W9-BQK-343

STRUCTURAL AND EVOLUTIONARY GENOMICS

NATURAL SELECTION IN GENOME EVOLUTION

New Comprehensive Biochemistry

Volume 37

General Editor

G. BERNARDI
Naples

ELSEVIER
Amsterdam - Lausanne - New York - Oxford - Shannon - Singapore - Tokyo

Structural and Evolutionary Genomics

Natural Selection in Genome Evolution

GIORGIO BERNARDI

Stazione Zoologica Anton Dohrn
Naples, Italy

2004
ELSEVIER
Amsterdam - Lausanne - New York - Oxford - Shannon - Singapore - Tokyo

ELSEVIER B.V.
Sara Burgerhartstraat 25
P.O. Box 211, 1000 AE Amsterdam, The Netherlands

Library of Congress Cataloging in Publication Data
A catalog record from the Library of Congress has been applied for.

ISBN: 0-444-51255-1

⊗ The paper used in this publication meets the requirements of ANSI/NISO Z39.48–1992 (Permanence of Paper).
Printed in The Netherlands.

* The picture on the dust cover is "Sky and Water I", a woodcut by M. C. Escher (1938). It can be seen not only as "*a powerful metaphor for the inseparability of life from life-supporting elements, air and water*" (Schattschneiden, 1990), but also as the transition from the oldest class of cold-blooded vertebrates, the fishes, to the youngest class of warm-blooded vertebrates, the birds.

For Gabriella

Preface

The main purpose of this book is to present our investigations in the areas of structural and evolutionary genomics, to critically review the relevant literature and to draw some general conclusions. Even if "functional genomics" is not included in the title, a number of functional implications derived from structural and evolutionary genomics will be discussed. While the majority of the book concerns genome organization, the last Parts present "*a long argument*" on the role of natural selection, "*the preservation of favourable variations and the rejection of injurious variations*" (Darwin, 1859), in genome evolution.

I intended to write this book for several years, but I hesitated mainly because firm conclusions on the role of natural selection in genome evolution had not yet been reached. Even if new results may modify the picture presented here, I now feel that its main features are correct, and that the time is ripe for publishing this overview.

Basically, the book presents experimental and conceptual advances in two major areas. The first one is genome organization. In spite of recent spectacular progress in genome sequencing, the remark that "*a large amount of detail is available, but comprehensive rules about the organization of genome have not yet emerged*" (Singer and Berg, 1991) still applies to the current literature. Our main discoveries, concerning the compositional compartmentalization of the vertebrate genome into a mosaic of isochores, the genome phenotypes, the genomic code, the bimodal distribution of genes and its correlation with functional properties, have led for the first time to a unified view of the eukaryotic genome as an integrated ensemble.

The second area is genome evolution. Our findings could not be accounted for by any of the current molecular evolution theories, since they were all based on single-nucleotide changes, and did not (and could not) take into consideration regional and compositional changes. We have been able to build a model of genome evolution, the neo-selectionist model, which accommodates not only some key features of the classical selection theory (essentially the selection of single-nucleotide changes in coding and regulatory sequences), but also those of the neutral theory (basically the random fixation of selectively neutral or nearly neutral changes in noncoding sequences). The neutral and nearly neutral changes certainly represent the majority of the changes in genome evolution, but they are finally controlled at the regional level by natural selection (essentially negative selection). In other words, the neo-selectionist model puts the neutral view of the genome into a new selectionist frame.

The book starts (**Part 1**) with a short history of the different views concerning the genome, a brief narrative of our early investigations, and a discussion of the molecular approaches that we used. **Part 2** deals with a small model genome, the mitochondrial genome of yeast, which shed light on the large genome in the nucleus.

In the central section of the book, **Parts 3** and **4** outline the compositional properties of the vertebrate genome, namely the compositional patterns of DNA molecules and of coding sequences, as well as the compositional correlations between coding and non-coding sequences, whereas **Parts 5, 6** and **7** discuss the most important properties of the vertebrate genome: the distributions of genes, of transposons and of integrated viral sequences in the genome and in chromosomes. This book is, however, not limited to the vertebrate genome, but also concerns other eukaryotic genomes, in particular plant genomes, as well as prokaryotic genomes (**Parts 8 and 9**).

The book ends with **Part 10**, which examines the correlations between gene composition and protein structure, **Part 11**, which considers how the organization of the vertebrate genome evolved in time, and **Part 12**, which discusses the general causes and mechanisms of this evolution. A recapitulation and our conclusions concerning the relative roles of natural selection and random drift in the evolution of living organisms are presented in the final sections.

The investigations reported here were carried out in the Centre de Recherches sur les Macromolecules of Strasbourg (1959–1969), in the Institut Jacques Monod of Paris (1970-2003) and in the Stazione Zoologica Anton Dohrn of Naples (since 1998). Summer visits at NIH as a Fogarty Scholar (1981–84), at Osaka University (1995) and at the National Institute of Genetics in Mishima (1996–2001) provided some pauses for reflection. I wish to thank here most warmly my hosts Maxine Singer, Gary Felsenfeld, Kenichi Matsubara and Takashi Gojobori.

The names and the contributions of the many people who participated in the investigations described in this book can be gathered from the references. I would like, however, to mention the names of those who either played a particularly important role in some phases of this work, or did more than the references suggest.

The first group comprises several people. My brother Alberto closely collaborated with me both in Strasbourg in the 1960's, on the preparation of DNases, exonucleases, phosphatases etc., which had never been prepared before, and later in Paris. In the early 1970's, Jean-Paul Thiéry, Gabriel Macaya and Jan Filipski set the foundations for the investigations that kept us busy for many years, while Dusko Ehrlich was the major contributor to our approach on the frequency of oligonucleotides in DNAs. My second son, Gregorio, started the computer analysis of DNA sequences in 1980, with the help of Jacques Ninio. My youngest son, Giacomo, initiated our investigations in molecular evolution in 1985 and has been collaborating with me since then. In the 1990's, Giuseppe D'Onofrio was responsible for pursuing further our investigations on both the organization and the evolution of the mammalian genome together with Simone Caccio, Oliver Clay, Kamel Jabbari, Dominique Mouchiroud, Hector Musto and Serguei Zoubak. In more recent years and until present, Giuseppe D'Onofrio, Oliver Clay, Kamel Jabbari and Héctor Musto were joined by Fernando Alvarez-Valin, Nicolas Carels, Stéphane Cruveiller and Adam Pavliček. Salvo Saccone was behind all the cytogenetic work in which compositional DNA fractions were used for *in situ* hybridization. Along the yeast mitochondrial research line, the major contributions came from Giuseppe Baldacci, Miklos de Zamaroczy, Godeleine Faugeron-Fonty, Regina Goursot, Gianni Piperno, Ariel Prunell, and Edda Rayko. The second group comprises Claude Cordonnier, Anne Devillers-Thiery, Audrey Haschemeyer and Alla Rynditch, who made investigations on hydroxyapatite chromatography, oligonucleotide frequencies, fish genomics and retroviral integrations, respectively. I certainly do not forget my faithful technicians Andréa Silvert and Henri Stebler, my draftman/photographer Philippe Breton, and Martine Brient, my secretary for almost thirty years.

I also wish to thank Fernando Alvarez-Valin, Giacomo Bernardi, Giuseppe D'Onofrio,

Regina Goursot, Kamel Jabbari, Adam Pavliček, Edda Rayko, and, especially, Oliver Clay and Héctor Musto for critical reading of sections of this book. Its preparation would have been impossible without the intelligent, competent and dedicated help of Gianna Di Gennaro and Romy Sole. I am grateful to Francisco Ayala, Takashi Gojobori, Daniel Hartl, Toshimichi Ikemura, Masatoshi Nei, Tomoko Ohta and Emile Zuckerkandl for their interest and encouragement. Last but not least, I wish to thank Dr. Arthur Koedam of Elsevier for his patience and understanding.

The first draft of this book was prepared at Hopkins Marine Biology Laboratory of Stanford University, Pacific Grove, in August 2001, thanks to the hospitality of George Somero. The book was written in the congenial atmosphere of the Stazione Zoologica Anton Dohrn, where it was completed in July 2003. Two notes were added in proof in early November 2003.

Finally, I would like to mention that some ideas presented in this book were developed during extensive travel and field work (essentially linked to specimen collection) in faraway places, often with my wife Gabriella and/or my son Giacomo. It was an honour, and a pleasure, to have, in some of these trips, the company of Professor Richard Darwin Keynes, FRS, the great grandson of Charles Darwin.

I would like to offer my sincere apologies to two groups of people. The first group comprises the colleagues whose work I am criticizing. I wish to make clear that criticisms were not just raised for polemical reasons, but because the analysis of a wrong experiment, or of a wrong viewpoint, can advance our understanding of a problem. Moreover, it is often instructive to present the background of wrong ideas against which new facts had to emerge. My feeling is that science makes progress, like evolution, more by negative selection (of wrong facts and views, which are abundant), than by positive selection (of good ideas, which are rare). Let me add that my personal opinion is that in science the principle "*Amicus Plato, sed magis amica veritas*" should prevail over any other consideration, diplomatic and otherwise.

The second group is that of the readers of this book. Covering over 40 years of work in one volume was not easy. For the sake of speeding up the preparation of this book, I did not hesitate to use *verbatim* quotations from our papers, especially the most recent ones. I hope the readers will excuse me for not having spent more time in polishing the style and smoothing out the jumps from one subject to another. They should, however, remember that this is not a textbook but a scientific monograph, that often deals with subjects at the border of our knowledge. Moreover, this book is focused on the general picture rather than on details, on the rule rather than on the exceptions. For this very reason, some subjects that are very important in themselves, were treated only in a cursory way, if their relevance to the main line of this book was marginal. I tried to be as clear as possible, while solving two problems, namely introducing methodological approaches which might not be generally familiar to the readers, and sketching a complex picture. Including all this information in the book was not a minor enterprise. This task was, however, made simpler by three factors. First, the main line of the book presents investigations carried out in a single laboratory. Second, the molecular biology approaches that we used provided results that could stand time (the buoyant density of DNA, for example, does not become obsolete over the years). Third, most of the data presented are very recent. In fact, some of the articles referred to will still be in press at the time this book will appear. This time will coincide with the 50th anniversary of the double helix paper by Francis Crick and Jim Watson, whom I salute and greet, and the 20th anniversary of the publication of *The Neutral Theory of Molecular Evolution* by Motoo Kimura, the great scientist to whose memory I pay homage.

Giorgio Bernardi

Contents

* Except where indicated otherwise, all figure legends are *verbatim* transcriptions of the original ones.

Part 1
Introduction

Introduction

1.1. The genome: a short history of different views

Since our starting point was the analysis of the organization of the eukaryotic genome, it may be appropriate to begin this introduction with a brief history of the different views concerning the genome. The term **genome** was coined over eighty years ago by Hans Winkler (1920), a Professor of Botany at the University of Hamburg, to designate the **haploid chromosome set**. Interestingly, the term genome was associated with eukaryotes from the beginning. The definition of Winkler was, however, a purely **operational definition**. This was in contrast to the older definition of **gene** (Johannsen, 1909), which was a conceptual definition. Indeed, the gene was defined as a unit of the genetic material localized in the chromosomes, and was originally supposed to be at the same time the ultimate unit of inheritance, of phenotypic difference and of mutation.

The term genome was not as successful as the term gene and, in fact, was forgotten for many years. Its utility became evident almost thirty years later, when Boivin et al. (1948) and Vendrely and Vendrely (1948) discovered that the amount of DNA per cell was a characteristic, constant feature of a given species and that somatic cells had a double amount of DNA compared to germ cells, two points confirmed and expanded later (Mirsky and Ris, 1949, 1951). The amount of DNA in haploid cells from organisms belonging to the same species was called **c-value** (Swift, 1950) for constant value, or **genome size** (Hinegardner, 1976; see Cavalier-Smith, 1985, and Petrov, 2001, for reviews). The identical functional potential of the genomes from all cells of a eukaryotic organism was demonstrated later by Gurdon (1962).

Between the end of the 1940's and the end of the 1960's, when the **prokaryotic paradigm** suggested that the eukaryotic genome was essentially made of genes, the word genome was considered to indicate the **sum total of genes**. In fact, the belief was widespread that, taking into account the different size of bacterial and human genomes, the human genome comprised one million genes.

The **large variability of genome sizes**, even among phylogenetically close species, and the discovery of **repeated sequences** led to the idea that genes only represented a part, and often a very small part, of the eukaryotic genome (see **Table 1.1**). The meaning of the word genome changed once more to indicate the **sum total of coding and non-coding sequences**. At this point, at the end of the 1960's, the term genome started its real career. Its increasing popularity accompanied the development of genome projects which began in the 1980's.

A crucial question is whether the eukaryotic genome is fully described by Winkler's definition (as proposed, often only implicitly, in all current textbooks of Molecular or Cell Biology, Genetics, Evolution), and by its subsequent modifications, or whether it is more than the sum of its parts. This dilemma may also be phrased differently, whether the component parts of the genome are endowed with simple **additive properties**, or with **co-operative properties**.

The first view, in which genes were visualized as distributed at random in the bulk of non-coding DNA, which would be "**junk DNA**" (Ohno, 1972) or, at least, "**selfish DNA**"

TABLE 1.1
Genome size, coding sequences and gene numbers in some representative organisms.

Organism	Genome size[a] Mb[b]	Coding sequences %	Gene number[a]	kb/gene[a, b]
Haemophilus	2	85	2,000	1
Yeast	12	70	6,000	2
Human	3,200	2	32,000	100

[a] in approximate figures
[b] kb, kilobases, or thousands of base pairs, bp; Mb, megabases, or millions of bp.

(Doolittle and Sapienza, 1980; Orgel and Crick, 1980), could be paraphrased from Mayr (1976) as the "**bean-bag view**" of the genome.

The second view of the eukaryotic genome as an **integrated ensemble**, defended here, is based on the notion that the genome is more than the sum of its parts, because structural, functional and evolutionary interactions occur among different regions of the genome and, more specifically, between coding and non-coding sequences.

To summarize, we have witnessed several different views of the eukaryotic genome: the purely **operational view** of Winkler, the **prokaryotic paradigm**, the **"bean-bag" view** and, finally, the **integrated ensemble view.** This latter view could, however, only be justified if one could define properties that are specific for the genome as a whole. The main achievements of our work were that we could define such genome properties and that we were able to build a coherent and comprehensive picture, which essentially emerged from an approach jointly based on molecular genetics and molecular evolution. Our main discoveries, concerning **the compositional compartmentalization of the vertebrate genome into a mosaic of isochores, the genome phenotypes, the genomic code, the bimodal distribution of genes and its correlation with functional properties** could not be accounted for by the classical selection theory or by the mutation-random drift theory, Kimura's neutral theory of evolution. This led us to investigate further the roles of natural selection and random drift in genome evolution, to propose a paradigm shift which could reconcile the neutral theory with our view of the dominant role played by natural selection in genome evolution and to formulate a **neo-selectionist model**.

1.2. Population genetics and molecular evolution

It has been stated (Li and Graur, 1991; Li, 1997; Graur and Li, 2000) that molecular evolution "*has its roots in two disparate disciplines: population genetics and molecular biology. Population genetics provides the theoretical foundation for the study of the evolutionary process, while molecular biology provides the empirical data*". In my opinion, this concept should be modified. First, population genetics has a number of intrinsic limitations that are best illustrated by its incapability to solve the neutralist-selectionist debate (see, for example, Hey, 1999; Kondrashov, 2000), to quote one example which is at the centre of the subject matter of this book. Second, molecular biology has played a role that is much more important than providing empirical data, which is rather the task of mapping, sequencing, etc. In fact, molecular biology, which arose in the middle of the past century from the disciplines of biochemistry and molecular structure, has revolutionized biology, one field after the other. One may wonder where we would be in genetics, immunology, virology, cell biology, if molecular biology had not invaded and pervaded those disciplines. Interestingly, starting with the epochal paper by Zuckerkandl and Pauling (1962), evolutionary studies are those that have undergone the deepest changes as a consequence of the development of molecular genetics, as strongly stressed by Kimura (1983), a population geneticist. Indeed, evolutionary genomics, which applies the molecular biology approach to the study of genome evolution, is progressively transforming the most speculative field of biology into the most rigorous one. This book shows how the structural genomics results obtained in our laboratory led not only to a better understanding of genome organization, but also to advances in evolutionary genomics which, in turn, opened the way to the solution of the neutralist-selectionist debate. Incidentally, it is because of the overwhelming role played by the molecular approach in our work that this book is published in a series called *New Comprehensive Biochemistry*.

1.3. Three remarks on terminology

I will not use in this book the expression **GC content**, preferring **GC level** or, simply, **GC** (which is defined as the molar ratio of guanosine and cytidine in DNA; see **Abbreviations**). Indeed, one can talk about a content only if there is a container. In the case of DNA, the nucleotides are not contained in DNA, they form DNA.

The counterpart of AT is not CG (as used by some authors) but GC, because what matters here is not the alphabetical order but the purine-pyrimidine order.

Finally, let me stress that by structural genomics I mean what used to be called nucleotide sequence organization or genome organization and not, as recently suggested (see, for example, Stevens et al., 2001; Baker and Sali, 2001), protein structure (although the latter also enters into the picture).

1.4. A brief chronology of our investigations

I will present here a short narrative of our research, concentrating on its early phases, because they are not dealt with in the book. My research career started in 1951, when I rang the bell of the Medicinska Nobel Institute, Department of Biochemistry in Stockholm, and was accepted by Professor Hugo Theorell as a summer student to start my thesis in biochemistry in view of a Medical Degree at the University of Padova. After the defence of my thesis, I spent two years in the Italian Air Force, during which time I kept in touch with the Department of Biochemistry of the University of Padova and started studying Physics (I later obtained a "Libera Docenza" in Physical Biochemistry in 1962 and a "Doctorat d'Etat ès-Sciences Physiques" in 1967 with Jacques Monod as the Chairman of the Jury). I then moved to the Biochemistry Department of the University of Pavia. My work there on the physico-chemical properties of mucopolysaccharides led me to visit the Centre de Recherche sur les Macromolécules of Strasbourg, directed by Professor Charles Sadron. As a consequence of this brief visit, in 1956 I left Italy to work in Strasbourg for several months. After six more months, during which I pursued investigations on mucopolysaccharides with Professor Frank Happey in Bradford, I joined, as a post-doctoral fellow, the National Research Council of Canada in Ottawa, where I worked on lipoproteins with Dr. W.H. Cook between 1957 and 1959. In those years, the return journey by boat from New York to Genoa took 12 days, a time long enough to think about the future and to make the decision to devote myself to the study of DNA. At that time, this was the research subject of only a handful of laboratories in the world (the Watson and Crick paper of 1953 was barely quoted in the 1950's). I knew that this was going to be a vast enterprise, but I could not imagine that my work in the field was going to span more than 40 years. It was, however, the best decision, because it allowed me to take an active role in an adventure that led us from the double helix to the human genome sequence through the golden era of molecular biology.

When I started working in the pleasant environment of the Centre de Recherche sur les Macromolécules in Strasbourg in 1959, it was clear to me that two tools were needed to understand the way the genome of eukaryotes was organized: enzymes that were able to cut DNA into large fragments and fractionation methods that were able to separate the fragments. I therefore embarked, at the age of 30, on two research projects along those lines.

As an enzyme, I chose acid DNase (which we had isolated for the first time), because preliminary experiments indicated that this enzyme led to the degradation of DNA into very large fragments of about 1 kb (Bernardi et al., 1960). The study of this and other DNases provided the first demonstration that these enzymes recognize short DNA sequences, contrary to the prevailing view (Laskowski, 1971, 1982) that they lacked specificity. It also indicated that acid DNase could cut both strands of native DNA at the same

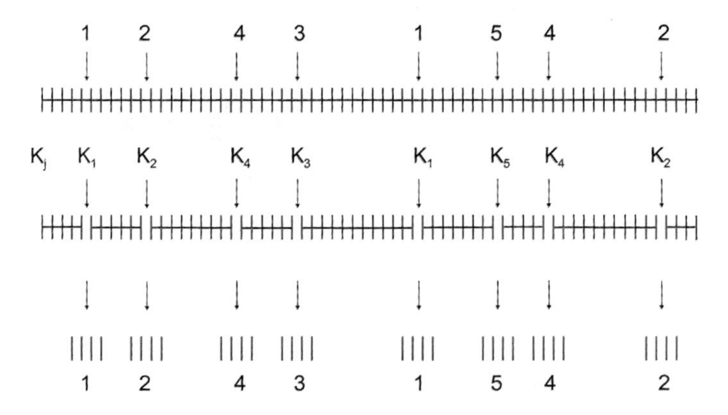

Figure 1.1. Analysis of termini: a number of sequences, shown as tetranucleotides and numbered 1 to 5, are recognized and split with different K_m (and/or V_{max}), indicated by K_1, K_2, etc. Terminal and penultimate nucleotides were isolated from the resulting oligonucleotides, and their base compositions (see Fig. 1.2) were determined. (From Bernardi et al., 1973).

5' WX ↓ YZ 3'

Figure 1.2. Scheme of a tetranucleotide split by a DNase at the position indicated by an arrow. Average nucleotide composition at the two terminal positions, W and Z, and at the two penultimate positions, X and Y, were determined using methods that we developed for their isolation and analysis. (From Bernardi et al., 1973).

time. In turn, this suggested that symmetrical sequences were recognized and split, so implying a symmetry in the DNase molecule, which was, in fact, an allosteric dimer (Bernardi 1965a). Even if these early studies (summarized in Bernardi, 1971) may need some revision, in many respects acid DNase was a prefiguration of the (class II) restriction enzymes which were discovered ten years later (Smith and Wilcox, 1970), and which made it obsolete for the purpose of cutting DNA into large pieces, because of their strict sequence specificity. One point which acid DNase, with its lower yet defined specificity, allowed us to assess was, however, the frequency of the sequences that the enzyme could recognize and split, namely the average composition of the terminal and penultimate nucleotides (the **termini**) on each side of the cuts (**Figs. 1.1** and **1.2**).

These nucleotides had compositions that were characteristic of the DNA under study (and of the DNase used). Since the percentages of the termini formed by DNases from bacterial DNAs were linearly related to their GC levels (**Fig. 1.3A**), a useful way to show the results from the DNAs under examination was to plot **difference histograms** like those of **Fig. 1.3B**. This approach extended the nearest neighbour analysis of dinucleotide frequencies (Josse et al., 1961) to a frequency approach involving the sequences, at least four

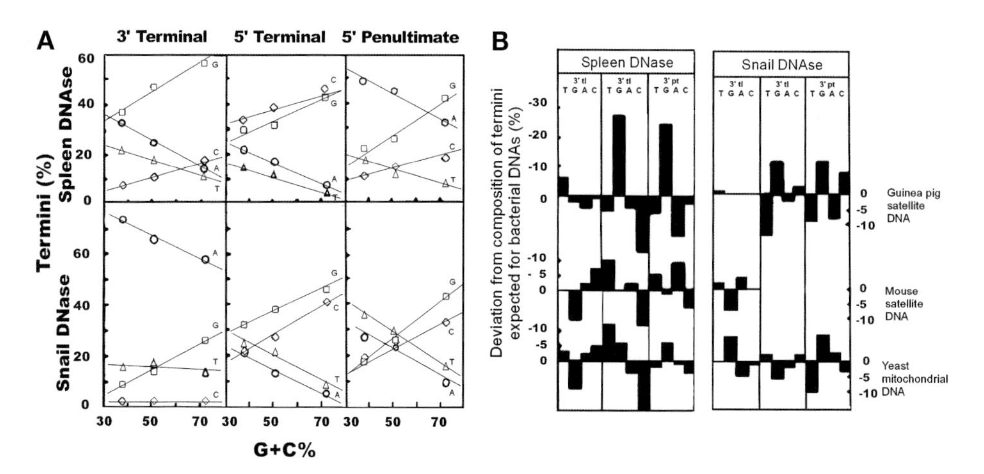

Figure 1.3. **A.** The percentages of the four nucleotides, A (circles), G (squares), C (diamonds) and T (triangles) in the 3'-terminal, 5'-terminal and 5'-penultimate nucleotides formed by the spleen and the snail DNase from bacterial DNAs (*Haemophilus influenzae*, 38% GC; *Escherichia coli*, 51% GC; *Micrococcus luteus* 72% GC) are plotted against the GC level of DNAs. Values obtained at an average chain length of 15 nucleotides were used. **B.** Deviation patterns of three repetitive DNAs. The histograms show the differences between the composition of termini formed from guinea pig satellite, mouse satellite and yeast mitochondrial DNAs by spleen and snail DNase and the compositions expected for bacterial DNAs having the same GC level; tl, terminal; pt, penultimate. (From Bernardi et al., 1973).

nucleotides long, that had been recognized and split by the enzyme (Bernardi et al., 1973). This allowed us to see specific patterns in different DNAs, not only in repetitive DNAs (see **Fig 1.3B**), but also in human "main-band" DNA and its major components (see **Fig. 3.8**). We later applied frequency methods to oligonucleotides from the mitochondrial genome of yeast (see **Part 2**). Interestingly, frequency methods involving di- to tetra-nucleotides have now been revived using self-assembly approaches and complete genomic sequences (Abe et al., 2003).

As a fractionation method, I developed chromatography of nucleic acids on hydroxyapatite, a calcium phosphate which had been used by Tiselius et al. (1956) for the fractionation of proteins. Previous observations (Bernardi and Cook, 1960a,b,c) that hydroxyapatite was particularly good as a chromatographic substrate for fractionating of phospholipoproteins characterized by different phosphorylation levels convinced me to try it on DNA. The main discovery was that hydroxyapatite could fractionate single- from double-stranded DNA (**Fig. 1.4** left panel), the former being eluted by a lower phosphate

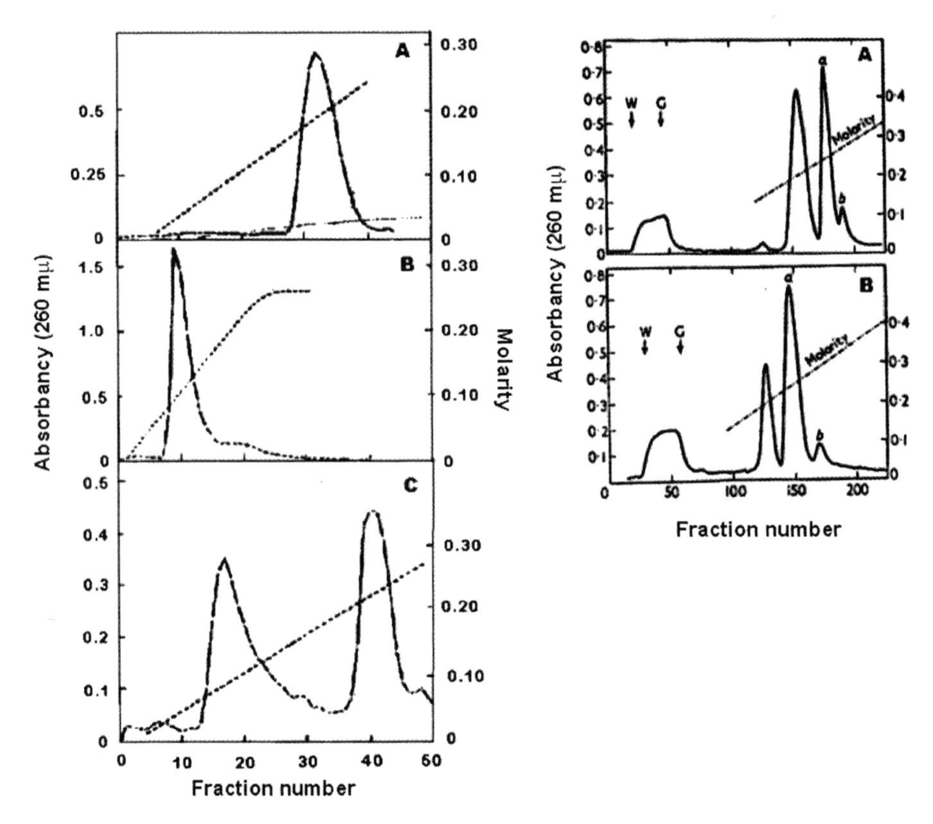

Figure 1.4. **Left panel**: Gradient elution by phosphate buffer in the presence of 1 per cent formaldehyde of bovine DNA: **A**, native DNA; **B**, heat-denatured (100°) DNA; **C**, a 1:1 mixture of native and heat-denatured (100°) DNA. (From Bernardi, 1965b). **Right panel**: Chromatography of DNA preparations **A** from wild-type yeast cells; **B** from a cytoplasmic petite mutant. The three peaks eluted by the phosphate gradient correspond to RNA, nuclear DNA (a) and mitochondrial DNA (b). W and G indicate washing and gradient. (From Bernardi et al., 1968)

buffer molarity compared to the latter (Bernardi, 1962, 1965b). The reason for this separation was that the random coil form of denatured DNA had fewer phosphate groups available for the binding to calcium sites of hydroxyapatite compared to the extended, relatively rigid, double-stranded, native DNA. This also explained the separations of compact, supercoiled from relaxed polyoma virus DNA (Bourgaux and Bourgaux-Ramoisy, 1967), of RNA from DNA (see for example **Fig. 1.4**, right panel), and, also, of denatured from native proteins (Bernardi et al., 1972a). The fractionation of single- and double-stranded DNA on hydroxyapatite provided an extremely powerful tool to study the kinetics of reassociation of denatured DNA and to demonstrate the existence of repeated sequences in eukaryotic DNAs (Britten and Kohne, 1968).

Not taking into consideration the case in which secondary (single-stranded DNA from ΦX 174 phage) or tertiary structures (twisted, circular DNA from polyoma virus) are grossly different and cause a different chromatographic behaviour, it is evident that differences in nucleotide sequences in otherwise similar DNAs (the similarity concerning the double-strandedness, the molecular weight, the linear or open circular configuration, and also the nucleotide composition) may be sufficient to determine different elution molarities. Indeed, DNAs containing short repetitive sequences may in general show particular elution molarities. For example, DNAs comprising alternating dAT:dAT and non-alternating dA:dT structures, like mitochondrial DNAs from *Saccharomyces cerevisiae* (Bernardi et al., 1968, 1970; see **Fig. 1.4** right panel, and **Part 2**), *Euglena gracilis* (Stutz and Bernardi, 1972; Fonty et al., 1975), *Ustilago cynodontis* (Mery-Drugeon et al., 1981) and chloroplast DNA from *Euglena gracilis* (Schmitt et al., 1981; Heizmann et al., 1981) showed high elution molarities compared to nuclear DNAs. In spite of such separations, in general the resolving power of hydroxyapatite was not good enough for fine fractionations of high molecular weight DNA according to base composition. The use of hydroxyapatite was, therefore, limited (not without problems) to reassociation kinetics experiments (see **Table 3.1**). Hydroxyapatite chromatography was, however, not abandoned. A long series of theoretical investigations started in our laboratory were carried out by Tsutomu Kawasaki, who summarized them in a book (Kawasaki, 2003).

In 1966 I turned, therefore, to equilibrium centrifugation of DNA in density gradients, in the presence of sequence-specific DNA ligands, as a fractionation approach based on the frequency of the oligonucleotides that could bind the ligand. This worked beautifully not only for separating satellite DNAs (Corneo et al., 1968), but also, to our surprise, for fractionating main-band bovine DNA into three major DNA components (in addition to several satellite and minor DNA components; Filipski et al., 1973). This work was done in Paris where I had moved from Strasbourg in 1970 to head the Laboratoire de Génétique Moléculaire at the new Institut de Recherches en Biologie Moléculaire (later called, at my suggestion, Institut Jacques Monod, after the name of its founder).

The finding of a striking, discontinuous, compositional heterogeneity at the macromolecular level in the bovine genome led us to investigate other eukaryotic DNAs (Thiery et al., 1976) and to show (Macaya et al., 1976) that the genome of vertebrates was a **mosaic of isochores** (for compositionally **equal landscapes**), namely of chromosomal DNA regions originally estimated as over 300 kb long, on the average, which were fairly homogeneous in composition, and belonged to different families characterized by different average GC levels. In 1984, we presented at the FEBS Meeting in Moscow the crucial discovery that the vertebrate genome was characterized by **compositional correlations** linking coding and non-coding sequences (Bernardi et al., 1985a,b). Indeed, this changed the previous definitions of the genome (see *Section 1.1*) into that of a system whose properties were not simply the sum of the properties of its constitutive parts.

In 1966, I also started a long-term project on the molecular genetics of yeast mitochondria. This model system provided important results as far as both the organization and the evolution of eukaryotic genomes are concerned. These findings will be briefly presented and discussed in **Part 2** of this book. In the 1960's we also performed a series of investigations on the physical chemistry of DNA (see, for instance, Froelich et al., 1963; Freund and Bernardi, 1963) and on transforming DNA from *Haemophilus influenzae* (Chevallier and Bernardi, 1965, 1968; Bernardi and Bach, 1968; Kopecka et al., 1973).

A subsequent major step in the history of our laboratory was its increasing involvement (starting in 1980) in both evolutionary aspects of genome organization and computer approaches to the study of nucleotide sequences. At the beginning of 1998, I moved to the Stazione Zoologica of Naples, where I started a Laboratory of Molecular Evolution, fulfilling, more than a century later, the dream of Anton Dohrn, my genial predecessor, who founded the Stazione in 1872 in order to carry out research with the aim of proving that Darwin was right.

1.5. Molecular approaches to the study of the genome

Among the molecular approaches used in our work on the genome organization, equilibrium centrifugation of DNA in density gradients played a key role and allowed us to analyse and fractionate DNA and to characterize the fractions so obtained. It is, therefore, useful to briefly illustrate it. Another good reason to devote some space to this approach is that it is not widely used and not well understood, as shown, for example, by the following statements: (i) "*Physical methods have suggested that G+C compositional heterogeneity is less marked in poikilothermic animals (Bernardi, 2000), and this is confirmed by our large-scale sequence analysis*" (Aparicio et al., 2002). (ii) "*Studies of the compositional patterns of the completely sequenced human genome (International Human Genome Sequencing Consortium, 2001) confirmed earlier indirect results (Bernardi et al., 1985) suggesting that an average GC content in both coding and non-coding DNA varies along the chromosome in a non-random way*" (Filipski and Mucha, 2002; our underlinings). Indeed, CsCl centrifugation, although it might be considered an indirect method, did not simply suggest, but actually demonstrated the points under consideration. Interestingly, the compositional heterogeneity of the vertebrate genome was quantified more than a quarter of a century before its "confirmation" by sequencing (Filipski et al., 1973, Thiery et al., 1976, Macaya et al., 1976; see Clay et al., 2003a for additional information).

Fig. 1.5 presents a simple scheme concerning the information that can be obtained by centrifuging DNA to equilibrium in an analytical CsCl density gradient. Scanning the DNA band formed in the centrifuge cell produces an absorbancy profile (at 260 nm) which allows estimating of the modal buoyant density ρ_0 (the density at the peak), the mean (average) buoyant density $<\rho>$, the asymmetry of the profile, $<\rho>-\rho_0$, and the heterogeneity, H.

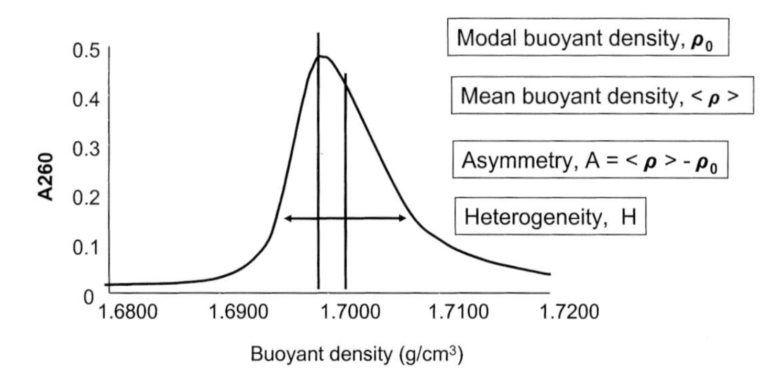

Figure 1.5. DNA profile obtained by analytical ultracentrifugation in CsCl gradient and deduced parameters.

More precisely (see Thiery et al., 1976), in order to calculate the **buoyant density**, ρ, at any point of abscissa r from the rotation axis, the relationship

$$\rho = \rho_\kappa - \frac{\omega^2}{2\beta_0}\left(r_\kappa^2 - r^2\right) \tag{1}$$

is used, where the subscript κ refers to the marker, ω is the angular velocity in radians s^{-1}, and β_0 was taken as equal to $1.19 \times 109\ cm^5\ g^{-1} s^{-2}$ (Ifft et al., 1961). Under such conditions, using phage 2C DNA (ρ=1.742 g/cm^3; Szybalski, 1968) as a density marker, a reproducible modal buoyant density, ρ_0 (density at the peak maximum, located at a distance ρ_0 from the rotation axis), of 1.7103 g/cm^3 is obtained for *E.coli* DNA.

The mean buoyant density, $<\rho>$, is calculated from the first moment of the band profile about the center of rotation:

$$< r > = \frac{\int_0^\infty cr\ \mathrm{d}\ r}{\int_0^\infty c\ \mathrm{d}\ r} \tag{2}$$

and from eqn (1), c being the DNA concentration at point of abscissa r.

The variance of the profile, $<\delta^2>$, is equal to the second moment about the mean

$$< \delta^2 > = m_2 = \frac{\int_{BAND} c\delta^2 \mathrm{d}\delta}{\int_{BAND} c\mathrm{d}\delta} \tag{3}$$

with $\delta = r - < r >$.

Integrals of eqns (2) and (3) can be calculated using Simpson's rule, with radial steps corresponding to 0.04 or 0.08 mm in cell dimensions.

The **asymmetry**, A, of CsCl main bands was determined as $A = < \rho > \rho_0$.

The intermolecular compositional **heterogeneity**, H, of CsCl main bands can be calculated according to Schmid and Hearst (1972)

$$H = \left[\left(\frac{G r_0{}^2 \omega^4 \langle \delta^2 \rangle}{RT\rho_0} - \frac{1}{M_3} \right) - \left(\frac{RT\rho_0}{\beta_0{}^2 a^2 G} \right) \right]^{1/2} \tag{4}$$

where G is the buoyancy factor, M_3, is the molecular weight of the dry Cs salt of DNA, R is the universal gas constant, T is the absolute temperature, and a is the slope, 0.098, of the relationship of Schildkraut et al. (1962) between **buoyant density and GC**,

$$\rho = 1.66 + a\%GC \tag{5}$$

It should be stressed that, while in vertebrate DNAs there is a significant correlation between asymmetry and heterogeneity (see **Fig. 4.2D**), this is not necessarily always the case. Indeed, while an asymmetric profile is always heterogeneous, a heterogeneous profile is not necessarily asymmetric (as when an asymmetry on the GC-rich side is compensated by an asymmetry on the GC-poor side of the profile).

Sedimentation velocities of DNAs can be determined as described by Prunell and Bernardi (1973), and **molecular weights** then calculated from the sedimentation coefficients $s_{20,w}$, using the relationship of Eigner and Doty (1965)

$$s_{20,w} = 0.034M^{0.405} \tag{6}$$

Molecular weight of DNA is currently expressed in bp (base pairs) and its multiples kb, Mb, Gb (see legend of **Fig. 1.1**). Older data are often given in daltons, 1kb being about equal to $6.18 \cdot 10^5$ daltons. Genome size is often given in picograms (1 pg = 10^{-12}g; 1 pg $\cong 0.98 \cdot 10^6$kb $\cong 6.02 \cdot 10^{11}$daltons).

When centrifuged to equilibrium in analytical CsCl density gradients, undegraded viral genomes show unimodal, narrow symmetrical bands, as expected for a population of compositionally identical DNA molecules (see **Fig. 1.6** for an example).

In the case of the much larger bacterial genomes (~ 2–5 Mb), DNA preparations consist

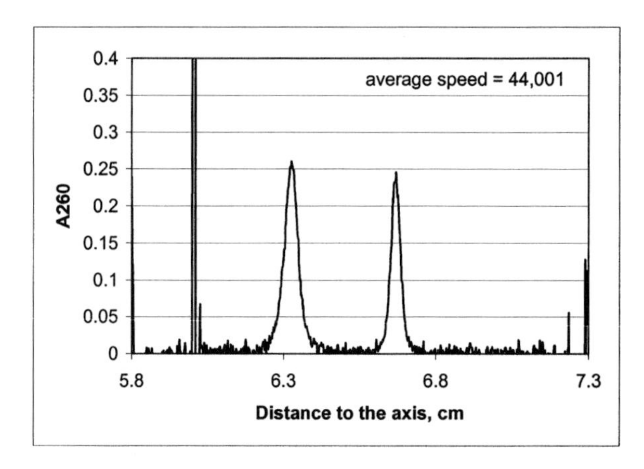

Figure 1.6. CsCl profiles of two viral DNAs, the DNAs from phage lambda (left peak) and phage 2C (right peak).

of fragments of different sizes. At average sizes above 50 kb, DNA fragments from *E.coli* DNA are essentially identical in composition (Yamagishi, 1970), hence the conclusion that bacterial genomes are characterized by a fairly homogeneous composition above a certain fragment size. This conclusion (to be discussed in detail in **Part 9**) was reinforced by the comparison with the then "standard" eukaryotic DNA, calf thymus DNA, which showed a broad, asymmetrical band (Meselson et al., 1957; Sueoka, 1959). In fact, the very strong asymmetry of the band was mostly due to the presence (see **Part 3**) of eight GC-rich satellite DNAs, six or seven (according to the resolution) of which were not separated from the "main-band" of bovine DNA (see **Figs. 3.5** and **3.6**). Satellite DNAs are components of eukaryotic DNAs (later shown to be made up of tandem oligonucleotides) that appear as separate bands accompanying the main band, or as shoulders of the main band (Kit, 1961; Sueoka, 1961), although some satellite DNAs remain cryptic, unseparated from the main band (see below).

When we started to look into the problem of fractionating eukaryotic DNAs, the first mammalian satellite DNA had just been isolated by preparative CsCl density gradient centrifugation from the mouse genome (Waring and Britten, 1966; Bond et al., 1967). The separation of mouse satellite (ρ=1.690 g/cm^3) from main-band DNA (ρ=1.700 g/cm^3) was not an easy one, because the two buoyant densities in CsCl were rather close. The guinea pig satellite was an even more difficult case because its buoyant density was 1.705 g/cm^3 *vs.* 1.700 g/cm^3 for the main band. We decided, therefore, to attempt the isolation of both mouse and guinea pig satellite DNAs by using centrifugation in Cs_2SO_4/Ag^+ density gradient, an approach just developed by Jensen and Davidson (1966) to separate the "AT satellite" from crab DNA. This approach was successful. Indeed, in such gradients the mouse satellite was well separated from the main band (by 45 mg/cm^3 instead of 10 mg/cm^3 in CsCl; see **Fig. 1.7** and **Table 1.2**). In the case of guinea pig, a major, heavy (GC-

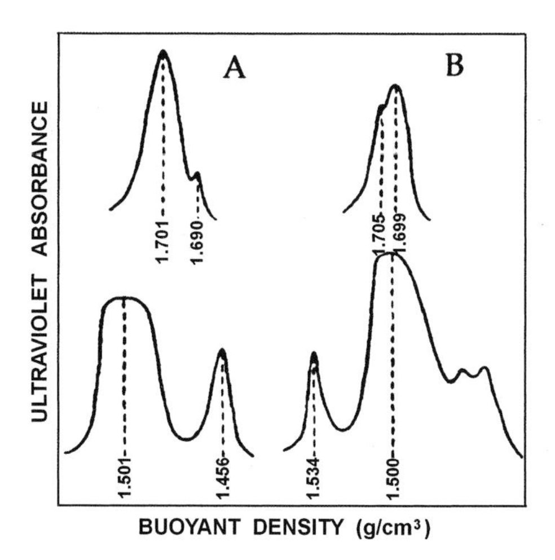

Figure 1.7. Microdensitometer tracings of **A** mouse and **B** guinea pig liver DNA, centrifuged to equilibrium in CsCl (upper tracings) and Cs_2SO_4/Ag^+ (lower tracings) density gradients. (From Corneo et al., 1968).

TABLE 1.2
Modal buoyant densities and GC levels of main-band and satellite DNAs from mouse and guinea pig[a]

Bands	Mouse DNA		Guinea Pig DNA	
	Main	Satellite	Main	Satellite
		ρ, g/cm^3		
CsCl	1.701	1.690	1.699	1.705
Cs_2SO_4/Ag^+	1.501	1.456	1.500	1.534
		GC, %		
CsCl buoyant density[b]	40.0	30.6	39.7	45.9
Melting temperature[c]	40.0	40.8	39.7	40.7
Nucleoside analysis	40.0	35.2	39.4	38.5

[a] From Corneo et al. (1968)
[b] Calculated according to Schildkraut et al. (1962)
[c] Calculated according to Marmur and Doty (1962)

rich), and two minor, light (GC-poor) satellite bands showed up (Corneo et al., 1968; see **Fig. 1.7**). While the former appeared as a shoulder on the CsCl profile, the latter were "cryptic" satellite DNAs that could not be detected in CsCl. As in the case of the mouse satellite, the major guinea pig satellite was better separated from main-band DNA in Cs_2SO_4/Ag^+(by 34 mg/cm^3 instead of only 6 mg/cm^3 in CsCl).

The mouse satellite and the guinea pig major satellite were then isolated by preparative ultracentrifugation in Cs_2SO_4/Ag^+ and their strands were separated by ultracentrifugation in alkaline CsCl. The satellite DNAs and their separated strands were enzymatically degraded to nucleosides (thus avoiding the losses associated with the chemical hydrolysis of DNA to bases) using, in succession, a series of enzymes isolated in our laboratory, spleen acid DNAse (Bernardi et al., 1966), spleen acid exonuclease (Bernardi and Bernardi, 1968) and spleen acid phosphomonoesterase II (Chersi et al., 1966). Nucleosides were then separated on Bio-Gel P-2 columns and quantified (Carrara and Bernardi, 1968; Piperno and Bernardi, 1971). Our analyses revealed that the two satellites from mouse and guinea pig only differed by 3% GC, instead of the 15% predicted by the relationship of Schildkraut et al. (1962). Moreover, the satellites were also characterized by an asymmetry of base compositions on the two strands. This was true not only for mouse satellite (as already seen by Flamm et al., 1967), but also, and much more so, for the major guinea pig satellite, which exhibited a "light strand" with less than 3% G. Finally, buoyant density in CsCl underestimated the GC level in mouse satellite, but overestimated it in guinea pig satellite (see **Table 1.2**). This was a clear indication that the basis for the separation was the differential binding of silver ions to the short sequences that made up the satellite DNAs under consideration (see **Fig. 1.8**). In fact, the high resolution depended upon such differential binding, since, *per se*, the resolving power of Cs_2SO_4 density gradients for DNA molecules of different base composition is much lower than that of CsCl gradients (Szybalski, 1968; Schmid and Hearst, 1972).

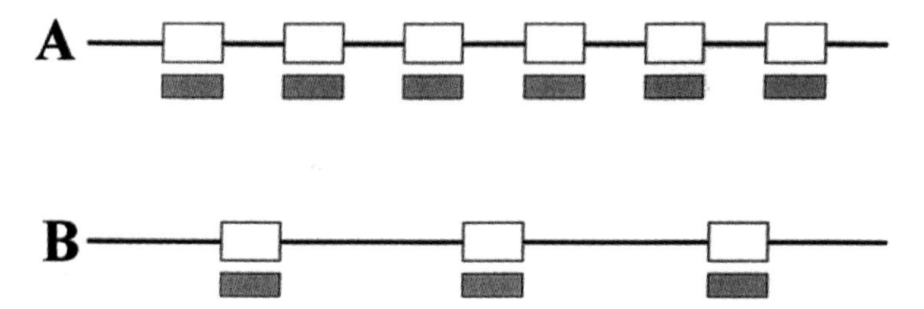

Figure 1.8. Scheme of the fractionation of complexes of DNA with sequence-specific ligands. Binding of the ligand (grey boxes) on DNA molecules depends upon the frequency of binding sites (oligonucleotides; open boxes). Two DNA fragments (**A** and **B**) are represented, which are characterized by different frequencies of such sites. The ligand is supposed to have a very strong affinity for the sites and to saturate them.

A novel observation was, however, that the base composition from the nucleoside analysis was very different from that calculated from CsCl buoyant density (Schildkraut et al., 1962), or from melting temperature (Marmur and Doty, 1962). Interestingly, while this was true for satellite DNAs, it was not for main band DNA (see **Table 1.2**). Having ruled out the presence of rare bases as the cause for the discrepancies, the explanation which was proposed was that "*satellite DNAs are conformationally slightly different from main-band DNAs, the differences being related to their peculiar nucleotide sequences*" (Corneo et al., 1968). Indeed, the sequence dependence of buoyant density in CsCl is most probably due to a different water binding by different short sequences in the DNA molecules (the bases themselves having identical "dry" densities). In connection with this explanation, it may be recalled that non-alternating poly (dA) (dT), alternating poly (dAT) (dAT) and poly (dG) (dC) were already known to show "anomalous" buoyant densities (Schildkraut et al., 1962), melting temperature (Marmur and Doty, 1962) and optical rotatory dispersion (Samejima and Yang, 1965). Indeed, their properties were different from those of the prokaryotic DNAs used to establish the relationships between base composition and physical properties. These anomalies were, generally, attributed to the extreme compositions of the synthetic polynucleotides, although Wells and Blair (1967) had shown a sequence-dependent behaviour of synthetic polynucleotides made up of repeated trinucleotides. Similar anomalous physical properties were also found (Bernardi et al., 1968, 1970; Bernardi and Timasheff, 1970) for yeast mitochondrial DNAs from wild-type cells (18% GC) and cytoplasmic "petite colonie" mutants (ranging from 4% to 18% GC), due to their abundance in short alternating and non-alternating AT sequences (see **Part 2**).

The **basic methods** that we routinely used to study genome organization were (see also **Table 1.3**):

(i) Analytical equilibrium ultracentrifugation in CsCl density gradients (see **Fig. 1.5**).

(ii) Analytical and preparative equilibrium centrifugation in Cs_2SO_4 density in the presence of sequence-specific DNA ligands (see **Fig. 1.7**); the two ligands that we used were Ag^+ ions at two different pH values and BAMD, bis(acetatomercurimethyl)dioxane, an organic mercurial first synthesized at the end of the 19th century (see Bünemann and Dattagupta, 1973). Two interesting features about BAMD are 1) that the compound is not sensitive to contaminating impurities, which is the case for Ag^+; and 2) that one can

TABLE 1.3
Methods for studying nucleotide sequences in DNAs[a]

Indirect methods
1. Sequence-dependent properties of double-stranded DNA
a. Buoyant density in CsCl
b. Melting temperature
c. Ligand binding
d. Hydroxyapatite binding
2. Sequence-dependent properties of single-stranded DNA
a. Hydroxyapatite binding
b. Reassociation kinetics
Direct methods
1. Frequency methods
a. Depurination
b. Nearest-neighbour analysis
c. Analysis of termini released by DNases
2. Mapping
3. Sequencing

[a] From Bernardi et al. (1973).

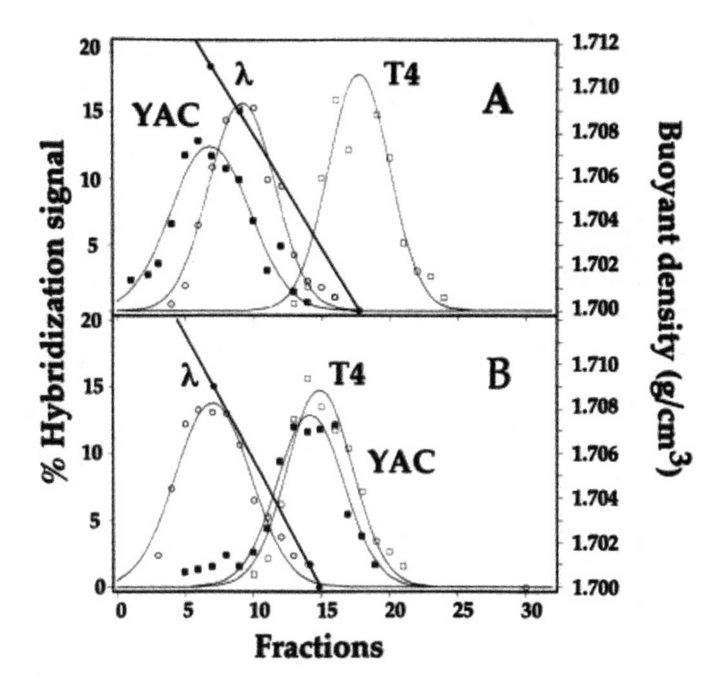

Figure 1.9. DNA distribution in a shallow CsCl gradient of **A** a GC-rich YAC (Yeast Artificial Chromosome) and **B** of a GC-poor YAC. λ and T4 bacteriophage DNAs were used as density markers. The intensities of the hybridization signals (left ordinate) and the buoyant densities (right ordinate) are plotted against the fractions collected from the gradient. (From De Sario et al., 1995).

shift the optimum resolution range by changing the ligand/nucleotide ratio (this is also true for Ag^{+}).

(iii) Preparative equilibrium centrifugation in shallow CsCl density gradients (De Sario et al., 1995; see **Fig. 1.9**); this approach, although not having quite the same resolving power as the DNA-ligand approach just mentioned, had the advantage of being much less laborious and of avoiding DNA losses, because dialysis to remove the ligand was not needed; DNA fractions could be alkali-denatured, loaded on filters, washed out of CsCl and hybridized with appropriate probes.

(iv) Hybridization of specific, labelled probes on fractionated DNA. This approach not

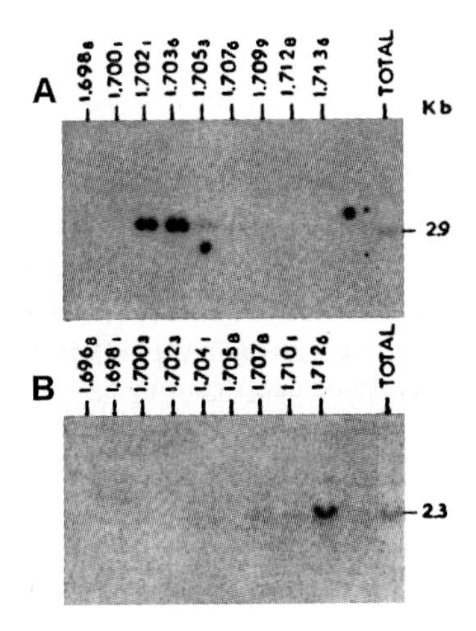

Figure 1.10. Gene localization in the human genome. 10 μg DNA from each fraction of a preparative Cs_2SO_4/BAMD gradient and of total human DNA were digested with EcoRI, electrophoresed in 0.8% agarose gels, transferred and hybridized with probes for **A** *c-mos* and **B** *c-sis* oncogenes. After hybridization, filters were washed under high-stringency conditions. *c-mos* was localized in relatively light (GC-poor) DNA fractions and *c-sis* in the heaviest (GC-richest) DNA fractions. Buoyant densities in CsCl are indicated for each fraction. (From Zerial et al., 1986a).

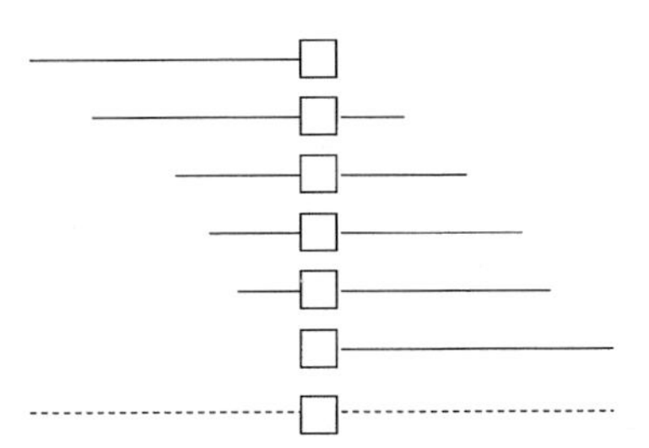

Figure 1.11. A schematic representation of a collection of random DNA fragments (assumed here to be equal in size) derived by random breakdown from a chromosomal region identified by a marker (box). The hybridization of a probe corresponding to the marker (*e.g.*, a gene) on DNA fractionated in a density gradient provides information on the average composition of a region having a size up to twice (broken line) the average size of the DNA molecules under consideration.

only localizes the sequence of interest in compositional fractions (**Fig. 1.10**), but also provides information on the average GC level of a DNA region having a size up to twice the average size of the target DNA molecules (see **Fig. 1.11**). PCR (polynucleotide chain reaction) was also used on DNA fractions in order to localize specific sequences.

(v) Hybridization of compositional DNA fractions on chromosomes (see **Part 7**) in the presence of an excess of unlabelled C_o t=1 DNA; (C_o t is the product of the initial DNA concentration by the reannealing time).

In conclusion, the main experimental approach that we followed to investigate genome organization was based on the most elementary properties of the genome, its nucleotide composition, and its oligonucleotide frequencies. This **compositional approach**, the only one that was possible before DNA sequencing, is still very useful, in particular for screening large numbers of genomes or genome fractions. Moreover, the compositional approach could be, and indeed was, easily transferred from DNA molecules to DNA sequences, when these became available. The usefulness for the compositional approach was threefold: (i) the compositional heterogeneity of the genome could not only be detected and assessed, but also used for fractionation purposes, so allowing us to study the properties of compositional DNA fractions; (ii) compositional features could be analysed in very large sequences, such as the human genome sequences; (iii) compositional changes resulting from the evolutionary process could be detected and compared.

Part 2
Lessons from a small dispensable genome, the mitochondrial genome of yeast

CHAPTER 1

The mitochondrial genome of yeast and the petite mutation

1.1. The "petite colonie" mutation

The mitochondrial genome of yeast is of special interest for two major reasons: (i) because, in contrast to its very compact counterpart in animal cells, it comprises abundant non-coding sequences (Bernardi et al., 1970); this situation is not unique to *Saccharomyces cerevisiae* since it is shared by other fungi and by *Euglena gracilis* (as shown by investigations presented in **Chapter 3**), nor to the mitochondrial genome since the chloroplast genome also shows it; in the case of *S. cerevisiae*, mitochondrial genome units are about five times larger than the units of animal mitochondrial genomes; (ii) because it is dispensable since *S. cerevisiae* can survive on fermentable carbon sources; mutants (the **cytoplasmic "petite colonie" mutants** of Ephrussi; see below) having undergone massive alterations in the nucleotide sequences of their mitochondrial genome (Bernardi et al., 1968; Mehrotra and Mahler, 1968) or having lost it altogether can therefore survive (dispensability also applies to the chloroplast genome of *E. gracilis*; see Heizmann et al., 1981).

In other words, although very small, the mitochondrial genome of yeast presents features that are common to those of the nuclear genome of eukaryotes. One can therefore study, for instance, the organization, the evolutionary origin and the function of non-coding sequences. Hence, our interest. Moreover, since the genome is dispensable, one can investigate genome changes which are usually incompatible with cell life.

In 1948 Boris Ephrussi gave the first account of investigations on the 'petite colonie' mutation in *S. cerevisiae* (Ephrussi, 1949), which were the starting point of **extra-chromosomal genetics.** It is difficult to describe the initial observation more clearly than in Ephrussi's own words: "*When a culture of baker's yeast, whether diploid or haploid, is plated, each of the cells gives rise in the course of the next few days to a colony. The great majority of these colonies are of very nearly identical size, but one usually finds also a very small number – say 1 or 2% – of distinctly smaller colonies* (**Fig. 2.1**). *These facts suggest that the population of cells which was plated was heterogeneous and that it may be possible to purify it by taking cells from either the big or the small colonies only. The results of such a selection show, however, that cells from the big colonies again and again produce the two types of colonies, while the cells from the small colonies give rise to small colonies only*" (Ephrussi, 1953). Besides describing the mutation and its irreversibility, this paper also reported a number of other fundamental observations: (i) that acriflavine treatment increased the number of 'petite' mutants from 1–2% to 100% (**Fig. 2.1**); (ii) that the mutants grew slowly because they could not respire, owing to the loss of their ability to synthesize a whole series of respiratory enzymes; and that in anaerobiosis wild-type cells and petite mutants grew at the same slow rate using fermentative pathways; (iii) that crosses of wild-type cells with petite mutants showed a **non-Mendelian segregation** of the mutation, in that they led either to wild-type progeny, or to both wild-type cells and petite mutants in different proportions

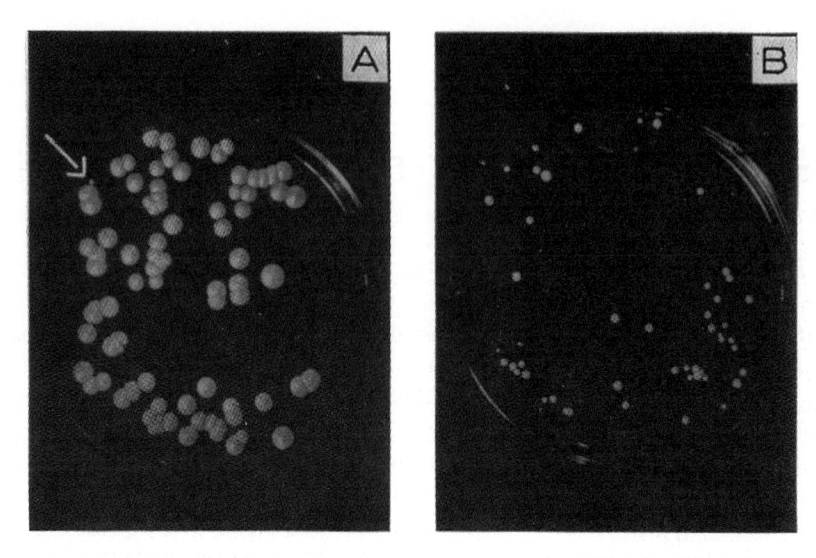

Figure 2.1. Colonies formed by baker's yeast on a solid medium. **A.** Colonies of a normal yeast, showing one small colony. **B.** Colonies formed by the same yeast grown in the presence of acriflavine prior to plating. (From Ephrussi, 1953).

depending upon the petite used in the cross; the petite mutants entering the cross were called **neutral petites** in the first case and **suppressive petites** in the second one (Ephrussi et al., 1955; see **Fig. 2.2**). The conclusion drawn by Ephrussi was that wild-type cells and petite mutants differed by "*the presence in the former and the absence in the latter of cytoplasmic units endowed with genetic continuity and required for the synthesis of certain respiratory enzymes*" (Ephrussi, 1953).

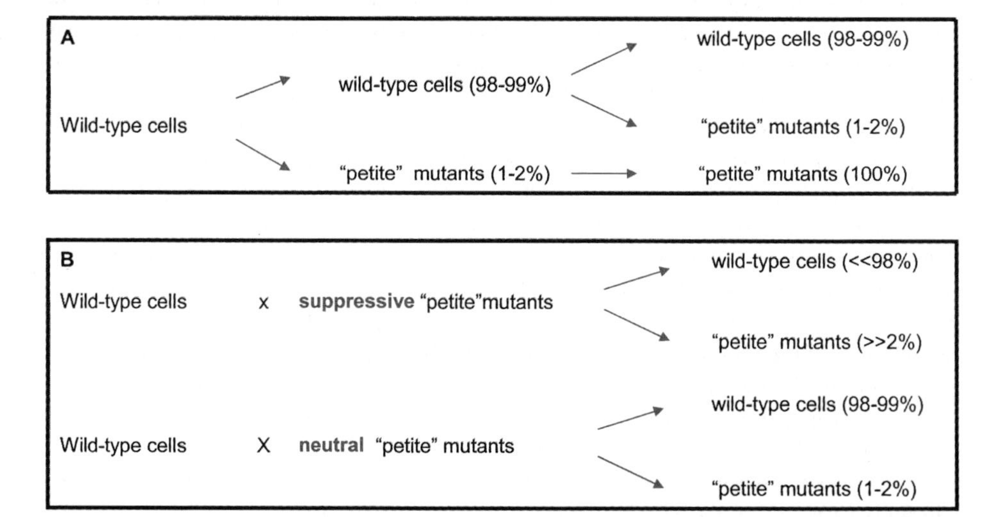

Figure 2.2. **A.** The generation of petite mutants from (cytoplasmic) wild-type cells. **B.** The results of crossing wild-type cells by petite mutants.

1.2. The petite mutation is accompanied by gross alterations of mitochondrial DNA

The cytoplasmic units postulated by Ephrussi in 1948 were only identified as mitochondrial genes 20 years later. Indeed the first hard facts about the molecular basis of the petite mutation were published by Bernardi et al. (1968) and by Mehrotra and Mahler (1968). These authors showed that mitochondrial DNAs from two genetically unrelated, acriflavine-induced, petite mutants had a grossly altered base composition (4% GC) compared to DNA from the parent wild-type strain (18% GC). These findings unequivocally established that massive alterations in the nucleotide sequences of the mitochondrial genome may accompany the petite mutation and be responsible for it. These conclusions were quickly confirmed by more detailed investigations (see the following sections), which probed the structure of mitochondrial DNA from both wild-type cells and petite mutants. Since we were interested in understanding the organization of the mitochondrial genome of yeast and the causes of these massive alterations, we decided to solve these problems using a strictly molecular approach, taking advantage of our ability to isolate mitochondrial DNA in large amounts by chromatography on hydroxyapatite (see **Fig. 1.4**). Wisely, we did not waste time using the classical genetic approach to study the petite mutation, and confined ourselves to spontaneous (*versus* ethidium-bromide-induced) mutants.

1.3. The AT spacers and the deletion hypothesis

Mitochondrial DNA from wild-type yeast cells was found (Bernardi et al., 1970; see **Fig. 2.3**) to be extremely heterogeneous in base composition, about half of it melting at a very low temperature and being almost exclusively formed by long stretches of short alternating AT:AT and non-alternating A:T sequences (the existence of which had already been predicted for the first petite genome investigated; Bernardi et al., 1968), and the rest melting over an extremely broad temperature range. Compared with mitochondrial DNA from wild-type cells, DNAs from three spontaneous suppressive petite mutants were shown to have lower GC levels, to lack a number of DNA stretches that melt at high temperature, and to renature very rapidly (**Fig. 2.4**). At that time, I interpreted those results as indicating that petite mutants had defective mitochondrial genomes, in which large segments of the parental wild-type genomes were deleted. I also suggested that such deletions arose by a mechanism (Campbell, 1969), involving illegitimate, unequal recombination events in the "**AT spacers**", which I supposed to contain sequence repetitions because of their extreme base composition. It was obvious that the loss of any known mitochondrial gene products (ribosomal RNAs, tRNAs, the sub-units of enzymes involved in respiration and oxidative phosphorylation) would have a pleiotropic effect and lead to a loss of respiratory functions. Incidentally, this could also happen as a consequence of mutations in nuclear genes that encode some sub-units of mitochondrial enzymes. These "**nuclear petite mutants**", that are characterized by a **mendelian inheritance**, will not be dealt with here.

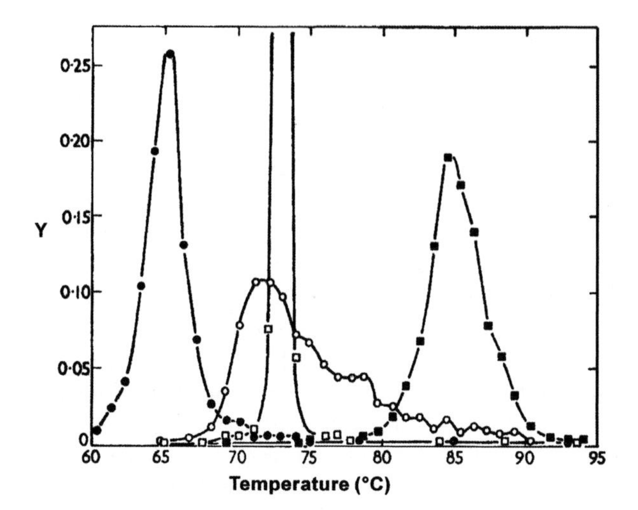

Figure 2.3. Differential melting curves obtained with poly (dAT : dAT) (solid circles), poly (dA : dT) (open squares), mitochondrial (open circles) and nuclear DNA from wild-type strain B (solid squares). The ordinate, Y, indicates the increment in relative absorbance per degree; Y_{max} of poly (dA:dT) had a value of 0.71. (From Bernardi et al., 1970).

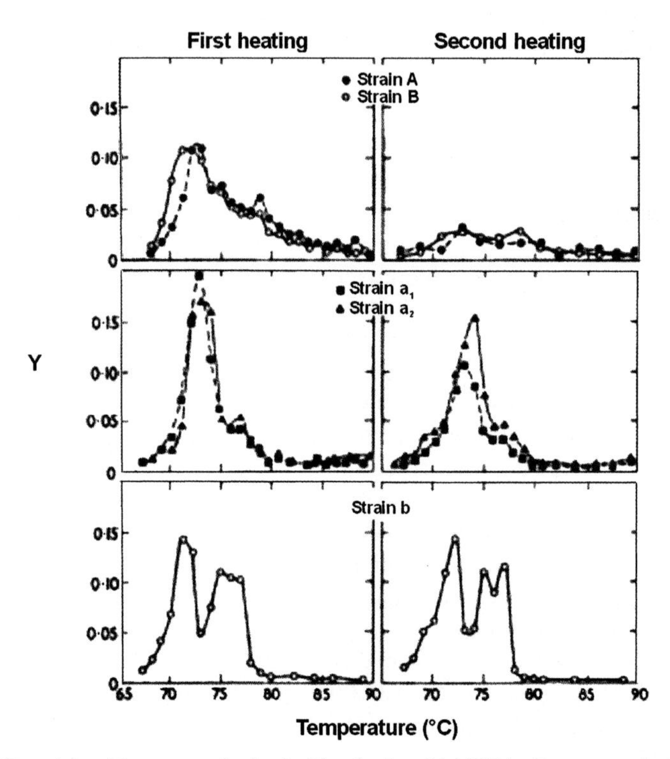

Figure 2.4. Differential melting curves obtained with mitochondrial DNAs. Every curve is an average from 2 to 5 replicate experiments. Strains A and B are two wild-type strains. Strains a_1, a_2 and b are spontaneous petite strains. See Fig. 2.3 for other indications. (From Bernardi et al., 1970).

1.4. The petite mutation is due to large deletions

Further work (Bernardi et al., 1972b; Prunell and Bernardi, 1977) showed that the AT spacers formed about 50% of the wild-type mitochondrial genome, had a GC content lower than 5%, were indeed repetitive in nucleotide sequence and were, therefore, likely to be endowed with sequence homology over stretches long enough to allow illegitimate site-specific recombination. In 1974, direct evidence was provided for both a deletion mechanism (Bernardi et al., 1975) and an accompanying amplification of the excised genome segment (Locker et al., 1974; Bernardi et al., 1975). Indeed, only a fraction of the restriction fragments of wild-type mitochondrial DNA were present in petite genomes and these were present in multiple copies per genome unit. These findings disposed of a number of strange *ad hoc* hypotheses put forward to explain the petite mutation and led to the scheme shown in **Fig. 2.5** in which the excised segment from the wild-type genome (depicted as circular) becomes the repeat unit of the petite genome. This may, in turn, undergo further deletions leading to secondary petite genomes, characterized by shorter repeat units. Incidentally, the analysis of restriction patterns of mitochondrial DNA from several strains of wild-type cells provided the first unequivocal estimate of the size of the mitochondrial genome unit, about 50 x 10^6 daltons or about 75 kb (Bernardi et al., 1975), an almost five-fold larger size than that of the mitochondrial genome units of animals, about 17 kb. The strain-related size variation of restriction fragments provided in fact the first example of what was later called "**restriction fragment length polymorphism** or **RFLP**"

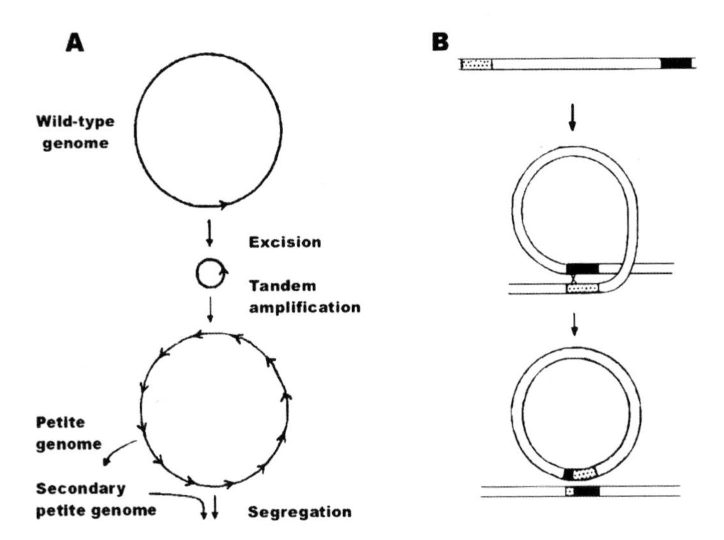

Figure 2.5. **A.** Scheme depicting the excision-amplification process leading to the formation of the genome of a spontaneous petite mutant. A segment of a wild-type mitochondrial genome unit is excised and tandemly amplified into a defective genome unit. This then replicates and segregates into the buds to form the genome of a petite mutant; the petite genome can undergo further excisions leading to the formation of secondary petite genomes. **B.** Scheme of the excision process leading to the formation of a petite genome unit. Dotted and black areas represent two direct sequence repeats. Excision is followed by a tandem amplification (see **A**). (From Bernardi, 1983)

(Botstein et al., 1980; see also other examples in recombinant mitochondrial genomes; **Section 2.1**).

1.5. The GC clusters

Another important advance in our knowledge of the organization of the mitochondrial genome was the discovery of short segments of mitochondrial DNA extremely rich in GC, the "**GC clusters**" (Bernardi et al., 1976; Prunell et al., 1977; Prunell and Bernardi, 1977). Operationally, two sorts of GC clusters can be distinguished, the (CCGG, GGCC) clusters, present in 60–70 copies per genome unit and recognizable because they are degraded by the restriction enzymes Hpa II and Hae III, and the GC-rich clusters which do not contain CCGG or GGCC sequences, but are often close to (CCGG, GGCC) clusters and to isolated CCGG sequences. A certain number of these clusters were likely to be homologous in sequence. Sequence analyses carried out in several laboratories (see below) indicated that the GC clusters were located in the middle of AT spacers and not at the ends of them, as first suggested as a working hypothesis (Prunell and Bernardi, 1977), so that an overall scheme of the organization of the mitochondrial genome of yeast is that given in **Fig. 2.6**. The predominant location of GC clusters in the middle of AT spacers suggests that they may fulfil a stabilizing role.

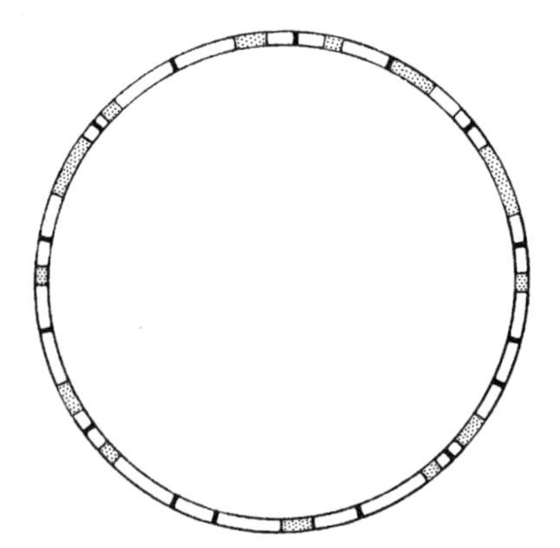

Figure 2.6. A schematic representation of the organization of a mitochondrial genome unit of yeast. Grey stretches represent genes, white stretches AT spacers, black bars GC clusters. (From Bernardi, 1979a).

1.6. The excision sites

The next step was the precise definition of the sequences involved in the excision process. The basic idea of the deletion model mentioned above was that the instability of the mitochondrial genome of yeast was due to the existence in each genome unit of a number of nucleotide sequences having enough homology to allow illegitimate, unequal recombination to take place. In this respect, clearly the GC clusters were at least as good candidates as the AT spacers.

Investigations on a number of spontaneous petite mutants showed that, indeed, most frequently the ends of the repeat units were formed by (CCGG, GGCC) clusters; less frequently, they appeared to correspond to GC-rich clusters and to AT spacer sequences (Faugeron-Fonty et al., 1979; Gaillard et al., 1980). For example, in the petite genomes of **Fig. 2.7**, excision of the repeat units occurred at, or very near, (CCGG, GGCC) clusters in the top four cases; and at GC clusters or AT spacers in the other two. Furthermore, it was shown that repeat units were organized in a perfect tandem (head-to-tail) fashion. Interestingly, secondary excision of simpler repeat units from the petite genomes originally derived from the parental wild-type genome appeared to take place at the same kind of sites used in the primary process, the end product being rather simple and stable petite genomes, such as those presented in **Fig. 2.7**. The spontaneous, cytoplasmic petite mutation should, therefore, be visualized as a **cascade of excisions**, which is slowed down, or stopped, only because sequences appropriate for excision are used up in the process, or because the DNA looping needed for recombination becomes impossible due to the short distance between potential recombination sites.

Such a simple excision mechanism was not found in petites induced by ethidium bromide, practically the only ones studied in other laboratories. These contained inverted

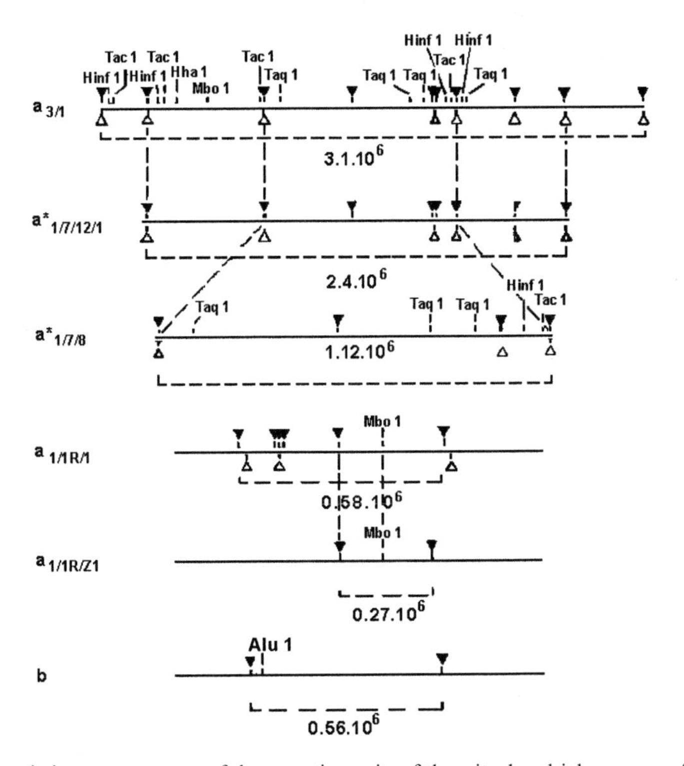

Figure 2.7. Restriction enzyme maps of the repeating units of the mitochondrial genomes of several spontaneous petite mutants. The molecular weights of the repeat units are indicated, along with the positions of *Hae* III (Δ), *Hpa* II (▼) and other restriction sites. In the case of $a_{3/1}$, five isolated *Hpa* II sites and *Hinc* II site are not shown. The broken lines indicate corresponding restriction sites in different repeat units. (From Bernardi, 1979a).

repetitions, multiple deletions, sequence rearrangements and, above all, were caused by much less specific excisions (Lewin et al., 1978). This is not surprising if one considers that the tremendous increase in petite formation upon mutagenization is accompanied by extensive genome fragmentation (Goldring et al., 1970) and that petites lacking mitochondrial DNA altogether (called ρ° **petites** in contrast to the ρ^{-} **petites**, which contain mitochondrial DNA) are frequently formed (Goldring et al., 1970; Nagley and Linnane, 1970). (By analogy, we called ϕ^{-} the bleached mutants of *Euglena gracilis* because, even if traditionally thought to be completely deprived of chloroplast DNA, they do contain a defective chloroplast genome, which is present at a very low copy number, and preferentially retains ribosomal RNA genes; Heizmann et al., 1981).

It was obvious that what was needed was detailed knowledge, at the nucleotide level, of the sequences involved in the excision of petite genomes. The simplest interpretation of the sequence data was that excision was due to a crossing-over process, and that the primary event in the spontaneous petite mutation was very similar to the excision of the lambda prophage from the *E.coli* chromosome, or to the dissociation of a bacterial transposon from its host plasmid; in this case, the GC clusters and sequences in the AT spacers play the same role as the insertion sequences delimiting a bacterial transposon.

1.7. Genomes without genes

The first determination of the nucleotide sequence of the repeat unit of the mitochondrial genome of the spontaneous petite mutant $a_{1/1R/Z1}$ (Gaillard and Bernardi, 1979; **Fig. 2.8**)

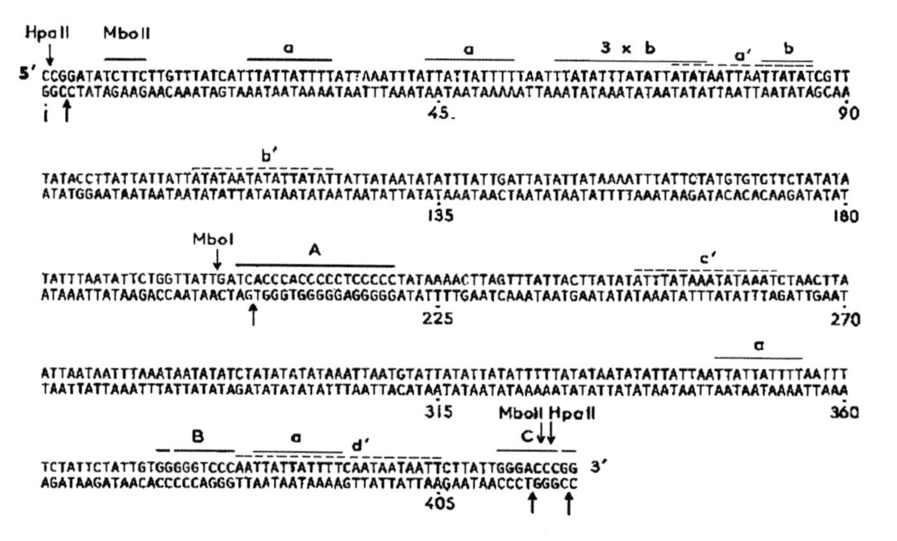

Figure 2.8. Nucleotide sequence of the repeat unit of the mitochondrial genome of spontaneous petite mutant $a_{1/1R/Z1}$ (see Fig. 2.7). A, B and C indicate GC-rich clusters; *a* and *b*, a repeated decanucleotide and a repeated hexanucleotide, respectively; *a'*, *b'*, *c'* and *d'*, palindromic sequences in the AT stretches. The restriction sites of *Hpa* II, *Mbo* I and *Mbo* II are indicated by arrows; the recognition site of the latter enzyme is also indicated. (From Bernardi, 1979a).

and preceding work (Cosson and Tzagoloff, 1979; Macino and Tzagoloff, 1979) provided complementary confirmations of several previous results, namely that AT spacers are made up of short alternating and non-alternating AT sequences (Bernardi and Timasheff, 1970; Ehrlich et al., 1972) and contain direct and inverted repeated sequences and palindromes (Prunell and Bernardi, 1977); that GC-rich clusters are largely contiguous to CCGG sequences and to (CCGG, GGCC) clusters (Prunell and Bernardi, 1977); and that the latter are to some extent endowed with homology (Prunell and Bernardi, 1977).The genome of $a_{1/1R/Z1}$ was of interest in three other respects: (i) it did not contain any gene, and was therefore a clear example of the lethality of the petite mutation as far as mitochondrial functions are concerned; incidentally, the surprising finding of **genomes without genes** can be taken as the best evidence that genetics and genomics are not the same thing; (ii) it replicates; in fact this is the only function left; since this genome was shown to be made up of a perfect tandem repetition of the basic unit (Faugeron-Fonty et al., 1979), the latter must contain a signal permitting the initiation of replication; (iii) it is excised, in all likelihood, from another petite genome, $a_{1/1R/1}$ (**Fig. 2.7**), and not directly from the wild-type genome; the complete sequence of the parental petite genome $a_{1/1R/1}$ (see **Fig. 2.12**) should, therefore, provide precise information as to the excision sites involved in the formation of $a_{1/1R/Z1}$.

To summarize, the first event in the spontaneous cytoplasmic petite mutation is the excision of a segment from one of the 25–50 mitochondrial **genome units (Fig. 2.9)** of a

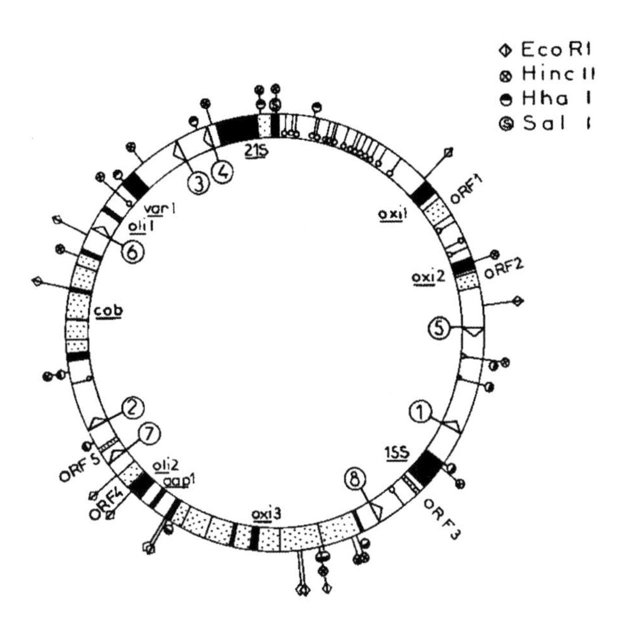

Figure 2.9. Physical and genetical map of the mitochondrial genome unit of wild-type yeast (strain A). Some restriction sites are indicated . Circled numbers indicate the location of *ori* sequences 1–8 (arrowheads point in the direction cluster C to cluster A; see Fig. 2.8). Black and dotted areas correspond to exons and introns of mitochondrial genes, respectively. Thin radial lines ending in small circles indicate tRNA genes. White areas correspond to long AT spacers embedding short GC clusters. (Modified from de Zamaroczy et al., 1979).

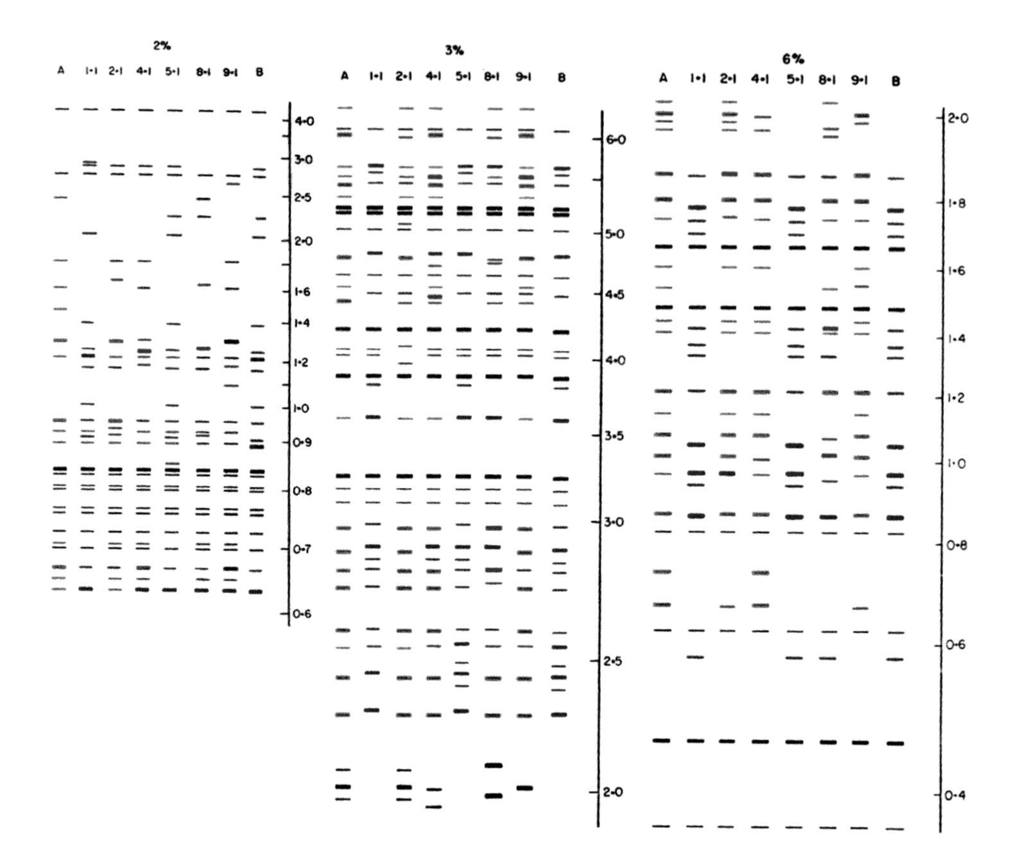

Figure 2.10. Scheme of the Hpa II band patterns on 0.5% agarose /2% polyacrylamide, 0.5% agarose/3% polyacrylamide, 6% polyacrylamide, as shown by the DNAs from parent strains A and B, and the diploid clones issued from the mass-mating experiment (1.1, 2.1, 4.1, 5.1, 8.1, 9.1). The molecular weights of the fragments are indicated (values in daltons are multiplied by 10^{-6} in the left-hand pattern, by 10^{-5} in the other two). Bands containing 1 or 2 (or more) fragments are indicated by a different thickness. This scheme only depicts the bands present in the best-resolved region of the gel and does not show the overlap region between different gels. Black bands are bands common to both parental strains; green and blue bands are bands only present in one of the two parents; red bands are bands differing in mobility or multiplicity from parental bands. Faint bands are indicated by broken lines (the faintest bands being indicated by 4 dashes, the others by 3). (From Fonty et al., 1978).

2.2. The canonical and the surrogate origins of replication of petite genomes

The mitochondrial genomes of the vast majority of spontaneous petites are exclusively derived from the tandem amplification of a DNA segment excised from any region of the parental wild-type genome (Faugeron-Fonty et al., 1979). Therefore, either the wild-type genome contains several origins of replication and at least one of them is present on the excised segment (Prunell and Bernardi, 1977), or sequences other than the origins of

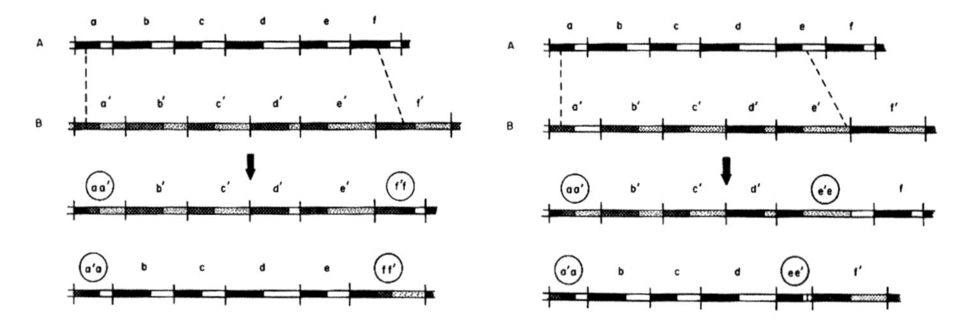

Figure 2.11. Scheme of crossing-over events leading to exchanges of DNA segments between the parental genomes A and B. Segments a′ to f′ and a to f indicate genetic units as split by HpaII in each segment, dark stretches correspond to genes, clear stretches to spacers (dotted in genome B) In the left-hand panel, crossing-overs take place between allelic genes. In the right-hand panel, one of the crossing overs takes place between allelic spacers and is unequal. (From Fonty et al., 1978).

replication of the wild-type genome are used as surrogate origins of replication. In fact, both situations have been found to occur, although with very different frequencies.

Considering that the first explanation was the more likely one, when we first sequenced (Gaillard and Bernardi, 1979) the repeat units of two petite genomes excised from the same region of the wild-type genome, we looked for a putative origin of replication in the segment shared by them and found a region characterized by two short GC clusters, A and B, flanking a palindromic AT sequence, *p*, and a short AT segment, *s*; and one long GC cluster, C, separated from B by a long AT segment, *l* (see **Figs. 2.8** and **2.12**). We also noticed (de Zamaroczy et al., 1981) that the potential secondary structure of the A-B region, the primary structure of cluster C and the general arrangement of the whole region were remarkably similar to those found in animal mitochondrial origins of replication (Crews et al., 1979; Gillum and Clayton, 1979; **Fig. 2.13**).

An *ori* sequence like the one just described was found in almost all the mitochondrial genomes of spontaneous petite mutants. Restriction mapping and hybridization of petite genomes on restriction fragments of wild-type genomes (de Zamaroczy et al., 1981; Bernardi, 1982b) provided evidence for the existence of eight *ori* sequences in the mitochondrial genome of wild-type cells. The primary structure of the *ori* sequences showed that they were extremely similar in sequence, particularly in their GC clusters (**Fig. 2.12**). All these canonical ***ori* sequences** have been precisely localized and oriented on the physical map of the genome (**Fig. 2.9**).

It should be noted: (i) that some *ori* sequences display one orientation and some the opposite one on the wild-type genome; (ii) that *ori* 2 and 7 as well as *ori* 3 and 4 are close to each other and tandemly oriented; (iii) that *ori* 4 is absent in a wild-type strain; (iv) that *ori* sequences containing the γ cluster were found only once (*ori* 4), or not at all (*ori* 6, *ori* 7) in extensive screenings of spontaneous petite genomes (see **Table 2.1**).

***Ori*° petites**, lacking a canonical *ori* sequence, were also found, although very rarely. An investigation of the mitochondrial genomes of eight such *ori*° petites (Goursot et al., 1982) has revealed that their repeat units contain, instead of canonical *ori* sequences, one or more ***ori*^{S}** or ***ori* "surrogate" sequences**. These *ori*^{S} sequences are a subset of GC clusters char-

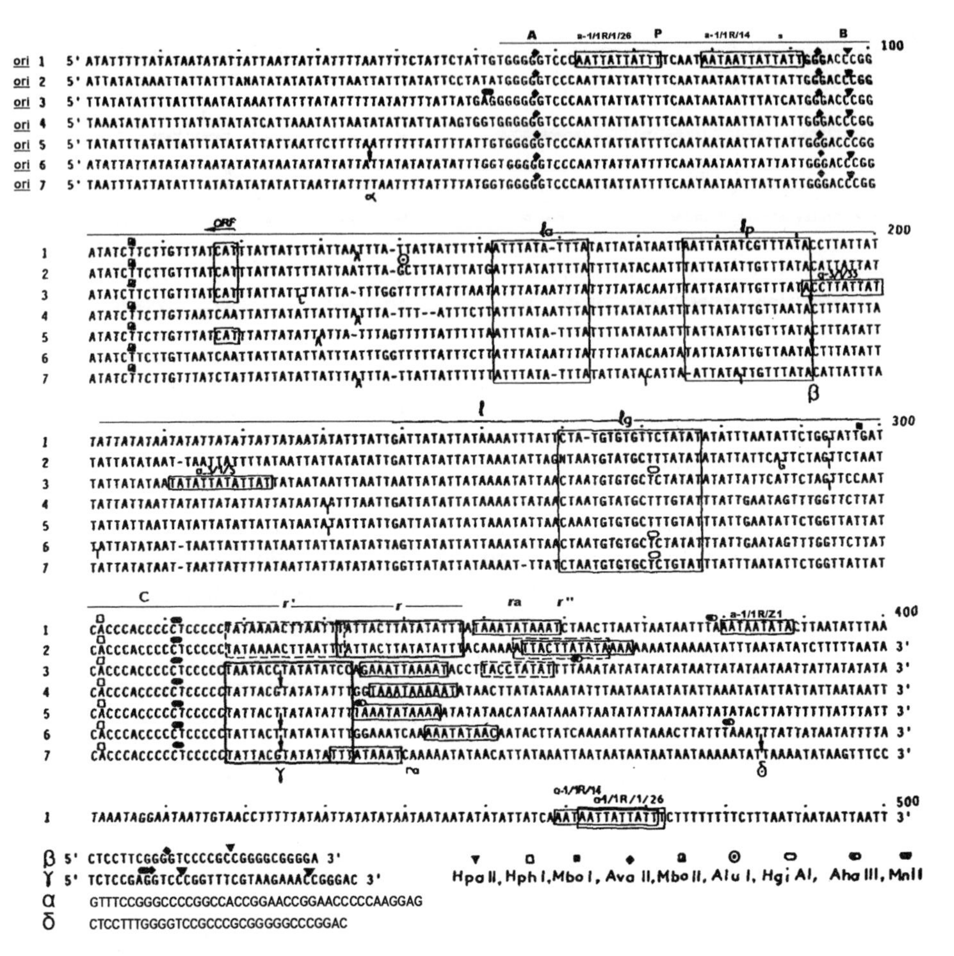

Figure 2.12. Primary structure of the *ori* sequences and their flanking regions. Regions of *ori* sequences are indicated by thick lines for GC clusters A, B and C and thin lines for AT stretches *p, s, l* and *r*. The positions of GC clusters α in *ori* 4, β in *ori* 4 and 6, γ in *ori* 4, 6 and 7, and δ in *ori* 7 are given; *r* sequences are indicated by heavy line boxes; *r* and *r* sequences by broken line boxes. Other boxes indicate sequences *l*a, *l*p, *l*g, *r*a, the excision sequences of the repeat units of petites a-1/1R/1/26, a-1/1R/14, a-3/1/33, a-3/1/5, and a-1/1R/Z1 and the initiation triplet of the open reading frames of *ori* 1, 2, 3 and 5. (From de Zamaroczy et al., 1983).

acterized by a potential secondary fold with two sequences ATAG and GGAG inserted in AT spacers; these sequences are followed by two AT base pairs, a GC stem (broken in the middle, and in most cases also near the base, by non-paired nucleotides) and a terminal loop (**Fig. 2.14**). This structure is reminiscent of that of GC clusters A and B from canonical *ori* sequences (**Fig. 2.13**). Like the latter, *ori*^{s} sequences are present in both orientations, are located in intergenic regions and can be used as excision sequences. *Ori*^{s} sequences are homologous with many other subsets of GC clusters (one of these subsets, the *ori*^{s}-like sequences, is shown in **Fig. 2.14**) some or all of which might perhaps also act as surrogate origins of replication, possibly still less efficiently than *ori*^{s} sequences.

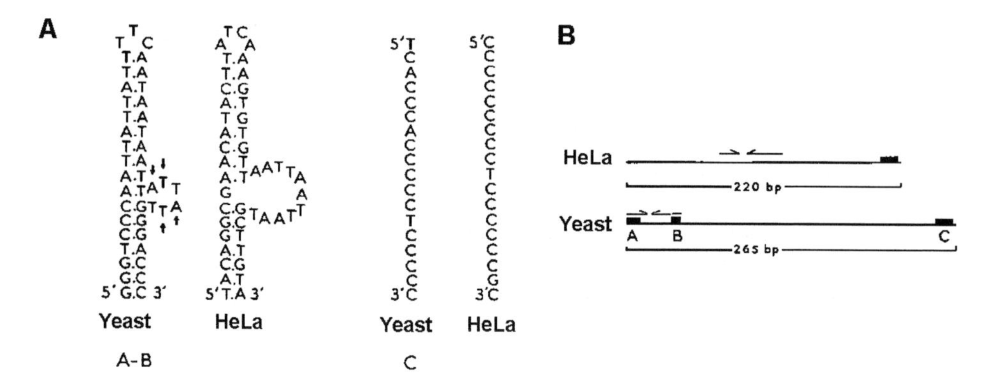

Figure 2.13. **A.** Comparison of *ori* sequences of mitochondrial genomes from yeast (de Zamaroczy et al., 1981) and HeLa cells (Crews et al., 1979). Homology of potential secondary structures is found for the inverted repeats in the cluster A - cluster B region; arrows indicate the base changes found in this region in different petite genomes. Homology of primary structure is found for cluster C. **B.** Comparison of the two *ori* sequences; the arrows indicate the inverted repeats of the A-B region, the broken line corresponding to the looped-out sequence; bp, base pairs. (From de Zamaroczy et al., 1981).

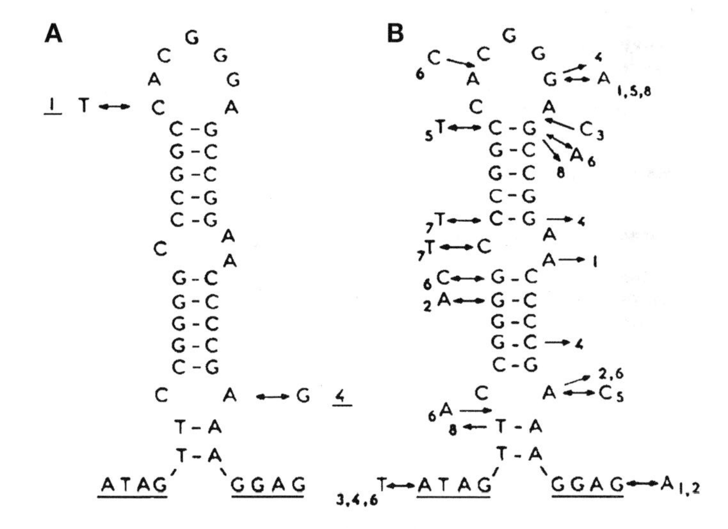

Figure 2.14. Potential secondary structure of: **A** the *ori*^{s} sequences; and **B** the *ori*^{s}-like sequences. All sequences are drawn in the same orientation ATAG → GGAG. Double-headed arrows indicate base-exchanges, arrows pointing towards, or away from, the structure indicate insertions and deletions, respectively. Numbers indicate the *ori*^{s}, or *ori*^{s}-like, sequences presenting these changes. (From Goursot et al., 1982).

2.3. The replication of petite genomes and the phenomenon of suppressivity

A functional evidence that *ori* sequences are indeed involved in the replication of the mitochondrial genome came from crosses of spontaneous petites, characterized in their mitochondrial genome and their suppressivity, with wild-type cells (de Zamaroczy et al., 1979; 1981; Goursot et al., 1980). In the suppressivity test, petite mutants are crossed with wild-type cells. The degree of suppressivity (or suppressiveness), namely the percentage of diploid petites in the progeny (which carry in every case the parental petite genome; Goursot et al., 1980; de Zamaroczy et al., 1981; Rayko et al., 1988) is determined by the replicative efficiency of the petite genome relative to the mitochondrial genome of the wild-type cells used in the cross (Bernardi et al., 1980a,b; Blanc and Dujon, 1980; de Zamaroczy et al., 1981). Crosses of spontaneous, highly suppressive petites having mitochondrial genomes formed by very short repeat units (400–900 base pairs) with wild-type cells produced diploids which harbored only the unaltered mitochondrial genomes of the parental petite, which was called **supersuppressive petite** (de Zamaroczy et al., 1979; Goursot et al., 1980). When petites with different degrees of suppressivity were used in the crosses, the genomes of diploid petite progeny had restriction maps identical to those of the parental haploid petites. Very few exceptions were found and these corresponded to new excision processes affecting one of the parental genomes.

There are two clear correlations between the *ori* sequence of the petite used in the cross and its degree of suppressivity. First, all other properties being comparable (namely, identical genetic background, the intact state of the *ori* sequence and the total amount of mitochondrial DNA per cell), the lower the overall density of *ori* sequences on the genome units, the lower the suppressivity. This can be called the **repeat-unit-size rule** (see **Section 2.5**). Second (see **Table 2.1**), partial or total deletion of the *ori* sequences, or their rearrangement, affect suppressivity: (i) ***Ori***$^-$ **petites**, in which the *ori* sequence is partially deleted, show a decreased suppressivity relative to *ori*$^+$ petites carrying intact *ori* sequences; the loss of cluster C with its flanking sequence has a much more dramatic effect than the loss of cluster A; (ii) *ori*^{r} petites, which show an inverted orientation of two *ori* sequences within the same repeat units (the latter having, in turn, an alternate inverted and tandem orientation) have a very low degree of suppressivity; (iii) *ori*^{o} petites, which lack the *ori* sequence altogether but contain *ori*^{s} sequences instead, have low to minimal

TABLE 2.1
Replication and transcription of petite genomes[a]

	Petites	Suppressivity[b]	Transcription		Petites	Suppressivity[b]	Transcription
A	Ori$^+$ petites			**B**	*Ori*$^-$ petites		
	ori 1	>95%	+		*ori* 1 A$^-$	80%	+
	ori 2	>95%	+		*ori* 1 C$^-$	n.d.	–
	ori 3	85%	+		*ori* 3 C$^-$	< 5%	–
	ori 4[c]		–	**C**	*Ori*^{o} petites		
	ori 5	90%	+		a-15/4/1/10/3	~ 1%	–
	ori 6[c]				a-3/1/B4[d]		–
	ori 7[c]			**D**	*Ori*^{r} petites	< 5%	n.d.

[a] Modified from Baldacci and Bernardi (1982).
[b] Values found for petite genomes having repeat units ~900 (*ori* 1, 2) or 1800 (*ori* 3, 5) base pairs long.
[c] *Ori* 4 was only found once, *ori* 6 and 7 were never found alone in the extensive screenings of spontaneous petite genome.
[d] Diploid.

degrees of suppressivity. These results provide a molecular basis for a replicative advantage being the explanation for suppressivity.

The first direct demonstration that the active *ori* sequences are indeed origins of DNA replication, as previously postulated on the basis of compelling but indirect evidence, was provided by Baldacci et al. (1984). These authors showed that in both *ori* 1 and *ori* 5, nascent DNA chains using as template the strand containing sequence *r* (the "*r* strand") start at the *r* end of cluster C, are elongated towards sequence *l*, and follow an RNA primer starting at sequence *r*. Nascent DNA chains copied on the "non-*r* strand" start within cluster C, are elongated towards sequence *r*, and follow an RNA primer starting in sequence *l* just before cluster C. *Ori* 1 and 5 are, therefore, used as sites for RNA-primed bidirectional replication of mitochondrial DNA (see **Fig. 2.15**).

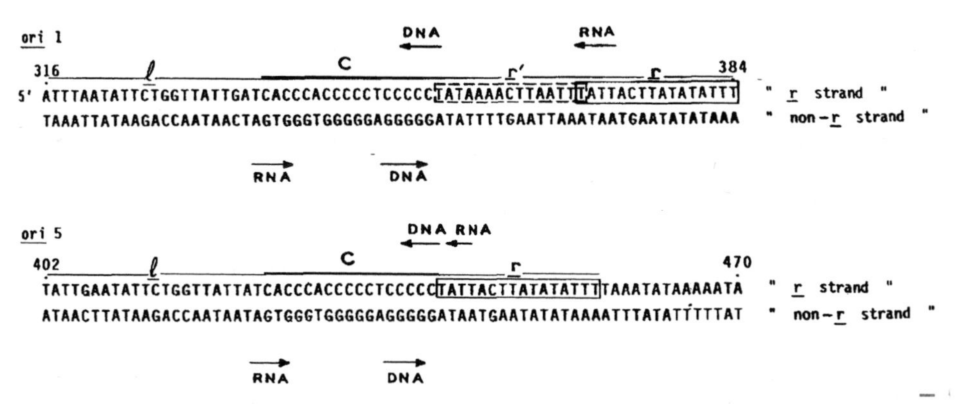

Figure 2.15. Initation of DNA replication. Cluster C and sequences *r* and *r'*) and the end of sequence *l* are shown. Arrows correspond to the position of starting points for RNA and DNA synthesis for each strand of the two *ori* sequences. (From Baldacci et al., 1984).

2.4. *The* ori *sequences as transcription initiation sites*

That *ori* sequences also act as transcription initiation sites is indicated by three results on petite genomes (Baldacci and Bernardi, 1982).

(i) Transcription initiation efficiency parallels replication efficiency. Petite genomes containing some canonical *ori* sequences (*ori* 1, 2, 5, and, to a lesser extent, *ori* 3) are transcribed very actively; others, containing *ori* 4, or deleted in their C clusters (but not those deleted in their A clusters), or lacking canonical *ori* sequences (*ori*° petites), are not (**Table 2.1**). Likewise, *ori* sequences containing a γ cluster (*ori* 4, 6 and 7) are probably not very efficient in DNA replication, as suggested by the fact that they are very rarely or never found in extensive screenings of spontaneous petites and may even be absent (*ori* 4) in some wild-type genomes (Faugeron-Fonty et al., 1984).

(ii) Transcription initiates next to the oligopyrimidine stretch of cluster C, at a sequence (**Fig. 2.12**) which is very similar to the transcription initiation sequences of rRNA genes (Osinga and Tabak, 1982), and proceeds from cluster C to cluster A. As already mentioned, the insertion of cluster γ in the middle of this sequence (as in *ori* 4) or the loss of cluster C (as in *ori* C^- petite genomes) is accompanied by a loss of transcriptional activity. This suggests that *ori* 2 and *ori* 5 might be among the initiation sites used for transcribing the sense strand, *ori* 1 and *ori* 3 among those used for the transcription of the other strand. A small number of other sequences largely homologous to the transcription initiation sites of tRNA genes have also been found. These might play a role in the multipromotor transcription (Levens et al., 1981) of the mitochondrial genome of wild-type cells which had been postulated by Prunell and Bernardi (1977).

(iii) Since transcriptionally active *ori* sequences (see above) are present in both orientations on the wild-type genome, it is very likely that both strands are transcribed, although the non-sense strand appears to be transcribed less accurately, or more slowly. Hybridization experiments with separated DNA strands have identified the template strand used in transcription as the strand which contains the oligonucleotide stretch of cluster C. This conclusion supports previous independent evidence (Coruzzi et al., 1981; Baldacci and Bernardi, 1982; Beilharz et al., 1982) and puts the transcription of the mitochondrial genome of yeast in line with that of the animal mitochondrial genome. Similarly, replication might proceed unidirectionally from some *ori* sequences (possibly *ori* 2 and 5 for one strand, and *ori* 1 and 3 for the other).

2.5. The effect of flanking sequences on the efficiency of replication of petite genomes

The mitochondrial genomes of progenies from 26 crosses between 17 cytoplasmic, spontaneous, suppressive, ori^+ petite mutants of *S. cerevisiae* were studied by electrophoresis of restriction fragments. Only parental genomes (or, occasionally, genomes derived from them by secondary excisions) were found in the progenies of the almost 500 diploids investigated; no evidence for intermolecular unequal mitochondrial recombination was detected. One of the parental genomes was always found to predominate over the other one, although to different extents in different crosses. This predominance appears to be due to a higher replication efficiency, which is correlated with a greater density of *ori* sequences on the mitochondrial genome (*i.e.* with a shorter repeat unit size of the latter).

Exceptions to this **'repeat-unit-size rule'** (see **Section 2.3**) were found, however, even when the parental mitochondrial genomes carried the same *ori* sequence (compare petites b11 and b7, $a^*_{1/7/8}$ and $a^*_{1/7/20}$, for instance). This indicates that noncoding, intergenic sequences flanking *ori* sequences also play a role in the **modulation of replication efficiency** (see **Fig. 2.16**). Since in different petites such sequences differ in primary structure, size, and position relative to *ori* sequences, this modulation is likely to take place through an indirect effect on DNA and nucleoid structure (Rayko et al., 1988; see **Fig. 2.21**).

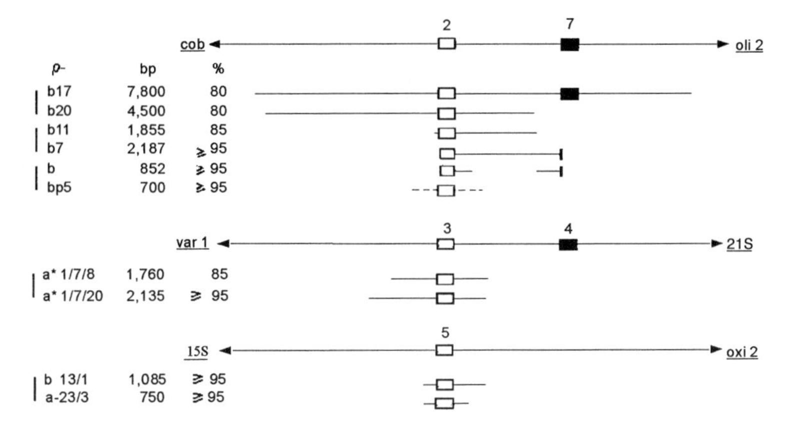

Figure 2.16. Repeat units of petites (ρ^-) harboring mitochondrial genomes carrying identical *ori* sequences and exhibiting suppressivities (%) and/or competitive abilities in petite x petite crosses which do not follow the repeat-unit-size rule (see Text). Repeat unit sizes (bp) and suppressivities are indicated. Open boxes indicate active *ori* sequences 2, 3 and 5, black boxes indicate inactive *ori* sequences 4 and 7 (de Zamaroczy et al., 1984). The *ori* sequences are all oriented with cluster C to the right of cluster A. Flanking genetic markers are indicated. Vertical bars on the left indicate the pairs of petites that were compared. (From Rayko et al., 1988).

2.6. The ori$^-$ *petites 14 and 26*

The effect of temperature on the replicative ability of the mitochondrial genomes was investigated on three isonuclear, spontaneous, cytoplasmic petite mutants of *S. cerevisiae,* al/1R/14, al/1R/1/26 and al/1R/Z1, which here will be referred to as petites 14, 26 and Zl, respectively (Goursot et al., 1988). The primary structure of the mitochondrial genomes of these petites had been previously determined (Gaillard et al., 1980; de Zamaroczy et al., 1981; 1983; 1984). Each one of these genomes arose from the excision of a DNA segment from the mitochondrial genome of yet another petite, a1/1R/1 (see **Fig. 2.17**). As is the rule for spontaneous, cytoplasmic petites (Gaillard et al., 1980; de Zamaroczy et al., 1983), these excisions (i) took place by unequal recombination events involving pairs of short direct repeats (shown in **Fig. 2.17**) and (ii) were followed by tandem amplification of the excised segments, which became the repeat units of the newly formed genomes of the secondary petites Z1, 14 and 26. While the repeat unit of the parental petite genome is 884 bp long, the repeat units of petites Z1, 14 and 26 are 416, 392 and 398 bp long, respectively. In the latter three cases, each repeat unit corresponds to about 0.5% of the wild-type mitochondrial genome and only contains about 100 bp in addition to one *ori* sequence, *ori*1. The amount of mitochondrial DNA in these petites being roughly the same as in wild-type cells, these petites contain about 50 times (200/4) more *ori* sequences than the parental wild-type cells.

While each repeat unit of the mitochondrial genomes of petites al/1R/1 and Z1 contains a complete *ori* sequence, those of petites 26 and 14 contain *ori* sequences which had lost 11 and 27 bp, respectively, in the excision process. In the case of petite 26, the deletion comprises GC cluster A; in the case of petite 14, it comprises also part of the neighbouring sequence *p* (de Zamaroczy et al., 1981; 1984; **Figs. 2.12** and **2.17**). In the complete *ori*

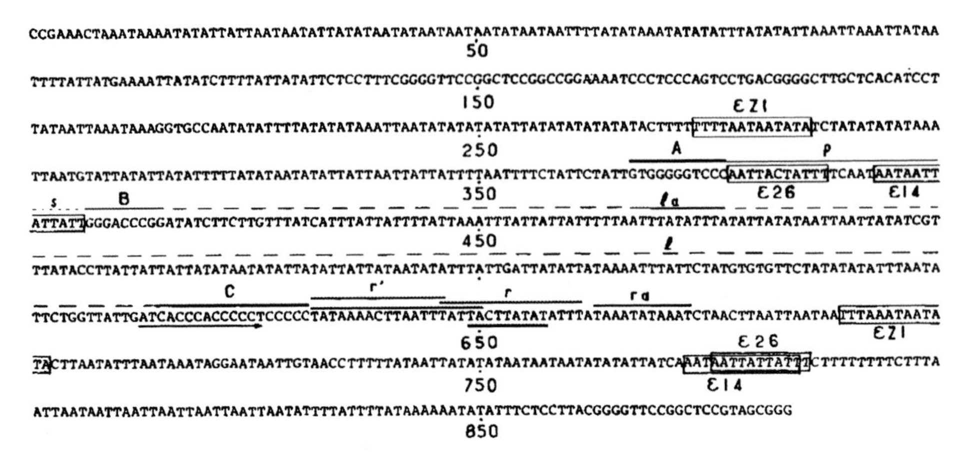

Figure 2.17. The repeat unit of the mitochondrial genome of primary petite a1/1R/1 (see Gaillard et al., 1980) is 884 bp in size and contains an *ori* sequence, *ori*1. This comprises three GC clusters, A, B and C (heavy overlines). Cluster B is separated from cluster A by two short AT sequences *p* and *s*, and from cluster C by a long AT sequence *l* (broken overline). Cluster C is followed by sequence *r* (light overline) which comprises, on the strand shown, a nonanucleotide (heavy underline) where transcription of the RNA primer for the other nascent DNA chain starts before cluster C (arrow pointing to the right) (Baldacci et al., 1984). Sequence r′ is followed by a sequence *ra* which might pair with sequence *l*a (within the *l* sequence) to form a hypothetical tertiary structure (de Zamaroczy et al., 1984; see Fig. 2.21). In the case of *ori*1, sequence *r* is preceded by a duplication, sequence r. Excision sequences (ϵ) of the secondary mitochondrial genomes from petites 14, 26 and Z1 (see Gaillard et al., 1980; de Zamaroczy et al., 1984) are indicated by boxes. (From Goursot et al., 1988).

sequences of *ori*$^+$ petites al/1R/1 and Z1, a secondary structure, the 'A-B fold', is possible (de Zamaroczy et al., 1981; 1984). This is a stem-and-loop structure, comprising 11 AT bp and 6 GC bp, which is formed by GC clusters A and B and the sequence in between (**Figs. 2.17–2.19**). In contrast, the formation of this structure is impossible in the *ori*$^-$ genomes of petites 14 and 26, which can form, however, '**replacement folds**' made of pure AT (**Fig. 2.19**), endowed, therefore, with a lower thermodynamical stability (especially in the case of petite 14) compared to the A-B fold (de Zamaroczy et al., 1984).

The *ori*$^-$ mitochondrial genome of petites 14 and 26, which lack a normal feature of the *ori* sequence, the A-B fold (a feature which is perfectly conserved in all *ori* sequences; de

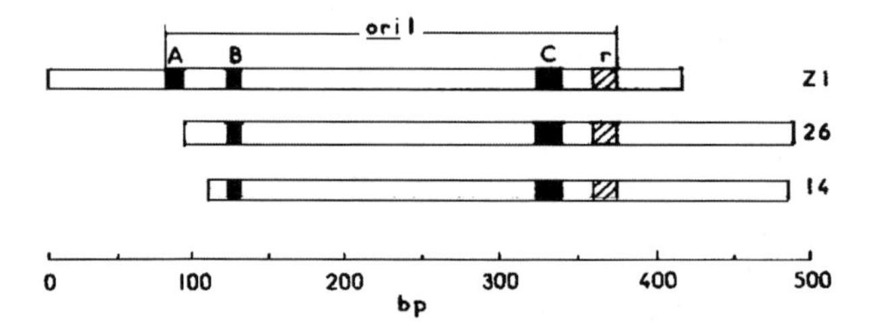

Figure 2.18. The repeat units of the mitochondrial genomes of petites Z1, 26 and 14. The *ori*1 sequence, with its main elements, GC clusters A, B, C (black boxes) and sequence *r* (hatched box), is also represented. (From Goursot et al., 1988).

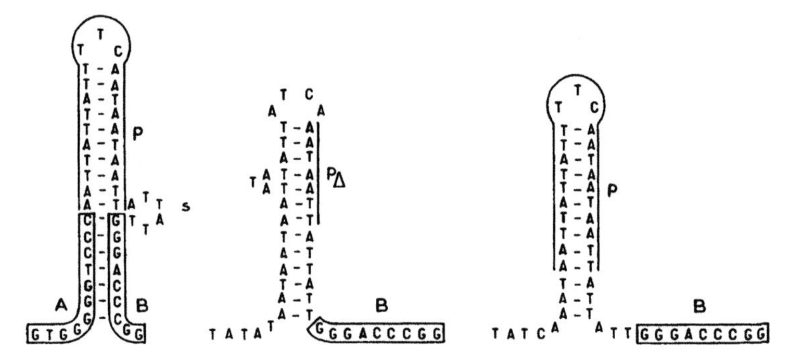

Figure 2.19. Potential secondary structure of the A-B fold of the *ori* sequence present in the repeat units of mitochondrial genomes from petites a1/1R/1 and Z1 (*ori*1) and of the replacement folds that can be formed in the *ori*1 sequence present in the mitochondrial genomes of petites 14 and 26. In the case of petite 14, the residual nucleotides from the partially deleted *p* stretch ($p\Delta$) can generate a hairpin structure with nucleotides from the preceding repeat unit, but the stem, only formed by 13 A/T nucleotides, carries different terminal and side loops compared to the A-B fold. In the case of petite 26, the upper part of the stem and terminal loop are identical to those of the A-B fold, but the lower part is replaced by three A:T pairs (three nucleotides are derived from the preceding repeat unit). The sequence involved in the structure shown (GC clusters A and B, sequences *p* and *s*) are those indicated in Fig. 2.13. (From Goursot et al., 1988).

Zamaroczy et al., 1981; 1984), do exhibit lower replicative abilities compared to the ori^+ mitochondrial genomes of petites al/1R/1 and Z1 (as well as of all ori^+ petite genomes having repeat units shorter than about 1000 bp; Rayko et al., 1988). This can be judged by the suppressivity test (Ephrussi et al., 1955). Indeed, when crossed with wild-type cells, ori^- petites 26 and 14 yield progenies comprising only 85% and 79% diploid petites, respectively, instead of 99% produced by crosses involving ori^+ petites al/1R/1 and Z1 (**Table 2.2**). Incidentally, much more dramatic decreases in suppressivity are associated with petite genomes deleted in the crucial C cluster region (de Zamaroczy et al., 1981), or lacking *ori* sequences altogether; as already mentioned, these latter ori^o petite genomes could replicate by making use of surrogate origins of replication, namely of ori^s sequences.

TABLE 2.2

Suppressivity of petites Z1, 26 and 14 at different temperatures[a]

Temperature[b] (°C)		Suppressivity (%)[c]		
	Petites:	Z1	26	14
23°		99	89	90
28°		99	85	79
33°		97	58	45

[a] From Goursot et al. (1988)

[b] Temperatures indicated were applied to precultures and to cultures used in crossings and in the incubation of diploids. Precultures in stationary phase were diluted ten times upon setting up the cultures used in crossings.

[c] Suppressivity was tested by crossing petite strains *(MATa, ade1)* against the wild-type strain B *(MATα, trp1, his1)*. The percentage of petite/total diploid colonies was determined by counting about 1000 colonies on minimal agar.

2.7. Temperature and the replicative ability of ori^{-} *petites 14 and 26*

In view of the presumed lower thermodynamical stability of replacement folds compared to the A-B fold, crosses of petites 14 and 26 with wild-type cells were also performed at both a higher (33°C) and a lower (23°C) temperature than the standard one, 28°C. The results of **Table 2.2** indicate that such temperatures decreased and increased, respectively, the replicative ability of the two mutants, whereas that of the control *ori*$^{+}$ petite Z1 did not show any significant change between these temperatures. The decrease of the replicative ability was stronger for petite 14 than for petite 26, as expected from the presumably lower thermodynamical stability of the replacement fold of the former. In the first case, the proportion of wild-type colonies obtained at 23°C and 33°C, respectively, was 10% and 55%, in the second 11% and 42%. The changes in replicative ability just described are immediate and reversible.

The results presented indicate that an environmental factor, **temperature, can reversibly affect the replicative ability of a genome** by altering its secondary (and possibly, its tertiary) structure. Indeed, (i) these changes cannot be ascribed to enzymes involved in DNA replication, as in temperature-sensitive mutants, since the petites discussed here are isonuclear, and lack mitochondrial protein synthesis, like all petites; (ii) the different effects of temperature on the replicative ability of petites Z1, 14 and 26 show an excellent correlation with those expected from the secondary structures of the postulated A-B fold and replacement folds (de Zamaroczy et al., 1981; 1984), an effect on tertiary structures being also possible.

It should be pointed out that the repeat unit of petite Z1 extends 80 bp to the left of cluster A and only 40 bp to the right of sequence *r*, whereas those of petites 26 and 14 extend 115 bp to the right of sequence r (**Fig. 2.18**); in all cases, however, these extensions to the left of *ori* sequences are just made of AT spacer. Effects of flanking regions of *ori* sequences on the replicative ability of the latter are known (Rayko et al., 1988; see **Section 2.5**), but they are small compared with the different suppressivities exhibited by petites 14 and 26 relative to petite Z1. There is, therefore, no doubt that these differences are due to the deletions in the *ori* sequences of petites 14 and 26.

Interestingly, the conclusion that DNA secondary structure is required for *ori* activity *in vivo* was also reached for bacteriophage G4, where a strong temperature-dependent impairment of replication was found after introducing by site-directed mutagenesis point mutations which destabilize intra-strand base-pairing in the *ori* sequence (Lambert et al., 1987).

From a general viewpoint, our results indicate the existence of **a novel type of environment-genome interaction**, in which reversible changes in higher-order DNA structures are induced with profound consequences on a basic genome function, such as replication. These changes consist in **genome transconformations** which, although non-inheritable as such, can be maintained for many generations in the presence of the appropriate environmental condition. Genome transconformations can provide, therefore, strong selective advantages or disadvantages and play an important role in evolution, in addition to classical mutations, which involve changes in the primary structure of DNA. Moreover, similar phenomena might (i) be induced by other environmental factors; (ii) affect other genome functions (*e.g.*, transcription); and also (iii) be operative in other organisms.

CHAPTER 3

The organization and evolution of the mitochondrial genome of yeast

3.1. The organization of the mitochondrial genome of yeast

The organization of the mitochondrial genome of yeast, as it has emerged from our work, is similar to that of the nuclear genome of eukaryotes in that (i) coding sequences are interspersed with non-coding sequences, and (ii) the non-coding sequences are similar to the interspersed repeated sequences and the fold-back sequences, identical or similar sequences being present in many copies in the genome; moreover, (iii) the genome contains two genes, *cox1* (cytochrome oxidase sub-unit 1) and *cytb* (apocytochrome b), that comprise several introns (in some strains), some of which are translated into maturases, reverse transcriptases or site-specific endonucleases; the 21S RNA gene also may comprise an intron (see Foury et al., 1998, for a review).

As already mentioned (see **Chapter 1, Section 1.1**), these genome features are not at all unique to the mitochondrial genome of yeast. Indeed, we have shown the presence of non-coding AT spacers in the mitochondrial genome of *Ustilago cynodontis* (Mery-Drugeon et al., 1981), a fungus belonging to the class of *Basidiomycetes* (in contrast of *S. cerevisiae* which is an *Ascomycete*), as well as in both the mitochondrial and chloroplast genomes of *Euglena gracilis,* a unicellular flagellate (Stutz and Bernardi, 1972; Fonty et al., 1975; Schmitt et al., 1981; Heizmann et al., 1981). A similar situation was also found in the mitochondrial genomes of other yeasts, *Torulopsis glabrata*, *Brettanomyces anomalus*, and *Kloeckera africana* (Clark-Walker and McArthur, 1978). Interestingly, another chloroplast DNA, from a higher plant, *Spinacia oleracea*, which is characterized by a higher GC level, 36.5%, also revealed a high degree of heterogeneity, 30% of the genome being only 22% GC and 10% of the genome being higher than 60% GC (Schmitt et al., 1981).

These results stress an important point, that mitochondrial genomes (to limit our discussion here to only one of the two organelle genomes) may range from the extremely compact structures that characterize not only most animal mitochondrial genomes, but also the mitochondrial genomes from some unicellular organisms (*e.g.*, *Schizosaccharomyces pombe*), to structures in which non-coding sequences form the majority of the genome, the best studied case being that of the mitochondrial genome of yeast. The monophyletic origin of the mitochondrial genome from a prokaryotic ancestor then raises two problems concerning **the mechanism of formation and the biological role of non-coding sequences** in the large mitochondrial genomes of yeasts.

Before we move to these subjects in the following sections, it should be stressed that the organization of the mitochondrial genome of yeast points to the fact that in eukaryotic genetic systems, where so much of the DNA is non-coding, there is a real need for a molecular approach, because approaches based on classical genetics or on the study of gene products suffer from serious intrinsic limitations and are unable to provide an overall picture of the genome. Incidentally, this organization disposed of a series of ideas, which were promoted for many years, and can be summarized as follows: (i) that the mitochondrial genome of yeast had a unique nucleotide sequence, namely a sequence lacking internal repetitions; this view had its origin in a misunderstanding of the renaturation kinetics of mitochondrial DNA; (ii) that the mitochondrial genome of yeast had, as a consequence, an informational content five times larger than that of animal mitochondrial genomes (which have a unit size of only 17 kb); (iii) that the unit size of the mitochondrial genome decreased in the evolution from unicellular organisms to animals. In conclusion, **the mitochondrial genome is a useful, simple model for the nuclear genome**.

3.2. The evolutionary origin of ori *sequences*

The extremely high homology of the eight canonical *ori* sequences indicates that they arose as the result of duplication and translocation events. More precisely, we proposed (see **Fig. 2.20**) that the canonical *ori* sequences derive from a primitive *ori* sequence (probably made of only a monomeric cluster C and its flanking sequences *r** and *r*) through (i) a series of duplications and inversions generating clusters A and B; and (ii) an expansion process producing the AT stretches of *ori* sequences. It is possible that the *ori* sequences are folded in a tertiary structure, as in the hypothetical model of **Fig. 2.21**.

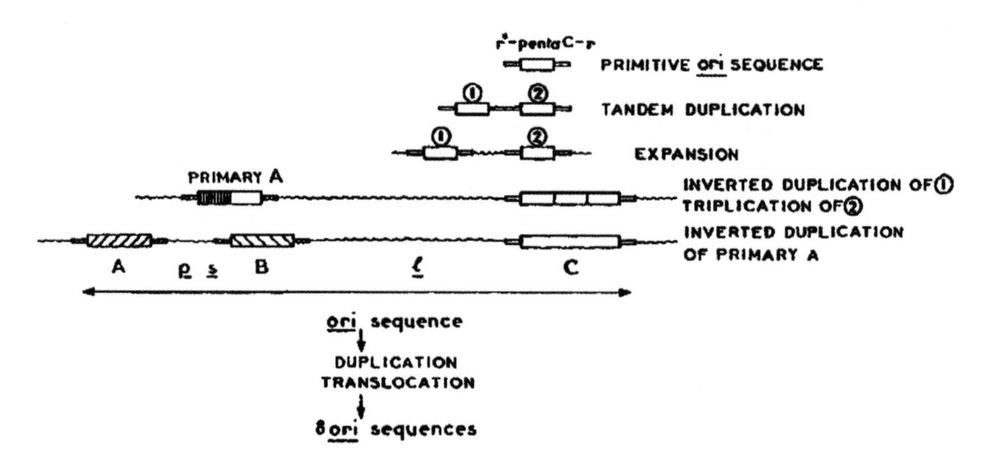

Figure 2.20. Hypothetical scheme for the evolutionary construction of *ori* sequences. (From de Zamaroczy and Bernardi, 1986a).

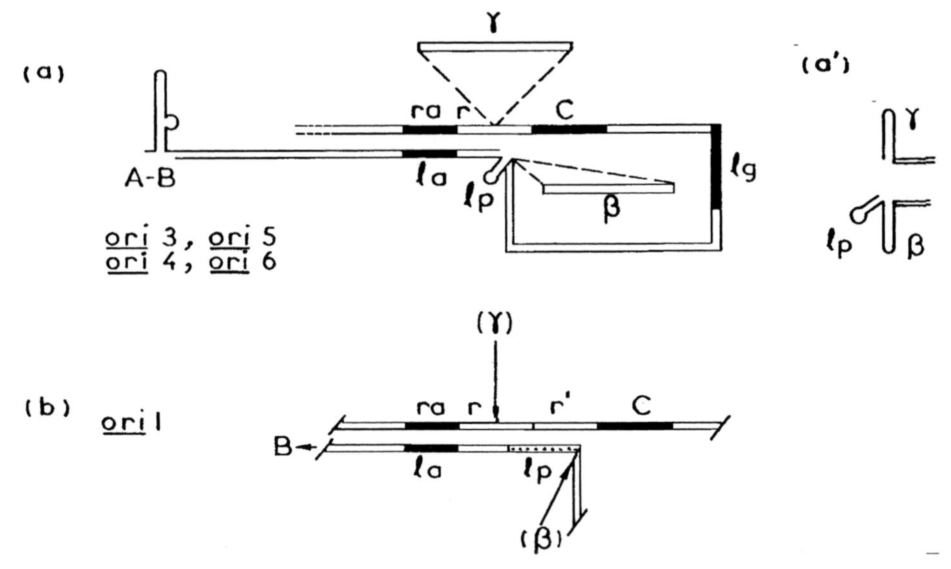

Figure 2.21. Hypothetical superfolding of *ori* sequences from the mitochondrial genome of yeast. (a, a') Superfolding of *ori* 3 and *ori* 5; base-pairing interactions take place between sequences *l*a and *r*a as well as between flanking sequences; *l*p sequence is supposed to loop out. In *ori* 4 and *ori* 6, the size of the loop (132 bp) between sequences *l*p and C would be considerably extended (by 70 bp) because of the insertion of clusters β and γ. This does not occur, however, if these clusters take the configuration shown in (a). (b) Folding of *ori* 1. In the case of *ori* 1, *l*p cannot fold upon itself, but can interact with sequence *r*; this interaction may also take place in *ori* 2. The overall result is no change in the size of the loop. Only in the case of *ori* 7, the loop would be larger by 30 bp, the size of cluster γ. (From de Zamaroczy and Bernardi, 1986b).

3.3. The evolutionary origin of the GC clusters

We studied the primary and secondary structures, the location and the orientation of the 196 GC clusters present in the 90% of the mitochondrial genome of *S. cerevisiae* which had been sequenced by 1986 (see de Zamaroczy and Bernardi, 1986b; the later completion of the sequence by Foury et al., 1998, did not change anything substantial in what follows). We found that (i) the vast majority of GC clusters is located in intergenic sequences (including *ori* sequences, intergenic open reading frames and the *var 1* gene; see next section) and in intronic closed reading frames (CRF's); (ii) most of them can be folded into stem-and-loop structures; (iii) both orientations are equally frequent. The primary structures of GC clusters permit to group them into eight families, seven of which are clearly related to the family formed by clusters A, B and C of the *ori* sequences. Most GC clusters apparently originated from primary clusters also derived from the primitive *ori* sequence in the course of its evolution towards the present *ori* sequences (de Zamaroczy and Bernardi, 1986a).

3.4. The evolutionary origin of the AT spacers and the var 1 *gene*

Intergenic sequences represent 63% of the mitochondrial 'long' (85 kb) genome of *S. cerevisiae*. They comprise 170–200 AT spacers that correspond to 47% of the genome and are separated from each other by GC clusters, ORFs, *ori* sequences, as well as by protein-coding genes. Intergenic AT spacers have an average size of 190 bp, and a GC level of 5%; they are formed by short (20–30 nt on the average) A/T stretches separated by mono- to tri-C/G. An analysis of the primary structures of intergenic AT spacers has shown that they are characterized by an extremely high level of short sequence repetitiousness and by a characteristic sequence pattern; the frequencies of A/T isostichs (oligonucleotides having the same size) conspicuously deviate from statistical expectations, and exponentially decrease when their (AT+TA)/(AA+TT) ratio, R, decreases. A situation essentially identical was found in the AT spacers of the small mitochondrial genome (19 kb) of *T. glabrata*. The sequence features of the AT spacers indicate that they were built in evolution by an **expansion process** mainly involving rounds of duplication, inversion and translocation events which affected an initial oligodeoxynucleotide (characterized by a particular R ratio) and the sequences derived from it. In turn, the initial oligodeoxynucleotide appears to have arisen from an ancestral promoter-replicator sequence which was at the origin of the nonanucleotide promoters present in the mitochondrial genomes of several yeasts (de Zamaroczy and Bernardi, 1987).

Common sequence patterns indicate that the AT spacers so formed gave rise to the *var1* gene (by linking and phasing of short ORFs), coding for the only mitochondrially encoded protein of the large ribosome sub-unit, to the DNA stretches corresponding to the untranslated mRNA sequences and to the central stretches of *ori* sequences. The case of *var 1* gene (Hudspeth et al., 1982) is of special interest in connection with the idea of the expansion of *ori* sequences. This gene is 10% GC and contains a 46 bp GC cluster accounting for 38% of total GC. Its similarity with spacer-cluster sequences is so striking that it suggests that this gene arose from an intergenic sequence only recently. Incidentally, *var 1* provided the first example of the generation of coding from non-coding sequences (de Zamaroczy and Bernardi, 1987), a subject of great current interest (see Jordan et al., 2003, and **Part 6, Section 2.3**).

3.5. The non-coding sequences: evolutionary origin and biological role

To sum up the preceding three sections, as far as the intergenic sequences are concerned, it is conceivable that they were derived from *ori* sequences by the expansion process proposed by Bernardi (1982a) on the basis of sequence comparisons (Bernardi and Bernardi, 1980). This might have taken place through three different mechanisms, all of which are likely to have played a role. First, a slippage of the replicase could occur at the *ori* sequences; this is a well-known phenomenon first studied in the reiterative replication of poly (dAT:dAT) by DNA polymerase I of *E. coli* (Kornberg et al., 1964). A second mechanism is unequal crossing-over; evidence for the high frequency of such a phenomenon in mitochondrial recombination is available (Fonty et al., 1978). A third mechanism is insertion. Almost all GC clusters are inserted in AT spacers; some rare ones are inserted in AT-rich regions of rRNA genes (Sor and Fukuhara, 1982) and even in a protein-coding gene, *var 1* (see above). Interestingly, these insertions are not only transcribed but, in the case of *var 1*, also translated.

Another interesting result (Bernardi, 1982a; de Zamaroczy and Bernardi, 1986a) was that the closed reading frames (CRF) of the intervening sequences of *oxi* 3 and *cob* genes share all the features of intergenic non-coding sequences. Indeed, when the relative amounts of di- to hexa-nucleotides were compared to those from random sequences having the same sizes and compositions, they were found to exhibit the same deviations as the intergenic noncoding sequences of the mitochondrial genome. In contrast, intronic open reading frames (ORFs) showed oligonucleotide patterns which were generally quite distinct from those of CRFs, although some similarities could be detected in some cases (especially for aI5α). The mitochondrial introns of yeast, therefore, are endowed with a mosaic structure, in which CRFs derive from mitochondrial intergenic sequences, whereas ORFs have a different origin (indicated as exogenous by other evidences).

It is evident that regulatory sequences acting as promotors, operators, sites for the initiation of replication, and sites involved in the processing of transcripts are present in the non-coding sequences of yeast mitochondrial DNA. Another function of the non-coding sequences is also well documented and has to do with illegitimate, unequal recombination. The excision of the spontaneous petite genomes just described is an example of these extragenic recombinational events. The same basic mechanism appears, however, to be more general and to account for: (i) the divergence of the mitochondrial genomes of wild-type yeast cells; it has been shown (Bernardi et al., 1975; Prunell et al., 1977) that different strains have mitochondrial genomes differing in the length of AT spacers, apparently the result of unequal crossing-overs in the sequences of spacers; (ii) similar changes in the mitochondrial genomes of the progeny arising from crosses of different wild-type strains (Fonty et al., 1978).

In conclusion, the evidence available at the present time appears to support the idea that the complex sequence organization of the mitochondrial genome of yeast corresponds to **the needs of very active and finely regulated replication, transcription, and recombination** processes. This seems to be achieved at the price of an **exceptional genomic instability**.

Twenty years ago, this genome instability was discussed as follows (Bernardi, 1983): The non-coding sequences of the mitochondrial genome of yeast are the source of three disadvantages for the genome. The main one is that the abundant direct repeats that they contain are potential excision sequences; as such, they are responsible for the extreme instability of the genome. The other two are that they increase replication time and energy expenditure. These disadvantages would quickly change wild-type yeast cells into suppressive and neutral petite mutants, if this intracellular selection were not counterbalanced by an intercellular selection in which the faster growing respiratory-competent wild-type cells compete out the respiratory-deficient petite mutants (incidentally, this accounts for the fact that in nature only wild-type yeast cells are found). Even if the disadvantages associated with the non-coding sequences do not lead, therefore, to the elimination of the mitochondrial genome, they should at least lead to the elimination of the non-coding sequences themselves. We know, however, that, although *S. cerevisiae* strains exist which lack a number of intervening sequences (introns) and also *ori 4*, in general non-coding sequences tend to be largely conserved. This indicates that the removal of non-coding sequences is selectively disadvantageous or, in other words, that non-coding sequences provide selective advantages which compensate for the disadvantages associated with them. This obviously raises the question of the nature of these advantages, namely of the physiological roles played by non-coding sequences.

It is clear from the genome map of the mitochondrial genome of yeast that the deletion of intergenic sequences, where most excision sequences used in the spontaneous mutation are located (de Zamaroczy et al., 1983), will frequently remove canonical *ori* sequences from the wild-type genome. Even if a wild-type genome lacking *ori 4* has been found, it is evident that in general such elimination will affect replication and also transcription. There is therefore a selective advantage in keeping *ori* sequences in the wild-type genome. As far as non-coding sequences outside *ori* sequences are concerned, it should first of all be stressed that the expansion process does not propagate non-sense sequences, but propagates instead sequences which have been highly selected and conserved in evolution and whose primary role is to interact specifically with enzymes involved in DNA replication and transcription. Thus, the expansion of *ori* sequences leads to the propagation of potential regulatory signals, which may be used in the regulation of gene expression, in repression (anaerobiosis or glucose can shut off transcription) and derepression, in the processing of primary transcripts, and in the regulation of nucleo-mitochondrial interactions. Another physiological role of non-coding sequences concerns recombination, since evidence exists (Fonty et al., 1978) that repeated and palindromic non-coding sequences are involved in mitochondrial site-specific recombination. Finally, it should be recalled that some non-coding sequences appear to be inserted into transcribed genes or even to be transformed into genes. **In summary**, a number of physiological roles apparently provide selective advantages compensating for the disadvantages inherent in the very existence of non-coding sequences. What we know about the non-coding sequences of the mitochondrial genome of yeast suggests that their conservation in the genome is due to selective advantages associated with their physiological roles; these sequences, or at least their majority, cannot, therefore, be considered "selfish DNA sequences" (*sensu* Doolittle and Sapienza, 1980; Orgel and Crick, 1980).

A final point on which the mitochondrial genome of yeast is relevant to the "selfish DNA" issue is the occurrence of functionless genomes in suppressive petites. Many of these genomes contain no gene, and yet replication, transcription and even transcript processing may still go on. In nature, as already pointed out, these genomes rapidly disappear since petites are competed out by the faster growing wild-type cells. Many of these genomes have such a replicative advantage over wild-type genomes that they could spread out through crosses with wild-type cells, if haploid. This does not occur in nature because parental wild-type cells and the derived petites have the same mating type. When isolated from competition with wild-type cells in the laboratory, however, petites not only survive, but frequently end up with very stable genomes which are the result of a selection on the basis of replication efficiency. These "**genomes without genes**" are practically made up of repeat units containing barely more than an *ori* sequence; replication is most efficient, the corresponding petites being supersuppressive, and transcription is only preserved because of the role played in replication. In other words, functionless genomes like the mitochondrial genomes of suppressive petites not only can exist and be quite stable, but they undergo a selection favoring those which are closest to the ultimate situation of being just a set of replication origins; this *in vivo* selection is very much the same found in the Qβ-replicase *in vitro* system by Mills et al. (1967). These "selfish genomes", exemplifying primordial self-replicating systems, will however be lost in the long run, when mitochondrial or nuclear mutations will inactivate the initiation of replication.

Part 3
The organization of the vertebrate genome

CHAPTER 1

Isochores and isochore families

1.1. The fractionation of the bovine genome

The success met by the Cs_2SO_4/Ag^+ ultracentrifugation in separating satellite DNAs from mouse and guinea pig (Corneo et al., 1968; see **Part 1, Section 1.3**) encouraged us, at the beginning of the 1970's, to approach the fractionation of another mammalian DNA, bovine DNA, which was known (Corneo et al., 1970) to comprise two satellites (at that time calf thymus DNA was the standard mammalian DNA because it could be easily prepared in gram amounts). This allowed us (Filipski et al., 1973) to resolve the bovine genome into seven DNA components (see **Fig. 3.1**), which comprised: (i) four satellite DNAs, classified as such on the basis of their narrow CsCl bands, sharp melting curves and fast reannealing properties; these 1.705, 1.710, 1.714 and 1.723 satellite components (DNA components are indicated by their modal buoyant density in CsCl) represented 4%, 1.5%, 7% and 1.5% of the DNA, respectively; the last two were the only known satellites at the time of our work; note that, at this point, only four satellite DNAs were resolved from the major components, and that the separation of the remaining four satellite DNAs required

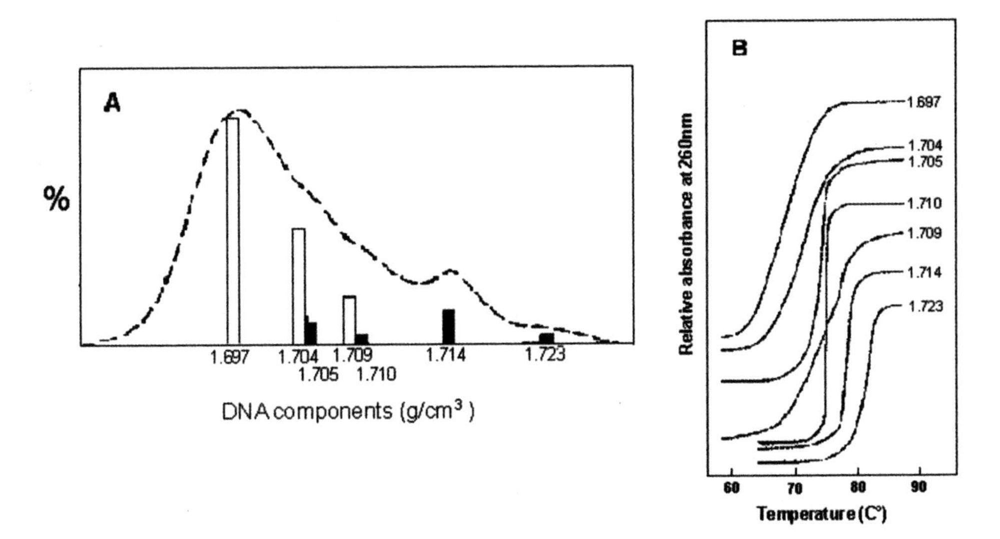

Figure 3.1. **A**. A histogram of the DNA components of the bovine genome. The height of each bar is proportional to the percentage of each component in DNA; solid bars correspond to the sharp-melting, open bars to the broad-melting components (see **B**). The broken line is an enlarged band profile of calf DNA in CsCl density gradient. **B**. Absorbance-temperature profiles for calf thymus DNA components in 0.1xSSC (standard saline citrate). Compare the very sharp transitions of 1.705, 1.714 and 1.723 g/cm^3 satellites with the broader ones of 1.697, 1.704 and 1.709 g/cm^3 major DNA components. (From Filipski et al., 1973).

a combination of different density gradient centrifugations (see below); and (ii) three major components, 1.697, 1.704 and 1.709, forming about 50%, 25% and 10% of the genome, respectively. Moreover, a number of minor components, formed together 4% of the genome; two of them, 1.719 and 1.699, were likely to correspond to ribosomal and mitochondrial DNAs, respectively (see **Table 3.1**).

Further work, in which we used both Ag^+ and another sequence-specific DNA ligand, BAMD, bis(acetato-mercuri-methyl)dioxane (Cortadas et al., 1977; Macaya et al., 1978; Kopecka et al., 1978; Meunier-Rotival et al., 1979), led to a complete picture of bovine satellite and ribosomal DNAs. Indeed, these investigations (i) precisely defined the relative amounts and buoyant densities of both satellite and minor components and revealed that as much as 23% of the bovine genome was formed by eight GC-rich satellite DNAs; (ii) showed that, as a consequence, the standard deviation of the main CsCl band of bovine DNA had been overestimated by Sueoka (1959), a mistake justified by the fact that the work was performed at a time when the very existence of satellite DNAs was not yet known (see also Elton, 1974); and (iii) stressed the much higher resolving power of density gradient centrifugation over the reassociation approach.

As far as the last point is concerned, while the reassociation approach could only resolve four classes of sequences, non-repetitive, intermediate, fast and very fast (Britten and Smith, 1968; Davidson and Britten, 1973), density gradient centrifugation could resolve (Macaya et al., 1978) four major, eleven minor (originally defined as DNA components representing less than 3% of the genome) and eight satellite components (see **Table 3.1**). Moreover, a comparison of the two sets of data indicated that most of the bovine satellite sequences reassociate in the intermediate class. This demonstrated that not all satellites are fast-reassociating, as originally suggested on the basis of the results obtained with mouse satellite DNA. The slow reassociation of the satellite DNA from the bovine genome was

TABLE 3.1

DNA components of the bovine genome[a]

Method of detection	Type of component	Number	Amount
Renaturation kinetics	Non-repetitive	1	55
	Intermediate	1	38
	Fast	1	2
	Very fast	1	3
	Total	4	98
Density gradient centrifugation	Major	4	73
	Minor	11	4
	Satellite	8	23
	Total	23	100

[a] From Macaya et al. (1978). In this paper the lightest major component of Filipski et al. (1973) had already been resolved into two components. Renaturation kinetics data are from Britten and Smith (1968).

due to the sequence divergence of the short repeats forming them, as shown by the sequencing of the major, 1.715 g/cm^3, satellite (Gaillard et al., 1981). In addition, this sequence and a restriction enzyme analysis allowed us to conclude that all eight satellite DNAs had a common evolutionary origin because they shared similar short basic repeats (Macaya et al., 1978; Kopecka et al., 1978). This conclusion was later confirmed by direct sequencing work (Pech et al., 1979). Further work demonstrated that, indeed, the 1.719 component was ribosomal DNA (Meunier-Rotival et al., 1979).

The demonstration of three well-defined major components by Filipski et al. (1973) forming the main band (**Fig. 3.1**) was an important novel finding, since it disproved the then generally accepted view that the bulk of the genome of higher organisms was formed by DNA molecules showing a continuous variation in GC level.

1.2. The fractionation of eukaryotic main-band DNAs

The fact that the three major components detected by Filipski et al. (1973) accounted for most or all of the main-band DNA of the bovine genome raised the possibility that they might be shared by all mammalian DNAs. Moreover, if the discontinuous compositional pattern was common to all mammalian genomes, one should explore more phyla to detect how far the compositional heterogeneity seen in the bovine genome could be found. Base composition, sedimentation coefficient, modal and mean buoyant density in CsCl were, therefore, investigated for DNAs from 25 eukaryotes ranging from human to yeast (see **Table 3.2** and **Fig. 3.2**). This information (Thiery et al., 1976) was subsequently extended to many more vertebrates (see **Part 4**). Moreover, four mammalian and two amphibian DNAs were fractionated by preparative Cs_2SO_4/Ag^+ density gradient centrifugation and the CsCl band profile of each preparative fraction was analyzed in terms of Gaussian curves (**Fig. 3.3**), to allow assessing the relative amount and the buoyant density of each DNA component (**Fig. 3.4**). This more laborious, but more informative, approach was later extended to several fishes (see **Parts 4 and 12**) and to two reptiles (see **Part 5**).

The main **conclusions** of Thiery et al. (1976) were the following: (i) all eleven mammalian DNAs analyzed exhibited the three major components 1.697, 1.704 and 1.709 g/cm^3, first observed in the bovine genome; similar components appeared in avian genomes and were, likewise, responsible for the trailing of the CsCl bands on the heavy, GC-rich, side; (ii) the DNAs of reptiles, amphibians and fishes showed a much lower skewness to the heavy side of their CsCl bands; (iii) essentially symmetrical bands in CsCl, neglecting satellite bands, were displayed by DNAs from three invertebrates (two echinoderms and *Drosophila*) and from three unicellular eukaryotes, *S. cerevisiae*, *E. gracilis*, *T. pyriformis* (see **Table 3.2**).

The work of Thiery et al. (1976), besides collecting a number of new observations on minor and satellite components, led to the first recognition of **major phylogenetic differences at the macromolecular level in the organization of eukaryotic genomes**. Minor differences in satellite DNAs may also occur in rather closely related species because even a single nucleotide change in a short basic repeat (for instance, a heptanucleotide) may lead to considerable changes in the buoyant density and compositional properties of the satellite.

TABLE 3.2
Properties of eukaryotic DNAs[a]

DNA source	Tissue	$S_{20,w}$	G+C (%)	ρ_0[b] (%)	$<\rho>$[b] (%)	$<\rho> - \rho_0$[b] (mg/cm^3)
ANIMALIA[c]						
CHORDATA						
MAMMALIA						
Homo sapiens (human)	Leucocytes	25		1.6985	1.7008	2.3
	Placenta	22.3	40.3	1.6990	1.7010	2.0
Bos taurus (calf)	Liver	25.2		1.7000	1.7039	3.9
	Thymus	26	43.2	1.6997	1.7033	3.6
Felis domesticus (cat)	Liver	32.7	42.8	1.7002	1.7031	2.9
Canis familiaris (dog)	Liver	31.3	44.1	1.7005	1.7040	3.5
Cavia porcellus (guinea pig)	Liver	24.8	40.1	1.6982	1.7010	2.8
Cricetus norvegicus (Chinese hamster)	Liver	22.6	42.8	1.7000	1.7015	1.5
Rattus sp. (rat)	Liver	27.6	41.8	1.7006	1.7021	1.5
Mus musculus (mouse)	Liver	24.5	40.3	1.7007	1.7008	0.1
					1.7020[d]	1.3[d]
	Thymus	21.6		1.7008	1.7016	0.8
Glis glis (garden dormouse)	Liver	35.8	40.3	1.6991	1.7004	1.3
Oryctolagus cuniculus (rabbit)	Liver	18.6	44.5	1.6999	1.7033	3.4
Erinaceus sp. (hedgehog)	Thymus	32.8	41.9	1.7004	1.7035	3.1
AVES						
Larus argentatus (sea-gull)	Liver	23.7	47	1.6998	1.7031	3.3
Gallus sp. (chicken)	Embryo	20	45	1.7001	1.7031	3.0
REPTILIA						
Iguana iguana	Erythrocytes	20.5	43.9	1.7015	1.7022	0.7
Testudo graeca	Liver	25.1	46	1.7027	1.7042	1.5
AMPHIBIA						
Pleurodeles waltlii	Testes	27.1	46.5	1.7041	1.7047	0.6
Xenopus laevis	Erythrocytes	35	40.9	1.6991	1.6997	0.6
PISCES						
Opsanus tau (toadfish)	Liver	20	42	1.7002	1.7004	0.2
Salmo irideus (salmon)	Testes	20.5	43.5	1.7028	1.7035	0.7
ECHINODERMATA						
ECHINOIDEA						
Strongylocentrotus purpuratus	Sperm	17	36.6	1.6989	1.6995	0.6
Paracentrotus lividus	Sperm	19	35.4	1.6972	1.6974	0.2
ARTHROPODA						
INSECTA						
Drosophila melanogaster	Embryo	27.4	39.1	1.7025	1.7014	–1.1

TABLE 3.2
Properties of eukaryotic DNAs[a] (continued)

DNA source	Tissue	$S_{20.w}$	G+C (%)	ρ_0[b] (%)	$<\rho>$[b] (%)	$<\rho> - \rho_0$[b] (mg/cm^3)
FUNGI						
ASCOMYCOTA						
Saccharomyces cerevisiae	Total	28	37.1	1.6995	1.6984	–1.1
			40[e]		1.7001	0.6[d]
PROTISTA						
EUGLENOPHYTA						
Euglena gracilis	Nuclear	16	49.9	1.7080	1.7080	0
CILIOPHORA						
Tetrahymena pyriformis	Macro nuclei	17.7	29.7	1.6910	1.6910	0
MONERA						
EUBACTERIAE						
Escherichia coli		34	51	1.7105	1.7101	–0.4

[a] From Thiery et al. (1976).
[b] The error on the ρ_0 values is ± 0.0005 g/cm^3; that on the $<\rho>$ values is ± 0.0002 g/cm^3.
[c] Systematic nomenclature according to Whittaker (1969).
[d] $<\rho>$ and $<\rho> - \rho_0$ values calculated for main band DNA, not taking into account the contribution of apparent satellites.
[e] DNA free of mitochondrial DNA and of the 1.705 g/cm^3 satellite (ribosomal DNA).

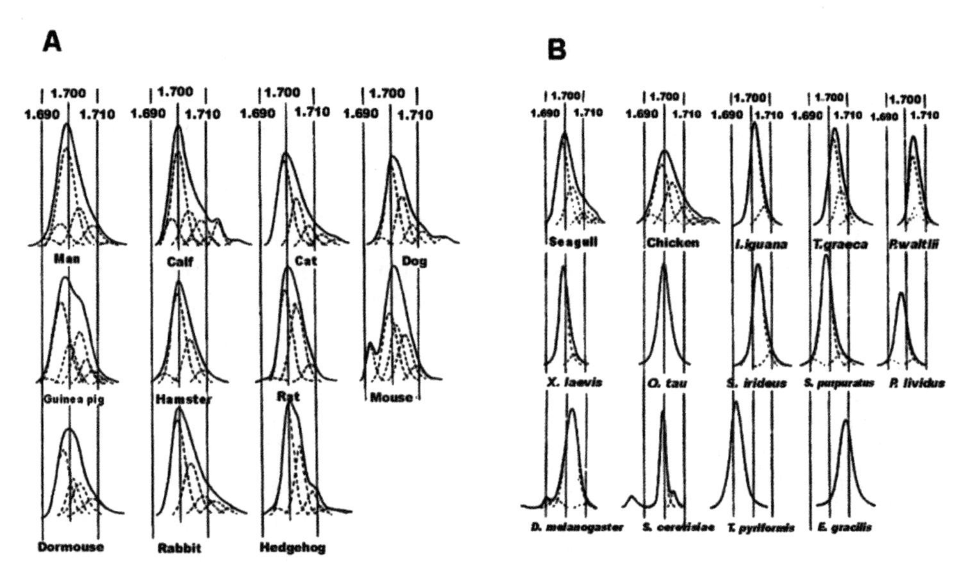

Figure 3.2. CsCl profiles of DNAs **A** from mammals and **B** from birds, reptiles, amphibians, fishes, invertebrates and unicellular eukaryotes. Buoyant densities are indicated at the top of the figures (From Thiery et al., 1976).

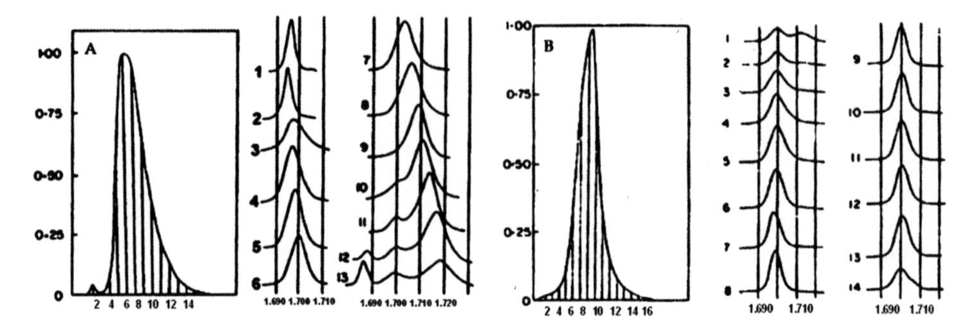

Figure 3.3. Profiles of DNAs in Cs_2SO_4/Ag^+ density gradients. CsCl profiles of fractions are shown on the right. **A** Human placenta DNA **B** Xenopus erythrocytes DNA. (From Thiery et al., 1976).

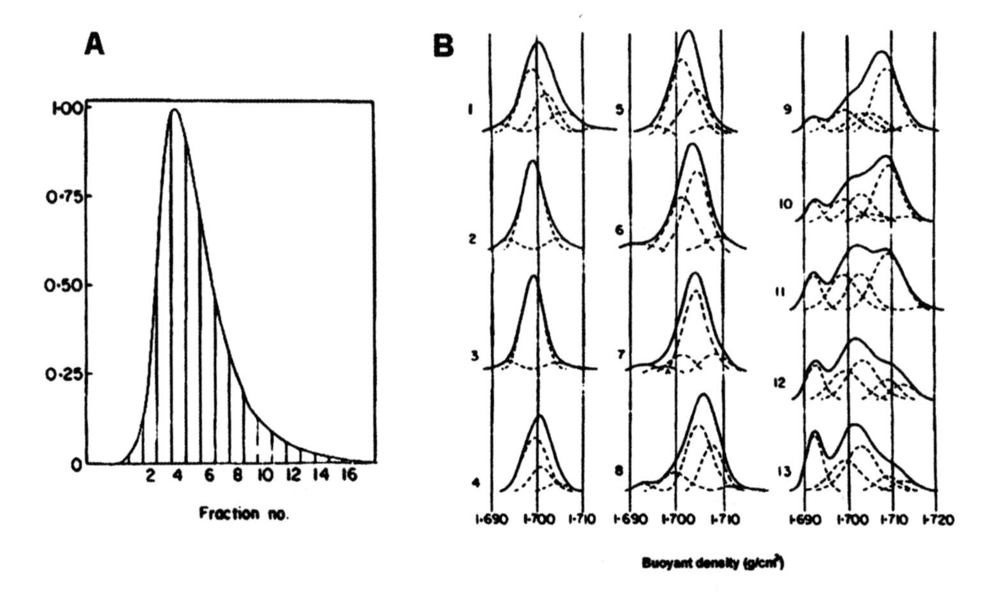

Figure 3.4. **A**. Profile of mouse liver DNA in a preparative Cs_2SO_4/Ag^+ density gradient. **B**. Analytical cesium chloride density gradient profiles of fractions from mouse liver DNA. Gaussian curves (broken lines) represent the different DNA components as resolved using a Dupont 310 curve resolver and the criteria given in the original paper (From Macaya et al., 1976).

1.3. Isochores and isochore families

The results of Thiery et al. (1976) established that gross compositional changes had occurred during the evolution of vertebrates. At the same time, they raised two problems, concerning the original size of the genome regions corresponding to the major DNA components and the physicochemical properties which were responsible for the fractionation.

(i) The problem of compositional homogeneity as related to fragment size was approached by investigating, in terms of buoyant densities and relative amounts of major components, DNAs degraded to about 3 kb, or prepared in such a way as to obtain fragment sizes higher than 300 kb (Macaya et al., 1976). In the first case, only a modest increase in molecular heterogeneity showed up. In the second, discontinuities corresponding to the buoyant densities of the major components appeared in the CsCl profile (**Fig. 3.5**). These results suggested a certain degree of compositional homogeneity in the major DNA components over a very broad molecular size range, 3 kb to over 300 kb. These long, compositionally fairly homogeneous segments originating the major DNA components by degradation during DNA preparation were later called **isochores** for (compositionally) **equal landscapes** (Cuny et al., 1981).

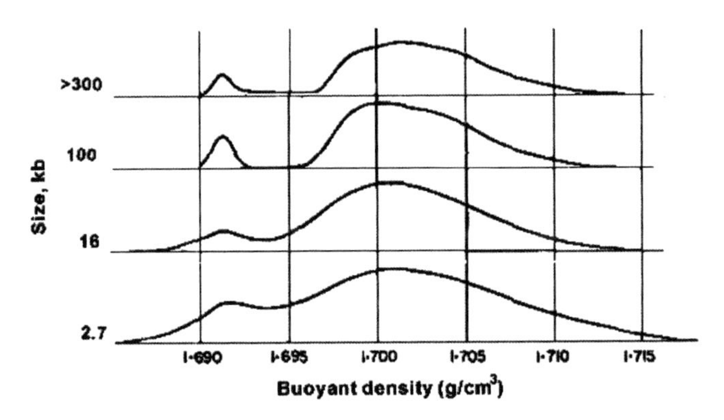

Figure 3.5. Analytical CsCl density gradient profiles of four mouse DNA samples of different molecular weight. Molecular weights were measured from sedimentation velocity experiments except for the highest one which was estimated from the CsCl profile (From Macaya et al., 1976).

The remarkable degree of compositional homogeneity of isochores was also demonstrated by using a completely different approach, namely by analyzing the distribution in preparative density gradients of human DNA fragments (average size 50–100 kb) that carried a particular gene. The first sequence so localized was the mouse β globin gene which was detected in the 1.701 g/cm^3 component, but not in the other components (Cuny et al., 1978). This was followed by the chicken ovalbumin gene found in the 1.7025 g/cm^3 component (Cortadas et al., 1979). **Fig. 1.10** shows that the *c-mos* and the *c-sis* oncogenes were comprised in fractions characterized by modal buoyant densities of 1.7021 and 1.7036 g/cm^3 in the first case and 1.7126 g/cm^3 in the second case. As shown by the *c-mos* example, DNA fragments containing a given, single-copy sequence may be present in more than one fraction, because of Brownian diffusion of DNA molecules in the gradient and/or because of the compositional heterogeneity of the chromosomal region where the sequence under consideration is located. Since the chromosomal segment represented by the population of overlapping fragments carrying the genes under consideration had a size of up to 100–200 kb (see **Fig. 1.11**), it could be concluded that base composition was fairly homogeneous over at least that size of DNA.

Other approaches provided additional information on isochores. (i) When the sequences of long genomic regions around the human α and β globin gene clusters became available,

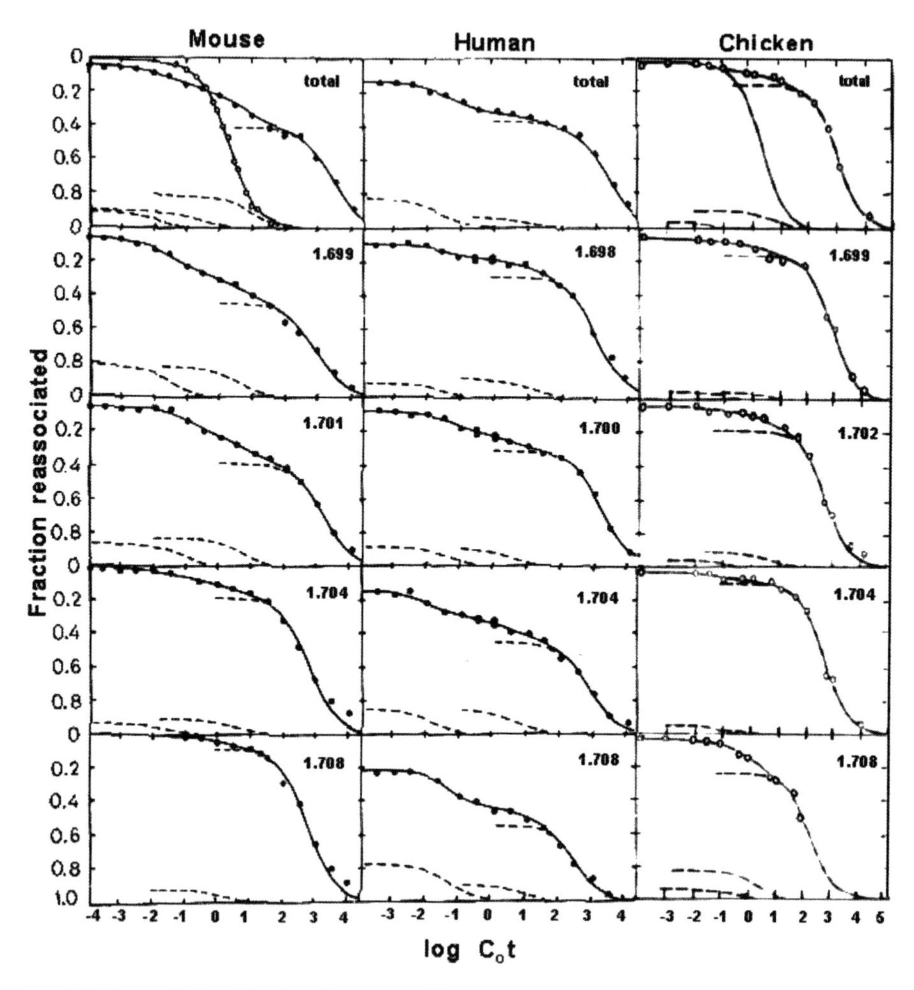

Figure 3.6. Reassociation kinetics of mouse and human DNAs and their major components. Results obtained with *E.coli* DNA (open circles; top left frame) are shown for the sake of comparison. The solid lines through the experimental points (solid circles) are the overall profiles resulting from the analysis of kinetic classes (broken lines). C_0t is the product of initial DNA concentration by reassociation time. (From Soriano et al., 1981). Chicken data are from Olofsson and Bernardi (1983).

consideration (as well as other genes tested later) in fact hybridized on fractions higher in buoyant density than the 1.708 g/cm^3 major component. An example of such hybridization is that for the *c-sis* oncogene shown in **Fig. 1.10**. This identified an additional "major" component, derived from another isochore family, which had been previously considered as a "minor component" because of its low relative amount in the human genome, and which was called H3 (Zerial et al., 1986a).

The compositional distribution of large DNA fragments from the human genome, which is a good representative of the majority of mammalian genomes (Sabeur et al.,

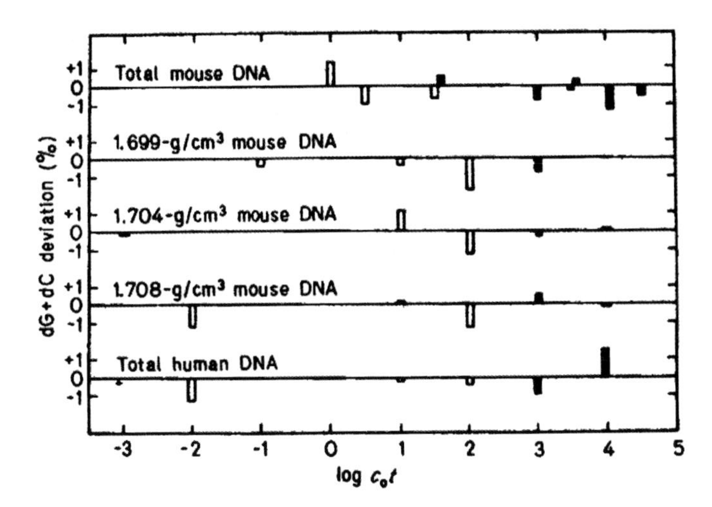

Figure 3.7. GC levels of DNA fractions denatured (white bars) or reassociated (black bars) at different C_0 t values. Data are presented as difference histograms with the GC level contents of total mouse DNA, its major components, and a total human DNA. Error is ± 1%. (From Soriano et al., 1981).

1993) and the most thoroughly studied, is characterized by the presence of five families of fragments derived from the corresponding isochores. In the human genome, the GC-poor isochore families L1 and L2 represent about 33% and 30% of the genome, the GC-rich H1 family corresponds to about 24% of the genome and the very GC-rich H2 and H3 families form 7.5% and 4–5% of the genome, respectively, the remaining DNA consisting of satellite and ribosomal DNAs (Zoubak et al., 1996). These percentages are slightly different from those presented in **Table 3.3**, which were determined on the basis of their yields from two subsequent large-scale runs of preparative centrifugation. **Figs. 3.9A** and **3.9B** summarizes the isochore organization of the human genome and its isochore pattern.

To sum up, the investigations presented in the preceding sections (Filipski et al., 1973; Thiery et al., 1976; Macaya et al., 1976) led to three major discoveries. The first one was the discovery of **isochores**, namely of long, compositionally fairly homogeneous regions forming most, or all, of the main bands of vertebrate genomes. For technical reasons, the average size of such fairly homogeneous regions originally could only be defined as larger than 300 kb. Incidentally, the sequences of satellite and ribosomal DNAs could also be visualized as isochores, because of their compositional homogeneity.

The second discovery was that isochores belonged to a small number of families characterized by distinct average base compositions and by other sequence features (see **Table 3.4**). The **major DNA components** were responsible for the discontinuities revealed by the CsCl analysis of Cs_2SO_4 ligand fractions (**Fig. 3.4**) and derived from **isochore families** that caused discontinuities in the CsCl profiles of very high molecular weight samples (**Fig. 3.5**).

The third discovery was that **isochore patterns** showed phylogenetic differences. The most striking difference concerned the genomes of warm- and cold-blooded vertebrates.

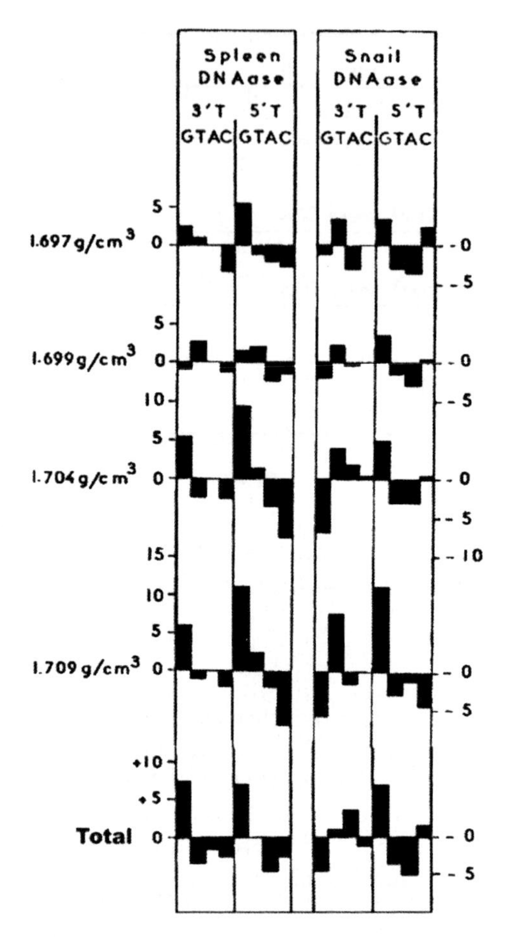

Figure 3.8. Deviation patterns of human DNA and its major components. The histogram shows the differences between the composition of the termini formed by spleen and snail DNase acting on human DNA and the compositions expected for bacterial DNAs having the same GC levels. The deviation patterns are different for total DNA and for its major components. (From Devillers-Thiery, 1974).

This difference was the starting point of evolutionary investigations on the vertebrate genomes, to be discussed in **Parts 11** and **12**.

Compositional patterns were called **genome phenotypes** (Bernardi and Bernardi, 1986a), because they are different in different vertebrate classes (see also **Part 4**). It should be stressed that, until the work of Thiery et al. (1976), the common belief was that, while DNA composition varied greatly among bacteria, animal species had a rather similar DNA composition (Kit, 1960). Although some small variations in modal buoyant densities and standard deviation of CsCl profiles had been detected in vertebrate DNAs, these could be due to different methylation levels (Kit, 1962).

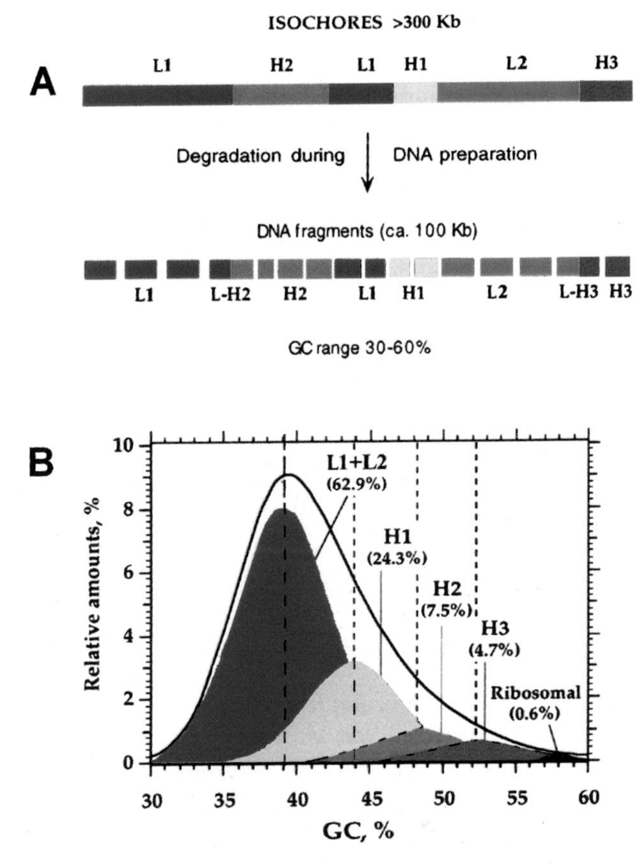

Figure 3.9. **A**. Scheme of the isochore organization of the human genome. This genome, which is typical of the genome of most mammals, is a mosaic of large DNA segments, the isochores, which are compositionally fairly homogeneous and can be partitioned into a small number of families, light or GC-poor (L1 and L2), and heavy or GC-rich (H1, H2 and H3). Isochores are degraded during DNA preparation to fragments of 50-100 kb in size. The GC range of these DNA molecules from the human genome is extremely broad, 30-60%. (From Bernardi, 1995.) **B**. The CsCl profile of human DNA is resolved into its major DNA components, namely DNA fragments derived from each one of the isochore families (L1, L2, H1, H2, H3). Modal GC levels of isochore families are indicated on the abscissa (broken vertical lines). The relative amounts of major DNA components are indicated. Satellite DNAs are not represented. (From Zoubak et al., 1996.)

1.4. Isochores and the draft human genome sequence

Early in 2001, two papers reported draft sequences of the human genome. Both of them dealt with, among other subjects, the broad genomic landscape and gene density. The paper by Venter et al. (2001) summarized these features in a Table (reproduced here as **Table 3.5**) in which the estimates from the human sequence were compared with our estimates (Zoubak et al., 1996; Bernardi, 2000a). Venter et al. (2001) found differences between the amounts of isochores "*observed*" by them and those "*predicted*", i.e., "*based* *on Bernardi's definitions of the isochore of the human genome*". These differences are, however, simply due to the fact that Venter et al. (2001) took as borders between isochore families GC levels different from ours. Indeed, the H3/H2 border was put at 48% GC instead of 50.8%, and the H1/L border at 43% instead of 41.5%. The differences in the relative amounts of genes located in different isochore families were due in part to the different estimates of isochore family borders (see above) and in part to an overestimate of genes located in GC-poor isochores (as judged by comparison with the data of Lander et al., 2001).

TABLE 3.5
Characteristics of G+C in isochores[a].

Isochore	G+C (%)	Fraction of genome		Fraction of genes	
		Predicted	Observed	Predicted	Observed
H3	>48	5	9.5	37	24.8
H1/H2	43–48	25	21.2	32	26.6
L	<43	67	69.2	31	48.5

[a] From Venter et al. (2001).

The second paper (Lander et al., 2001; also referred to as International Human Genome Sequencing Consortium, 2001) studied "*the draft genome sequence to see whether strict isochores could be identified*" and failed to find any. They concluded that their results "*rule out a strict notion of isochores as compositionally homogeneous*" and that "*isochores do not appear to deserve the prefix 'iso'.*"

Since the terminology "*strict isochores*" was misleadingly used by the authors to denote sequences that cannot be distinguished from random (uncorrelated) sequences (in which every nucleotide is free to change), their failure to identify in the human genome sequences as homogeneous as random sequences (masquerading as "*strict isochores*") could have been predicted easily on three accounts.

First, for over 40 years, since at least the work of Rolfe and Meselson (1959; see also Elton, 1974), random sequences had been known to be much more homogeneous than the least heterogeneous genomic DNAs, namely bacterial DNAs (viral DNAs are not considered here because, if intact, they are perfectly homogeneous; see **Fig. 1.6**). In turn, bacterial DNAs are much less heterogeneous than mammalian DNAs (even if satellite and minor components are neglected).

Second, early indications that **standard deviations of CsCl profiles** of major components from the bovine genome were comparable to those of bacterial DNAs of the same size had already been obtained by Filipski et al. (1973). Moreover, unfractionated DNAs from many fishes were compositionally more homogeneous than bacterial DNAs having the same fragment size and the same composition (Hudson et al., 1980; see **Figs. 3.10, 3.11**). Finally, Cuny et al. (1981) quantified the compositional heterogeneity of the major DNA components (namely the compositional families of 50–100 kb DNA molecules derived from isochore families; see **Tables 3.1, 3.3** and **Figs. 3.10, 3.11**) from human and mouse

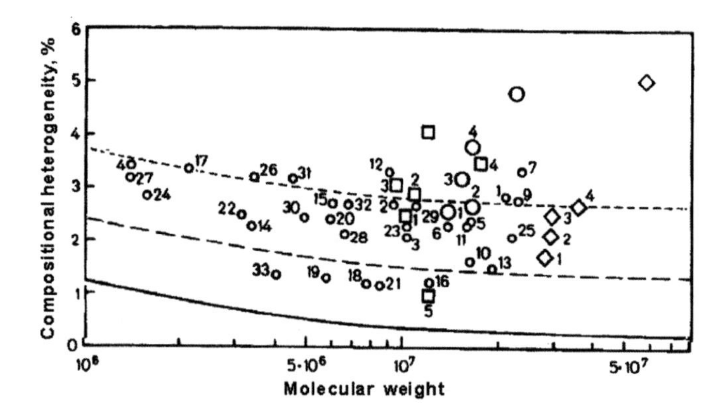

Figure 3.10. Intermolecular compositional heterogeneity (H) plotted as a function of molecular weight (10^6 daltons is about 1500 bp) for DNAs of 33 fish species. Numbering of species is from Table 2 of Hudson et al. (1980). The solid line represents the dependence of H on fragment size for a DNA of random sequence. The broken lines show this dependence for two bacterial DNAs: *E.coli* (51% dG+dC, upper dashed line) and *H.influenzae* (38% dG+dC, lower dashed line). Data for human (○), main-band mouse (□), and chicken (◇) DNAs and their components (1 to 4) are also given. Data for random DNA, bacterial, mammalian and avian DNAs are from Cuny et al., 1981. (From Hudson et al., 1980).

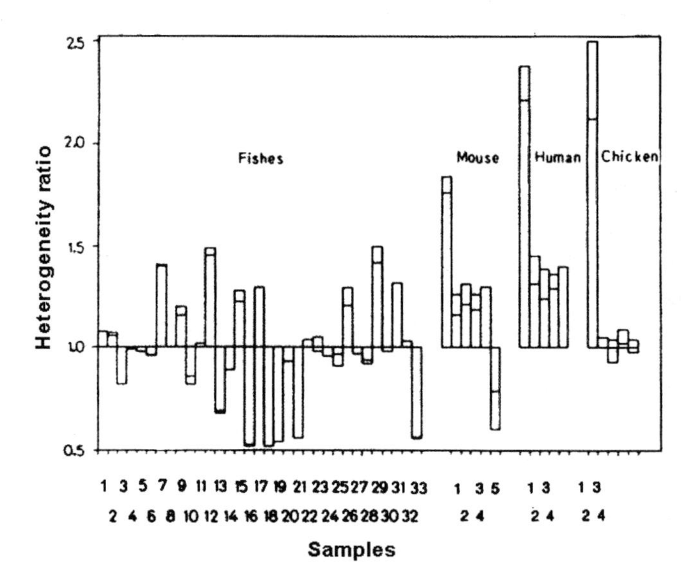

Figure 3.11. Comparison of heterogeneity ratios of DNAs from 33 species of fishes. Ratios were obtained by dividing intermolecular compositional heterogeneity values from each species of fish by the corresponding value of bacterial DNA having the same GC level and the same molecular weight (standard error ± 0.2). Numbering of samples is from Table 2 of Hudson et al. (1980). GC levels for fish DNAs were calculated from $<\rho>$ values given in the same table; values given by nucleoside analysis were also used when available to calculate ratios; in this case ratios were always higher than those calculated from $<\rho>$ and are also shown in the figure. Results of Cuny et al. (1981) for mouse main-band DNA, total human and chicken DNAs, and their components (1 to 4), including mouse satellite (5), are also shown (From Hudson et al., 1980).

and found that they were only about 30% more heterogeneous (by comparison of standard deviations) than bacterial DNAs having the same size and composition, but much more heterogeneous than random sequences. On this basis, Cuny et al. (1981) defined isochores as **"fairly homogeneous regions"**. It should be stressed that heterogeneity was overestimated by Cuny et al. (1981), because it was measured on pooled Cs_2SO_4/Ag^+ fractions obtained from two subsequent large-scale centrifugations, which entailed some extent of cross-contamination between DNA components. **Figs. 3.10** and **3.11** show that, in the case of chicken DNA (Cortadas et al., 1979), the heterogeneity of major components was barely different from those of reference bacterial DNAs (see above), and that unfractionated fish DNAs are, on the average, indistinguishable from bacterial DNAs, that satellite DNAs are more homogeneous than bacterial DNAs and that degradation of mouse DNA components down to 7.5 kb fragments only led to a modest increase in heterogeneity.

Third, "*strict isochores*" cannot exist in any natural DNA (i) because coding sequences are made up of codons, in which the compositions of the three positions are correlated with each other (D'Onofrio and Bernardi, 1992); (ii) because non-coding sequences are compositionally correlated with the coding sequences that they embed (Bernardi et al., 1985b; Clay et al., 1996; see **Chapter 3** below); and (iii) because interspersed repeats are characterized by their own specific sequences. More detailed discussions of this problem were presented by Clay and Bernardi (2001a,b), Clay et al. (2001) and Clay (2001).

In summary, the conclusion of Lander et al. (2001) that "*isochores*" are not "*strict isochores*", *i.e.* are not as homogeneous as random sequences is correct, but it is something we have known for at least 20 years. To take random sequences as a reference for homogeneity and looking for the same level of homogeneity in human DNA sequences was a mistake. Furthermore, it raised doubts about the very existence of isochores in readers who were not familiar with this issue, and this was unfortunate.

Around the same time, other laboratories also missed the point that "*strict isochores*" cannot exist in natural DNA and also took random sequences as references for compositional homogeneity (Häring and Kypr, 2001; and also, in part, Nekrutenko and Li, 2000; these papers have been commented in detail in Clay and Bernardi 2001a,b; Clay et al., 2001).

Along another line, Lander et al. (2001) did not accept the idea that the human genome is a mosaic of isochores, namely that the large-scale compositional heterogeneity is discrete or discontinuous, rather than continuously drifting, as generally believed thirty years ago and as proposed in a model by Fickett et al. (1992). This proposal was uncritically, yet enthusiastically, accepted by some authors: "*From sodium chloride* (sic!) *centrifugation experiments, Bernardi et al. (1985) defined three major classes of genomic fragments with low, median, and high GC content, respectively and called them isochores. This description now appears to be artificial. Analyses of the complete human genome (International Human Genome Sequencing consortium, 2001) have dismissed the underlying hypothetical picture of*

Figure 3.12. A colour-coded compositional map of the human chromosomes, representing 100 kb moving window plots that scan the draft human genome sequence of Lander et al. (2001). Colour codes span the spectrum of GC levels in 5 steps, from ultramarine blue (GC-poorest isochores) to scarlet red (GC-richest isochores). Grey vertical lines and bands correspond to the 5,000 gaps present in the euchromatic part of the chromosomes (Modified from Pavliček et al., 2002a.)

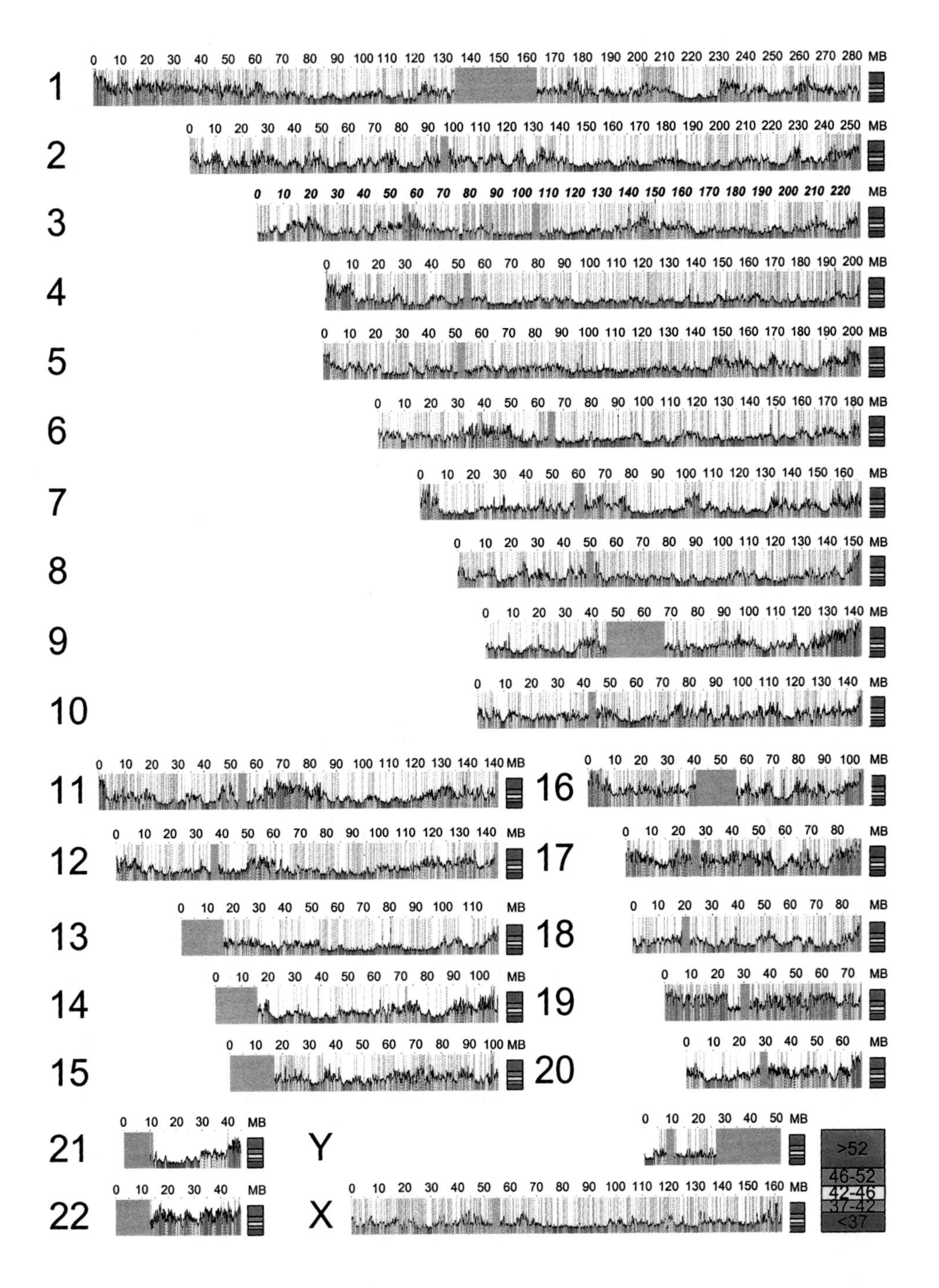

sharp boundaries between long homogeneous fragments: GC content turns out to vary continuously, and somewhat erratically, along chromosomes" (Galtier et al., 2001; our underlining; see below for the *erratic* nature of compositional variation). The authors' rejection of the discontinuous compositional heterogeneity was due to their being apparently unaware of the detailed investigations that led to this conclusion (Filipski et al., 1973; Thiery et al., 1976; Macaya et al., 1976) and to the conclusion that isochore patterns are conserved in mammals and birds (see **Part 11, Chapter 2**). Moreover, Lander et al. (2001) seem to have overlooked the evidence that GC-rich isochores arose at the transition between cold- and warm-blooded vertebrates from regions that were much lower in GC level and, in fact, very similar to the GC-poor isochores of the human genome (see **Section 1.5** below and **Part 11**). Obviously, this emergence from the more homogeneous compositional spectrum of cold-blooded vertebrates was the primary source for a discontinuous distribution (see **Fig. 12.5**).

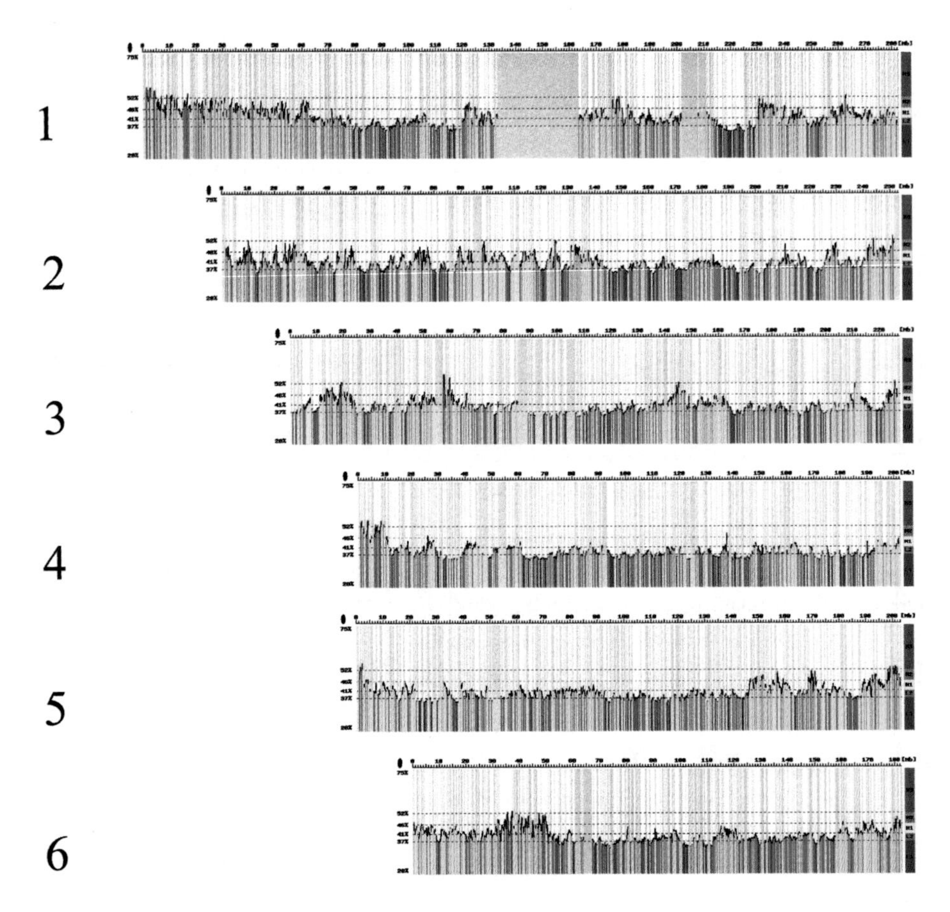

Figure 3.13. A colour-coded compositional map of human chromosomes 1-6. Other indications as in Fig. 3.12. (From Pavliček et al., 2002a.)

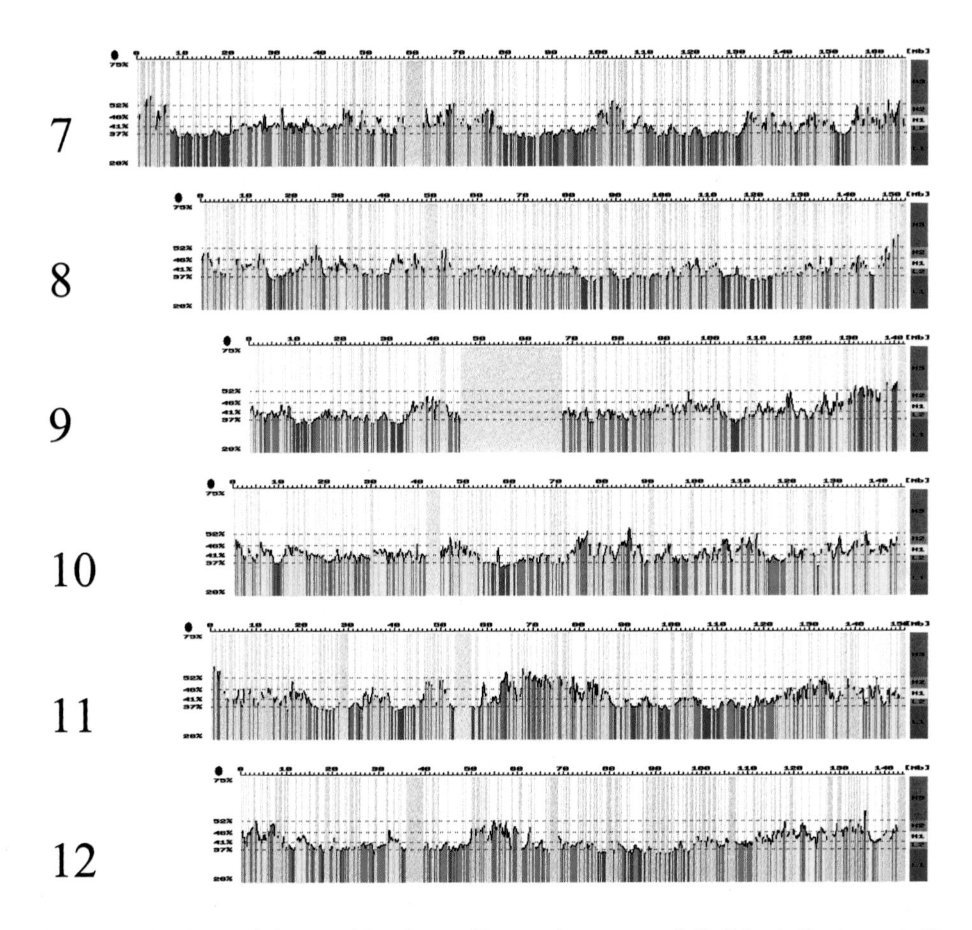

Figure 3.14. A colour-coded compositional map of human chromosomes 7-12. Other indications as in Fig. 3.12. (From Pavliček et al., 2002a.)

The denial of the very existence of isochores, if correct, would have major consequences which, apparently, were not realized by Lander et al. (2001) . The first one would be the denial of a compositionally discontinuous sequence organization and the return to the continuous compositional spectrum for the human genome (see above) that was the predominant view until the work of Filipski et al. (1973). The second consequence would be the denial of "*an important level of genome organization, insofar as gene density (Zoubak et al., 1996), gene length (Duret et al., 1995), and patterns of codon usage (Sharp et al. 1995), as well as the distribution of different classes of repetitive elements (Soriano et al., 1983; Duret et al., 1995), are all correlated with GC content*" (Fullerton et al., 2001). Other properties that could be added to the list are replication timing, recombination frequency (Fullerton et al., 2001), chromosomal banding, and stability and transcription of integrated sequences. In other words, the second consequence would be the denial of "*a fundamental level of genome organization*" (Eyre-Walker and Hurst, 2001).

Very interestingly, the best graphical display of the mosaic organization of the human

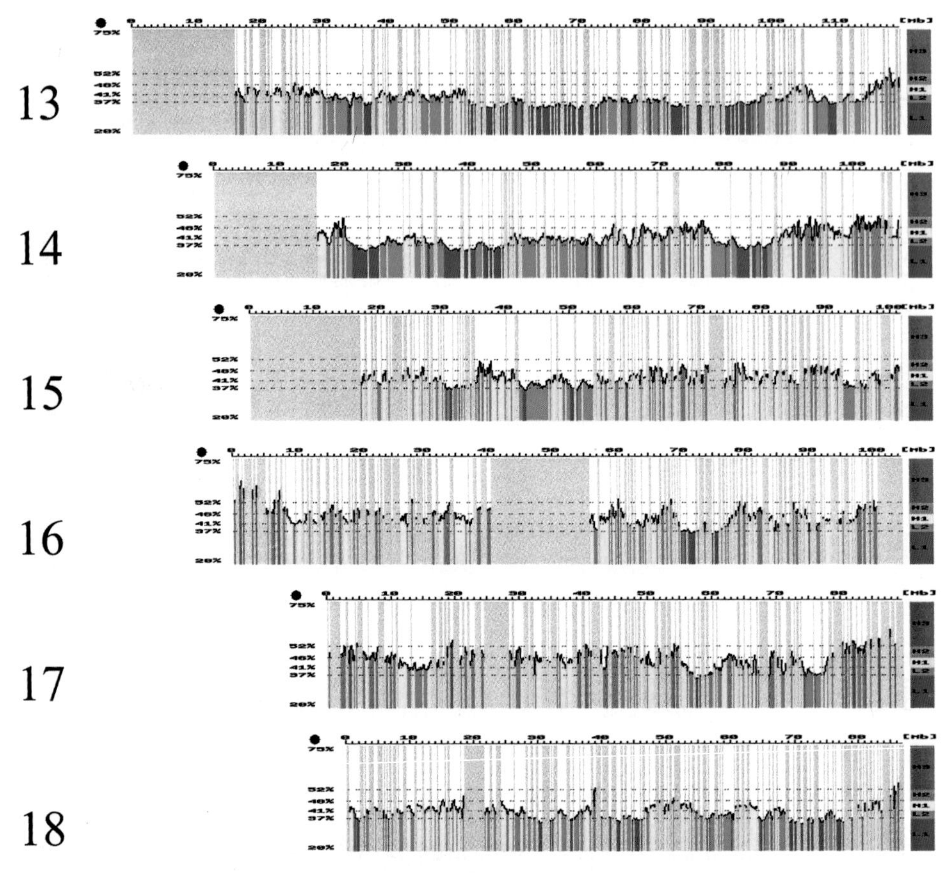

Figure 3.15. A colour-coded compositional map of human chromosomes 13-18. Other indications as in Fig. 3.12. (From Pavliček et al., 2002a.)

genome available at present is that of the compact compositional maps of human chromosomes (**Figs. 3.12–3.16**) derived from the data of Lander et al. (2001) which confirm, a quarter of a century later, the isochore structure of mammalian genomes (Macaya et al., 1976). Indeed, apart from the presence of about 5,000 gaps (grey bars) in the euchromatic regions of most chromosomes (in fact of all chromosomes, except chromosomes 21 and 22, which had been sequenced previously by two other groups, Hattori et al., 2000, and Dunham et al., 1999), the striking feature of the map is undoubtedly the large proportion of the genome represented by long GC-poor regions, uninterrupted by GC-rich regions. The next most notable observation is the scarcity of GC-poor regions in many of the blocks characterized by GC-rich regions. As far as these two points are concerned, the results fit with the previous estimates of the relative amounts of GC-poor and GC-rich isochores (see **Fig. 3.8**) and with the very high yields of major DNA components from the human genome (Cuny et al., 1981). They contradict the suggestion (Eyre-Walker and Hurst, 2001) that an isochore structure accounts for "*only some parts*" of the genome. The third observation is the increasing compositional fluctuations when moving from GC-

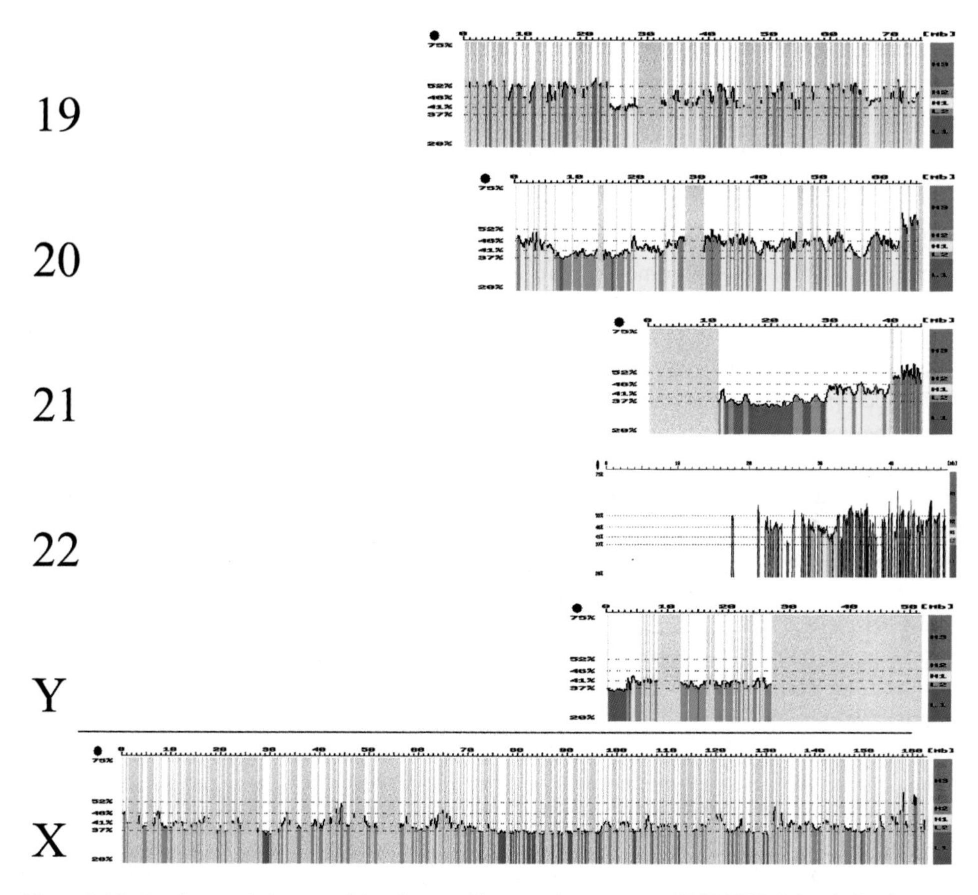

Figure 3.16. A colour-coded compositional map of human chromosomes 19-22 X,Y. Other indications as in Fig. 3.12. (From Pavliček et al., 2002a.)

poor to GC-rich isochores. This had already been noticed in previous work (Cuny et al., 1981; De Sario et al., 1996; 1997) and was confirmed by a detailed analysis of chromosomes 21 and 22 (Saccone et al., 2001). A working hypothesis on this point is presented in **Part 7, Chapter 3**.

Finally, it should be noted that the considerations presented in this section were not affected by the recent advances towards completing the human genome sequence (see, for example, Heilig et al., 2003; Mungall et al., 2003).

1.5. Other misunderstandings about isochores

Another article (Nekrutenko and Li, 2000) also exemplifies some common misunderstandings about isochores, but, in addition, proposes new criteria and measures of heterogeneity that are relevant to the results presented here. Indeed, the authors propose a new heterogeneity parameter (apparently as an alternative to the standard deviation), namely the average magnitude of the difference between GC levels of successive segments of length l in an n-segment partitioning (or covering by partially overlapping windows) of a DNA sequence of interest (after appropriate normalization):

$$\overline{\Delta GC} := \frac{1}{n-1}\sum_{i=2}^{n} |GC_i - GC_{i-1}|$$

In fact, the quantity $\overline{\Delta GC}$, independently used by us in a comparative analysis of the human and *Drosophila* genome (Jabbari and Bernardi, 2000), has a nice graphical interpretation. If one plots the GC level of each segment i against the GC level of the preceding segment i - 1, one obtains, in human and even in *Drosophila* DNA, a good correlation, with the points of the scatter plot lining up near the main diagonal of slope 1 (see **Part 9**; **Fig. 9.7**). Such plots are called "time-delay plots" in physics (Ruelle, 1989), or "**phase plots**", since they can show the phase of a periodic fluctuation. They can be used to detect and display correlations along DNA sequences, if nonoverlapping windows are chosen. It can be easily seen that $\overline{\Delta GC}$ is simply the mean vertical (or horizontal) distance of the points from the main diagonal; the greater the (absolute) GC differences between successive segments, the farther the points from the diagonal, and the higher the heterogeneity.

It is, therefore, natural that Nekrutenko and Li (2000) also suggest a criterion, based on the differences $|GC_i - GC_{i-1}|$, for the detection of isochore boundaries. Clearly, the points representing pairs of adjacent segments within a single GC-poor isochore will cluster around one part of the diagonal, those representing an adjacent GC-rich isochore will cluster around another part of the diagonal, and the few segment pairs spanning a sharp boundary between two adjacent isochores will be found far from the diagonal. A natural question is how far the points need to be from the diagonal, *i.e.*, how big $|GC_i - GC_{i-1}|$ needs to be, before it is indicative of an isochore boundary. Any such critical distance must depend on the variation in GC level that can be expected within a single isochore (i.e., depend on the fragments' GC distribution), and on some chosen significance level such as 5%, 1% or 0.1%.

Unfortunately, the authors did not resolve this problem. Their first normalization is incorrect: not only did they choose a binomial distribution, *i.e.*, an uncorrelated DNA sequence, as the expected value for normalizing (which, as argued above, and as subsequently realized also by the authors, is inappropriate), but they neglected that the absolute difference $|GC_i - GC_{i-1}|$ has a standard deviation that is $2/\sqrt{\pi} \approx 1.128$ times larger than that of GC_i, i.e., than that of a simple binomial distribution or of its normal approximation. A factor $\sqrt{2}$ arises because the variance of a distribution of differences of two random variables is, in the correlation-free case considered here, the sum of the variances of their distributions; an additional factor $\sqrt{2/\pi}$ is introduced by taking absolute values (and using the approximate normality of the binomial distributions of interest; see Shiryayev, 1984, pp. 237 ff., or Bourbaki, 1969, p. 74). As a result, the authors' 400,000 simulations for different fragment lengths and overlaps (Nekrutenko and Li, 2000, Table 1), using randomly generated, uncorrelated sequences, repeatedly give values close to 1.12 – 1.13, namely $2/\sqrt{\pi}$, instead of 1 as expected. This apparent "*behavior*" or "*baseline value*" of their compositional heterogeneity index is, therefore, simply an artefact of incorrect normalization. In addition, since the authors use the binomial distribution, *i.e.*, sequences of independent nucleotides, as their first standard for homogeneity, they find no long homogeneous regions and are obliged to search for a more appropriate standard. (A similar two-step reasoning, amounting to a rediscovery of the results of Cuny et al., 1981, can be discerned also in Lander et al., 2001, and in Häring and Kypr, 2001, a study that has been discussed in more detail by Clay and Bernardi, 2001a.)

The second attempt by Nekrutenko and Li (2000) to compare $|GC_i - GC_{i-1}|$ with a control distribution, namely by assuming that intra-isochore fluctuations in human should not exceed typical intra-chromosomal fluctuations in yeast, the smallest ones observed so far in an eukaryotic genome (see **Part 9**), is no less arbitrary and extrinsic, but it does lead to somewhat longer "homogeneous" regions than when random sequences are used as a yardstick. This partial success leads the authors rather quickly to a bold proposal, namely to a "*new definition of an isochore as any genomic fragment longer or equal to 100 kb such that when it is divided into a series of overlapping 10-kb windows, no two windows can differ by >7% GC*". In addition to the underestimate of intra-isochore fluctuations imposed by their yeast threshold, there is a second problem inherent in their approach: the probability of witnessing at least one unlikely event (such as 4 consecutive heads when tossing a coin) will always increase with the number of trials, so that working outwards from an "*isochore seed*" until a given fluctuation limit is reached will make it increasingly difficult for long isochores to meet the authors' criterion. It does not come as a surprise, therefore, that their method results in many very short homogeneous fragments, and in no long ones. The short ones are duly eliminated by the artificially imposed lower bound of 100 kb, and banished to regions of DNA purported to contain no isochores at all, while long "isochores" remain conspicuously absent. For example, in chromosome 21, where most moving window plots immediately reveal a strikingly homogeneous, long, very GC-poor region covering more than 7 Mb (De Sario et al., 1997; Hattori et al., 2000; Oliver et al., 2001; Li, 2001; see **Fig. 3.16**), the authors' histogram shows only two "isochores" longer than 350–370 kb (with lengths of $\approx$ 430 and $\approx$ 490 kb, respectively). Clearly, such a redefinition of isochores cannot be considered tenable, since it could furnish, at best, only the shortest isochores in the human genome.

Recent results show that isochores and their boundaries can be recognized not only by color-coded moving window plots (Pavlíček et al., 2002a; see **Figs. 3.12–3.16**), but also by entropic segmentation (Oliver et al., 2001; Li, 2001), and by simple windowless method (Zhang and Zhang, 2003a,b). Interestingly, this approach was applied by Macaya and Bernardi thirty years ago, but its use on total DNA fragments about 50 kb in size was not (and could not be) as successful as when applied to DNA sequences. Statistical correlations in DNA sequences have been investigated by Bernaola-Galvan et al. (2002) and isochore chromosome maps of the human genome have been published by Oliver et al. (2002). The principle and the applications are described in **Part 9, Chapter 5.3**. Furthermore, if traditional statistical procedures are correctly applied, they reveal the existence of a significant relative homogeneity in GC within isochores (Li et al., 2003; this paper critically discusses the calculations of Lander et al., 2001). Interestingly, the isochore structure of the mammalian genome has reached the sequencing community, as witnessed by the number of **compositional plots** presented by the Mouse Genome Sequencing Consortium (2002).

CHAPTER 2

Compositional patterns of coding sequences

Fig. 3.17A presents a schematic view of the isochore patterns of the vertebrate genomes best studied by density gradient centrifugation. The relative amounts, modal buoyant densities and GC levels of isochore families are shown. Apart from the major differences between the *Xenopus* and human patterns, the latter showing GC-rich isochore families that are absent in *Xenopus*, it is remarkable that the human pattern is also slightly yet significantly different from the chicken pattern, which exhibits an additional very GC-rich family H4, and from the mouse pattern, which lacks the GC-rich family H3. These differences will be discussed in more detail in **Part 4**.

Very remarkably, the differences in isochore patterns of vertebrate genomes were paralleled by differences in base composition (**Fig. 3.17B**) of coding sequences (Bernardi et al., 1985b, 1988; Perrin and Bernardi, 1987; Bernardi and Bernardi, 1991) which only represent a very small minority of the genomes under consideration. A first comparison could be made by looking at the compositional distribution of coding sequences, or of their average third codon positions, GC_3. This showed striking differences not only between human and *Xenopus*, but also between human and chicken, and between human and mouse. Indeed, the compositional distribution of coding sequences from *Xenopus* is narrower and centered on a lower value compared to the human distribution. The distribution of chicken genes, also seen as GC_3 values, reached a higher level, in fact 100%, compared to the human distribution. Finally, differences could also be seen between human and mouse, the latter distribution being narrower.

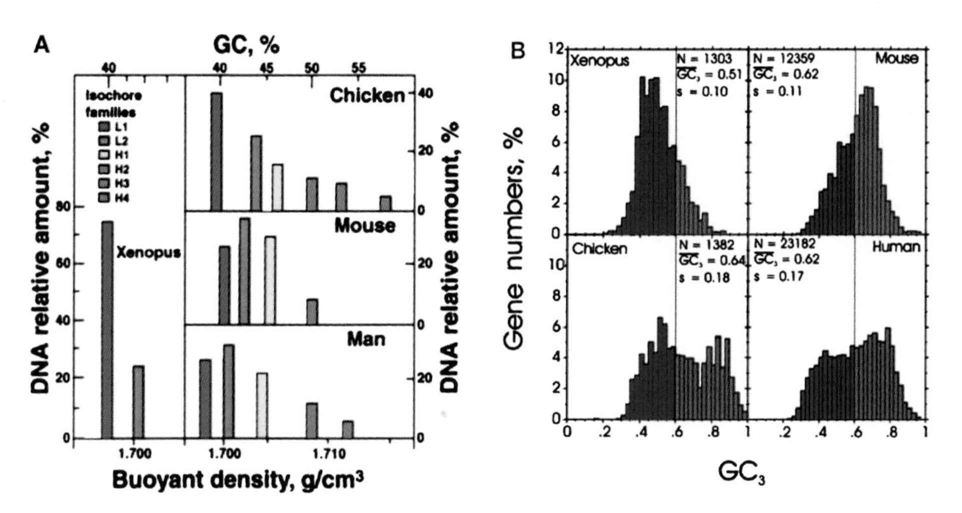

Figure 3.17. **A.** Isochore families from *Xenopus*, human, mouse and chicken, as deduced from density gradient centrifugation. **B.** Compositional patterns of coding sequences (represented by GC_3 values averaged per coding sequence) for *Xenopus*, mouse, chicken and human. (Modified from Bernardi, 1995).

Compositional correlations between coding and non-coding sequences

The similar compositional features exhibited by isochore families and by coding sequences of the genomes of **Fig. 3.17** raised the question of the possible correlation between the composition of the isochores, which cover a 30 to 60% GC range in the case of the human genome, and the composition of the coding sequences embedded in them. This question could be answered by localizing coding sequences of known primary structure in compositional DNA fractions (or in isolated major DNA components).

As already mentioned, hybridization of appropriate probes on compositional fractions of genomic DNA (**Fig. 1.10**) allows localizing sequences in such fractions, defining the GC levels of those fractions and exploring chromosomal regions up to twice the size of the DNA fragments making up the DNA samples (see **Fig. 1.11**). When a number of coding sequences, essentially from human, mouse and chicken, were localized by hybridization of appropriate probes, and their GC levels were plotted against GC levels of the major components or compositional fractions in which they had been localized, linear correlations were found (Bernardi et al., 1985a,b; preliminary reports were published by Bernardi, 1979b; 1984). Linear correlations were also found when plotting GC values for entire genes (exons + introns; **Fig. 3.18**) and for introns, or GC_3 values, against those of the corresponding isochores. These results were of great relevance because they showed compositional correlations between the coding sequences (that represent 2–3% of the genome) and the non-coding sequences (introns and intergenic sequences, that represent 97–98% of the genome), in which, or next to which, they were located. It should be noted that a correlation between GC levels in third positions and GC levels of flanking sequences was independently reported by Ikemura (1985) for some vertebrate genes.

When GC and nucleotide levels of the three codon positions of prokaryotic, viral and vertebrate genes were plotted against GC levels of the corresponding genome or (in the case of vertebrate genes) of exons, linear correlations with high correlation coefficients were found in all cases (Bernardi and Bernardi, 1985; 1986a,b; **Fig. 3.19**). Incidentally, the results on prokaryotic GC correlations (top frames of **Fig. 3.19**) were later confirmed on a small gene sample by Muto and Osawa (1987).

The localization of coding sequences was further studied when long genomic sequences from human DNA became available. In this case, direct comparisons could be done between the GC levels of coding and non-coding sequences and more precise estimates of the correlations could be obtained (Aïssani et al., 1991; Clay et al., 1996; Zoubak et al., 1996). **Figure 3.20A** shows that linear correlations hold between the GC_3 levels of coding sequences and the GC levels of the isochores in which coding sequences are located. Interestingly, GC_3 values of GC-poor coding sequences and their flanking sequences show very similar values, whereas GC_3 values of GC-rich coding sequences are increasingly higher above the diagonal, essentially because GC_3 values depart more and more from the intergenic sequences. Linear correlations also hold (**Fig. 3.20B**) between the GC levels of coding sequences and the GC levels of the introns of the same genes (Bernardi et al., 1985b; Aïssani et al., 1991; Clay et al., 1996). The values of introns were systematically lower than those of exons, a point commented upon in **Part 12**, **Chapter 2**.

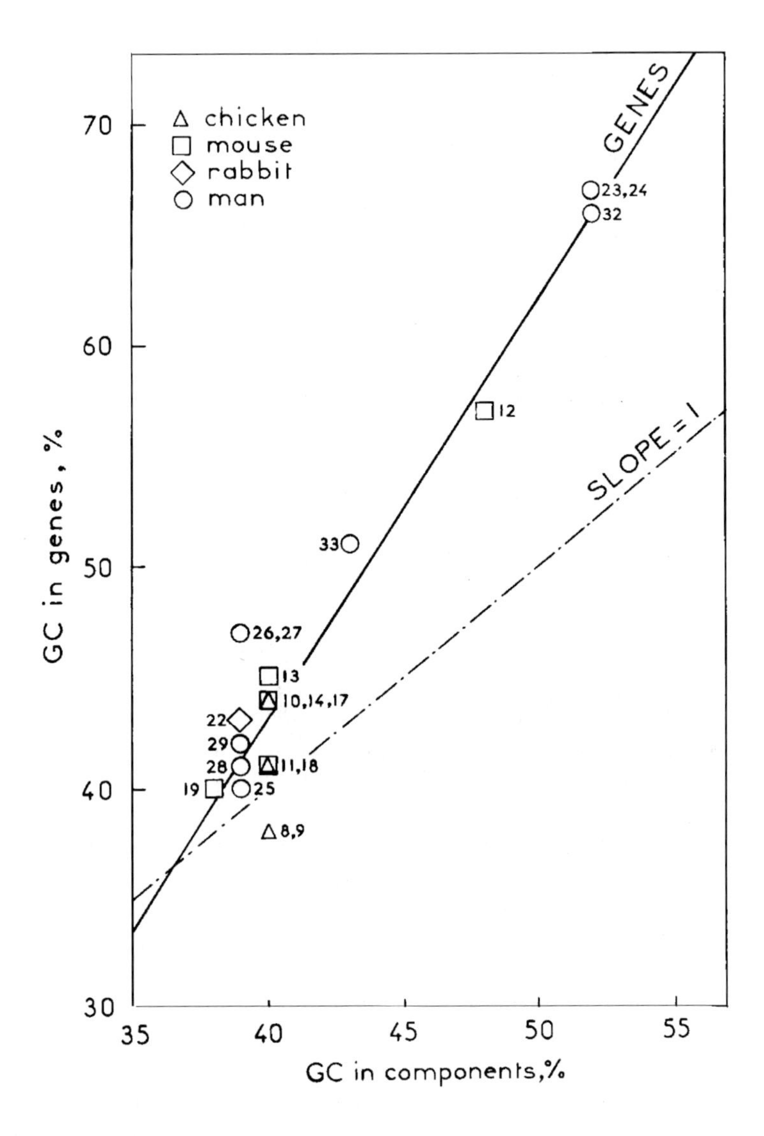

Figure 3.18. Plot of GC content of genes against the GC levels of DNA components in which they are located. The numbers indicate genes (see the original article). The line was drawn using the least-square method. The unit slope line corresponds to the coincidence in GC contents of genes and major components in which genes are located. (Modified from Bernardi et al., 1985b).

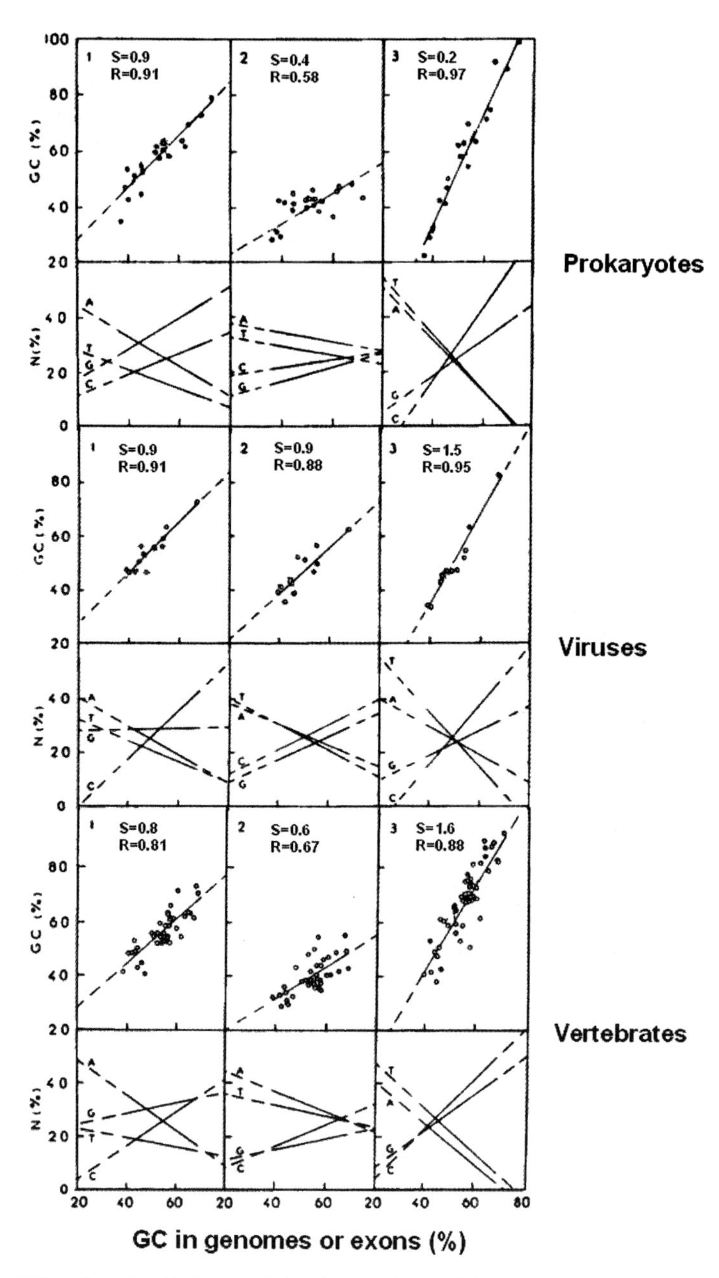

Figure 3.19. GC and nucleotide levels of the three codon positions (1, 2, 3) of prokaryotic, viral and vertebrate genes plotted against GC levels of the corresponding genomes (or exons in the case of vertebrates; filled circles are average GC_3 levels of genes belonging to the same compartment of a given genome). For details about the genes studied, see the original paper. (From Bernardi and Bernardi, 1986a).

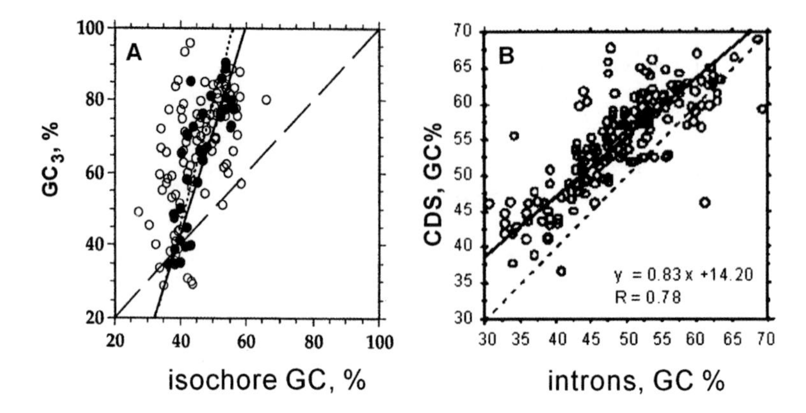

Figure 3.20. **A.** Correlation between GC_3 of coding regions of genes and the GC level of fractions or Yeast Artificial Chromosomes (YACs) in which the genes were localized (filled circles), and of 3' flanking sequences farther than 500 bp from the stop codon (open circles). The corresponding, essentially coincident orthogonal regression lines, indicated here by a single dashed line, are given by $GC_3 = 3.45 \times GC-94.6$ (R = 0.82, N = 32; fractions and YACs) and $GC_3 = 3.39 \times GC-88.2$ (R=0.56, N=103; far 3' sequences). The relation obtained by comparing the independent Gaussian decomposition of GC and GC_3 distribution is indicated by a solid line, and given by $GC_3 = 2.92 \times GC-74.3$. This solid line happens to be almost indistinguishable from the regression line for those genes that were known to contain no CpG islands, $GC_3 = 2.89 \times GC-71.9$ (R = 0.87, N = 13; fractions and YACs) and $GC_3 = 2.90 \times GC-66.4$ (R = 0.67, N = 30; far 3' sequences). (From Zoubak et al., 1996). **B.** Correlation between GC levels of human coding sequences and the GC levels of the corresponding introns. The diagonal (unity slope line) is also shown. (Modified from Clay et al., 1996).

We have already reported that the use of the orthogonal regression as done in **Fig. 3.20A**, *i.e.*, regression along the principal eigenvector of the variance-covariance matrix, gives a better description of the scatter diagrams than linear regression, since the orthogonal regression line is obtained by minimizing the sum of squares of the orthogonal distances parallel to the ordinate (see D'Onofrio et al., 1991; Mouchiroud et al., 1991). Indeed, as stressed by Cruveiller et al. (2003), bivariate distributions often have an approximately linear shape and are well characterised by their major axis (also called orthogonal regression line, or principal axis). In passing, it shoud be mentioned that traditional linear regression lines (such as are used to described unilateral dependence relationships) do not provide a satisfactory characterisation, since they do not follow the points when scatterplots are characterised by step slopes, but instead systematically slice the scatterplots at a lower angle (Jolicoeur, 1990; Harvey and Pagel, 1991; Mouchiroud et al., 1991; Clay et al., 1996).

As a final remark, one could say that the compositional correlations between coding and non-coding sequences and between the three codon positions (the genomic code) have shown, to paraphrase Galileo, that the book of the genome is written in a mathematical language.

Part 4
The compositional patterns of vertebrate genomes

CHAPTER 1

The fish genomes

1.1. Compositional properties: a CsCl analysis

There were several reasons to make the vertebrate genome the central subject of our investigations on structural and evolutionary genomics: the vast knowledge available on the biology of vertebrates, the existence of a good paleontological record, and the fact that vertebrates are a small taxon, which share many basic properties. But, obviously, the decisive factor was the discovery of massive compositional changes at the transitions between cold- and warm-blooded vertebrates (Thiery et al., 1976). On the one hand, this raised important questions as to how this could have occurred and which were the causes; on the other it provided an experimental access to its analysis.

As far as fishes are concerned, the work was done in four steps involving a total of 2 (Thiery et al., 1976), 34 (Hudson et al., 1980), 39 (Pizon, 1984), 122 (Bernardi and Bernardi, 1990a) and 201 (Bucciarelli et al., 2002) species, respectively. The reasons for exploring such a large number of fish genomes was (i) to obtain a representative sample of different orders; fishes include a vast array of distantly related vertebrates with about 25,000 species, corresponding to about half of the extant vertebrate species (Nelson, 1994); (ii) to extend as much as possible comparisons of fish orders, families, and genera, in order to investigate the compositional differences that arose over evolutionary time (see Bernardi and Bernardi, 1990b); and (iii) to draw some general conclusions about the organization of fish genomes.

Table 4.1 presents a classification of fishes derived from Nelson (1994) in order to indicate the taxonomic position of the species studied. This classification puts the fishes in an order that reflects their postulated evolutionary relationship, ranging from the most ancient splits in vertebrate evolution, *Chondrichthyes* (sharks, rays), *Sarcopterygii* (Lungfish, Coelacanths and Tetrapods) and *Actinopterygii* (ray-finned fishes), to the most recent teleost orders, *Pleuronectiformes* (flounders) and *Tetraodontiformes* (puffers).

Our study included representatives from 3 out of 9 orders of *Elasmobranchi* (sharks and rays), both orders of dipnoan lungfishes, and both orders of chondrosteans (sturgeons and bichirs). We also studied 19 out of 38 teleostean orders, which represent all but 4 (minor) superorders of the subdivision *Teleostei*, a group comprising about 23,600 species (96% of all extant fishes). This leaves for further studies two subclasses, *Holocephali* (chimaeras), and *Coelacanthimorpha* (gombessas). **Table 4.2** lists all species studied with their properties.

Figure 4.1 displays the analytical CsCl profiles of the fish DNAs studied by Bernardi and Bernardi (1990a). Only three DNA samples (4, 11, 12; all of them from *Chondrichtyes*) showed sizable amounts of resolved satellite bands, which were on the heavy, GC-rich side of the main band. Minor satellite bands were found in other DNA samples, mainly on the

TABLE 4.1
Classification of fishes[a]

Class	Subclass	Infraclass	Division	Subdivision[b]	Superorder	Order[c]
Chondrichthyes	**Holocephali*					
	Elasmobranchi				*Euselachi*	*Lamniformes*
						Squaliformes
						Rajformes
Sarcopterygii	**Coelacanthimorpha*					
	Dipnoi				*Ceratodontimorpha*	*Ceratodontiformes*
						Lepidosireniformes
	(Tetrapoda)					
Actinopterygii	*Chondrostei*					*Polypteriformes*
						Acipenseriformes
	Neopterygii		*Teleostei*	*Osteoglossomorpha*		*Osteoglossiformes*
				Elopomorpha		*Anguilliformes*
				Clupeomorpha		*Clupeiformes*
				Euteleostei	*Ostariophysi*	*Cypriniformes*
						Characiformes
						Siluriformes
					Protacanthopterygii	*Salmoniformes*
					**Stenopterygii*	
					Cyclosquamata	*Aulopiformes*
					**Scopelomorpha*	
					**Lampridiomorpha*	
					**Polymixiomorpha*	

TABLE 4.1
Classification of fishes[a] (continued)

Class	Subclass	Infraclass	Division	Subdivision[b]	Superorder	Order[c]
					Paracanthopterygii	*Gadiformes*
						Ophidiiformes
						Batrachoidiformes
						Lophiiformes
					Acanthopterygii	*Cyprinodontiformes*
						Syngnathiformes
						Dactylopteriformes
						Scorpaeniformes
						Perciformes
						Pleuronectiformes
						Tetraodontiformes

[a] From Nelson, 1994.
[b] Asterisks indicate groups not investigated.
[c] Only orders listed in Table 4.2 are presented.

TABLE 4.2

Properties of DNAs from fishes[a]

Sample number	Order	Family	Species	MW kb	ρ_0 g/cm^3	$<\rho>$ g/cm^3	A mg/cm^3	H %GC	GC %	c pg
1	*Lamniformes*	*Scyliorhinidae*	*Scyliorhinus stellaris*	29.55	1.7053					6.2
2		*Carcharinidae*	*Mustelus mosis*	34.55	1.7044	1.7047	0.3	2.9	45.6	
3			*Scoliodon terranovae*	13.98	1.7030	1.7035	0.5	2.9	44.4	3.6
4			*Carcharinus galapagensis*	15.24	1.7033	1.7037	0.4	2.3	44.7	
5		*Sphyrnidae*	*Sphyrna lewini*	1.57	1.7030	1.7042	1.2	4.6	45.1	3.5
6	*Squaliformes*	*Squalidae*	*Squalus acanthias*	48.50	1.7050	1.7059	0.9	4.3	46.8	7.2
7		*Squatinidae*	*Squatina dumerili*	29.55	1.7050	1.7072	2.2	5.0	48.2	
8	*Rajiformes*	*Rajidae*	*Raja erinacea*	33.25	1.7035	1.7041	0.6	3.7	45.0	3.5
9			*Raja stellulata**	11.70	1.7014					
10		*Dasyatidae*	*Gymnura altavela*	44.44	1.7015	1.7022	0.7	2.7	43.1	
11		*Myliobatidae*	*Myliobatis freminvillei*	33.25	1.7029	1.7039	1.0	4.4	44.8	4.9
12		*Torpedinidae*	*Torpedo marmorata*	45.36	1.7011					7.0
13			*Torpedo ocellata*	49.46	1.7012					7.5
14	*Ceratodontiformes*	*Ceratodontidae*	*Neoceratodus forsteri*		1.7031	1.7027	–0.4		43.5	
15	*Lepidosireniformes*	*Protopteridae*	*Protopterus sp.*	55.15	1.7005	1.7001	–0.4	2.5	40.9	50.0
16	*Polypteriformes*	*Polypteridae*	*Polypterus senegalus*	28.84	1.6995	1.7008	1.3	2.3	41.6	
17	*Acipenseriformes*	*Acipenseridae*	*Acipenser sturio*			1.7019			42.8	1.6
18	*Osteoglossiformes*	*Pantodontidae*	*Pantodon buchholzi*	5.53	1.7022	1.7023	0.1	2.0	43.2	0.8
19		*Notopteridae*	*Notopterus notopterus*	27.91	1.7015	1.7021	0.6	3.6	42.9	
20		*Mormyridae*	*Gnathonemus petersii*	24.58	1.7029	1.7033	0.4	2.5	44.2	1.2
21	*Anguilliformes*	*Anguillidae*	*Anguilla anguilla*	13.68	1.7018	1.7031	1.3	2.3	44.0	0.9
22			*Anguilla rostrata*	35.07	1.7011	1.7015	0.7	3.5	42.6	1.4
23	*Clupeiformes*	*Clupeidae*	*Brevoortia tyrannus*	20.52	1.7021	1.7024	0.3	2.5	43.4	0.8–1.4
24			*Sardinella anchovia*	20.14	1.7022	1.7022	0.0	3.3	43.1	
25	*Cypriniformes*	*Cyprinidae*	*Carassius auratus*	24.36	1.6970	1.6971	0.1	1.8	37.9	1.7
26			*Cyprinus carpio*	72.03	1.6963	1.6965	0.2	2.0	37.2	1.7
27			*Brachydanio rerio*	49.46	1.6959	1.6962	0.3	2.6	36.9	1.8
28			*Labeo bicolor*	18.65	1.6959	1.6965	0.7	2.0	37.2	1.3
29		*Cobitididae*	*Acanthophtalmus semicinctus*		1.6975	1.6978	0.3		38.6	

TABLE 4.2

Properties of DNAs from fishes[a] (continued)

Sample number	Order	Family	Species	MW kb	ρ_0 g/cm^3	<ρ> g/cm^3	A mg/cm^3	H %GC	GC %	c pg
30	*Characiformes*	*Characidae*	*Astyanax mexicanus*		1.6995	1.6998	0.3		40.6	1.1–2.1
31	*Siluriformes*	*Callichthydae*	*Corydoras aeneus*		1.6979	1.6980	0.1		38.8	4.4
32	*Salmoniformes*	*Salmonidae*	*Salmo salar*	10.99	1.7028	1.7035	0.7		44.4	
33			*Salmo fario*	45.97	1.7030	1.7039	0.9	1.9	44.8	
34			*Onchorhynchus keta*	30.51	1.7038				44.7	
35			*Onchorhynchus kisutch*	31.24	1.7033	1.7036	0.3	3.0	44.5	3.0
36			*Onchorhynchus mykiss*	28.14	1.7024	1.7026	0.2	3.0	43.5	3.2
37			*Onchorhynchus nerka*	27.68	1.7035				44.4	
38			*Coregonus autunnalis migr.*	3.49	1.7034	1.7035	0.1	3.1	44.4	
39	*Aulopiformes*	*Synodontidae*	*Synodus foetens*	32.24	1.7025	1.7029	0.4	2.5	43.8	1.2
40			*Synodus intermedius*	1.31	1.7021	1.7027	0.6	5.6	43.6	
41			*Trachinocephalus myops*	31.00	1.7034	1.7045	0.9	3.5	45.4	
42	*Gadiformes*	*Gadidae*	*Urophycis chuss*	5.86	1.7070	1.7063	–0.7	4.2	47.2	
43			*Urophycis regius*	22.29	1.7065	1.7068	0.3	3.3	47.8	0.9
44		*Merluccidae*	*Merluccius bilinearis*	52.43	1.7041	1.7042	0.1	2.5	45.1	0.9
45	*Ophidiiformes*	*Ophidiidae*	*Ophidion holbrooki*	6.43	1.7018	1.7027	0.9	4.2	43.6	0.7–0.8
46	*Batrachoidiformes*	*Batrachoidae*	*Opsanus tau*	23.73	1.6996	1.7001	0.5	2.4	40.9	2.8
47			*Porichthys notatus**	46.00	1.6979					2.2
48			*Porichthys porosissimus*	42.34	1.6998	1.7000	0.2	2.3	40.8	1.7
49	*Lophiiformes*	*Lophiidae*	*Lophius americanus*	13.08	1.6996	1.6998	0.2	3.5	40.6	1.0
50	*Cyprinodontiformes*	*Aplocheilidae*	*Aplocheilus dayi*		1.7002	1.7008	0.6		41.6	
51			*Aphyosemion amieti*	13.38	1.7016	1.7020	0.4	1.8	42.9	
52			*Aphyosemion australe*	65.34	1.7066	1.7071	0.5	2.1	48.1	0.6
53			*Aphyosemion cameronense*	15.40	1.7009	1.7012	0.3	2.7	42.0	
54			*Aphyosemion elegans*	13.23	1.7008	1.7007	–0.1	1.8	41.5	
55			*Aphyosemion hertzogii*	36.95	1.7013	1.7014	0.1	2.0	42.2	
56			*Aphyosemion marmoratum*	25.22	1.7018	1.7021	0.3	3.7	43.0	
57			*Aphyosemion punctatum*	6.43	1.7009	1.7011	0.2	1.7	41.9	0.6
58			*Aphyosemion schelii*	17.59	1.7011	1.7014	0.3	1.4	42.2	

TABLE 4.2

Properties of DNAs from fishes[a] (continued)

Sample number	Order	Family	Species	MW kb	ρ_0 g/cm^3	$<\rho>$ g/cm^3	A mg/cm^3	H %GC	GC %	c pg
117		*Pomacanthidae*	*Holocanthus passer**	37.00	1.6999					
118		*Kyphosidae*	*Hermosilla azurea**	20.08	1.6997	1.7001	0.4	3.9	40.9	
119		*Labridae*	*Bodianus diplotaenia**	34.00	1.6997	1.6996	0.0	2.2	40.5	
120			*Bodianus rufus**	44.00	1.6990	1.6994	0.4	2.2	40.2	
121			*Clepticus parrae**	46.30	1.6990	1.6988	–0.1	2.7	39.6	
122			*Coris julis*	22.70	1.7006	1.7015	0.9	2.9	42.3	1.2
123			*Gomphosus varius**	36.00	1.6972	1.6977	0.5	2.5	38.4	
124			*Halichoeres garnoti**	60.00	1.6985	1.6987	0.2	2.6	39.5	
125			*Halichoeres nicholsi**	30.00	1.6977					
126			*Halichoeres pictus**	40.50	1.6979					
127			*Halichoeres poeyi**	51.00	1.6986	1.6988	0.2	2.6	39.6	
128			*Pseudodax moluccanus**	19.00	1.6995	1.7000	0.5	3.0	40.8	
129			*Semicossyphus pulcher**	28.16	1.6995					
130			*Symphodus cinereus*	24.79	1.7002	1.7016	1.4	3.2	42.4	1.2
131			*Symphodus mediterraneus*	17.94	1.7006	1.7015	0.9	3.1	42.3	0.6
132			*Symphodus ocellatus*	20.14	1.7004	1.7012	0.8	3.2	42.0	1.1
133			*Thalassoma amblycephalum**	31.80	1.6971					
134			*Thalassoma bifasciatum**	25.80	1.6982	1.6979	–0.2	3.5	38.7	1.0
135			*Thalassoma grammaticum**	29.80	1.6974	1.6979	0.5	3.8	38.7	
136			*Thalassoma hardwicke**	62.00	1.6975	1.6976	0.1	2.5	38.3	
140			*Thalassoma hebraicum**	24.60	1.6970	1.6972	0.2	3.2	37.9	
137			*Thalassoma lucasanum**	25.00	1.6976					
138			*Thalassoma purpureum**	63.00	1.6976	1.6975	–0.2	2.3	38.2	
139			*Thalassoma trilobatum**	26.30	1.6971					
141			*Xyrichthys novacula*	28.14	1.6998	1.7001	0.3	2.0	40.9	
142		*Labrisomidae*	*Dialommus fuscus**	53.00	1.7001	1.7007	0.5	2.7	41.5	
143		*Pomacentridae*	*Chromis atrovirens**	27.00	1.6986					
144			*Chromis chromis*	17.76	1.7005	1.6993	–1.2	4.4	40.1	1.2
145			*Chromis cyanea**	17.57	1.6982	1.6987	0.5	5.2	39.5	

TABLE 4.2

Properties of DNAs from fishes[a] (continued)

Sample number	Order	Family	Species	MW kb	ρ_0 g/cm^3	$<\rho>$ g/cm^3	A mg/cm^3	H %GC	GC %	c pg
146			*Chromis multilineata**	38.50	1.6983	1.6984	0.1	3.1	39.2	
147			*Dascyllus aruanus**	32.40	1.6991					
148			*Dascyllus flavicaudus**	29.90	1.6985					
149			*Microspathodon chrysurus**	30.00	1.6979	1.6984	0.5	3.4	39.2	
150			*Microspathodon dorsalis**	35.00	1.6985	1.6989	0.4	4.1	39.7	
151			*Stegastes dorsopunicans**	18.77	1.6990	1.6990	–0.1	3.8	39.8	
152			*Stegastes planifrons**	21.00	1.6991	1.6992	0.1	3.0	40.0	
153		*Paracirrhytidae*	*Paracirrhytices forsteri*	22.23	1.7006	1.7015	0.9	3.4	42.3	
154		*Scaridae*	*Calotomus carolinus**	56.00	1.6967					
155			*Hypposcarus harid**	39.00	1.6976	1.6980	0.4	2.6	38.8	
156			*Nicholsina denticulata**	62.00	1.6965	1.6964	–0.1	3.0	37.2	
157			*Scarus coelestinus**	78.00	1.6974	1.6976	0.2	2.5	38.4	
158			*Scarus ghobban**	77.00	1.6973	1.6982	0.9	2.9	39.0	
159			*Scarus gibbus*	12.79	1.6984	1.6988	0.4	1.6	39.6	1.9–2.1
160			*Scarus psitticus**	50.00	1.6973	1.6977	0.4	2.7	38.5	
161			*Scarus schlegeli**	24.20	1.6970	1.6964	–0.6	3.7	37.2	
162		*Zanclidae*	*Zanclus cornutus**	31.80	1.6975	1.6979	0.4	2.4	38.7	
163		*Lutjanidae*	*Lutjanus synagris*	2.62	1.7010	1.7022	1.2	4.2	43.1	0.9–1.4
164		*Lethrinidae*	*Lethrinus nebulosus*	11.53	1.6999	1.7001	0.2	1.7	40.9	
165		*Cichlidae*	*Alcolapia alcalicus grahami*	8.64	1.7007	1.7012	0.5	3.6	42.0	
166			*Cichlasoma meeki*		1.7007	1.6994	–1.3		40.2	1.4
167			*Orechromis spilurus*	6.15	1.7003	1.7006	0.3	2.2	41.4	0.9
168			*Oreochromis aureus*	9.11	1.7003	1.7008	0.5	2.3	41.6	1.2
169			*Oreochromis mossambicus*	9.71	1.7009	1.7010	0.1	3.0	41.8	1.0
170			*Oreochromis niloticus*	7.65	1.7002	1.7009	0.7	2.0	41.7	0.9
171			*Symphysodon discus*	24.36	1.7004	1.7005	0.1	2.5	41.3	1.5
172			*Tilapia buettikoferi*	25.88	1.7005	1.7009	0.4	2.5	40.8	
173		*Sphyraenidae*	*Sphyraena barracuda*	8.64	1.7015	1.7017	0.2	1.8	42.6	0.8–1.2
174			*Sphyraena ensis*	8.76	1.7024	1.7027	0.3	2.7	43.6	

TABLE 4.2
Properties of DNAs from fishes[a] (continued)

Sample number	Order	Family	Species	MW kb	ρ_0 g/cm^3	$<\rho>$ g/cm^3	A mg/cm^3	H %GC	GC %	c pg
175		*Nototheniidae*	*Pagothenia borchgrevinki*	15.56	1.7002	1.7010	0.8	2.5	41.8	
176			*Trematomus bernacchii*	4.25	1.7011	1.7020	0.9	3.1	42.9	1.9
177			*Trematomus centronotus*	1.82	1.7009	1.7017	0.8	4.1	42.5	2.0
178			*Trematomus hansoni*	13.38	1.6999	1.7002	0.3	2.4	41.1	1.8
178			*Trematomus nicolai*	4.71	1.7010	1.7017	0.7	3.7	42.5	
180			*Trematomus newnesi*	1.53	1.7005	1.7018	1.3	4.5	42.6	2.1
181			*Dissosticus mawsoni*	9.84	1.7003	1.7007	0.4	2.5	41.5	
182		*Bathydraconidae*	*Gymnodraco acuticeps*	17.24	1.7007	1.7012	0.5	3.5	42.1	1.9
183		*Blenniidae*	*Ophioblennius atlanticus*	7.33	1.7007	1.7011	0.4	2.8	41.9	0.8–1.0
184		*Callyonimidae*	*Synchiropus splendidus*	11.12	1.7073	1.7076	0.3	2.7	48.6	
185	*Pleuronectiformes*	*Pleuronectidae*	*Siacium papillosum*	2.62	1.7024	1.7025	0.1	4.7	43.4	0.7–1.0
186			*Limanda aspera*	6.62	1.6995	1.7002	0.7	3.5	41.0	0.6–1.0
187			*Limanda ferruginosa*	10.34	1.7012	1.7021	0.9	4.7	42.9	
188			*Psettichthys melanostictus*	80.00	1.7002	1.7006	0.3	2.7	41.4	
189			*Pleuronichthys californicus**	49.00	1.6996	1.6992	–0.4	2.7	40.0	
190	*Tetraodontiformes*	*Balistidae*	*Melichthys vidua**	12.79	1.7014	1.7022	0.8	2.5	43.1	
191			*Balistes capriscus*	17.07	1.7023	1.7024	0.1	2.4	43.3	0.7
192			*Monacanthus tuckeri**	37.80	1.7042	1.7046	0.4	3.6	45.5	
193			*Rhinecanthus aculeatus*	8.19	1.7047	1.7048	0.1	1.5	45.7	
194			*Stephanolepis hispidus*	3.73	1.7037	1.7043	0.6	4.4	45.2	0.7
195			*Aluterus schoepfi*	1.44	1.7038	1.7039	0.1	5.4	44.8	0.6
196		*Ostraciidae*	*Acanthostracion quadricornis*	51.10	1.7001	1.7008	0.7	2.7	41.6	
197		*Tetraodontidae*	*Lagocephalus laevigatus*	9.35	1.7022	1.7026	0.4	3.5	43.5	0.4–0.5
198			*Arothron diadematus*	10.09	1.7025	1.7036	1.1	3.0	44.5	
199			*Arothron meleagris*	26.77	1.7034	1.7058	1.4	4.8	46.7	
200			*Sphoeroides annulatus*	18.47	1.7047	1.7053	0.6	3.6	46.2	
201		*Diodontidae*	*Diodon holocanthus*	6.24	1.7009	1.7012	0.3	2.1	42.0	0.9

[a] Asterisks indicate values from Bucciarelli et al. (2002). All other values are from Bernardi and Bernardi (1990a). See this reference for additional information and for the sources of *c* values. The sources of samples are given in Bernardi and Bernardi (1990a) and in Bucciarelli et al. (2002). See also legend of Fig. 4.1.

heavy side and much more rarely on the light side. Finally, some other samples showed poorly resolved satellite on the heavy side.

Figure 4.2 displays the distribution of the modal buoyant densities of the fish DNAs studied. The modal buoyant density range covered was 1.696–1.708 g/cm^3. In the case of *Actinopterygii*, the average ρ_0 was 1.7014 g/cm^3 and the corresponding standard deviation, σ, was 2.2 mg/cm^3; most values were between 1.699 and 1.704 g/cm^3. In the case of *Chondrichthyes*, the average ρ_0 was 1.7035 g/cm^3, and σ was 1.1 mg/cm^3; most values were in the 1.702–1.704 g/cm^3 range. Because the distribution of ρ_0 values could be biased by the over-representation of species from the same genera or families of fishes, a histogram was constructed in which only one DNA sample per genus was taken into account when the other species within the genus had the same buoyant density. This histogram (**Fig. 4.2A**) was not, however, remarkably different from that including all species (see Bucciarelli et al., 2002).

The distribution of DNA CsCl band asymmetries for all fish species investigated are presented in the histogram of **Fig. 4.2B**. The majority of fishes showed asymmetry values around 0.5 mg/cm^3 (A = 0.65 mg/cm^3; σ = 0.54 mg/cm^3 in the case of *Actinopterygii*).

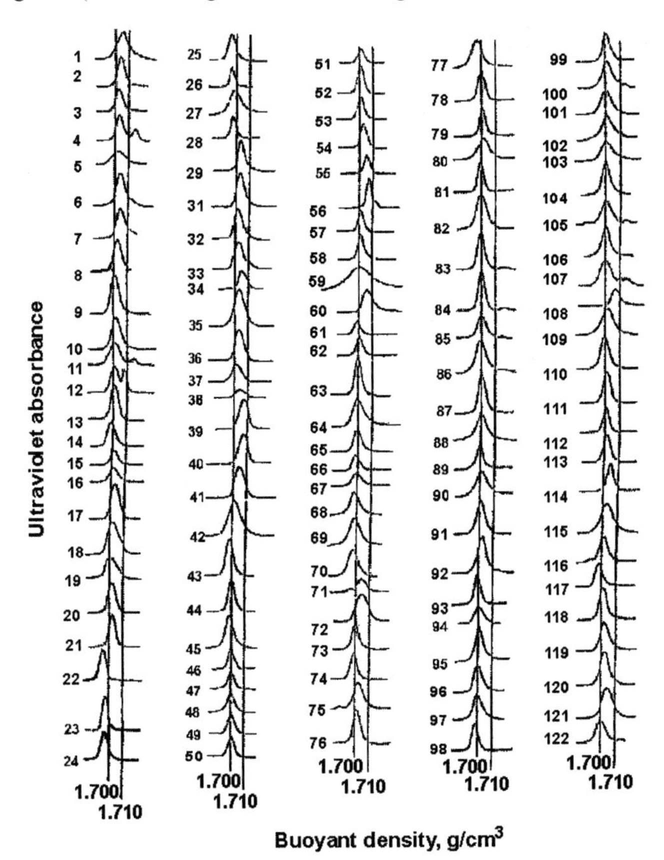

Figure 4.1. CsCl band profiles of fish DNAs. See Table 3 of the original paper for the numbering of samples. Interestingly, even the very low molecular weight samples (see Table 4.2) yielded acceptable profiles, because of the presence of large-size molecules in DNA preparations and of the compositional homogeneity of fish DNAs. (From Bernardi and Bernardi 1990a).

Intermolecular compositional heterogeneities (H) basically reflect the spread of GC levels among different fragments in a given DNA preparation. (see **Part 1**, **section 1.3**) **Fig. 4.2C** displays the distribution of H values for different fish species. Values ranged from 0.5% GC for *Aphyosemion striatum* (sample 59) to 4.7% GC for *Arothron meleagris* (sample 199; the higher value of 4.9% for *Squatina dumerili*, sample 7, was due to a shoulder on the heavy side of the main band, indicative of a satellite). The average heterogeneity, $\overline{H}$, of DNAs from *Osteichthyes* was equal to 2.6% GC ($\sigma = 1.1$% GC). Values higher than 3% GC were often due to poorly resolved satellites (see, for instance, sample 7). Artefactual reasons, like low molecular weight of DNA samples, may also account for some high values of H. A plot of CsCl band asymmetry *versus* intermolecular compositional heterogeneity (**Fig. 4.2**) shows a linear relationship with a significant slope ($p < 0.001$), stressing the contribution of asymmetry to heterogeneity (see **Part 1**, **Section 1.5**). In most cases, compositional heterogeneities were equal to, or lower than, those of comparable bacterial DNAs (see **Figs. 3.10** and **3.11**).

In conclusion, an almost two-fold increase in the number of fish DNAs investigated, reaching a total of 200 species, including species from a number of orders and families not previously explored, did not lead to any significant differences in the range and average values of modal buoyant densities, asymmetry and heterogeneity compared to previous

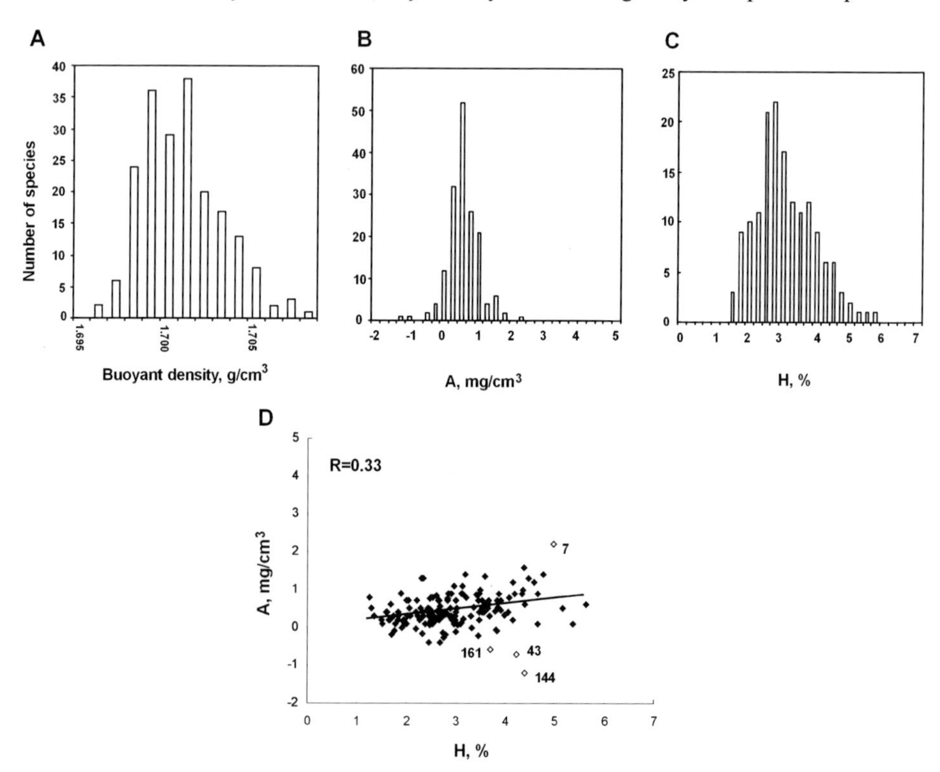

Figure 4.2. The number of fish species is plotted against **A** modal buoyant densities (only species characterized by different modal buoyant densities within a genus were used), **B** CsCl band asymmetry and **C** intermolecular compositional heterogeneity of DNAs. **D**. Plot of CsCl band asymmetry against compositional heterogeneity of DNAs from fishes (four outliers, numbered according to Table 4.2, were not taken into consideration because the anomalous asymmetries were due to satellite DNAs). The correlation coefficient, R, is given. (From Bucciarelli et al., 2002).

investigations (Bernardi and Bernardi, 1990a). This suggests that these values are unlikely to change with further increases in the fish samples explored.

1.2. Compositional properties: a Cs_2SO_4/BAMD analysis

Fractionation in Cs_2SO_4/BAMD was used to analyze in more detail some fish genomes. The analyses of DNA fractions, presented as histograms of relative DNA amounts *versus* modal buoyant densities in CsCl in **Fig. 4.3**, provided information on cryptic, or poorly detectable, satellites, which all showed up as GC-rich satellites. When not resolved from the main band, cryptic satellites were sometimes detected by restriction enzyme digestions. These analyses were done for different reasons.

(i) Some analyses were used to confirm the correlation proposed by Cuny et al. (1981; see also **Part 7**) between chromosomal banding patterns and compositional heterogeneity of DNA (Medrano et al., 1988). *Anguilla anguilla* and *Epinephelus guttatus* exhibit different degrees of asymmetry (1.3 and 0.2 mg/cm^3, respectively) and compositional heterogeneity (2.3% and 1.7% GC, respectively). These species were shown to be characterized by strong differences in Giemsa and Reverse banding of their metaphase chromosomes. Indeed, practically no banding pattern was found in *E. guttatus*, whereas a rather distinct pattern could be detected in *A. anguilla*. Another species, *Labeo bicolor*, exhibiting intermediate values of asymmetry (0.7 mg/cm^3) and heterogeneity (2.0% GC) showed a banding pattern of intermediate intensity and reciprocity. Interestingly, the other two fishes explored, *Salmo gairdneri* (or *Onchorynchus mykiss*) and *Cyprinus carpio*, also showed a remarkable

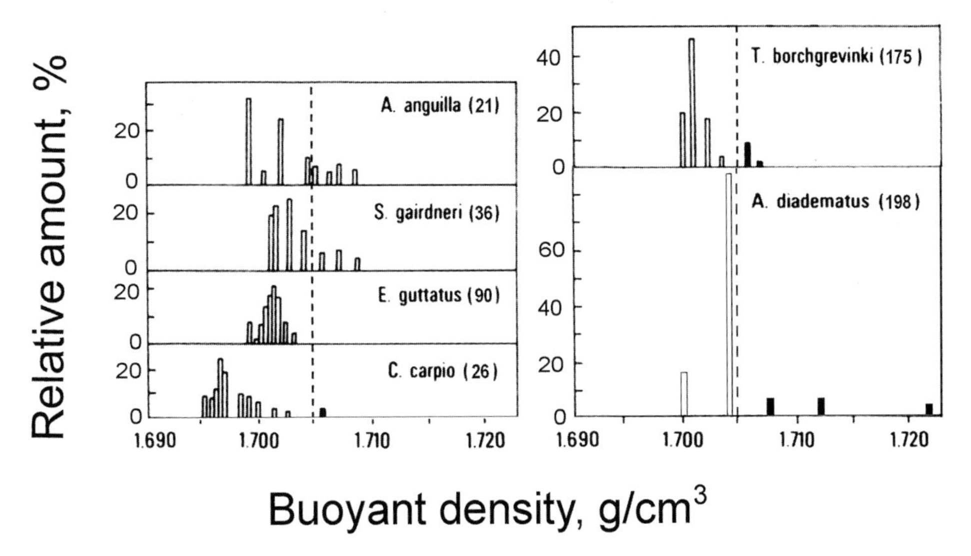

Figure 4.3. Histograms showing the relative amounts and the modal buoyant densities in CsCl of DNA fractions obtained by preparative Cs_2SO_4/BAMD density gradient centrifugation from some fish species. Black bars correspond to satellite components. The vertical broken line at 1.705 g/cm^3 is shown to provide a reference. Numbers in parentheses refer to Table 4.2. (From Bernardi and Bernardi 1990a).

compositional heterogeneity in their DNAs. Yet, the chromosomal banding patterns were very different. In *S. gairdneri* the compositional spectrum was similar to that of *A. anguilla* and, expectedly, the Giemsa banding pattern was quite distinct. In the case of *C. carpio*, which is characterized by one of the GC-poorest DNAs of all vertebrates, chromosomes are extremely compact. (ii) In the case of the genome of *Pagothenia* (formerly *Trematomus*) *borchgrevinki*, a typical Antarctic fish (sample 175), the CsCl bands for the six fractions obtained show (Hudson et al., 1980) that 85% of the DNA is in the range 1.700-1.702 g/cm^3 (fractions 1-3) pointing to a very homogeneous genome. Among higher density fractions, the most abundant, fraction 5, showed, upon digestion with restriction enzyme HaeIII, three strong bands indicative of a satellite component. (iii) *Arothron diadematus* was investigated because the order to which it belongs (*Tetraodontiformes*) is characterized by a very small genome (c = 0.4-0.5 pg), in fact, the smallest among vertebrate genomes. In this case, renaturation kinetics showed that this DNA is the vertebrate DNA lowest in repeated sequences (Pizon et al., 1984, see **Fig. 4.4**). The very homogeneous *A. diadematus* DNA could be resolved into several components, characterized by buoyant densities of 1.700, 1.704_5, 1.708, 1.712 and 1.723 g/cm^3, and representing 15%, 73%, 4%, 4% and 2.5%, respectively, of total DNA. The last component comprised a satellite DNA and ribosomal DNA. A family of interspersed repeats, possibly related to the *Alu* family of warm-blooded vertebrates, showed an extremely specific genomic distribution, being present only in the 1.708 g/cm^3 component, which it matched in base composition (see **Fig. 4.3**). Later work on another tetraodontiform, *Takifugu rubripes*, showed that these small genomes are also characterized by short introns (Brenner et al., 1993).

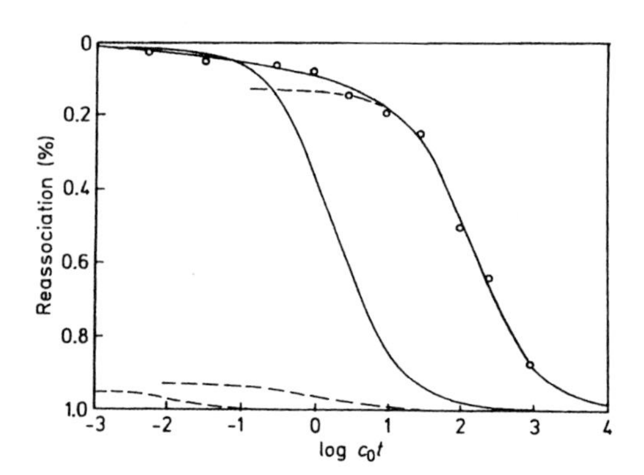

Class	Amount %	Complexity kb
Foldback	1	
Fast	5	1.4
Intermediate	7	2.2×10^2
Slow	87	3.9×10^5

Figure 4.4. Reassociation kinetics of *A. diadematus* DNA. Results with *Escherichia coli* DNA (–) are shown for comparison. The solid line through the experimental points (o) is the overall profile resulting from the analysis of kinetic classes (Form Pizon et al., 1984; see Soriano et al., 1981).

1.3. Compositional properties: an analysis of long sequences

The very recent availability of long DNA sequences from three fishes, a cyprinid, *Brachidanio rerio* (*Danio* or zebrafish), and two *Tetraodontiformes*, *Takifugu rubripes* (*Fugu* or pufferfish) and *Tetraodon nigroviridis*, made it possible to push further the compositional

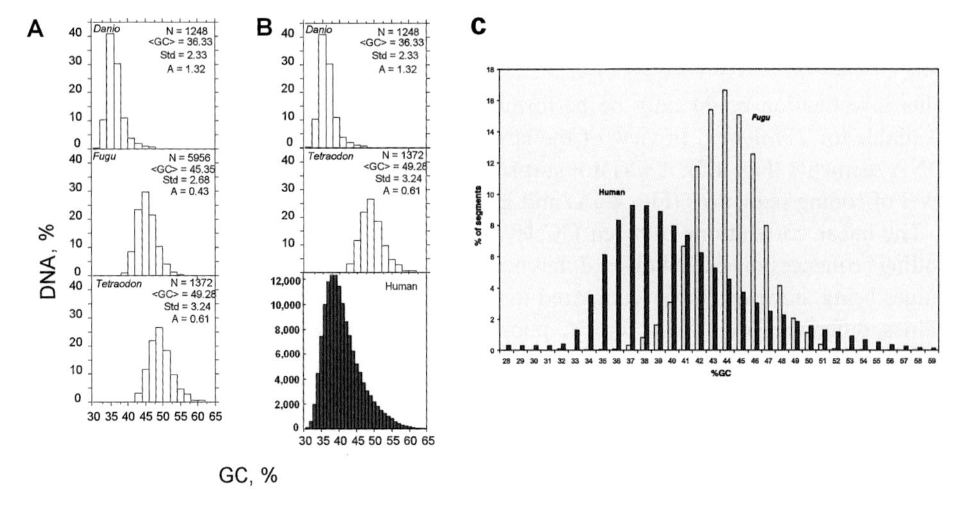

Figure 4.5. **A**, **B**. Compositional distribution of 20 kb DNA segments from *Danio*, *Fugu*, *Tetraodon* and human. (From Jabbari and Bernardi, 2004a; human data are from Lander et al., 2001). **C**. Distribution of GC content in the *Fugu* and human genomes. Sliding windows of 50 kb were used; similar conclusions were derived with windows of 25 and 100 kb. (From Aparicio et al., 2002).

analysis of fish genomes (Jabbari and Bernardi, 2003a) and to compare them with the compositional patterns of 20 kb segments from the human genome (in all cases, satellite and ribosomal sequences were absent from the sequences). The differences observed in **Fig. 3.17** between the genomes from warm-blooded (human) and cold-blooded (*Xenopus*) are found again in the *Danio*/human comparison (**Fig. 4.5**). The case of the other two fish genomes is of special interest because the genomes from the most recent order *Tetraodontiformes* are exceptionally small (0.4 pg/haploid cell), very poor in repeated sequences (Pizon et al., 1984), characterized by short introns (Brenner et al., 1993) and GC-rich in most species (Bernardi and Bernardi, 1990a). In spite of these specific features, these genomes share with those of all other species an extremely narrow compositional distribution of large DNA segments compared to those of warm-blooded vertebrates (**Fig. 3.18**).

Two additional conclusions could be drawn from **Fig. 4.5**: (i) a positive asymmetry (*i.e.*, a skew on the GC-rich side) seems, indeed, to be a property of main-band DNA and is not due to satellites, since the latter were excluded from the large DNA segments analyzed; this conclusion provides a final confirmation of what could be deduced 1) from the CsCl profiles; in fact, it was most unlikely that all the positive asymmetries of the CsCl profiles were systematically due to cryptic, GC-rich satellites; and 2) from the Cs_2SO_4/BAMD fractionation which showed (**Fig. 4.3**) that GC-rich satellites were not the only, or even the main, contributors to the asymmetric distributions of DNA molecules; (ii) a remarkable difference in average GC level was found between *A. diadematus* (**Table 4.2**) and *Fugu*, which showed values of 44.5% and 45.3%, on the one hand, and *T. nigroviridis*, which showed a value of 49%, on the other hand. This surprising difference for the genomes of fishes belonging to the same subfamily, *Tetraodontinae* (Tyler, 1980), will be discussed in **Part 12**.

1.4. Compositional properties of coding sequences and introns

This investigation could only be performed on *Fugu* and *Danio*, no annotation being available for *Tetraodon*. In view of the large, 9%, difference in average GC level of long DNA segments (**Fig. 4.5**), it was not surprising to find a 6% difference in the average GC level of coding sequences (**Fig. 4.6A**) and in that of GC_3 (**Fig. 4.6B**).

The linear correlations between GC levels of introns and those of the corresponding coding sequences or GC_3 showed differences which increased with increasing GC, intron values being, however, lower compared to coding sequence. In the comparison with GC_3 values, differences were small for GC-poor genes, but they increased very much for GC-rich genes (not shown).

1.5. Compositional correlations

Fig. 4.6C shows the compositional correlation between GC_3 and flanking sequences (20 kb in size) for *Fugu*. This was used to assess gene density (**Fig. 5.7**), by the approach developed for estimating gene density in the human genome (see **Part 3**). Interestingly, the GC levels of flanking sequences corresponding to the highest GC_3 values are only about 47%, *versus* 55% in the case of human isochores (see **Fig. 3.20**).

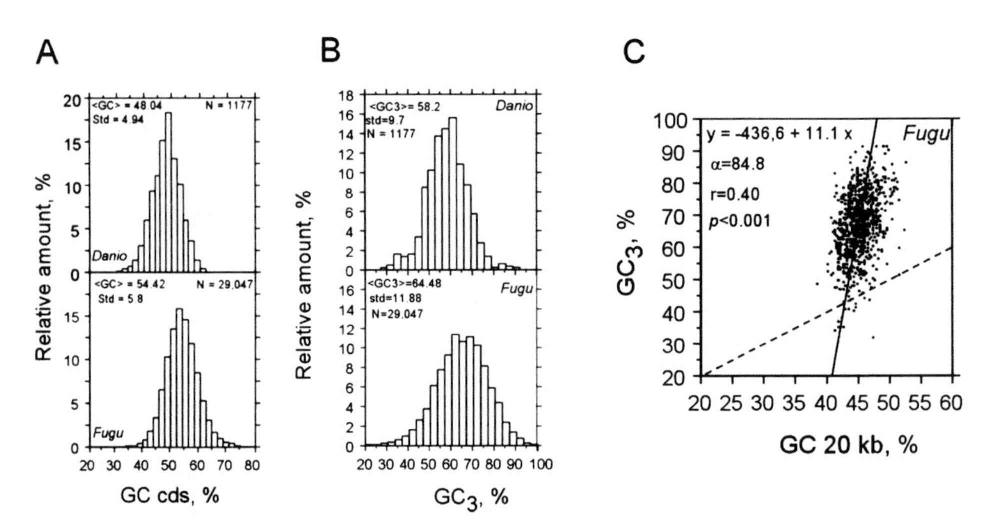

Figure 4.6. Relative amounts of coding sequences from *Danio* and *Fugu* are plotted against their GC (**A**) and GC_3 (**B**) levels. **C.** Plot of GC_3 against 20 kb segments of *Fugu* genome. (From Jabbari and Bernardi, 2004a).

CHAPTER 2

Amphibian genomes

In the case of amphibians, the basic compositional properties (Thiery et al., 1976; Bernardi and Bernardi, 1990a) were similar to those just described for fishes (see **Table 4.3** and **Fig. 4.7**). The modal buoyant densities of DNAs from anurans range from 1.6991 g/cm^3 for *Xenopus laevis* (sample 1 of **Table 4.3**), a strictly aquatic species from the ancient superorder *Pipoidei*, to 1.7013, 1.7028 and 1.7029 g/cm^3 for *Leptodactylus pentadactylus*, *Bufo paracnemis*, and *Rana sp.* (samples 2–4), respectively; these three species belong to the more recent, not strictly aquatic, superorder *Ranoidei*. The ρ_0 value of DNA from the urodele *Pleurodeles waltlii* (sample 5) is equal to 1.7041 g/cm^3.

The CsCl profiles of DNAs from amphibians (**Fig. 4.7**) show no satellite bands; cryptic satellites may exist, however. For instance, in the case of *X. laevis*, the CsCl analysis of fractions from preparative centrifugation in Cs_2SO_4/Ag^+ showed (see **Fig. 3.3**) the presence of minor components banding at 1.705 and 1.706 g/cm^3 and of a minute amount of a 1.712 g/cm^3 component (Thiery et al., 1976).

On the other hand, the Cs_2SO_4/Ag^+ fractionation of *P. waltlii* DNA revealed the presence of a component banding at 1.706 g/cm^3 and representing about 25% of DNA, whereas the majority of DNA bands at 1.704 g/cm^3 (Thiery et al., 1976). It is possible that the heavier component is a satellite, as no such component was found after HindII+III digestion (Macaya et al., 1976), as if it had been degraded to small fragments.

All amphibian DNAs are characterized by low asymmetries (ranging from 0.4 to 0.8 mg/

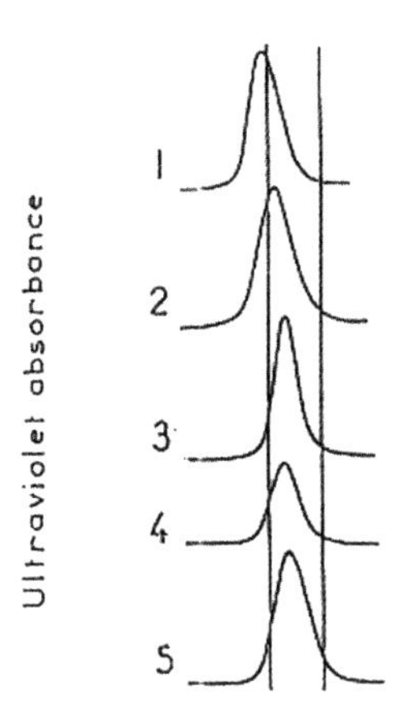

Figure 4.7. CsCl band profiles of DNAs from amphibians. See Table 4.3 for the numbering of samples. (From Bernardi and Bernardi 1990a).

CHAPTER 3

Reptilian genomes

Extant reptiles comprise three main groups: (i) turtles and (ii) crocodiles, two homogeneous orders accounting for a relatively small number of species; and (iii) lepidosaurs, or squamates, consisting of lizards and snakes, which are morphologically more diversified and richer in species (see Olmo et al., 2002). Reptiles were studied in three steps (Thiery et al., 1976; Bernardi and Bernardi, 1990a; Hughes et al., 2002) involving a total of 2, 13 and 43 species, respectively. Because of differences in the classification of reptiles, in the experimental setup (Beckman Optima XL *versus* Spinco model E ultracentrifuges) and in the slightly different criteria used to define CsCl band borders (and therefore asymmetries and heterogeneities), the older results (Bernardi and Bernardi, 1990a) and the more recent ones (Hughes et al., 2002) will be presented separately. The advantage of this separate presentation is that each data set is perfectly coherent. Moreover, the older set was assessed exactly like the fish and amphibian sets investigated by Bernardi and Bernardi (1990a; see **Chapters 1** and **2**).

As far as the older results are concerned, DNAs from reptiles exhibited modal buoyant densities that were in the range 1.6992–1.7047 g/cm^3 (**Table 4.4A** and **Fig. 4.8A**). Shoulders on the CsCl bands can be seen in at least three DNAs, those from *Tarentola mauretanica*, *Natrix maura* and *Bothrops neuwiedi* (samples 5, 9, 13 of **Table 4.4A**). Moreover, cryptic satellites might be responsible for the GC-rich tails exhibited by the CsCl band profiles of most DNAs.

The asymmetries of CsCl bands of DNAs from reptiles were often close to those shown by DNAs from amphibians. They were, however, larger in several cases (**Table 4.4A**). In five cases out of the six in which heterogeneity was determined, values were close to those found in DNAs from amphibians, with a higher value for *Testudo graeca* (sample 3; **Table 4.4A**). Again, the presence of cryptic heavy satellites may be responsible, at least in part, for the higher asymmetry and heterogeneity values of DNAs from reptiles. Such presence is suggested by the existence of components melting at high temperature in DNAs from some reptiles (Olmo, 1981). Asymmmetry and heterogeneity of DNAs from reptiles are in the same range as those of most fish DNAs, with some exceptions.

A crocodile (*Crocodylus niloticus*) and a turtle (*T. graeca*), two GC-rich genomes (see **Table 4.4A**) were studied in more detail, namely by Cs_2SO_4/BAMD fractionation and CpG island analysis (Aïssani and Bernardi, 1991a). Fractionation results showed the presence of sizable amounts of heavy fractions in both DNAs (see **Fig. 5.9**). Methylation data (Jabbari et al., 1997) and the analysis of a few coding sequences (Hughes et al., 1999) from these reptilian genomes will be presented in **Parts 5** and **12**, respectively.

The more recent results (Hughes et al., 2002; see **Table 4.4B** and **Fig. 4.8B**) are characterized by ρ_0 values comprised between 1.6974 and 1.7032 g/cm^3 (covering a lower range than that previously reported), a by higher asymmetry, and heterogeneity values.. An interesting point of the new data is that often several species from the same sub-family

TABLE 4.4A

Properties of DNAs from reptiles[a]

Order	Family		Species	$s_{20,w}$ S	ρ_0 g/cm^3	$<\rho>$ g/cm^3	A mg/cm^3	H %	GC %
Crocodylia	*Crocodylidae*	1	*Crocodylus niloticus*	37.7	1.7046	1.7051	0.5	3.2	46.0
Chelonia									
Suborder	*Emydidae*	2	*Pseudemys nelsoni*	34.4	1.7022	1.7023	0.1	1.3	43.2
Cryptodira	*Testudinidae*	3	*Testudo graeca*[b]	25.1	1.7027	1.7042	1.5	4.5	45.1
Squamata									
Sauria	*Iguanidae*	4	*Iguana iguana*[b]	20.5	1.7015	1.7022	0.7	2.7	43.1
	Gekkonidae	5	*Tarentola mauretanica*	23.9	1.7047	1.7057	1.0	2.7	46.6
	Scincidae	6	*Chalcides ocellatus*		1.7039	1.7050	1.1		45.9
	Tefidae	7	*Ameiva ameiva*		1.7014	1.7018	0.4		42.6
	Varanidae	8	*Varanus griseus*		1.7024	1.7029	0.5		43.8
Ophidia	*Colubridae*	9	*Natrix maura*	23.1	1.7002	1.7018	1.6	3.4	42.6
		10	*Tomodon dorsatus*		1.7014	1.7015	0.1		42.3
		11	*Dromicus porcilogyrus s.*		1.7006	1.7007	0.1		41.5
	Viperidae	12	*Cerastes cerastes*		1.6995	1.7007	1.2		41.5
	Crotalidae	13	*Bothrops neuwiedi p.*		1.6992 (1.7028)[c]	1.7029			43.8

[a] From Bernardi and Bernardi, 1990a

[b] Data from Thiery et al. (1976). H values from Bernardi and Bernardi (1990a)

[c] Satellite band

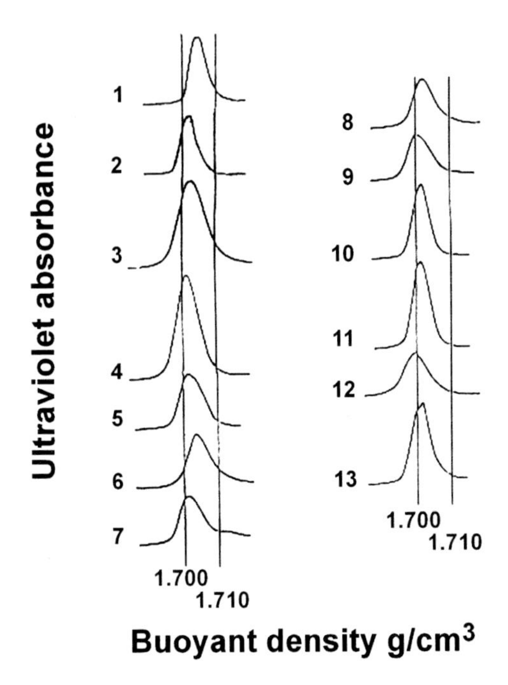

Figure 4.8A. CsCl band profiles of DNAs from reptiles of Table 4.4A. See this Table for the numbering of samples. (From Bernardi and Bernardi, 1990a).

were studied and showed different asymmetries of CsCl profiles. In this case, obviously the lowest values should be the most representative, since changes within a subfamily are most likely due to changes in satellite DNAs (or interspersed repeats).

Fig. 4.9 presents the asymmetry and the modal buoyant density values from **Table 4.4B** (when more than one species per genus or subfamily was available the lowest asymmetry values was used). As far as asymmetry is concerned, all values (except for one, close to the borderline) are below the lower boundary of asymmetry of mammals and birds, with many values below the upper boundary of fishes and amphibians. It should be noted that asymmetry values of **Table 4.4B** tend to be higher than those of **Table 4.4A**. In terms of modal buoyant density, 1.701 g/cm^3 seems to be the borderline between snakes and turtles on the lower GC side and lizard and crocodiles on the upper side.

TABLE 4.4B

Properties of DNAs from reptiles[a]

Order Suborder	Superfamily Family	Subfamily Species	ρ_0 (g/cm^3)	$<\rho>$ (g/cm^3)	$<\rho>-\rho_0$ (mg/cm^3)	H (GC%)	MW (kb)
Squamates							
Lacertilia	*Gekkota*						
	Gekkonidae	*Gekko gecko*	1.7032	1.7046	**1.4**	5.2	16
	Scincomorpha						
	Lacertidae	*Lacerta viridis*	1.7019	1.7031	**1.2**	3.5	>60
		Podarcis muralis	1.7019	1.7025	**0.6**	3.1	>60
	Diploglossa						
	Anguidae	*Anguis fragilis*	1.7019	1.7032	**1.3**	3.2	>60
Serpentes	*Colubroidea*						
	Colubridae	*Colubrinae*					
		Boiga dendrophila	1.6997	1.7007	1.0	4.3	>60
		Elaphe obsoleta	1.6984	1.6993	**0.9**	3.7	>60
		Gonyosoma oxycephalum	1.6988	1.7005	1.7	5.1	27
		Lampropeltis triangulum polyzona	1.6983	1.7011	2.8	6.8	36
		Rhinocheilus leconteï	1.6996	1.7015	1.9	6.0	21
		Xenodontinae					
		Clelia rustica	1.6989	1.6995	**0.6**	2.8	>60
		Boodontinae					
		Lamprophis olivaceus	1.6995	1.7022	**2.7**	6.7	45
		Dipsadinae					
		Pliocercus euryzona	1.7010	1.7014	**0.4**	3.5	54
		Sibon longifrenis	1.7005	1.7015	1	4.3	57
		Sibon nebulata	1.7000	1.7009	0.9	3.9	41
	Elapidae	*Bungarinae*					
		Aspidelaps lubricus	1.6984	1.7020	3.6	7.6	25
		Walterinnesia aegyptiae	1.6983	1.6995	**1.2**	3.5	>60
		Notechinae					
		Notechis scutatus	1.6980	1.7000	**2**	4.4	>60

TABLE 4.4B

Properties of DNAs from reptiles[a] (continued)

Order **Suborder**	**Superfamily** **Family**	**Subfamily** **Species**	ρ_0 **(g/cm^3)**	$\langle\rho\rangle$ **(g/cm^3)**	$\langle\rho\rangle$-ρ_0 **(mg/cm^3)**	**H (GC%)**	**MW (kb)**
	Viperidae	*Viperinae*					
		Bitis gabonica gabonica	1.6982	1.7003	2.1	5.5	15
		Vipera ammodytes	1.6980	1.6991	**1.1**	3.5	60
		Crotalinae					
		Crotalus lepidus	1.6985	1.7006	**2.1**	5.5	29
		Tropidolaemus wagleri	1.6995	1.7026	3.1	7.3	23
	Henophidiae						
	Boidae	*Boa constrictor*	1.6982	1.6997	**1.5**	3.5	26
	Pythonidae	*Python molurus*					
		bivittatus	1.6974	1.6999	2.5	5.3	34
		Python curtus	1.6983	1.7005	**2.2**	6.2	18
Turtles							
Cryptodires	*Emydidae*	*Trachemys scripta elegans*	1.7006	1.7028	**2.2**	4.7	>60
		Trachemys scripta elegans	1.7007	1.7031	2.4	4.9	>60
Pleurodires	*Pelomedusidae*	*Pelusios subniger*	1.7002	1.7026	**2.4**	5.6	35
Crocodiles	*Alligatoridae*	*Alligator mississippiensis*	1.7017	1.7037	**1.9**	4.5	>60
	Crocodylidae						
		Crocodylus acutus ♀	1.7022	1.7040	1.8	4.3	>60
		Crocodylus acutus ♂	1.7026	1.7044	1.8	4.5	60
		Crocodylus niloticus	1.7021	1.7036	**1.5**	4.1	>60
Birds	*Phasianidae*	*Coturnix coturnix*	1.6990	1.7018	2.8	6.4	>60
		Gallus gallus	1.6996	1.7026	3.0	6.4	25
Amphibians	*Ranidae*	*Rana esculenta*	1.7024	1.7028	0.4	3.1	33
	Pipidae	*Xenopus laevis*	1.6973	1.6980	0.7	2.9	55

[a] The classification given for reptiles is that of the Reptile EMBL database (http://www.embl-heidelberg.de/~uetz/LivingReptiles.html). Sources of samples are given in the original paper. Asymmetry values in bold are the values used in Fig. 4.9, where only the lowest values from each subfamily or genus with more than one species were used. Some results with avian and amphibian DNAs are also presented. (From Hughes et al., 2002).

CHAPTER 4

Avian genomes

Table 4.5 shows a classification of birds and the species investigated. Eight species from eight different orders, including both paleognathous and neognathous birds, were explored, six of them by the Cs_2SO_4/BAMD approach followed by analytical centrifugation in CsCl. Avian DNAs showed a high degree of similarity in their isochore patterns. Major DNA components reached higher GC levels in avian DNAs (see **Table 4.6** and **Fig. 4.10**) compared to mammalian DNAs; moreover, the former exhibited 10–12% of their DNAs having a buoyant density of 1.710 g/cm^3 or higher, whereas the latter only showed 6–10% of their DNAs in that buoyant density range (Kadi et al., 1993).

TABLE 4.5

A classification of birds and the species investigated in the present work[a]

Superorder and order	Family	Species investigated	
Paleognathi			
Struthioniformes	*Struthionidae*	*Struthio camelus*	Ostrich
Rheiformes	*Rheidae*	*Rhea americana*	Rhea
Neognathi			
Sphenisciformes	*Spheniscidae*	*Spheniscus demersus*	Penguin
Anseriformes	*Anatidae*	*Cairina moschata*	Duck
Galliformes	*Phasianidae*	*Gallus gallus*	Chicken
Charadriiformes	*Laridae*	*Larus argentatus*	Herring gull
Columbiformes	*Columbidae*	*Columba livia*	Pigeon
Passeriformes	*Ploceidae*	*Passer domesticus*	Sparrow

[a] (From Kadi et al., 1993). The classification of birds is from Perrins and Middelton (1985), except for the introduction of superorders (Sibley and Monroe, 1990).

Interestingly, significant linear correlations were found between coding sequences (and their first, second and third codon positions), flanking regions (5' and 3'), and introns, as is the case in the human genome. These compositional correlations were not limited to GC levels but extended to individual bases. Finally, an analysis of coding sequences has confirmed a correlation among GC_3, GC_2, and GC_1 (Musto et al., 1999).

TABLE 4.6
Buoyant densities of avian DNAs[a]

DNA	ρ_0 g/cm^3	GC-rich DNA[b] %
Ostrich	1.7001	9
Rhea	1.6999	9
Penguin	1.7006	
		10
Duck	1.6991	12.5
Chicken	1.6997[b]	
		12.4
Seagull	1.6998[c]	
Pigeon	1.7007	13
Sparrow	1.7002[d]	

[a] From Kadi et al. (1993). See the original paper for additional information.
[b] % of DNA having a buoyant density of 1.710 g/cm^3 or higher (neglecting satellite and minor components).

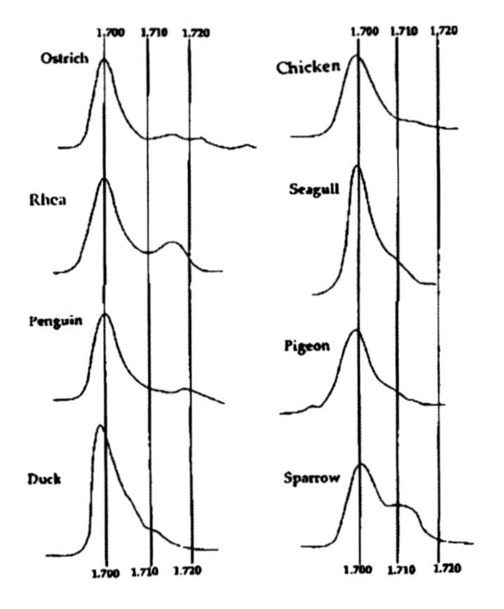

Figure 4.10. Analytical CsCl density profiles of unfractionated DNA preparations. (From Kadi et al., 1993).

Mammalian genomes

In contrast with avian DNA, mammalian DNAs showed differences in their compositional patterns. The first indication in this direction came from the early work of Thiery et al. (1976) which showed a lower degree of asymmetry for the DNAs from mouse and rat (as well as for those from hamster and garden dormouse) compared to those of other mammals (see **Table 3.2**). Investigations by Salinas et al. (1986) on the mouse genome and by Zerial et al. (1986a) on the human genome demonstrated important differences between the isochore patterns of these genomes. Indeed, the mouse genome lacked a GC-rich DNA component, H3, which was present in the human genome. Initially, it was difficult to decide whether this GC-richest component H3 was the result of a GC increase in the sequences forming a genome compartment (the isochore family H3), or of an invasion of GC-rich repetitive sequences, like Alu sequences (which are much more numerous in human than in mouse). The first explanation was unequivocally shown to be correct by comparing the compositional properties of orthologous genes from human and *Xenopus* (see **Part 11**) and by investigating the effect of removing (by masking) repeated sequences from the genome (see **Part 6**).

Table 4.7 presents a classification of mammals and the species investigated. **Figs. 4.11** and **4.12** show the ρ_0 values and the CsCl profiles of the DNAs under consideration here and **Table 4.8** lists the parameters obtained from such profiles. The work of Sabeur et al. (1993) on 20 species belonging to nine of the 17 eutherian orders demonstrated the existence of a "**general distribution**" in species belonging to eight mammalian orders. The human isochore pattern was typical of this general distribution. A "**myomorph distribution**" was found in *Myomorpha* (such as mouse and rat; see above), but not in the other rodent infraorders *Sciuromorpha* and *Histricomorpha*, which shared the "general distribution". Two other distributions were found in a megachiropteran (but not in a microchiropteran, which, again, shared the general distribution) and in pangolin (a species from the only genus of the order *Pholidota*), respectively. The main difference between the general distribution and the other distributions was that the former contains sizable amounts, 6–10 %, of GC-rich isochores (detected as DNA molecules equal to, or higher than, 1.710 g/cm^3 in buoyant density), which are scarce, or absent, in the other distributions. It should be stressed that while the "myomorph distribution" is now understood (see **Part 11**), the other special distributions have not been investigated further. We know, however, that they are different for the "myomorph distribution" (see Sabeur et al., 1993).

TABLE 4.7

A classification of mammals and the species investigated[a]

Order, sub-, and infraorder	Family	Species investigated	
Monotremata			
Marsupialia			
Insectivora	*Erinaceidae*	1. *Erinaceus europeus*	Hedgehog
	Soricidae	2. *Crocidura russula*	Shrew
	Talpidae	3. *Talpa europea*	Mole
Dermoptera			
Chiroptera			
Megachiroptera	*Pteropodidae*	4. *Pteropus sp.*	Fruit bat
Microchiroptera	*Vespertilionidae*	5. *Myotis myotis*	Bat
Primates	*Hominidae*	6. *Homo sapiens*	Man
Edentata			
Pholidota	*Manidae*	7. *Manis sp.*	Pangolin
Lagomorpha	*Leporidae*	8. *Oryctolagus cuniculus*	Rabbit
Rodentia			
Sciurognathi			
Sciuromorpha	*Sciuridae*	9. *Sciurus vulgaris*	Squirrel
		10. *Marmotta monax*	Woodchuck
Myomorpha	*Muridae*	11. *Rattus norvegicus*	Rat
		12. *Mus musculus*	Mouse
	Cricetidae	13. *Cricetus norvegicus*	Hamster
	Spalacidae	14. *Spalax sp*	Mole rat
	Gliridae	15. *Glis glis*	Dormouse
Histricognathi			
Caviomorpha	*Caviidae*	16. *Cavia porcellus*	Guinea pig
Cetacea			
Carnivora	*Canidae*	17. *Canis familiaris*	Dog
	Felidae	18. *Felis domesticus*	Cat
Pinnipedia			
Tubulidentata			
Proboscidea			
Hyracoidea			
Sirenia			
Perissodactyla	*Equidae*	19. *Equus caballus*	Horse
Artiodactyla	*Bovidae*	20. *Bos taurus*	Calf

[a] From Nowak and Paradiso (1983). Rodents are classified according to Colbert and Morales (1991).

TABLE 4.8

Buoyant density properties of mammalian DNAs[a]

Order		Species	ρ_0 (g/cm^3)	$<\rho>$ (g/cm^3)	$<\rho>-\rho_0$ (mg/cm^3)	GC-rich DNA[b] (%)
Insectivora	1.	Hedgehog	1.7004	1.7035	3.1	
	2.	Shrew	1.6976	1.7001	2.5	**7**
	3.	Mole	1.6987	1.7011	2.4	**9**
Chiroptera	4.	Fruit Bat	1.6965	1.7002	3.7	<1
	5.	Bat	1.7000	1.7024	2.4	**10**
Primates	6.	Man	1.6984	1.7006	2.2	**8**
Pholidota	7.	Pangolin	1.6997	1.7026	2.9	<1
Lagomorpha	8.	Rabbit	1.6998	1.7039	4.1	**7**
Rodentia	9.	Squirrel	1.6982	1.7006	2.4	**9**
	10.	Woodchuck	1.6983	1.7026	4.3	
	11.	Rat	1.7008	1.7014	1.6	<1
	12.	Mouse	1.7006	1.7009	0.3[c]	2
	13.	Hamster	1.7000	1.7021	2.1	<4
	14.	Mole Rat	1.7000	1.7019	1.9	2
	15.	Dormouse	1.6991	1.7004	1.3	
	16.	Guinea Pig	1.6989	1.7018	2.9	**6**
Carnivora	17.	Cat	1.6986	1.7019	3.3	**7**
	18.	Dog	1.6990	1.7017	2.7	**6**
Perissodactyla	19.	Horse	1.6990	1.7030	4.0	**8**
Artiodactyla	20.	Calf	1.7000	1.7039	3.9	**7**

[a] From Sabeur et al. (1993). See the original paper for additional information.

[b] This column lists rounded-out figures from Table 3 of Sabeur et al. (1993). Values equal to, or higher than 6% (typical of the "general distribution") are shown in bold type.

[c] The light satellite DNA was not taken into account in calculating $<\rho>$.

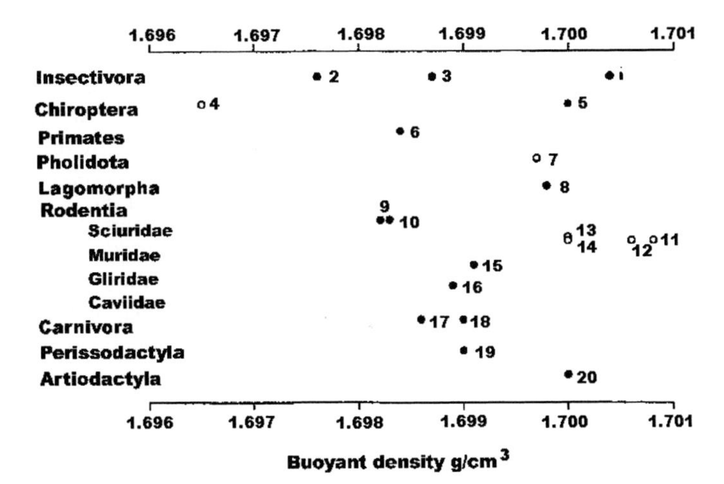

Figure 4.11. Modal buoyant density (ρ_0) values for mammalian DNAs. For the numbering of samples see Table 4.8. (From Sabeur et al., 1993)

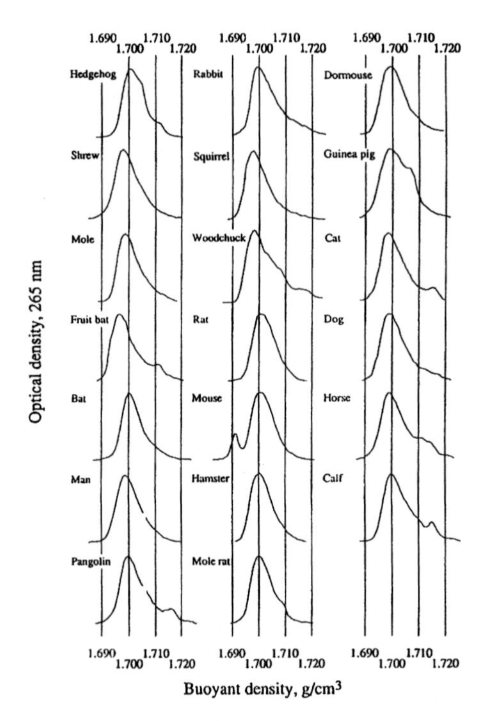

Figure 4.12. Analytical CsCl density profiles of unfractionated DNA preparations. The hedgehog and calf profiles are from Thiery et al. (1976), the woodchuck profile from F. Kadi (personal comm.). Profiles were normalized to the same height. (From Sabeur et al., 1993).

Buoyant density profiles of high-molecular-weight DNAs centrifuged in CsCl gradients, *i.e.*, compositional distribution of 50- to 100-kb genomic fragments, have revealed a clear difference between the murids and most other mammals, including other rodents. Sequence analyses have revealed other, related, compositional differences between murids and nonmurids. The study of CsCl profiles of 17 rodent species representing 13 families showed that the modal buoyant densities obtained for rodents span the full range of values observed in other eutherians (**Figs. 4.11** and **4.13**; **Table 4.9**). Scatterplots of these and related CsCl profile parameters show groups of rodent families that agree largely with established rodent taxonomy, in particular with the monophyly of the *Geomyoidea* superfamily and the position of the *Dipodidae* family within the *Myomorpha*. In contrast, while confirming and extending previously reported differences between the profiles of *Myomorpha* and those of other rodents, the CsCl data question a traditional hypothesis positing *Gliridae* within *Myomorpha*, as does the recently sequenced mitochondrial genome of dormouse (Reyes et al., 1998).

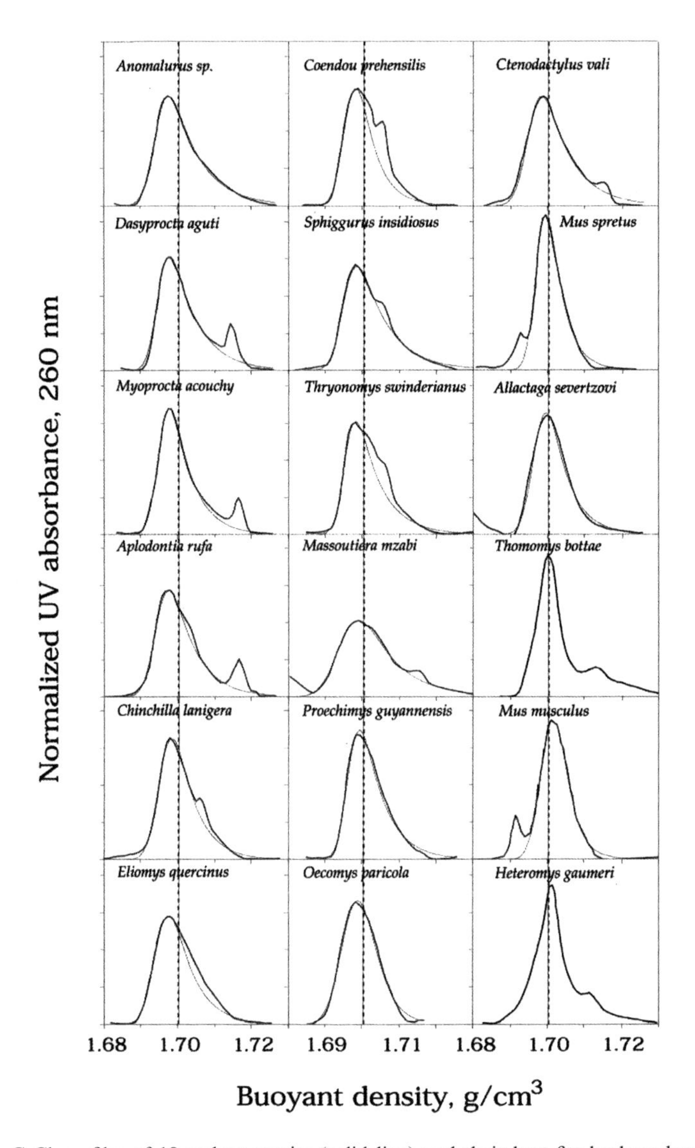

Figure 4.13. CsCl profiles of 18 rodent species (solid line) and their best fits by broadened exponential distributions (thin dashed lines). The abscissa shows modal buoyant density in g/cm^3; the ordinate shows relative amounts of DNA. All profiles are normalized to unit area and shown in three columns in order of increasing modal buoyant density. No fits are shown for the two *Geomyoidea* profiles (see the original paper). The dashed vertical line, at 1.7015 g/cm3, indicates the mean buoyant density of the 18 species. (From Douady et al., 2000).

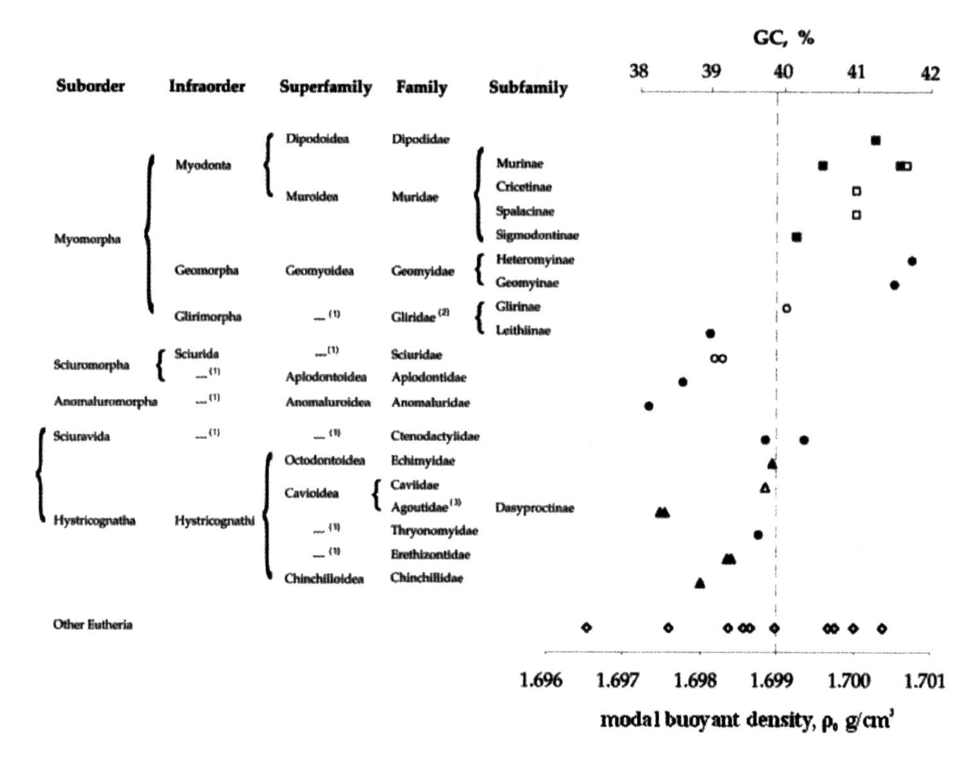

Figure 4.14. Modal buoyant density and GC values of rodent and other eutherian DNAs. Closed and open symbols refer to the data of Douady et al. (2000) and those of Sabeur et al. (1993), respectively. Triangles, squares, circles and diamonds denote caviomorphs, muroids, other rodents, and other mammals, respectively. Each row shows the values grouped at the family or subfamily level. The systematic arrangement follows McKenna and Bell (1997). The grouping of the *Ctenodactylidae* (*Sciuravida* suborder) with the *Hystricognatha* suborder (leftmost bracket) is supported by results from nuclear DNA sequence comparisons (von Willebrand factor gene; Huchon et al., 2000). For clarity, the two identical values observed in each of the *Erethizontidae* (1.6984 g/cm^3) and *Dasyproctidae* (1.6976 g/cm^3) have been artificially separated. With one exception (from *Leithiinae*), all the *Myomorpha* studied have modal buoyant density, ρ_0, equal to 1.699 g/cm^3(dashed line); only one other rodent species (from *Ctenodactylidae*) has ρ_0 in this range. Other Eutheria are, from left to right: *Pteropus sp.* (megabat), *Crocidura russula* (shrew), *Homo sapiens* (man), *Felis domesticus* (cat), *Talpa europea* (mole), *Canis familiaris* (dog), *Manis sp.* (Asian pangolin), *Oryctolagus cuniculus* (rabbit), *Bos taurus* (cow), and *Erinaceus europaeus* (hedgehog). (1), Not named or not appropriate; (2) also called *Myoxidae*; (3), also called *Dasyproctidae*.

Part 5
Sequence distribution in the vertebrate genomes

CHAPTER 1

Gene distribution in the vertebrate genome

1.1. The distribution of genes in the human genome: the two gene spaces

The first indication that coding sequences are unevenly distributed in the human genome came from the observation that the majority of the sequences first tested were localized in the GC-richest isochores of the H3 family (Bernardi et al., 1985b), which only represents 4-5% of the genome. A quantitative assessment could be made later, on 1,610 (Mouchiroud et al., 1991) and 4,270 (Zoubak et al., 1996) genes, using the following approach. The correlation of GC_3 levels of coding sequences with the GC levels of the isochores in which genes were located (see **Fig. 3.20A**) allows positioning of the GC_3 histogram of human genes relative to the distribution of DNA molecules, namely to the CsCl profile. Once this positioning is done, the ratio of each histogram bar by the corresponding point on the CsCl profile provides the relative concentration of genes (Mouchiroud et al., 1991; Zoubak et al., 1996). **Fig. 5.1A** shows that gene concentration increases from a very low average level

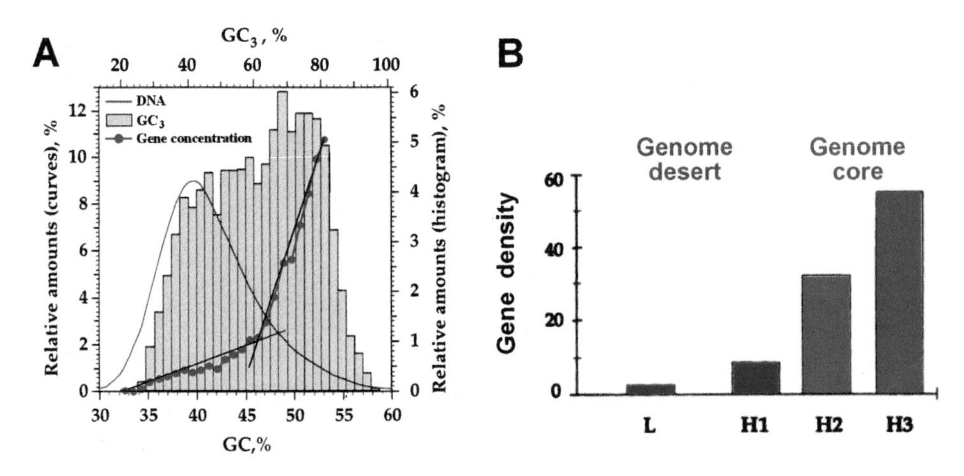

Figure 5.1 **A**. Profile of gene concentration (red dots) in the human genome, as obtained by dividing the relative numbers of genes in each 1.5% GC_3 interval of the histogram of gene distribution (yellow bars) by the corresponding relative amounts of DNA deduced from the CsCl profile (blue line). Values beyond the last one are unreliable because the corresponding DNA amounts are very low. The positioning of the GC_3 histogram relative to the CsCl profile is based on the correlation of GC_3 *vs.* GC of the isochores embedding the corresponding genes shown in Fig. 3.20A. (Modified from Bernardi, 2000a.) **B**. Density of genes in isochore family. Relative numbers of sequences over relative amounts of isochore family are presented (From Zoubak et al., 1996).

in L1 isochores up to a 20-fold higher level in H3 isochores. **Fig. 5.1B** presents the data of **Fig. 5.1A** in the form of a gene density histogram.

Since it had been tacitly assumed until then that genes were uniformly distributed in eukaryotic genomes, the **strikingly non-uniform gene distribution** in the human genome (and, for that matter, in the genomes of all vertebrates; see below) came as a big surprise. Moreover, a closer look at the gene concentration line of **Fig. 5.1A** reveals a clear break in the slopes of the GC-poor and the GC-rich isochores at 60% GC_3 of coding sequences and at 46% GC of isochores . The break defines **two gene spaces** in the human genome: a GC-rich, gene-rich space corresponding to the H2 and H3 isochores and representing about 12% of the genome, which was called the **genome core** (Bernardi, 1993a), and a GC-poor, gene-poor space corresponding to the L1, L2 and H1 isochores and representing about 88% of the genome, which was called the **genome desert** or the **empty quarter** (from the classical name of the Arabian desert; Bernardi, 2000a). About half of the human genes are located in the small genome core, the other half being located in the large genome desert. The properties of these two gene spaces are very distinct (see the following section).

The plot of gene density as a function of GC in the human genome, as determined for 9,315 genes by Lander et al. (2001) is essentially identical (see **Fig. 5.2**) to that published by us for 1,610 and 4,270 genes ten and five years before, respectively (Mouchiroud et al., 1991; Zoubak et al., 1996). Most surprisingly, Lander et al. (2001) did not mention our previous results, in spite of the fact that they were reported again in a review paper (Bernardi, 2000a) quoted in their reference list. As a consequence, some authors, apparently not familiar with the literature, are now quoting the paper by Lander et al. (2001) as the original report about the non-uniform gene distribution in the human genome. (Fortunately, this mistake was not made again by the Mouse Genome Sequencing Consortium, 2002). Interestingly, we observed that the gene density values of Lander et al. (2001) fall on

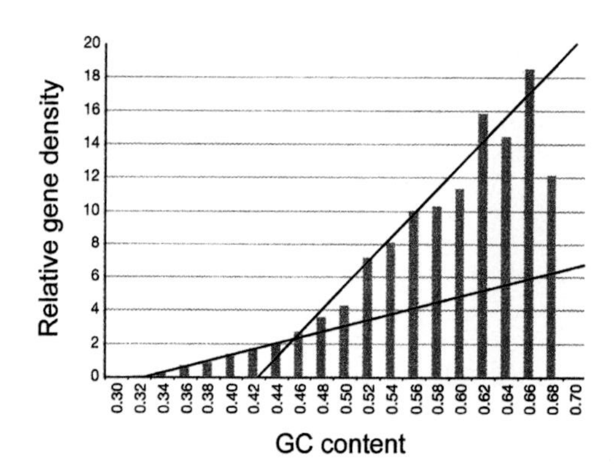

Figure 5.2. Gene density as a function of GC content. Values are less accurate at higher GC levels because the denominator is small. For 9,315 known genes mapped to the draft genome sequence, the local GC content was calculated in a window covering either the whole alignment or 20,000 bp centered around the midpoint of the alignment, whichever was larger. Ns (undefined nucleotides) in the sequence were not counted. GC content for the genome was calculated for adjacent nonoverlapping 20,000 bp windows across the sequence. Both the gene and the genome distribution were normalized to sum one. The two slopes added here to the original figure concern the genes from GC-poor and GC-rich isochores, respectively. (Modified from Fig. 36b of Lander et al., 2001; from Bernardi, 2001).

two straight lines crossing each other at about 46% GC (see **Fig. 5.2**). Along the same line, an independent assessment of gene density in chromosomes 21 and 22 (Saccone et al., 2001) also showed two slopes crossing each other at about 45% GC (see **Part 7**). The explanation given by Lander et al. (2001) for the results of **Figs. 5.1** and **5.2** is that "*the correlation between gene density and GC appears to be due primarily to intron size, which drops markedly with increasing GC content*" (see **Fig. 5.3**). This is an astounding explanation, since introns only represent 4-6% of the genome and cannot possibly account for the correlation.

1.2. Properties of the two gene spaces

The "genome core" is endowed with several specific features, which are quite distinct from those of the "genome desert". **Table 5.1** summarizes the properties of the two gene spaces and provides indications as to where they are presented and/or discussed in this book. Here we will only comment on some points concerning gene expression.

As reviewed by Pesole et al. (2001), regulation of gene expression is achieved through a series of complex mechanisms which can be basically divided into two distinct steps. The first one involves the control of transcription mediated by *cis*-acting DNA elements such as promoters, enhancers, locus control regions and silencers to produce a mature mRNA. This mechanism has been well characterized for many genes. The second step covers the post-transcriptional control of mRNA nucleo-cytoplasmic transport, translation efficiency, subcellular localization and stability. This step has been less comprehensively characterized, although it is known to be mediated by *cis*-acting RNA elements generally located in 5' and 3' mRNA untranslated regions (5'UTRs and 3'UTRs; Sonenberg, 1994; McCarthy and Kollmus, 1995; Pesole et al., 1997; 2000; Bashirullah et al., 1998; van der

TABLE 5.1
Properties of the two gene spaces of the vertebrate genome

Gene space:	**Genome core**	**Genome desert**	**Ref.**
Genome, %	~ 12%	~ 88%	**Fig. 3.10**
Genes, %	~ 50%	~ 50%	**Fig. 5.1**
Gene density	High	Low	**Figs. 5.1, 5.2, 5.9**
CpG	High	Low	**Fig. 5.12**
Methylation	High	Low	**Fig. 5.15**
CpG islands	Abundant	Scarce	**Fig. 5.9**
Introns	Short	Long	**Fig. 5.3, ***
Untranslated sequences	Short	Long	*
Consensus start sequence	RCC **atg** R	AAR **atg** R	*
Replication	Early	Late	**Part 7**
Recombination	High	Low	**Part 7, ***
Transcription	Strong	Weak	*
Chromatin	Open	Closed	**Part 7, ***
Chromosomal location	Distal	Proximal	**Part 7**

* See this Section

Velden and Thomas, 1999). Unlike DNA-mediated regulatory signals whose activity is essentially associated with their primary structure, the biological activity of regulatory patterns acting at the RNA level relies on a combination of primary and secondary structure elements assembled in a consensus structure generally recognized by specific RNA-binding proteins.

(i) **Introns**. Small numbers and sizes of introns are generally viewed as advantageous features for genes that are transcribed in an extensive way. The short and relatively rare introns of the highly expressed genes which are preferentially located in GC-rich isochores (see **Part 7**) fit with this view. In the case of GC-poor genes, the abundance and the large size of introns would be favourable for alternative splicing, an important mechanism of expression regulation in tissue-specific genes (Bell et al., 1998).

Although apparently not noticed by Lander et al. (2001), their results on intron size (see **Fig. 5.3)** confirm the existence of two classes of genes characterized by distinctly different intron sizes (Carels and Bernardi, 2000) in that they show a sharp transition with a mid-point at about 45% GC, again in agreement with the boundary between the genome core and the empty quarter.

(ii) **Transcription**. Ten years ago it was proposed that "*the H3 isochore family presumably has the highest level of transcription because of its very high concentration of genes, especially housekeeping genes*" (Bernardi, 1993a). Subsequent analyses on the correlation between gene distribution and gene expression levels (Gonçalves et al., 2000) claimed that the majority of the widely expressed (housekeeping) genes were localized mainly in GC-poor isochores, whereas tissue-specific genes were localized in the GC-rich ones. The authors drew this conclusions by analyzing the base composition and the distribution of genes with or without retropseudogenes, the former being, in their view, more widely expressed than the latter. However, in the human genome, using a different algorithm,

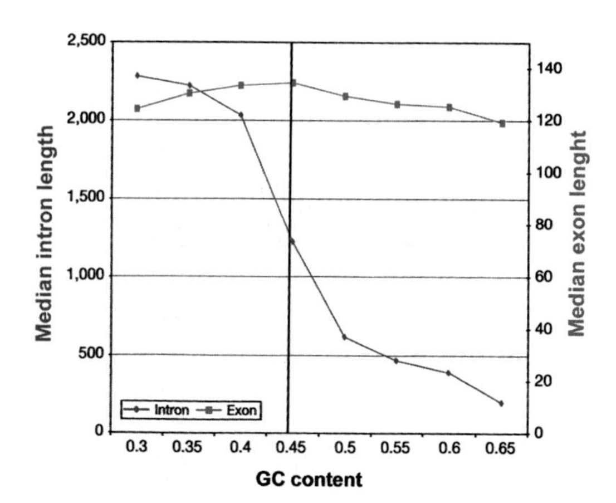

Figure 5.3. Dependence of mean exon and intron lengths on GC content. For exons and introns, the local GC content was derived from alignments to finished sequence only, and were calculated from windows covering the feature or 10,000 bp centered on the feature, whichever was larger. The vertical straight line added here to the original figure marks the midpoint of the transition. (Modified from Fig. 36c of Lander et al., 2001; from Bernardi, 2001).

the propensity for retrotransposition was found to be unaffected by the GC content of the genes (Venter et al., 2001). Studying the origin of CpG islands, Ponger et al. (2001) confirmed that tissue-specific genes were mainly localized in GC-rich isochores (64% in H3), whereas the distribution of widely expressed genes was independent of the isochore context.

It should be stressed that, in both papers dealing with the correlation between gene distribution and expression patterns (Gonçalves et al., 2000; Ponger et al., 2001), the gene partition was performed according to the criteria defined in Mouchiroud et al. (1991). In the last decade, however, several results based on theoretical and experimental approaches led to an improvement of the gene partition criteria (Saccone et al., 1993, 1999; Zoubak et al., 1996; Federico et al., 2000). Using this partition criteria and two independent datasets, the analysis of the gene frequencies in GC-poor and GC-rich regions led to the conclusions that tissue-specific and widely expressed genes: 1) show no compositional differences at GC_3 level; and 2) follow the general gene distribution, reaching the highest frequency in the GC-rich isochores (D'Onofrio, 2002; see **Fig. 5.4**). Needless to say, these conclusions are in complete contradiction with those of Gonçalves et al. (2000) and Ponger et al. (2001; see also note added in proof in **Part 12**, Chapter 6).

(iii) **Chromatin structure**. It has been known for a long time that the open chromatin structure is characterized by accessibility to DNases (Kerem et al., 1984) and that CpG island chromatin is characterized by a larger spacing of nucleosomes, absence of histone H1 and acetylation of histones H3 and H4 (Tazi and Bird, 1990). Since CpG island density parallels gene density (Aïssani and Bernardi, 1991a,b; Jabbari and Bernardi, 1998; see **Fig. 5.11**), one should conclude that chromatin structure is largely open in the genome core, and largely closed in the empty quarter. In fact, it is quite possible that **open chromatin regions are more abundant, even in absolute amounts, in the scarce GC-rich, gene-rich isochores** than in the abundant GC-poor, gene-poor isochores.

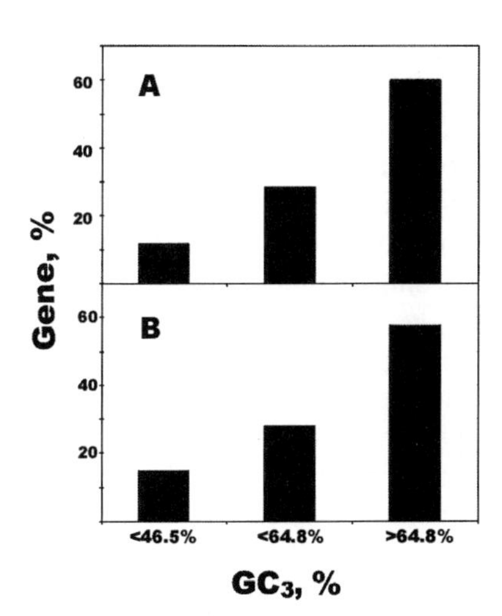

Figure 5.4. Distribution of **A** widely expressed (housekeeping) genes and **B** tissue-specific genes. The three GC_3 classes were the result of averaging the boundaries defined by theoretical (Zoubak et al., 1996) and experimental results (Saccone et al., 1993, 1999; Federico et al., 2000). (From D'Onofrio, 2002).

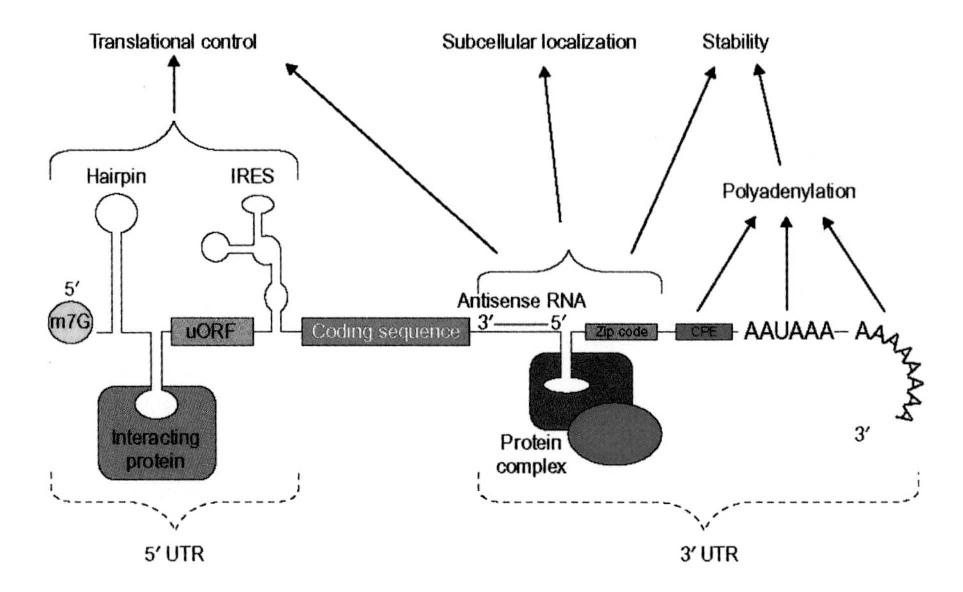

Figure 5.5. The generic structure of a eukaryotic mRNA, illustrating some post-transcriptional regulatory elements that affect gene expression. Abbreviations (from 5' to 3'): UTR, untranslated region; m7G, 7-methyl-guanosine cap; Hairpin, hairpin-like secondary structures; uORF, upstream open reading frame; IRES, internal ribosome entry site; CPE, cytoplasmic polyadenylation element; AAUAAA, polyadenylation signal. (From Pesole et al., 2001)

(iv) **Recombination** has been shown to be more frequent in GC-rich compared to GC-poor isochores (see Nachman, 2002 for a review, and **Part 7**, **Chapter 4**).

(v) **Post-transcriptional control of gene expression** (see **Fig. 5.5**). 1) The efficiency of AUG start codon recognition in translation initiation is modulated by its sequence context. This context is different in genes located in different isochores. In particular, of the two main consensus start sequences, RCC**atg** R is five-fold more represented than AAR**atg** R in genes from the GC-rich H3 isochores compared to genes from the GC-poor L isochore. Furthermore, genes located in GC-rich isochores have shorter 5' UTRs and stronger avoidance of upstream ATG compared to genes located in GC-poor isochores (Pesole et al., 1999). This suggests that genes requiring highly efficient translation are located in GC-rich isochores and genes requiring fine modulation of expression are located in GC-poor isochores. Interestingly, the most represented initiator context in mammalian genes, as well as in warm-blooded vertebrates genes in general, is GCC**atg** G, whereas in cold-blooded vertebrates it is AAA**atg** G (a context very similar to that observed in GC-poor human isochores), as expected for genomes in which GC-rich isochores are very scarce or absent (Pesole et al., 1999). 2) Both 3' and 5' UTR (untraslated regions), but especially the former, are shorter as their GC level increases, in parallel with GC_3 (Pesole et al., 1997, 2000). It was also observed that the average GC of 5'UTRs is higher than that of 3'UTRs, this difference being more marked in warm-blooded vertebrates, 60% GC for 5'UTRs *vs.* 45% for 3'UTRs, probably because of the frequent presence of CpG islands (Pesole et al. 2001).

To sum up, an ample body of data has conclusively demonstrated that the genomes of mammals (and, more generally, of warm-blooded vertebrates; see the following section) are not only **structural mosaics**, made up of compositional compartments, the isochores, but also **functional mosaics**.

1.3. The distribution of genes in the vertebrate genomes

The obvious question as to how widespread among vertebrates is the distribution of genes exhibited by the human genome was approached experimentally in several ways. (i) Hybridization of human DNA corresponding to H3 isochores (in an excess of unlabeled Alu sequences in order to avoid hybridization of repeated sequences) on compositional fractions from other vertebrate genomes. Single-copy human DNA derived from H3 isochores hybridized on the GC-richest compositional fractions representing 10-15% of all mammalian and avian genomes tested, indicating a similar compositional distribution of the highest concentration of genes (Cacciò et al., 1994; **Fig. 5.6**). More interestingly, hybridization of human single-copy sequences from H3 isochores was also found in the 10-15% GC-richest compositional DNA fractions of cold-blooded vertebrates (see **Fig. 5.6**). (ii) Prediction of ORFs (open reading frames) in long DNA sequences. In *Brachidanio rerio*, the estimate of ORF density per 10 kb is 1.7 (sd 0.99) for 895 kb of GC-poor (35.1% GC) regions, and 3.5 (sd 1.82) for 617 kb of GC-rich (39.7% GC) regions, showing again an increased gene density in the GC-rich regions. (iii) Extending the approach of **Fig. 5.1** to

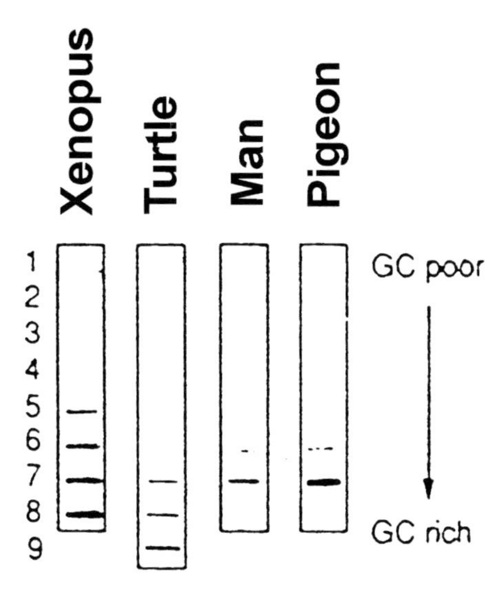

Figure 5.6. Slot blot hybridisation patterns obtained using as a probe the gene-richest fraction of human DNA on compositional fractions from vertebrate genomes. Species names are indicated at the top. The modal buoyant densities of the fractions increase from the top to the bottom, as indicated at the left of the figure, the last fractions often containing, however, satellite DNAs. Fraction numbers are indicated at the left of the figure. The amounts of DNA in the fractions are variable. In all cases, however, hybridization-positive fractions represent 10-15% of DNA. (Modified from Cacciò et al., 1994).

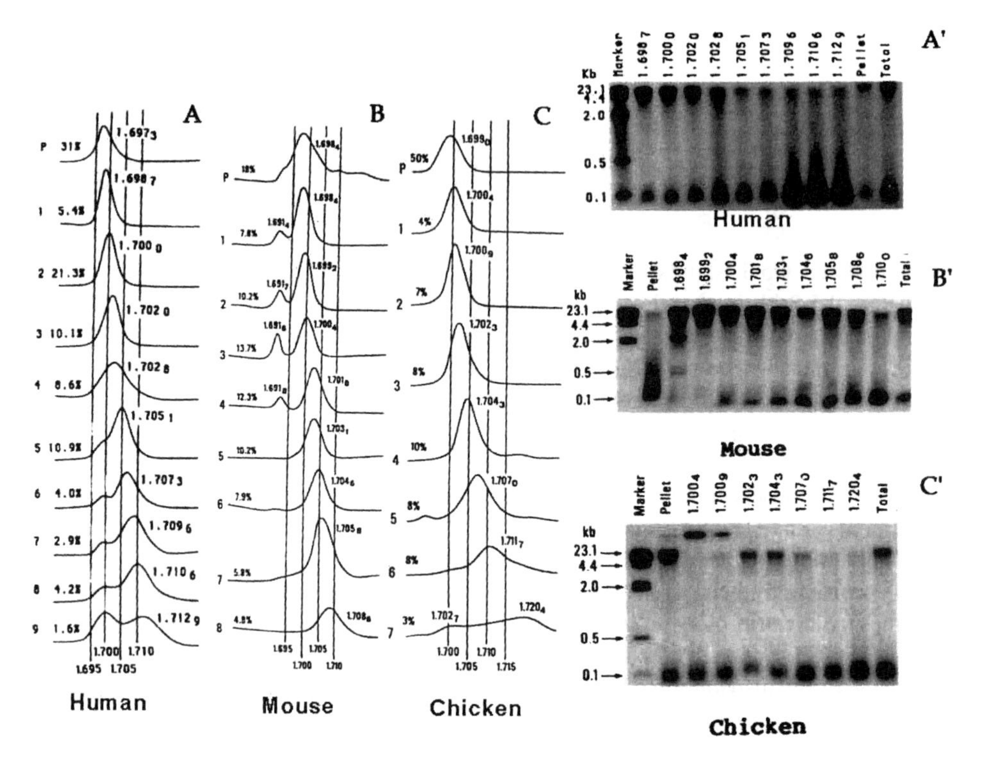

Figure 5.8. CsCl profiles of compositional DNA fractions from **A** human placenta, **B** mouse liver and **C** chicken erythrocyte, as obtained from preparative centrifugation in Cs_2SO_4/BAMD density gradient. The *rf* (ligand/nucleotide molar ratio) values used were 0.12 for mouse and 0.14 for human and chicken. Modal buoyant densities and relative amounts of the fractions are indicated. P indicates the pellet. Notice the satellite peak (centered at about 1.700 g/ml) in the last fractions of human, and the satellite peak (1.691-1.692 g/cm^3) in mouse DNA fractions 1-4. **A', B'** and **C'** display autoradiograms of terminally labelled (Cooper et al., 1983; 1.2% agarose gels were used) *Hpa* II fragments from DNA fractions. (From Aïssani and Bernardi, 1991a).

(i) The frequency of CpG islands in long human sequences increases with increasing GC and almost parallels gene frequency (**Fig. 5.11**). (ii) GC and CpG levels of CpG islands are positively correlated with the GC levels of the long sequences in which they are located. (iii) CpG levels of CpG islands increase with their own GC levels.

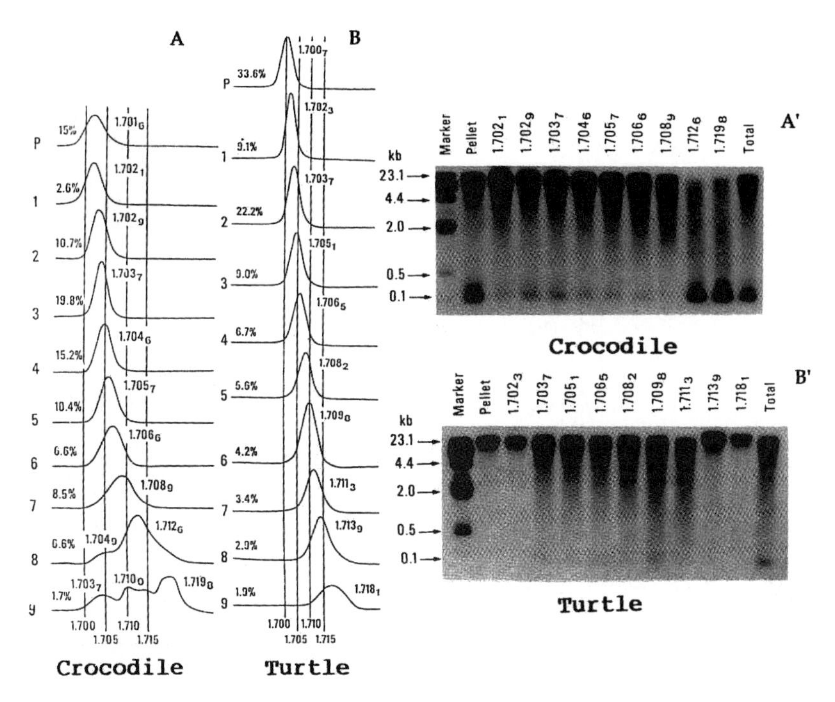

Figure 5.9. CsCl profiles of compositional DNA fractions from **A** crocodile (*Crocodylus niloticus*) and **B** turtle (*Testudo graeca*) erythrocytes, as obtained from preparative centrifugation in Cs_2SO_4/BAMD density gradient. The *rf* values used in the fractionations of crocodile and turtle DNA were 0.14 and 0.16, respectively. Modal buoyant densities and relative amounts of the fractions are indicated. P, pellet. **A'** and **B'** show autoradiograms of terminally labelled *Hpa* II fragments from DNA fractions. (See also legend of Fig. 5.8; from Aïssani and Bernardi, 1991a).

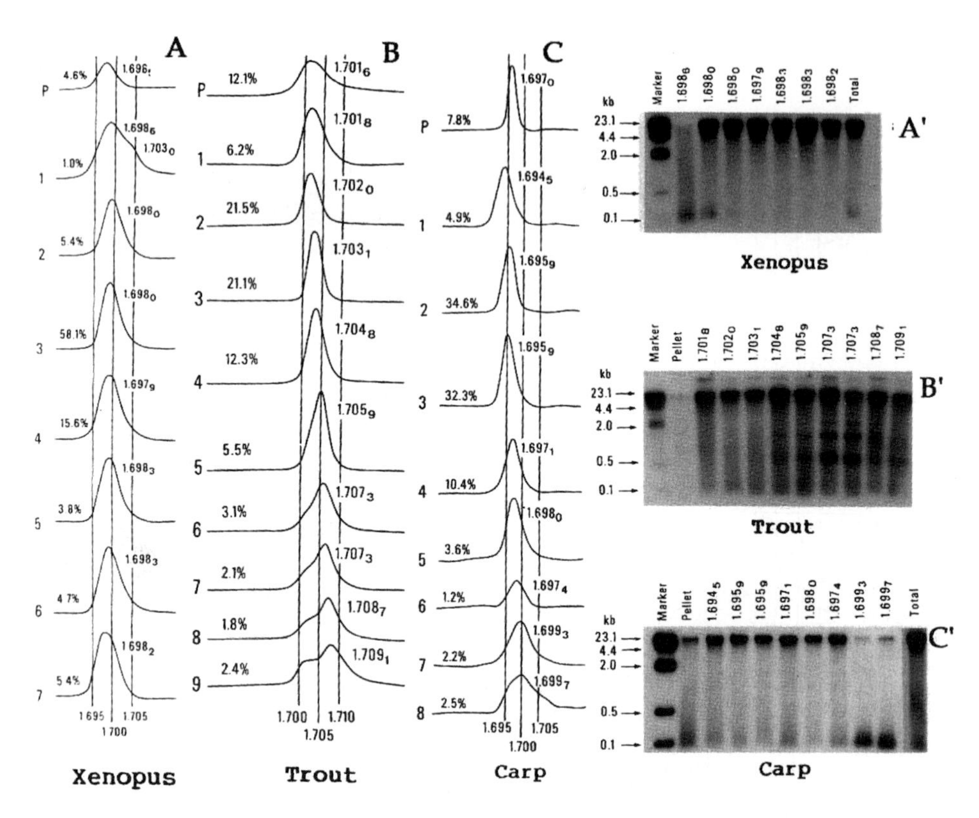

Figure 5.10. CsCl profiles and *Hpa* II fragments of compositional DNA fractions from Xenopus, trout and carp DNAs. CsCl profiles of **A** *Xenopus laevis,* **B** trout (*Onchorhynchus mykiss*), and **C** carp (*Cyprinus carpio*) liver, as obtained from preparative centrifugation in Cs_2SO_4/BAMD density gradient (Bernardi and Bernardi, 1990a); *rf* values of 0.14 were used. Modal buoyant densities and relative amounts are indicated. P, pellet. Satellite peaks or shoulders are visible in the CsCl profiles of fraction 1 of *Xenopus* DNA, fractions 6-9 of trout DNA and fractions 7 and 8 of carp DNA. Panels **A'**, **B'** and **C'** show autoradiograms of *Hpa* II fragments from DNA fractions. In panel **C'**, the pellet was very poorly labelled. (See also legend of Fig. 5.8; from Aïssani and Bernardi, 1991a).

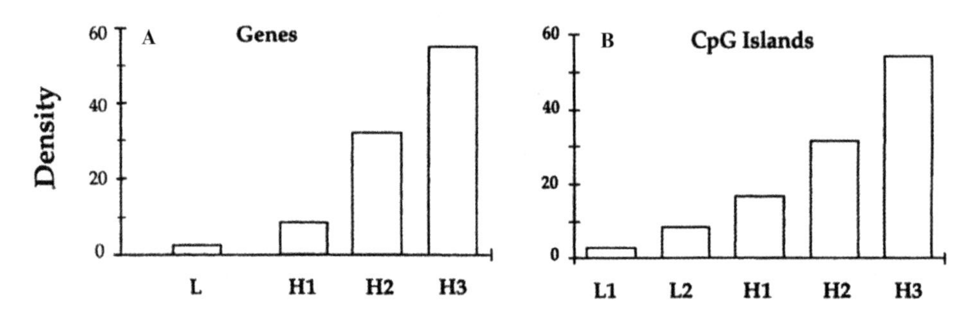

Figure 5.11. Density of **A** genes and **B** CpG islands in isochore families. Relative numbers of sequences over relative amounts of isochore families are presented in the histograms. (From Jabbari and Bernardi, 1998).

CHAPTER 3

The distribution of CpG doublets and methylation in the vertebrate genome

3.1. CpG doublets

It is well known that vertebrate DNAs show a characteristically low frequency of CpG doublets, whereas all other doublets are present at about the frequency expected from a random distribution of nucleotides (Josse et al., 1961; Swartz et al., 1962). Two explanations were proposed to account for this outstanding feature of the vertebrate genome, the so-called CpG shortage.

(i) The first explanation (Subak-Sharpe et al., 1966) was based on the observation that the CpG shortage is also exhibited by the genomes of small vertebrate viruses (like polyoma and SV40), which use essentially all their genetic information for directing protein synthesis. It was proposed (Subak-Sharpe et al., 1967, 1974) that the genomes of these viruses derived from polypeptide-specifying DNA of the host cells and, therefore, exhibited the same CpG shortage, which was visualized as reflecting constraints from the translation apparatus. In agreement with this proposal, CpG shortage was absent in tRNA, 5sRNA and rRNA genes. The lack of CpG shortage in the genome of intermediate and large viruses (such as adenovirus and herpes simplex virus, respectively) was attributed to their capacity to modify the host translation apparatus. Since only a very small proportion of vertebrate DNA is actively involved in protein coding and since a CpG shortage was found in all compositional fractions of guinea pig DNA, as obtained by gradient centrifugation, by differential renaturation or from dispersed and condensed chromatin, it was further concluded that "*the bulk of vertebrate DNA derives from, and maintains the gross sequence characteristic of polypeptide-specifying DNA*" (Russell et al., 1976).

This hypothesis was already somewhat compromised by the discovery that α globin mRNAs from rabbit (Salser, 1977) and human (Forget et al., 1979) carry a large number of CpG doublets, and by the different levels of CpG in both coding and non-coding sequences from the human α and β globin gene clusters (Lennon and Fraser, 1983). Subsequent investigations (Bernardi et al., 1985b; Bernardi, 1985; Bernardi and Bernardi, 1986a) definitely demonstrated that CpG shortage is not related to polypeptide coding. In fact, CpG shortage (i) is strong in GC-poor genes, but becomes increasingly weaker in GC-rich genes from warm-blooded vertebrates; and, moreover, (ii) decreases with increasing GC in non-coding (intronic and intergenic) sequences. In other words, CpG levels in both coding and non-coding sequences from warm-blooded vertebrates are correlated with the GC levels of the corresponding isochores. These findings not only contradicted the explanation proposed by Subak-Sharpe and co-workers for the CpG shortage, but also demonstrated that CpG shortage is not uniform throughout the genome, at least in the case of warm-blooded vertebrates, in contrast with the conclusion of Russell et al. (1976). Finally, as far as the different CpG levels exhibited by the genomes of small and large vertebrates viruses are concerned, it was shown that these differences were not related to genome size, but simply to the GC levels of the corresponding genomes. Indeed, a plot of CpG *vs* GC for viral genomes exhibited a positive linear relationship (Bernardi and Bernardi, 1986a) which was very close (**Fig. 5.12**) to that previously found for vertebrate genes (Bernardi et al., 1985b).

(ii) An alternative explanation for the CpG shortage in vertebrates was based on the consideration that the C residues in CpG doublets are highly methylated in vertebrates (Sinsheimer, 1955; Doskocil and Sorm, 1962; Grippo et al., 1968; Van der Ploeg and Flavell, 1980; Gruenbaum at al., 1981; Kunnath and Locker, 1982), to the extent of 50-90% in different mammalian DNAs (Gruenbaum et al., 1981). It was proposed that mCpG represents a hot-spot for mutation (Salser, 1977; Coulondre et al., 1978) since it can be deaminated to TpG. Even in the presence of a repair system that specifically corrects mispaired T/G back to C/G (Kramer et al., 1984), the evolutionary effect will be a progressive loss of CpG dinucleotides, leading to a CpG shortage, and to a concomitant increase in TpG (and of its complementary doublet CpA on the other strand). Indeed, a scenario which was put forward for the evolution of DNA methylation in vertebrates (Cooper and Krawczak, 1989) is that the vertebrate genome originally was strongly methy-

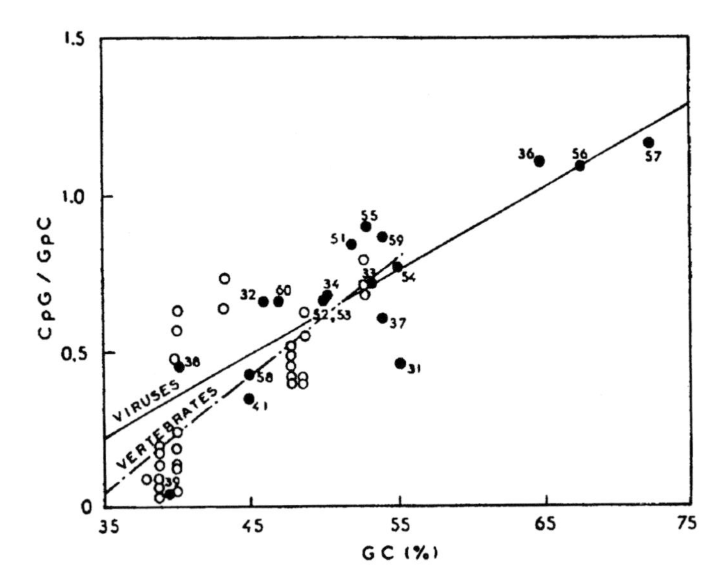

Figure 5.12. CpG/GpC ratios for viral genomes, plotted against genomic GC levels (filled circles). Numbers refer to genomes listed in Table 1 of the original paper. Except for frog virus 3 (point 60), all viruses were from warm-blooded vertebrates. The dashed-and-dotted line and the open symbols correspond to a similar plot (Bernardi et al., 1985b) obtained for vertebrate genes or exons. (From Bernardi and Bernardi, 1986a).

lated (and GC-rich) and that the methylated CpG doublets (as well as the GC levels) subsequently underwent a monotonous decay. This explanation as well as the alternative hypothesis of an equilibrium between CpG depletion and CpG formation (Sved and Bird, 1990), which also predicts a continuous decrease in CpG during vertebrate evolution, were, however, also shown to be wrong by investigations from our laboratory (see **Sections 3.2** and **3.3**).

3.2. Two different CpG levels in vertebrate genomes

Table 5.2 displays the CpG frequency and the observed/expected ratios for the CpG dinucleotides, as well as the genomic GC levels. The data show that the genomes of fishes and amphibians ranging from *Chondrychthyes* to *Anura* have a similar CpG o/e (observed/expected) ratio, averaging 0.37, which is significantly higher than that, averaging 0.26, found for nine genomes from eight mammals and one bird. The fact that mammals from six orders share similar CpG o/e values and the star-like phylogeny of mammals (see **Fig. 11.3**) suggest that this value was also that of their last common ancestor (see **Part 11, Section 3.3** for additional comments). Unfortunately, no data are available for reptiles.

TABLE 5.2
GC and CpG levels of some vertebrate genomes[a]

	GC	CpG	CpG o/e	Reference[b]
Squalus acanthius	45.3	1.8	0.35	4
Salmo salar	43.1	1.7	0.37	1
Latimeria sp.	41.7	1.8	0.41	4
Rana catesbeiana	45	1.7	0.33	5
average	43.78	1.75	**0.37**	
Oryctolagus cuniculus	44.3	1.3	0.25	1
Rattus norvegicus	43.9	1.2	0.25	2
Mus musculus	42.2	1	0.23	1
Bos taurus	44	1.4	0.29	1
Phocoena phocoena	41.4	1.3	0.33	3
Homo sapiens	42	1	0.25	1
Halichoerus sp.	39	0.9	0.24	3
Cavia porcellus	40	1	0.26	3
Gallus gallus	44.2	1.1	0.23	1
average[c]	42.33	1.13	**0.26**	

[a] (From Jabbari et al., 1997)
[b] References: (1) Swartz et al. (1962); (2) Skalka et al. (1966); (3) Russell (1974); (4) Russell and Subak-Sharpe (1977); (5) McGeoch (1970).
[c] *Bos taurus* was excluded from the average because of large amounts (23%) of GC-rich satellites (Macaya et al., 1978).

3.3. Two different methylation levels in vertebrate genomes

Investigations aimed at understanding the biological significance of DNA methylation in vertebrates have been carried out along two lines.

(i) Studies on the correlation between promoter methylation and gene expression led to a number of interesting insights. Siegfried and Cedar (1997) recognized three different strategies by which methylation could affect gene expression: (1) methylation of CpG may interfere directly with the binding of specific transcription factors to DNA; this does not appear to affect gene-specific transcription factors, but rather ubiquitous factors; (2) the direct binding of specific factors to methylated DNA may lead to gene repression; (3) methylation may cause repression by altering chromatin structure; indeed, DNA methylation affects positioning of nucleosomes and influences the sensitivity of DNA to DNAses.

(ii) In contrast, analyses of overall DNA methylation levels in DNA from different vertebrate and different tissues (Vanyushin et al., 1970, 1973; Pollock et al., 1978; Ehrlich et al., 1982; Gama-Sosa et al., 1983a,b; Serrano et al., 1993) did not lead to any conclusion. A re-examination of the general problem of DNA methylation in vertebrates has, however, led to new insights.

Table 5.3 presents the GC and 5mC levels of the 42 vertebrate genomes studied by Jabbari et al. (1997), as well as those reported in the literature for 45 additional species. *c* values (haploid genome sizes) and the tissues from which DNAs were extracted are also indicated. It should be noted that, even in the sample of **Table 5.3**, the largest studied so far, vertebrate classes are not represented in a balanced way, fishes and mammals being, by far, predominant.

A summary of the results of **Table 5.3** indicates (**Table 5.4**) that the average methylation values in fishes (1.70%) and amphibians (1.98%) are roughly twice as large as those found in mammals (0.88%), birds (1.02%) and reptiles (1.00%). Since average GC levels are close to 42.5% for all five vertebrate classes, this means that approximately 9% of all cytosines are methylated in fishes and amphibians compared with approximately 4.5% in mammals, birds and reptiles.

Fig. 5.13 shows that two statistically significant correlations ($p<0.0001$) hold between 5mC and genomic GC levels, one for fishes and amphibians and another one for mammals and birds. It should be noted that (i) a few mammalian values deviate from the correlation shown by mammals/birds; deviating points concern, beside the bovine somatic cells mentioned above, *Erinaceus europeus, Meriones unguiculatus, Cricetus norvegicus, Manis* sp. (lower values) and *Pteropus poliocephalus* (higher value), these deviations being possibly due to under- or over-methylated satellites, respectively; (ii) reptilian values show a larger scatter compared with values from mammals/birds (see **Table 5.4** and **Fig. 5.13**) to which they are close; (iii) one of the fishes, the only chondrichthyan studied, showed an especially low value compared with other fishes, 1.15%.

The correlation between 5mC and genomic GC levels is understandable if one recalls that CpG frequency is linearly correlated with the GC content of the genome (Bernardi, 1985; Aïssani and Bernardi, 1991a,b) and that CpG is the major site of methylation, methylation on other dinucleotides (Woodcock et al., 1987) and on CpNpG trinucleotides (Clark et al., 1995) being extremely limited. A similar correlation was also found in plant

TABLE 5.3.

5mC, GC and *c*-values in vertebrate DNAs[a].

Order	Family	Species	Tissue	GC (mol%)	5mC (mol.%)	5mC ref.[b]	*c*-value
Fishes							
Cyclostomata							
Petromyzontiformes	*Petromyzontidae*	*Lampetra fluviatilis*	blood	50.9	2.09	1	
Chondrichthyes		*Petrolamiops longimanus*	sperm	45.4	1.16	1	
Osteichthyes							
Acipenseriformes	*Acipenseridae*	*Acipenser guldenstadtii*	sperm	43.8	1.79	1	
		Huso huso	sperm	43.4	1.81	1	1.8
		Acipenser ruthenus	sperm	43.0	1.87	1	
		Acipenser stellatus	sperm	42.9	1.87	1	
Antheriniformes	*Orizatidae*	*Oryzias latipes*	liver	40.8	1.85	5	1.1
Clupeiformes	*Clupeidae*	*Clupea harengus*	sperm	45.2	1.87	1	0.77
Cypriniformes	*Cyprinidae*	*Cyprinus carpio*	sperm	38.9	1.56	1	1.7
			blood	38.9	1.38	1	
		Brachydanio rerio	liver	37.3	1.27	1	1.8
Siluriformes	*Siluridae*	*Silurus glanis*	lood	41.1	1.64	1	
	Ictaluridae	*Ictalurus nebulosus*	liver	41.3	1.92	5	1.2
Salmoniformes	*Esocidae*	*Esox lucius*	sperm	43.5	1.87	1	0.85
			liver	44.2	1.98	1	
			blood	43.5	1.47	1	
	Salmonidae	*Osmerus eperlanus*	Sperm	45.4	2.01	1	
		Oncorhynchus sp.	sperm	44.0	1.64	1	3
			liver	44.1	2.13	1	
		Oncorhynchus tschawitscha	sperm	41.9	1.73	1	3.3
		Oncorhynchus mykiss	liver	42.9	1.45	5	2.8
Gadiformes	*Gadidae*	*Gadus morhua*	blood	50.0	2.24	1	0.4
		Eleginus navaga	blood	48.2	2.15	1	
Scorpaeniformes	*Cottidae*	*Cottus sp.*	sperm	41.0	1.70	1	1

TABLE 5.3.
5mC, GC and *c*-values in vertebrate DNAs[a]. (continued)

Order	Family	Species	Tissue	GC (mol%)	5mC (mol.%)	5mC ref.[b]	*c*-value
Perciformes	*Scombridae*	*Trichiurus japonicus*	sperm	41.3	1.50	1	
	Thynnidae	*Thunnus thynnus*	sperm	37.0	1.54	1	1
	Cichlidae	*Oreochromis grahami*	liver	41.5	1.29	6	
		Oreochromis niloticus	liver	41.0	1.26	6	0.95
		Tilapia buttikofferi	liver	40.3	1.46	6	
	Mugilidae	*Mugil cephalus*	sperm	39.5	1.87	2	0.99
			liver	39.5	1.80	2	
Pleuronectiformes	*Pleuronectidae*	*Pleuronectes flesus*	blood	44.9	1.70	1	
		Liopsetta glacialis	blood	45.2	1.84	1	0.73
Tetraodontiformes	*Tetraodontidae*	*Sphaeroides sp.*	sperm	41.3	1.50	1	0.5
Amphibia							
Anura	*Leptodactylidae*	*Odontophrynus americanus*	blood	41.9	1.63	6	
	Pipidae	*Xenopus laevis*	liver	42.2	1.40	1	3.1
	Ranidae	*Rana temporaria*	liver	45.1	2.60	1	4.2
			blood	45.1	2.61	1	
		Rana pipiens	liver	47.3	2.30	1	5.8
Reptilia							
Crocodylia	*Crocodylidae*	*Crocodilus niloticus*	blood	44.4	1.06	6	
Chelonia	*Testudinidae*	*Testudo horsfieldi*	blood	45.9	1.30	1	
			liver	44.8	1.50	1	
		Testudo graeca	blood	45.7	0.77	6	5.4
Squamata	*Varanidae*	*Varanus griseus*	blood	43.7	1.15	1	2.4
Ophidia	*Crotalidae*	*Bothrops neuwiedi*	blood	40.0	1.01	6	
		Bothrops jararac	blood	39.7	0.97	6	
	Boinae	*Boa constrictor amarali*	blood	40.4	0.61	6	3.15
Aves							
Rheiformes	*Rheidae*	*Rhea americana*	blood	44.0	0.98	6	
Sphenisciformes	*Spheniscidae*	*Spheniscus demersus*	blood	43.3	1.01	6	1.6
Anseriformes	*Anatidae*	*Cairina moschata*	blood	41.5	0.64	6	1.35
Galliformes	*Phasianidae*	*Gallus gallus*	blood	44.2	1.04	1	1.25

TABLE 5.3.

5mC, GC and *c*-values in vertebrate DNAs[a]. (continued)

Order	Family	Species	Tissue	GC (mol%)	5mC (mol.%)	5mC ref.[b]	*c*-value
Columbiformes	*Columbidae*	*Columba livia*	blood	44.0	1.02	6	1.3
Passeriformes	Corvidae	*Corvus corone*	blood	45.4	1.15	1	1.7
Ralliformes	*Rallidae*	*Fulica atra*	blood	49.3	1.33	1	1.7
Mammalia							
Monotremata	*Ornithorhynchidae*	*Ornitorhyncus anatinus*	blood	45.6	1.32	6	3.78
	Tachyglossidae	*Tachyglossus aculeatus*	blood	45.9	0.99	6	3.57
Eutheria							
Insectivora	*Erinaceidae*	*Erinaceus europeus*	thymus	45.5	0.53	6	
	Soricidae	*Crocidura russula*	liver	41.4	0.66	6	
Dermoptera	*Cynocephalidae*	*Cynocephalus variegatus*	liver	40.6	0.89	6	
Megachiroptera	*Pteropodidae*	*Pteropus sp.*	liver	40.5	0.96	6	
	Pteropodidae	*Pteropus poliocephalus*	liver	41.5	1.39	6	
Microchiroptera	*Hipposideridae*	*Hipposideros galeritus*	whole animal	41.4	0.86	6	
	Rhinolophidae	*Rhinolophus creaghi*	whole animal	41.5	0.81	6	
	Vespertilionidae	*Myotis lucifugus*	whole animal	43.5	0.97	6	3.2
	Phyllostomidae	*Chiroderma salvinii*	liver	41.4	0.86	6	
	Nycteriidae	*Nycteris thebaica*	whole animal	42.9	0.88	6	
	Molossidae	*Chaerephon pumila*	liver	41.4	0.86	6	2.9
Primates	*Hominidae*	*Homo sapiens*	liver	42.6	0.88	3	3.5
			sperm	42.6	0.84	3	
			blood	42.6	0.96	3	
	Cercopithecidae	*Macaca mulatta*	liver	41.3	0.86	4	3.15
		Macaca fascicularis	liver	41.3	0.96	4	
		Cercopithecus aethiops	liver	41.3	0.98	4	
	Cebidae	*Saimiri sciureus*	liver	41.3	0.85	4	
	Tupaiidae	*Tupaia montana*	liver	41.9	1.08	6	
	Lemuridae	*Hapalemur griseus*	liver	41.4	0.93	6	
Pholidota	*Manidae*	*Manis sp.*	liver	42.4	0.59	6	
Lagomorpha	*Leporidae*	*Oryctolagus cuniculus*	liver	44.3	0.86	6	3.2
Rodentia	*Sciuridae*	*Sciurus vulgaris*	liver	39.5	0.61	6	5.1

TABLE 5.3.
5mC, GC and *c*-values in vertebrate DNAs[a]. (continued)

Order	Family	Species	Tissue	GC (mol%)	5mC (mol.%)	5mC ref.[b]	*c*-value
	Cricetidae	*Cricetus norvegicus*	liver	40.7	0.37	6	4
	Spalacidae	*Spalax sp.*	liver	38.4	0.58	6	
	Caviidae	*Cavia porcellus*	liver	39.7	0.74	6	2.95
	Muridae	*Rattus norvegicus*	liver	43.9	0.94	6	3.8
		Mus musculus	liver	42.2	0.95	4	3.25
			sperm	42.2	0.83	4	
		Meriones unguiculatus	nd	45.2	0.70	2	
	Procaviidae	*Procavia capensis*	liver	41.0	0.70	6	3.64
Cetacea	*Balenopteridae*	*Balenoptera physalus*	liver	41.3	0.94	6	
	Physiteridae	*Physeter macrocephalus*	liver	41.9	1.06	6	
	Phocoenidae	*Phocena phocena*	liver	41.4	0.97	6	
Carnivora	*Canidae*	*Canis familiaris*	liver	41.1	0.67	6	3.3
	Felidae	*Pantera uncia*	liver	41.5	0.94	6	
Perissodactyla	*Equidae*	*Equus caballus*	liver	42.8	1.04	6	2.95
Artiodactyla	*Bovidae*	*Bos taurus*	liver	44.0	1.40	4	3.35
			kidney	44.0	1.30	4	
			sperm	44.2	0.75	4	
	Suidae	*Sus scrofa*	kidney	44.0	1.20	4	2.77
			sperm	42.1	0.77	4	
	Ovidae	*Ovis aries*	liver	42.6	1.13	4	2.9
			sperm	42.0	0.76	4	

[a] (From Jabbari et al., 1997)
[b] 5mC values are from (1) Vanyushin et al. (1970, 1973); (2)Pollock et al. (1978), Gama-Sosa et al. (1983b); (3) Ehrlich et al. (1982); (4) Gama-Sosa et al. (1983a); (5) Serrano et al. (1993) and (6) present work. GC values are from Bernardi and Bernardi (1990a) for fishes, amphibians and reptiles, Kadi et al. (1993) for birds and Sabeur et al. (1993) for mammals. *c*-values are from literature quoted in the above papers and from Bachmann (1972) and Venturini et al. (1986)

TABLE 5.4.

5mC and GC levels in different classes of vertebrates[a]

Class	Species	GC%	sd[b]	5mC	sd[b]
Fishes[c]	28	42.44	2.87	1.70	0.26
Amphibians	4	44.13	2.56	1.98	0.56
Reptiles	7	42.75	2.63	1.00	0.26
Birds	7	44.53	2.41	1.02	0.21
Mammals	39	42.07	1.68	0.88	0.20
Vertebrates	85	42.55			

[a] From Jabbari et al. (1997)

[b] sd, standard deviation

[c] Only *Osteichthyes* were taken into consideration.

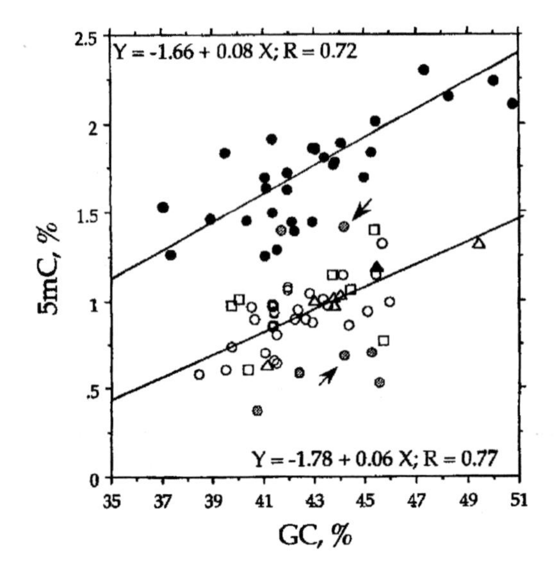

Figure 5.13. Plots of 5mC levels against GC levels for the genomes of fishes/amphibians (solid circles), reptiles (open sqares), mammals (open circles) and birds (open triangles). When data from different tissues were available, mean values were used. Deviating points of mammals are in grey (two of them, indicated by arrows, corresponding to DNAs from bovine somatic and germ cells, respectively) and were not taken into account in drawing the regression lines of mammals and birds; the single deviating point from fishes is shown as a solid triangle. Correlation coefficients, slopes and intercepts are given for fishes/amphibians and mammals/birds, respectively. (From Jabbari et al., 1997).

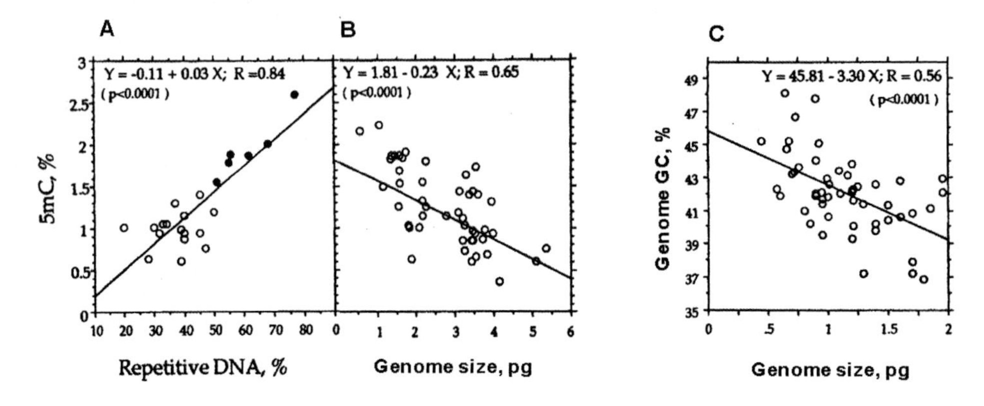

Figure 5.14. **A**. Plot of 5mC against the percentage of repetitive DNA sequences of the corresponding genomes (from Ginatulin, 1984). Solid circles correspond to fishes/amphibians, open circles to mammals/birds. **B**. Plot of 5mC levels against the corresponding haploid genome size of vertebrates; data from Table 5.3 were used to construct the plot. **C**. GC levels of fish DNAs are plotted against the corresponding haploid genome sizes; data are from Table 3 of Bernardi and Bernardi (1990a). *c*-values were used only up to 2pg, higher values being due to polyploidization. In all cases the correlation coefficients, slopes, intercept and *p*-values are shown. (From Jabbari et al., 1997).

genomes (Matassi et al., 1992; see **Part 8**) and in isochore families from individual vertebrate genomes (human, mouse, chicken and *Xenopus*; see below).

The two different methylation levels in vertebrate genome might be related to the different relative amounts of repetitive DNA sequences since (i) repetitive DNA sequences, as isolated by reassociation experiments, are more methylated than single-copy DNA sequences, at least in mammals (Romanov and Vanyushin, 1981; Gama-Sosa et al., 1983b); and (ii) the proportion of repetitive DNA sequences is larger in fishes/amphibians (59-77%) than in reptiles/mammals/birds (27-41%) (see Olmo et al., 1989, for a review). As expected from these premises, when the relative amount of repetitive DNA sequences (from Ginatulin, 1984) was plotted against the methylation level of the corresponding genomes, a positive, statistically significant correlation was found (see **Fig. 5.14A**).

There are, however, some problems with the relationship of **Fig. 5.14A**. (i) While the correlation persisted when plotting only the results from five fishes and one amphibian, a non-significant correlation was found when plotting only data from birds and mammals. (ii) At least one of the fish points of **Fig. 5.14A** is very doubtful. Indeed, *Clupea harengus*, with a *c* value of only 0.77 pg (see **Table 5.3**), was estimated to comprise 62% repetitive sequences, leaving only 0.29 pg for non-repetitive DNA. Now, the work of Pizon et al. (1984) has shown that in *Arothron diadematus*, one of the Tetraodontid fishes that have the smallest genome size of all vertebrates, 0.45 pg, repetitive sequences amount to 13%, leaving 0.39 pg for non repetitive DNA. This result casts a serious doubt on the *C. harengus* results, and decreases the points supporting the correlation to only five. More importantly, the results of Pizon et al. (1984) led to the conclusion that the reduction in genome size, such as occurred in *Tetraodontiformes*, largely took place at the expense of repetitive sequences. Conversely, this conclusion confirms the suggestion (Olmo et al., 1989) that genome size increases not due to polyploidization largely take place by an increase of repetitive sequences. (3) The genomes of fishes characterized by the lowest *c* values and by the lowest amount of repetitive sequences (*Tetraodontiformes* and *Gadidae*) show a methylation level that is similar to those of fishes with a genome size at least twice as large. Along the same lines, there is no substantial difference in methylation level between the two classes of warm-blooded vertebrates, in spite of the fact that birds have a genome size close to one-half that of mammals and a much larger amount of slow-reassociating DNA (84% in chicken DNA *vs.* 62% in human DNA; Soriano et al., 1981; Olofsson and Bernardi, 1983). It should, therefore, be concluded that the difference in methylation level between fishes/amphibians and reptiles/mammals/birds is not correlated with the amounts of repetitive sequences within either one of the two sets of genomes.

This conclusion is confirmed by investigations concerning the correlation between methylation and *c* values, the haploid genome sizes. As discussed above, *c* values are related to the amounts of repetitive sequences. In fact, they have the advantage of providing a larger data set compared with reassociation data. As shown in **Fig. 5.14B**, a negative correlation holds between 5mC levels and *c* values. This negative correlation can be understood on the basis of the following considerations. A negative correlation holds between GC and genome size in the case of fish genomes (Bernardi and Bernardi, 1990a; see **Fig. 5.14C**). This indicates that genome expansion (in a *c* value range where polyplodization is absent) takes place by an increase in the size of intergenic regions, which, as a general rule (Clay et al., 1996), are GC-poorer than coding sequences and, as a consequence, have fewer methylatable sites. Therefore, increasing the genome size means increasing GC-poorer regions of the genome which leads to a relative decrease of methylation, so accounting for the results of **Fig. 5.14B**.

5-Methylcytosine (5mC) levels were also determined in compositional DNA fractions corresponding to different isochore families from the genomes of *Xenopus*, chicken, mouse and human, four vertebrates which show different isochore patterns (Cacciò et al., 1997). The results obtained indicate that: (i) positive correlations exist between the 5mC levels and the GC levels of isochores within any given genome (**Fig. 5.15**); (ii) DNA from *Xenopus* isochore families is twice as methylated as DNA from the isochores having the same GC levels from mouse, human and chicken; (iii) higher methylation levels are associated with higher GC levels of main-band fractions; in some cases plots display deviating points that correspond to hypermethylated fractions (usually satellite DNAs). A similar result was obtained for plant genomes by Matassi et al. (1992; see **Fig. 8.18**).

In conclusion, our findings completely contradicted the explanations for CpG shortage proposed by Subak-Sharpe et al (1966) and Russell et al (1976), as well as those proposed by Cooper and Krawczak (1989) and Sved and Bird (1990). An alternative explanation based on our data on two different CpG and methylation levels in cold- and warm-blooded vertebrates will be presented in **Part 11, Section 3.5**.

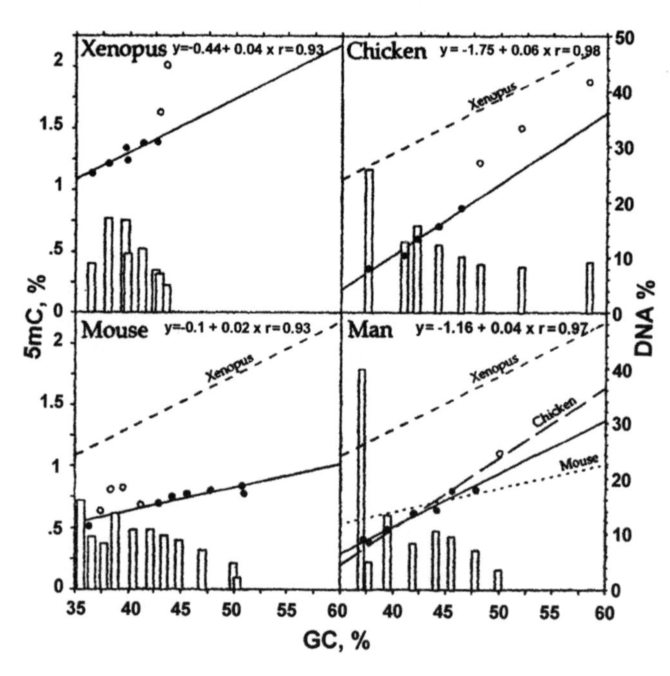

Figure 5.15. Plot of 5mC *vs* GC for compositional DNA fractions from *Xenopus*, chicken, mouse and man (left-hand scale). The histograms indicate 5mC levels. Least square lines through the points (and their equations) are shown; only filled circles were used in drawing the lines. The *Xenopus* (broken) line is also shown in the other diagrams for comparison. Likewise, the mouse and chicken lines are shown in the human diagram. (From Cacciò et al., 1997).

Part 6
The distribution of integrated viral sequences, transposons and duplicated genes in the mammalian genome

CHAPTER 1

The distribution of proviruses in the mammalian genome

1.1. The integration of retroviral sequences into the mammalian genome

The integration of retroviral genomes as proviruses into the genomes of host cells is a critical step in the life cycle of retroviruses (Temin, 1976; Varmus, 1984). Indeed, the replication of retroviruses is dependent upon integration (see Goff, 1992; Brown, 1997, for reviews), and several pathogenic effects of retroviruses are associated with their integration. For example, integration may inactivate host cell genes by disruption, activate them by the action of viral promotors and enhancers, or modulate their activity (reviewed in Kung et al., 1991; Tsichlis and Lazo, 1991; Athas et al., 1994; Jankers and Berns, 1996). Integration may be also associated with the modification of cellular genes and with the formation of new transforming viruses by acquisition of cellular gene sequences (reviewed in Bishop, 1980; Neil, 1983; Hughes, 1983). These events may lead to cell transformation and cancer development. It is evident, therefore, that retroviral integration is a very important phenomenon. Retroviral integration is, in fact, a special case of integration of foreign DNA into the genome of host cells. Obviously, a good understanding of this case has a crucial importance in gene therapy, especially when retroviruses are used as vectors (reviewed in Bushman, 1995; Hodgson, 1996; Anderson, 1998).

Integration of retroviral genomes (see **Fig. 6.1**) is a site-specific process as far as the viral sequences are concerned, since **long terminal repeat (LTR)** sequences are involved (reviewed in Varmus and Brown, 1989; Grandgenett and Mumm, 1990). As for the host genome, local features of integration loci (nucleosomal *versus* non-nucleosomal DNA, bent *versus* unbent DNA etc...), as well as the role of integrase in target site selection (reviewed in Holmes-Son et al., 2001), have been investigated, but while "*it is not clear to what extent local effects can account for the pattern of integration over the whole genome*", it has been concluded that "*the distribution of integration sites across the whole genome remains to be adequately characterized*" (Brown, 1997). Investigations carried out in our laboratory since the late 1970's, taking into account the isochore organization of the genome of warm-blooded vertebrates, have led, however, to an understanding of this distribution.

1.2. The bimodal compositional distribution of retroviral genomes

Compositionally, retroviral genomes belong in two classes (Zoubak et al., 1992; Tsyba et al., 2003): a GC-poor class and a GC-rich class (**Fig. 6.2A,B**). The first, minor, class comprises all lentiviruses and spumaviruses (which have no oncogenes but contain genes for regulatory proteins), oncoviruses of B type and some oncoviruses of D type, like MMTV (which also do not contain oncogenes). The second, major, class includes all

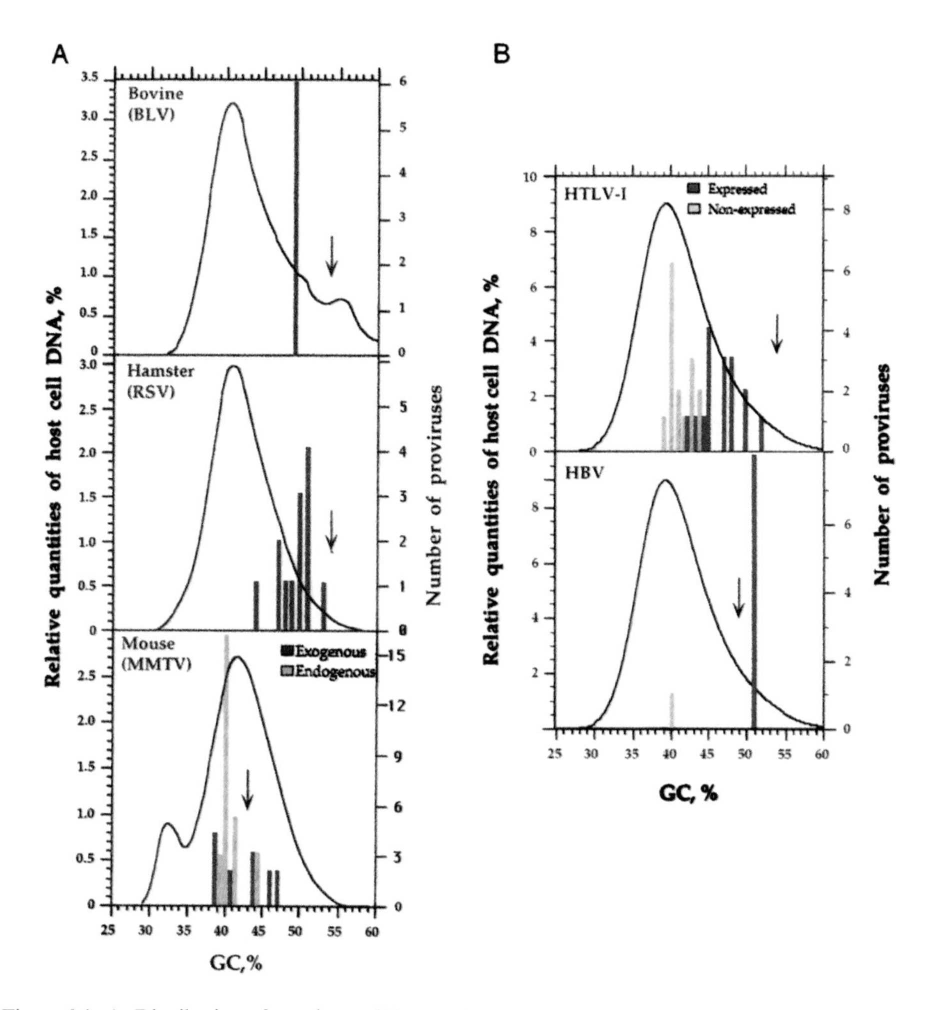

Figure 6.4. **A.** Distribution of proviruses (BLV, RSV, MMTV) in the genome of calf (Kettmann et al., 1979), hamster (Rynditch et al., 1991) and mouse (Salinas et al., 1987). The CsCl profiles of host cell DNAs (with buoyant densities converted into GC levels) are shown along with the localization of proviral sequences (open bars). Arrows indicate the GC levels of the viral genomes. **B.** Distribution of integrated viral sequences in the human genome: HTLV-1 (Zoubak et al., 1994) and HBV (Zerial et al., 1986b). The profile of human DNA is shown by the solid line. Arrows indicate the GC level of the viral sequences. (From Rynditch et al., 1998).

RSV. In this case, the localization of integrated retroviruses was investigated in the genomes of six well-characterized clones of hamster tumor cells. Twelve out of thirteen proviral sequences were observed in DNA fragments present in compositional fractions corresponding to the GC-richest 20% of the hamster genome, with a peak at 50% GC (**Fig. 6.4**). More precisely, GC-rich RSV (54% GC) from the four cell lines derived from different tumors containing one or two complete or defective copie(s) were found in DNA fractions having GC levels equal to 50–53% (Rynditch et al., 1991). The direct sequencing of extended DNA stretches flanking the provirus in one such cell line confirmed its integration in GC-rich DNA regions (Machon et al., 1996). In the case of cell lines comprising both complete and defective proviral copies, complete copies were present in fractions centered at about 50% GC, whereas defective copies were found in these fractions as well as in GC-poorer ones.

HTLV-1. The localization of this GC-rich provirus (54% GC) was investigated (**Fig. 6.4**) in five immortalized cell lines, containing a total of 22 integrated complete or defective proviruses, and in seven T-cell clones obtained from patients with TSP/HAM (tropical spastic paraparesis/HTLV-1-associated myelopathy), in which case each clone comprised 1–3 integrated proviral sequences for a total of 18 proviruses (Zoubak et al., 1994). These forty proviruses were found in the 39–54% GC range of the human genome, whereas no HTLV-1 sequences were found in isochores having a GC level lower than 39% (in spite of these isochores representing as much as 40% of the human genome).

MMTV. GC-poor MMTV proviral sequences (43.3% GC) were localized in the DNAs from the livers of five inbred strains of mice and from GR cells derived from primary implants of mammary tumors (Salinas et al., 1986). Twelve sets of endogenous MMTV sequences (corresponding to 27 proviruses) were characterized by restriction patterns and chromosomal localizations. Exogenous sequences were also localized and produced strong hybridization signals (the total number was 13), or weak hybridization signals corresponding to proviral integrations in a minor cell population (the total number was 6). While endogenous sequences were present in the GC-poor 43% of the host genome, exogenous sequences showed a broader distribution, being present in the GC-poor 60% of the genome (**Fig. 6.4**), therefore still within the L isochores. Interestingly, endogenous sequences were centered at 40% GC, whereas exogenous sequences showed a broader distribution centered at 43% GC.

To sum up, four retroviruses, three of C-type (BLV, RSV, HTLV-1) and one of D-type (MMTV), were localized in four mammalian species. The data show that integrated retroviral sequences are not spread at random in the host genomes, but are found in some host genome compartments approximately matching the viral sequences in composition. This conclusion is stressed by a presentation of the results in terms of proviral density in host DNA (**Fig. 6.5**). Expectedly, in none of the cases investigated were the proviral sequences found in satellite or ribosomal DNA.

Results similar to those just described for retroviral sequences were obtained for the localization of integrated GC-rich sequences from human hepatitis B virus (HBV), a DNA virus (Zerial et al., 1986b). Eight out of the nine viral sequences integrated in the hepatoma Alexander cell line were localized in DNA segments having a GC level of 51% (**Fig. 6.4**). Likewise, integration of a GC-rich recombinant SV40-adenovirus 5 was observed in the H3 isochore family (Romani et al., 1993).

It should be stressed that our conclusions are not contradicted by results indicating, for instance, that "*HTLV-1 integration is not random at the level of the nucleotide sequence. The virus was found to integrate in A/T rich regions*" (Leclercq et al., 2000), because these investigations dealt, in fact, with a different problem, concerning the short (40-nucleotide long) host sequences flanking the provirus. This very local composition may obviously be present in any isochore. It should also be stressed that our results concern the localization of stably integrated proviral sequences and not the primary events of integration (which will be discussed below).

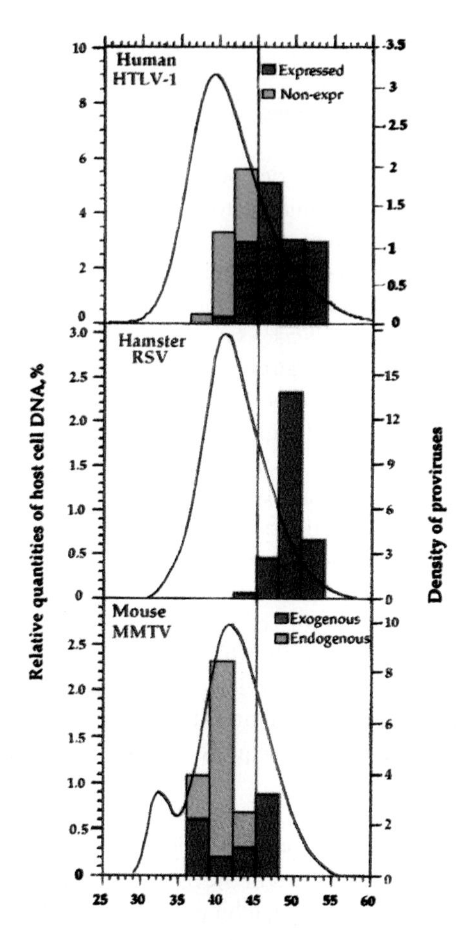

Figure 6.5. Data on HTLV-1, RSV and MMTV are presented in terms of number of proviruses in each 3% GC interval divided by the corresponding relative amounts of host DNA deduced from the CsCl profile. (From Rynditch et al., 1998).

1.4. An analysis of integration sites near host cell genes

Another approach that we used to localize integrated retroviral sequences should be mentioned. Since host cell genes are "isochore markers", in that their composition is correlated with that of the isochores in which they are embedded, an analysis of proviruses located near host cell genes can be done in order to localize proviruses (Rynditch et al., 1998). This was done for 64 MuLV (Murine Leukemia Virus) sequences for which information about the GC level of several genes in the neighborhood of integration sites was available. This analysis of host genes located close to MuLV proviruses showed that these genes are distributed in a GC range centered around 54% GC (**Fig. 6.6**), a value very close to the GC level of MuLV (53%). This means that in the case of insertional mutagenesis, it is possible to observe an isopycnic localization of integrated proviral sequences.

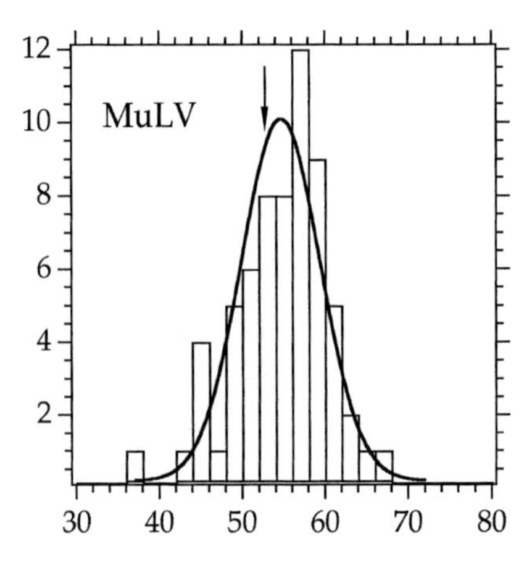

Figure 6.6. Compositional distribution of genes located in the neighborhood of 64 MuLV proviruses. The arrow indicates the GC level of the provirus. (Modified from Rynditch et al., 1998, by the addition of 36 proviruses to the original set of 28 proviruses)

1.5. Correlation between the isochore localization of integrated retroviral sequences and their transcription

The results on the compositional distribution of HBV sequences (Zerial et al., 1986b) provided the first information about a possible correlation between the isopycnic localization of integrated viral sequences in compositional compartments of the host genome and their transcription. Indeed, the eight HBV sequences integrated in the GC-richest regions of the genome of the Alexander cell line were expressed, whereas the single one located in a GC-poorer region of the genome was not (see **Fig. 6.4**). Likewise, all RSV transcribed sequences were localized in the GC-richest isochores (**Fig. 6.4**), whereas the proviral sequences whose expression was not detected were distributed in isochores with a lower GC content (Rynditch et al., 1991).

A more detailed analysis was possible on integrated HTLV-1 sequences, in which case 40 sequences from 12 cell lines and T-cell clones were investigated (Zoubak et al., 1994). Indeed, the transcriptionally active HTLV-1 sequences were found to be localized in the GC-richest regions which corresponded to the H2 and H3 isochore families of the human genome, whereas the transcriptionally inactive HTLV-1 sequences were localized in GC-rich regions corresponding to the H1 isochore family (**Fig. 6.4**).

In the case of MMTV, at least two loci, MTV-1 and MTV-2, located in GC-poor isochores (37–38%) are expressed as infectious viruses in GR (Bentvelzen et al., 1970) and C3H (Van Nie and Verstaelen, 1975) strains of mice and in the GR mammary tumor cell line. Moreover, most mouse strains contain at least one copy of endogenous MMTV provirus (Marcus et al., 1981) expressed not only in mammary glands during lactation (Choi et al., 1987), but also in several organs at a lower level (Henrard and Ross, 1988). These data indicate that MMTV sequences can, indeed, be expressed when integrated in GC-poor isochores.

*1.6. Integration in "*open*" chromatin and/or near CpG islands*

Several observations suggested that retroviral localization is correlated with the level of transcription and recombination, and with the degree of DNA accessibility in chromatin (Vijaya et al., 1986; Robinson and Gagnon, 1986; Rohdewohld et al., 1987; Scherdin et al., 1990; Mooslehner et al., 1990; Fincham and Wyke, 1991).

In bursal lymphomas, ALV (Avian Leukemia Virus) integrants were found near each of five major DNase I hypersensitive sites which are immediately upstream of the coding sequences for *c-myc* (Schubach and Groudine, 1984; Robinson and Gagnon, 1986). The integration of MoMuLV (Moloney Murine Leukemia Virus) sequences in the α-1 collagen gene has been localized near a DNase I hypersensitive site (Breindl et al., 1984). Likewise, MoMuLV integrations in seven regions of chromosomal DNA were mapped near DNase hypersensitive sites (Vijaya et al., 1986). Three of these regions, containing *c-erbB*, *c-myc* and *dsi-1*, were targets for multiple tumor-inducing integrations. Interestingly, the other four unselected integrations also occurred near DNase I hypersensitive sites, suggesting that retroviral sequences preferentially integrate near hypersensitive sites in all cases. This observation has been confirmed by the analysis of chromatin structure of MoMuLV proviral integration sites in early embryonic cells and in differentiated fibroblasts, where integration occurs within a few hundred base pairs of a DNase I hypersensitive chromatin region (Rohdewohld et al., 1987). It is well known that DNase I hypersensitive sites correlate with gene expression and are preferentially located in regions with an open chromatin structure (Edmondson and Roth, 1996).

These results prompted the analysis of transcription of cellular sequences flanking the integrated proviruses. In three out of five randomly chosen mouse strains harboring one copy of MoMuLV in their germ lines, the provirus was integrated in the vicinity of DNA regions transcribed in an embryonal stem cell line and in an embryonal carcinoma cell line (Mooslehner et al., 1990). Among nine sequences randomly chosen for MoMuLV integration in NIH3T3 mouse fibroblasts, at least six were in transcriptionally active regions and/or contained CpG-rich islands (Scherdin et al., 1990). Finally, the proviral integration sites of transcribed and non-transcribed RSV in rat DNA were examined (Fincham and Wyke, 1991). In this case, the transcribed sequences showed a tendency to integrate close to the 3' ends of CpG islands, whereas non-transcribed sequences did not.

A more recent study (Carteau et al., 1998) of 61 HIV-1 integration site sequences after a short (14 hrs) experimental infection of a human T-cell line led, however, to the conclusion of no clear bias for integration in transcription units, but that centromeric alphoid repeats were absent at integration sites. An analysis of 48 such sites in terms of genome localization (Elleder et al., 2002) showed, however, that 54% of them were located in genes (introns) and 69% in transcribed sequences, in line with the older results mentioned above. As far as the isochore localization is concerned, as shown in **Fig. 6.7**, while the HIV-1 integrations are mainly found in isochores L2, H1 and H2, their highest density (and, therefore, their preference of integration) is in the genome core (isochores H2 and H3), which is characterized by an open chromatin structure (see **Parts 4, 5 and 7**). This is a most interesting result in view of the fact that HIV-1 has a very GC-poor genome.

The conclusions of Elleder et al. (2002) have been confirmed by a more extensive study on the sites of HIV-1 integration in the human genome (Schröder et al., 2002). Needless to say, the gene dense-regions in which the preferential localization of these early integration sites occurs coincide (although apparently not noticed by the authors) with the GC-richest isochores (see **Fig. 6.8**). An interesting aspect of the work of Schröder et al. (2002) was that these authors not only investigated HIV-1 integration *in vivo* (as already mentioned), but also *in vitro*. Using naked genomic DNA from Sup T1 cells as an integration target and HIV-1 preintegration complexes (*i.e.* replication intermediates that can be isolated from infected cells and which contain the viral cDNA, integrase, and other viral and cellular proteins), they found an integration pattern which did not show the specificity for genes

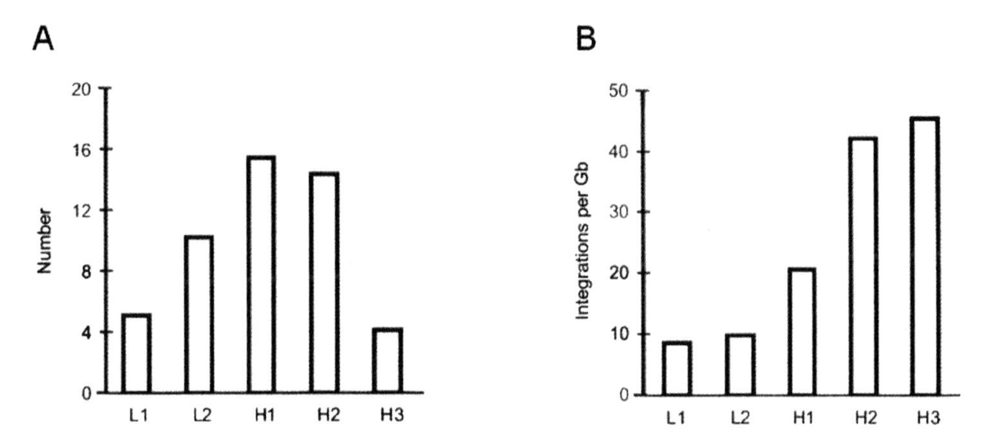

Figure 6.7. A. Numbers of HIV-1 sequences in each isochore family, the GC intervals are L1 (<37%), L2 (37–41%), H1 (41–46%), H2 (46–52%) and H3 (>52%). The size of the genome belonging to individual isochore families in 100 kb windows is 571 Mb for L1, 977 Mb for L2, 706 Mb for H1, 321 Mb for H2 and 85 Mb for H3. **B.** The vertical axis shows the probability of targeting the individual isochore family. This is calculated as the number of integrations per Gb whose 100 kb surrounding belongs to the corresponding isochore family. (Modified from Elleder et al., 2002).

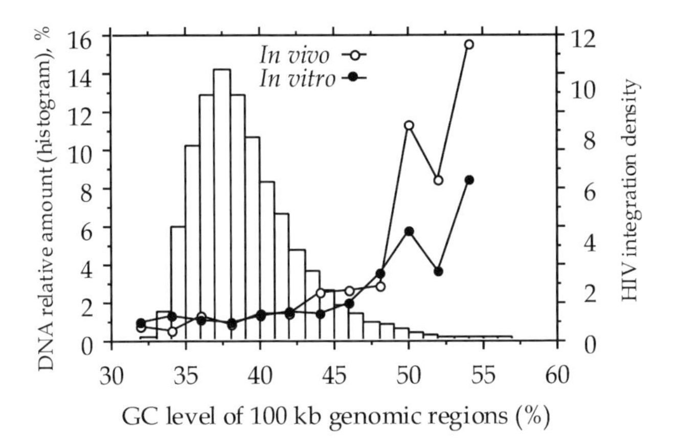

Figure 6.8. Distribution of 500 *in vivo* and 110 *in vitro* HIV-1 integration sites in the human genome based on the data from Schröder et al. (2002). The curves represent the integration density (% integrations/ % DNA). (From Tsyba et al., 2004).

isopycnic integration. It should be noted, however, that the examples of non-isopycnic integration practically only concern a GC-poor provirus, MMTV, being integrated next to GC-rich genes, a point which is now better understood in view of the results on HIV-1 (see preceding section). Indeed, the opposite case, namely that of GC-rich proviruses activating GC-poor oncogenes, has only been found once (in the case of MuLV activating *c-ki-ras*), in spite of the fact that the number of GC-rich proviruses explored was much larger than that of GC-poor proviruses.

These conclusions have some implications concerning both the compositional evolution of retroviral genomes and gene therapy: (i) the bimodal compositional distribution of retroviral sequences may be visualized as the result of their compositional coevolution with the sequences of the host genome; it is conceivable that the compositional transitions, which led to the formation of GC-rich isochores of warm-blooded vertebrates, also led to similar compositional transitions (acting on their integrated forms) of what are now the GC-rich retroviral sequences (Zoubak et al., 1992); the results on duplicated genes reported in **Chapter 3** reinforce this explanation; and (ii) the reasons for the poor expression of retroviral constructs used in gene therapy (Anderson, 1998) should be investigated not only at the level of the promotors present in the constructs, but also at the level of the integration sites which are used by the constructs; in many cases, constructs are characterized by compositional features which are never found in the host genome.

CHAPTER 2

The distribution of repeated sequences in the mammalian genome

2.1. *Alu and LINE repeats in human isochores*

Alus are short (~300 bp) GC-rich, non-autonomous elements, derived from 7SL RNA. They make up to 10% of the human genome (Lander et al., 2001). Mutational databases show that Alus may transpose into the introns of human genes (Schmid, 1988; Batzer and Deininger, 2002). Full-length LINEs are long (6–8 kb) GC-poor sequences encoding an RNA binding protein and a reverse transcriptase/endonuclease. LINE1 elements represent the most abundant group of LINEs (see Ostertag and Kazazian, 2001, for a review) and correspond to 17% of the human genome (see **Fig. 6.10**).

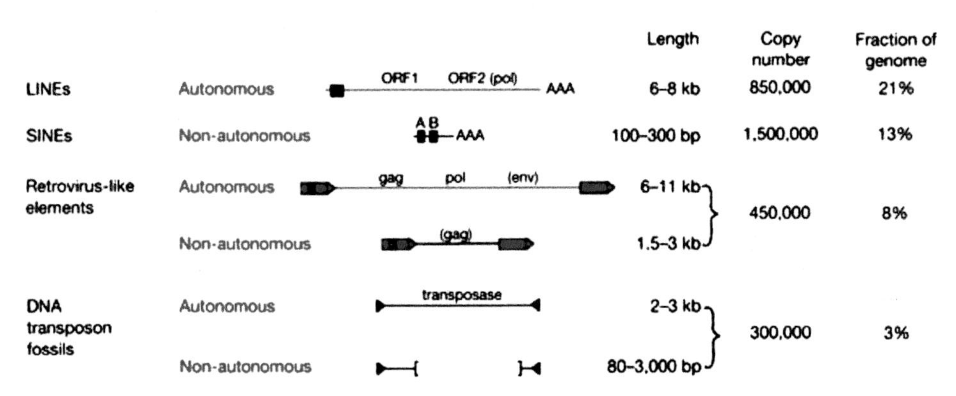

Figure 6.10. Classes of interspersed repeat in the human genome. (From Lander et al., 2001).

Compositional fractionation of human DNA not only allowed us to characterize the genome in terms of isochore families and gene densities, but also to investigate the distribution of interspersed repeated sequences. Reassociation kinetics provided the first indication that the distribution of intermediate repetitive sequences was different in DNA fractions from different isochore families (Soriano et al., 1981; Olofsson and Bernardi, 1983b). Indeed, as shown in **Fig. 3.7**, the relative amounts of interspersed repeats and unique sequences were strikingly different in the different major components of human, mouse and chicken genome. In a more direct approach, hybridisation of appropriate probes on the major DNA components (**Fig. 6.11**) showed that the GC-poor LINE-1 and the GC-rich Alu families have their highest densities in GC-poor and GC-rich isochores of the human and mouse genome, respectively (Meunier-Rotival et al., 1982; Soriano et al., 1983; Zerial et al., 1986b). These results, later confirmed by assessments based on sequences from data banks (Smit, 1996; 1999; Jabbari and Bernardi, 1998), showed that

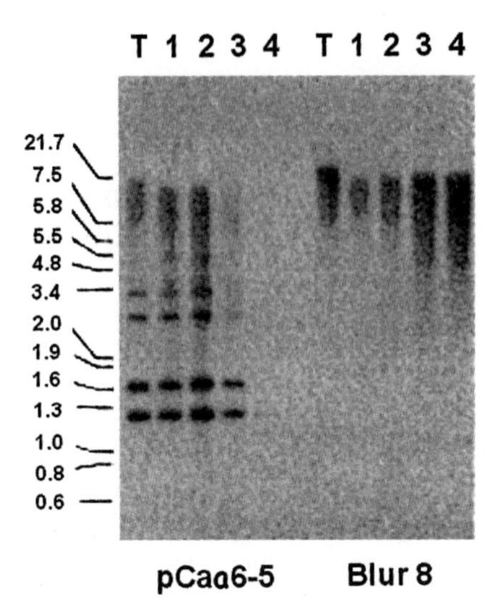

Figure 6.11. Autoradiograms of cloned, repeated sequences pCaα6–5 (a LINE 1 probe) and Blur 8 (an Alu 1 probe) hybridized on nitrocellulose transfers from a gel electrophoresis on 0.6% agarose gel of *Kpn* I digests of unfractionated human DNA (lanes T) and its major components 1.698 (lanes 1), 1.700 (lanes 2), 1.704 (lanes 3), and 1.708 (lanes 4) loaded in equal amounts (1 μg). (Modified from Soriano et al., 1983).

the "ubiquitous" distribution of Alu sequences (Houck et al., 1979), was far from homogeneous.

The location of the long interspersed mouse LINE-1 repeats (originally called Bam H1 repeats) in the two light isochore families L1 and L2 of mouse showed a compositional match with the isochores in which they were embedded, and was strikingly different from that of protein-encoding genes, indicating that the repeats were not involved, at least in a simple way, in the regulation of gene expression (Meunier-Rotival et al., 1982). Subsequent work indicated not only that the human LINE-1 repeats (which were originally called Kpn I repeats) were also located in light isochores, but also that the short interspersed SINE repeats (of the B1 family of mouse and of the Alu-1 family of human) were most frequent in the GC-rich isochores. These results showed that the genome distribution of interspersed repeats is non-uniform, different for the two major families of repeats, and similar in two mammalian species. Moreover, the base composition of both classes of repeats was correlated with that of the isochores in which they had their highest concentration, and both sorts of repeats were shown to be transcribed (Soriano et al., 1983).

Recents assessments have quantified the distribution of Alus and LINES in the human genome and provided information on the specific distribution of four categories of repeats separated according to their divergence, Δ, from the consensus sequence (see **Table 6.1**). Very young Alus with $\Delta < 2\%$ (1,563 of the analyzed 280,809 Alus) were found to be twice as frequent in L1 compared to H3 isochores. Alus with Δ = 2–4% showed no preference, and Alus with $\Delta > 4\%$ have the typical bias toward the GC-rich part of the genome. The oldest and most abundant class of Alus Δ>6% have a more than threefold higher density in H3 compared to L1 isochore, and correspond to the general distribution of Alu sequences in the human genome.

TABLE 6.1
Alu and LINE distribution in the human genome[a]

DFC	L1	L2	H1	H2	H3
Δ < 2%	0.077	0.081	0.068	0.046	0.033
Δ 2%–4%	0.062	0.052	0.069	0.073	0.069
Δ 4%–6%	0.2	0.22	0.36	0.47	0.56
Δ > 6%	4.9	7.7	13.3	18.1	16.5
All Alus	**5.2**	**8.1**	**13.8**	**18.7**	**17.2**
Δ < 2%	0.1008	0.0818	0.0410	0.0103	0.0011
Δ 2%–4%	1.41	1.27	0.67	0.17	0.012
Δ 4%–6%	1.1	1.2	0.48	0.11	0.031
Δ > 6%	19.9	15.8	10.1	6.8	4.3
All LINEs	**22.5**	**18.3**	**11.3**	**7.1**	**4.4**

[a] From Pavliček et al. (2001). Isochore family intervals are: less than 37%, 37% to 41%, 41% to 46%, 46% to 52%, and more than 52% GC. Repeat densities were calculated for 10 kb-long non overlapping sections using RepeatMasker. Alus and LINEs are divided into four categories corresponding to less than 2%, 2–4%, 4–6% and more than 6% divergence (Δ) from the family consensus (DFC) sequence. The general Alu and LINE distributions are also shown (bold figures).

Alus are CpG-rich and contain about 30% of all CpGs in the genome (Hellmann-Blumberg et al., 1993). Since the high rate of mutation of CpGs could affect age estimation based on the divergence from the consensus, Alu classification based on diagnostic sites was also used as another criterion (Batzer et al., 1996; Jurka, 2000; **Fig. 6.12** presents a recent phylogenetic tree of Alu sequences). The young family AluYa5 (Deininger and Slagel, 1988) has a genomic distribution similar to the least divergent Alus (**Fig. 6.13**). The family AluYb8 is already more dense in GC-rich isochores, and AluY and the other families (members of AluS and AluJ group) are biased toward GC-rich isochores.

In the case of LINES, the preference for GC-poor isochores is much higher (90 times more in L1 than in H3 isochores) for young LINEs of L1HS family, the only interspersed repeats that still actively transpose in the human genome (Lander et al., 2001), compared to old LINEs (5 times), which practically show the general distribution (**Fig. 6.14**). The general distribution of both LINES and Alus are shown in **Fig. 6.15**.

We also calculated the Alu and LINE1 distribution on human chromosomes 21 and 22 (Dunham et al., 1999; Hattori et al., 2000). The densities of Alu and LINE1 co-vary with the GC level, LINEs showing their highest density in GC-poor isochores and Alus being in GC-rich isochores. Insertions of the young Alu family AluYa5 seemed to be independent of the general Alu distribution on chromosomes 21 and 22 (**Figs. 6.16A**; **6.16B**). The distribution of LINE L1PA2 family is strongly correlated with regions of low GC content. To test this point, the Alu (or LINE) density around each Alu (or LINE) in the database was analyzed within 20 kb flanking sequences. We found that the density of Alus surrounding each Alu insertion is higher for old families. Similar conclusions hold for LINEs, older LINEs being preferentially located within LINE clusters. We concluded that both Alus and LINEs integrate both inside and outside clusters of older elements and that there are chromosomal regions that are more favorable for their accumulation (GC-poor for LINEs

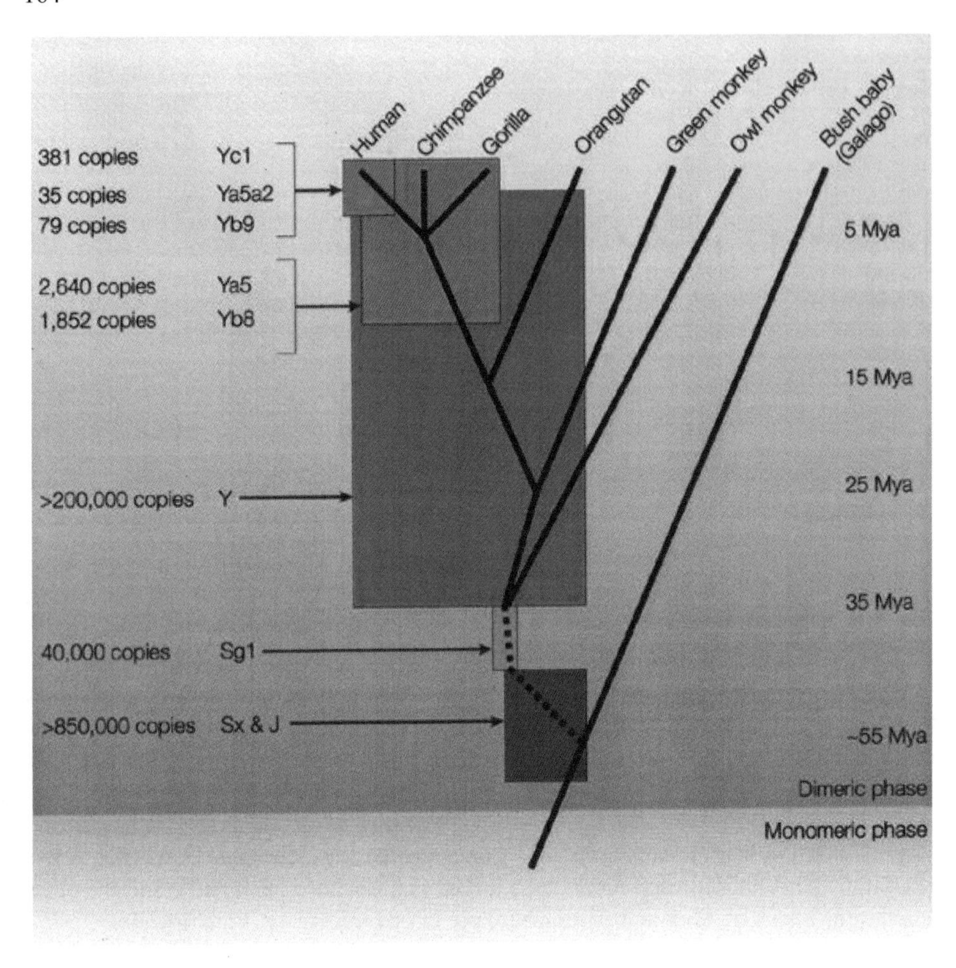

Figure 6.12. The expansion of Alu subfamilies (Yc1, Ya5a2, Yb9, Y, Sg1, Six and J) is superimposed on a tree of primate evolution. The expansion of the various Alu subfamilies is colour coded to denote the time of peak amplification. The approximate copy numbers of each Alu subfamily are also noted. Mya, million years ago. (From Batzer and Deininger, 2002).

and GC-rich for Alus). Remarkable exceptions are the GC-rich telomeric regions of the chromosomes, which are relatively free of Alus and rich in LINEs.

A final, important point is the following. Although the repeat density distribution is so different for Alus and LINES (as shown for the general distribution in **Fig. 6.15A**), the actual numbers of repeats follow rather similar trends (**Fig. 6.15B**). The largest number of Alus and LINES are present in L2 and the lowest in H3 isochores, as expected from the relative amounts of these isochores families. Evident differences exist in L1 isochores, where LINES are more abundant than Alus and in H1-H3 isochores where the opposite is true.

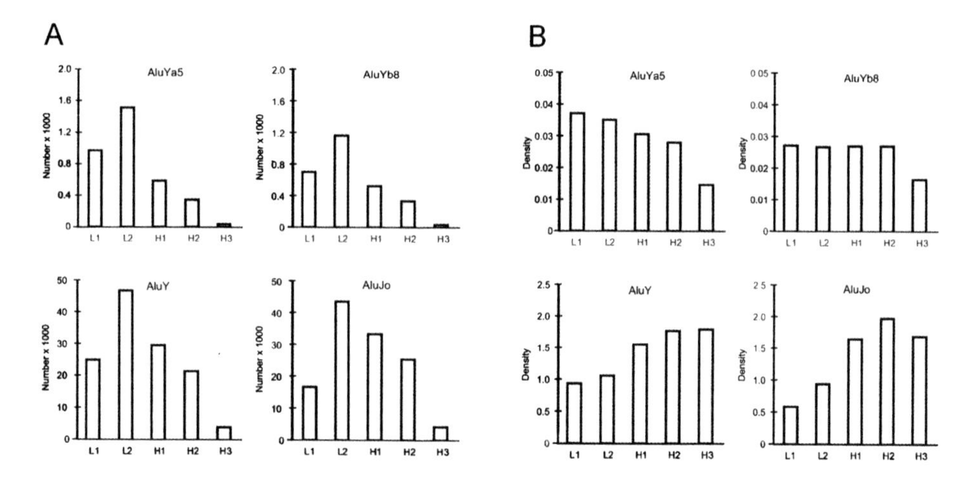

Figure 6.13. Human genome distribution of Alu sequences in terms of numbers of sequences (**A**) and of density as calculated in 100 kb-long nonoverlapping segments (**B**). Isochore family intervals are as in Table 6.1. Distributions concern AluYa5 family (mean Δ = 2.26%); AluYb8 family (mean Δ = 5.23%); AluY family (mean Δ = 7.15%); AluJo family (mean Δ = 16.8%). (Updated from Pavliček et al., 2001).

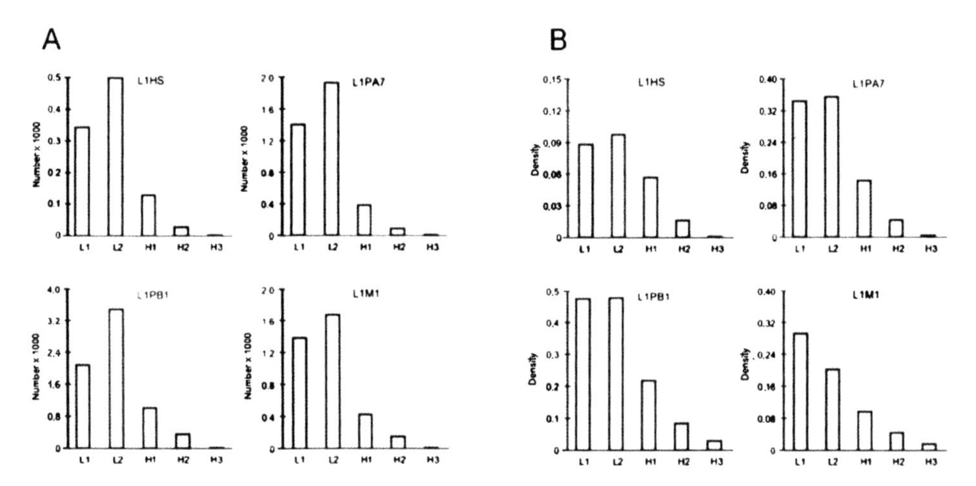

Figure 6.14. Human genome distribution of LINE 1 sequences in terms of numbers of sequences (**A**) and of density as calculated in 100 kb-long non-overlapping segments. (From Pavliček, 2002, personal comm.)

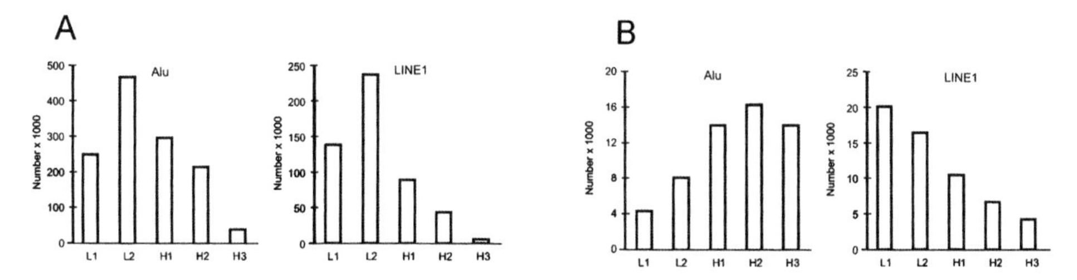

Figure 6.15. General human genome distribution of Alu and LINE 1 sequences in terms of numbers of sequences (**A**) and of density as calculated in 100 kb-long non-overlapping segments. (From Pavliček, 2002, personal comm.)

TABLE 6.3
Proposed explanations for the densities of Alus and LINEs in GC-rich and GC-poor isochores

Integration into:		Favored by:	Compositional match	Consequence
Alus:	**GC-rich isochores**	more open chromatin	+	**Stability**
	GC-poor isochores	more acceptor sites	–	**Elimination (slow)**
LINES:	**GC-rich isochores**	more open chromatin	–	**Elimination (rapid)**
	GC-poor isochores	more acceptor sites	+	**Stability**

2.3. Repeated sequences in coding sequences?

The contribution of transposable elements (TEs), including Alus, to human coding sequences has recently been reported to be high, 4% (1.3% Alus) out of 13,799 sequences (Nekrutenko and Li, 2001; Li et al., 2001). This is surprising, because previous examinations had revealed only very few repeats, and almost no Alus, in coding sequences (Brosius, 1999; Smit, 1999, Lander et al., 2001). Since extreme caution about input data has been suggested (Tugendreich et al., 1994; Zietkiewicz et al., 1994), we examined the database of Nekrutenko and Li (2001) and found that many (~30%) of its TE-containing sequences or their protein products are defined as '*hypothetical*', and 63% (421/669 sequences) are annotated as '*predicted, without experimental evidence or records without final NCBI revision*'. Such a dataset is likely to contain several sequences that remain untranscribed, and more that remain untranslated. Not even experimental validation (Iyer et al., 2001), let alone computer prediction of functional genes, is foolproof: the errors in coding sequence databases such as those used in Nekrutenko and Li (2001) may well amount to 1-2% or more.

Essentially all reported coding regions derived from Alus, or containing alternatively spliced Alus, have been detected at the RNA (cDNA) level, instead of at the protein level (Brosius, 1999; Hoenika et al., 2002). In eukaryotic cells, there is a significant turnover of RNA, and several steps of quality control exist for the synthesized RNA in both nucleus and cytoplasm (Jackson et al., 2000; Wilusz et al., 2001; Maquat and Carmichael, 2001; Moore, 2002; Iborra et al., 2001). mRNAs with an aberrant 3'P end are generally retained and/or degraded at their site of transcription (Hilleren et al., 2001) and the majority of stable RNA polymerase II transcripts remain in the nucleus as 'junk' RNA, so they never reach the cytoplasm (Jackson et al., 2000). The minority of transcripts that are successfully exported from the nucleus undergo additional check(s) during their translation. For example, there are specialized degradation mechanisms for transcripts having premature stop codons or lacking terminal codons, which prevent the creation of aberrant, potentially pathogenic proteins (Wilusz et al., 2001; Moore, 2002; Frischmeyer et al., 2002). Thus, even detection of a transcript at the mRNA (cDNA) level cannot guarantee that these mRNAs are ever translated into stable proteins. As has been summarized in the light of growing evidence (Pradet-Balade et al., 2001), '*mRNA abundance is a poor indicator of the levels of the corresponding protein*', yet '*it is the proteome that determines cell phenotype*': the transcriptome does not faithfully represent the proteome. Furthermore, to become a viable protein, a transcript must (after its accurate translation and possible post-translational modification) resist degradation until it can serve its functional role at the site of its required action. These facts underline the importance of detection at the protein level, for elucidating whether SINEs or other repeats contribute to true coding sequences in humans or mice.

The most accurate sources of proteins are 3D structure databases and direct amino acid sequencing. Out of 781 non-redundant human proteins from a 3D database or determined at the amino acid level and compared to human repeats in RepBase (Jurka, 2000) using TFASTX (Pearson et al., 1997), we found no Alu-related protein domain (the best hit has an E-value of 0.5). Twenty-eight apparently significant hits with E-values under 0.01 were detected, but mainly from protein-coding elements (DNA transposons and LINE1). When cDNAs encoding these 28 proteins were extracted and searched by RepeatMasker (Smit and Green), no interspersed repeats were detected. In addition, the similarity regions that had been reported by TFASTX were also found in other vertebrate orthologs. In summary, we did not detect any repeat sequence in our dataset of 781 protein sequences.

In 1994, it was pointed out (Tugendreich et al., 1994) that a discovery of a translated Alu element(s) in a functional part of a functional human protein '*would represent the first report of its kind and would have important evolutionary implications*'. Despite the 7 years since this challenge, confirmed cases of Alu-containing sequences that encode a functional protein still remain extremely elusive.

The paucity of documented examples is a good indication that proteins are unlikely to utilize domains encoded by Alus for functional ends. The reluctance to accept this view is understandable, given the huge proportion of interspersed repeats in the human genome (around 45%; Smit, 1999) : in principle, at least some of them might have been recruited for functional purposes at the protein level. The great majority of previously detected repeat-derived coding sequences comes, however, from protein-coding repeats, and particularly from DNA transposons (Smit, 1999; Boeke, 1997). LINEs are less common in coding sequences and only a few Alus had been identified prior to the analysis of Nekrutenko and Li (2001) and Li et al. (2001). Since SINEs are derived from RNA genes without protein-coding capacity, the lack of Alu-encoded proteins is consistent with the notion that new domains arise from existing sequences encoding functional proteins (for example, by exon shuffling) and that the *de novo* creation of coding sequences from non-coding DNA is rare (although not impossible; see **Part 2**, **Section 3.4**).

The relative frequencies for the TE classes found by Nekrutenko and Li (2001) are similar to genome-wide repeat proportions, *i.e.* to expectations under random sampling of sequences or random errors in predicting exons. In contrast, our findings are in good agreement with previous reports (Smit, 1999; Lander et al., 2001) and the above arguments that repeat-derived protein-coding sequences, especially those corresponding to Alus and other SINEs, should be rare. Indeed, Alus are derived from 7SL RNA, part of the signal recognition particle on ribosomes (Ullu and Tschudi, 1984), and the strong selection for such 7SL-like secondary structures that they apparently experience (Boeke, 1997) would not leave much freedom for Alu RNAs to fulfil other roles: in particular, it would be difficult to harness them to simultaneously encode functionally important proteins.

It should be emphasized that we are not questioning the presence, or even a proposed abundance (Nekrutenko and Li, 2001), of alternatively spliced transcripts containing Alus. We are questioning the notion that such transcripts will be translated to yield functional proteins, except possibly in one or two extremely rare cases. In spite of anecdotal reports of involvement of Alu-derived or -sequestered amino acid sequences in molecular recognition or binding (*e.g.* of the HPK1 Alu in activating AP1; Nekrutenko and Li (2001), or of a group of Alu-derived peptides in binding tau; Hoenika et al., 2002), it is not at all clear that such possibly fortuitous involvement reflects a functional role.

In summary, the available examples suggest that transposable elements could occasionally correspond to parts of rare, atypical proteins that arise by alternative splicing and subsequent translation, resulting in aberrant products with potentially pathological effects. In general, however, the presence of SINEs in a putative or predicted human coding sequence still appears to be a good indication that it will seldom, if ever, be translated into a functional protein *in vivo*.

CHAPTER 3

The distribution of duplicated genes in the human genome

The reason for dealing with the distribution of duplicated genes in the human genome in this part of the book is that many gene duplications are very ancient (see below) and that most duplicated genes have been transposed to other locations since.

Thirty years ago, Ohno (1970) proposed that two complete genome duplications took place at the origin of vertebrates, and that this tetraploidization was followed by the diversification of the functions of duplicated genes. Ohno's hypothesis is also called the 2R hypothesis (for two rounds of genome duplication).

More precisely, the model proposed two duplications before and after the emergence of the jawless fishes, approximately 500 and 430 Mys ago (Ohno, 1970, 1999; Lundin, 1993; Holland et al., 1994; Sidow, 1996; Katsanis et al., 1996; Spring, 1997; Bailey et al., 1997). Two rounds of complete genome duplication should lead to a 1:4 ratio between invertebrate and mammalian coding sequences. The *Hox* clusters fit the model well since they occur in one copy in invertebrates and in four copies in most vertebrate genomes. In humans, potentially quadruplicated regions have been analyzed in detail, on chromosomes 1, 6, 9 and 19, by Endo et al. (1997), and on chromosomes 4, 5, 8 and 10, by Pebusque et al. (1998). Phylogenetic analysis of paralogous genes has, however, challenged this view (see Wolfe, 2001, for a review).

Since many gene duplications in the human genome are ancient duplications going back to the origin of vertebrates (see above), the question may be asked about the fate of such duplicated genes at the compositional transitions from cold- to warm-blooded vertebrates, when about half of the genes increased their GC levels. The compositional distribution of the 1,111 pairs of duplicated genes investigated by Jabbari et al. (2003) is very similar to the

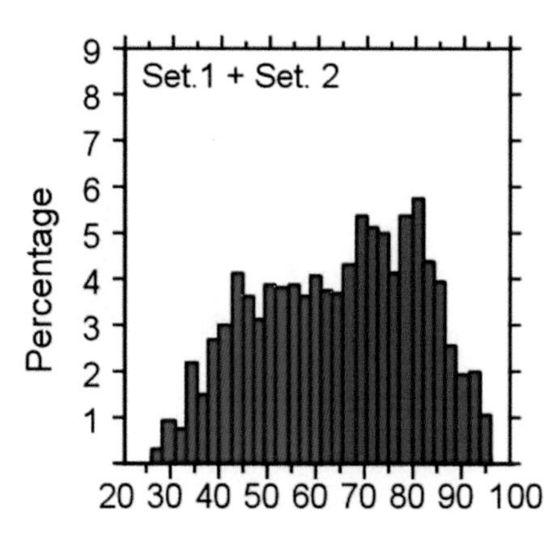

Figure 6.18. Distribution of GC_3 values of the 1,111 human gene pairs used (complete data set). (From Jabbari et al., 2003b).

general distribution of human genes (**Fig. 6.18**), where GC_3-poor (<60% GC_3) and GC_3-rich (>60% GC_3) genes are present in comparable amounts.

Two possibilities might then be considered (**Fig. 6.19A**), namely that the compositional transitions affected either (i) one copy of each gene pair only (or, at least, preferentially); or, in addition, (ii) none or both of the two copies, with approximately the same probability. Ideally, the two possibilities could be distinguished if each copy is put in one sub-set according to its composition. Indeed, the two distributions would be different if one copy kept the ancestral GC-poor composition, whereas the second copy, which had supposedly acquired a different function, had undergone the compositional change. In contrast, in the other case the distribution just mentioned would be blurred by the copies which both changed or maintained their composition. As shown in **Fig. 6.19B**, a comparison of the GC_3-poor copies with the GC_3-rich copies indicates that, indeed, one copy preferentially underwent the compositional transition, suggesting a different structure/function of one copy relative to the other one (Jabbari et al., 2003). Expectedly, we are not quite in the ideal case presented above. This can be understood because some recently duplicated genes are present in the data set. Indeed, if only genes showing a Nei-Gojobori (1986) distance higher than 0.7, or if only genes located on different chromosomes are taken into account (another way of eliminating recent tandem duplications), the difference between the two histograms is increased, with over 45% of the copies present in the GC-rich class.

One should now consider that (i) the two copies are generally located very far from each other in the genome (in fact they are most frequently located on different chromosomes; see **Fig. 6.20A**); (ii) that the compositional transition specifically affected the "ancestral genome core" (the gene-dense compartment of the ancestors of present-day mammals and birds; see **Part 12**); and (iii) that the GC-enriched copy is characterized by shorter introns compared to the GC-poor copy (Rayko et al., 2003; see **Fig. 6.20B**); the average intron size shortening by a factor of 2 is probably accompanied by an increased expression. Then, one should conclude that the "major line" of events following the ancestral gene duplications (disregarding a number of phenomena, like recent duplications, because they do not blur the events concerning the majority of duplicated genes) can be described as a succession of

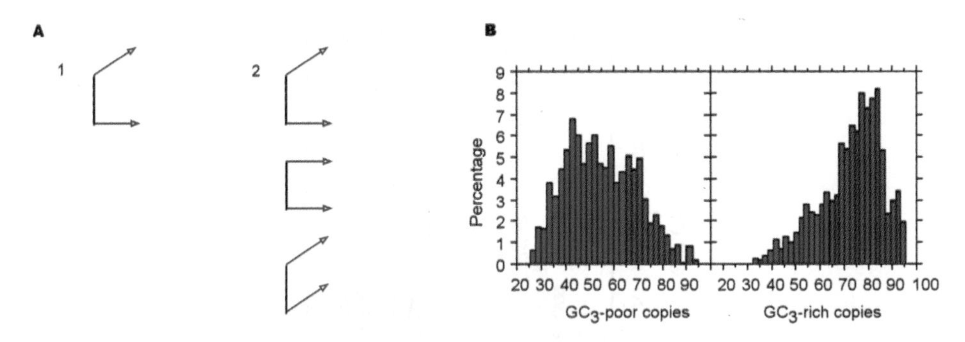

Figure 6.19. **A.** Model of two extreme situations hypothesized for duplicated genes at the transitions from cold- to warm-blooded vertebrates. 1) One copy of each pair preferentially undergoes the transition (red arrow), the other copy maintaining its original low GC-level (blue arrow). 2) In addition to situation 1, both copies may undergo the transition or maintain their original low GC level. **B.** Distribution of GC_3 values of the 1,111 human gene pairs used (sets of GC_3-poor and GC_3-rich copies). (From Jabbari et al., 2003b).

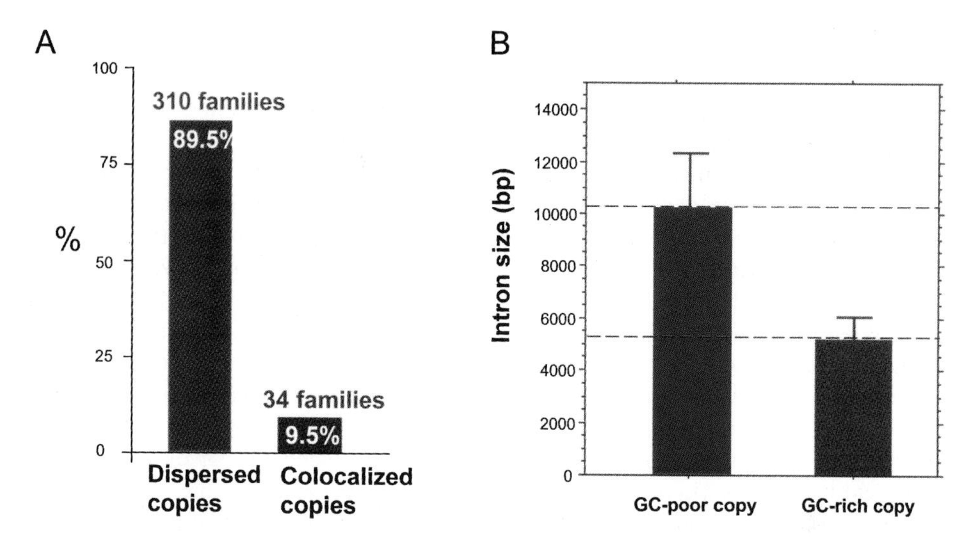

Figure 6.20. **A.** Chromosomal distribution of human gene families. Most families are localized on different chromosomes (dispersed copies), a minority being on the same chromosome. **B.** Average intron size for the GC poor and GC-rich copies of human duplicated genes. All pairs (111) for which genomic sequences were available were analyzed. (From Rayko et al., 2004).

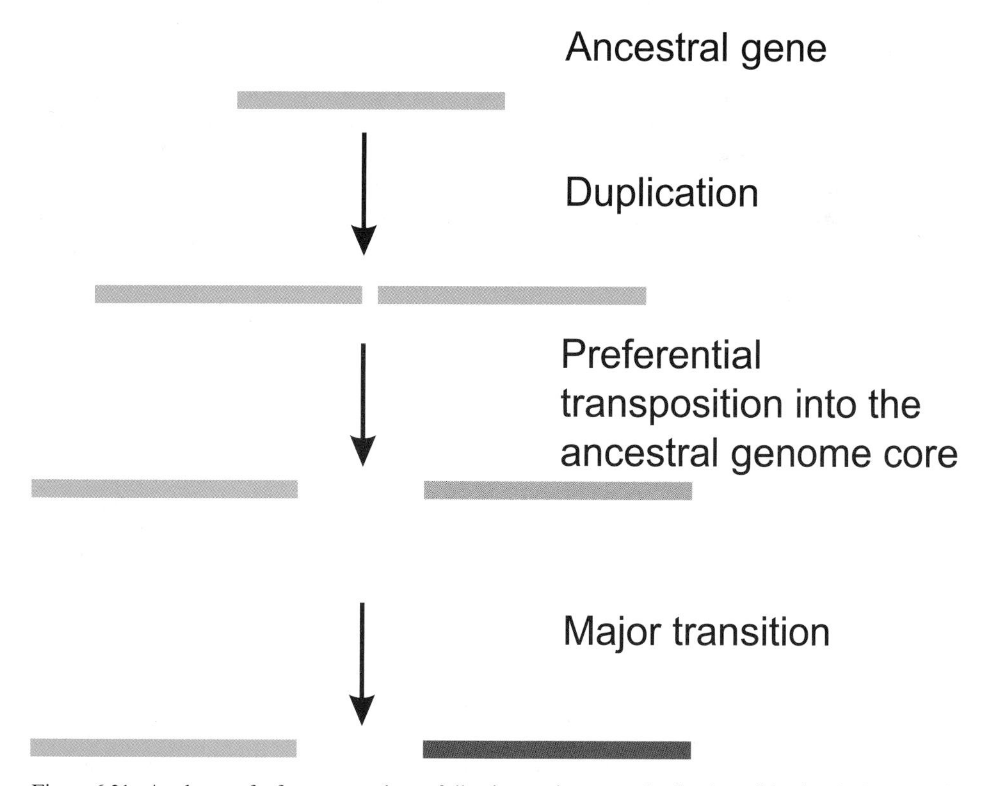

Figure 6.21. A scheme of a frequent pathway following ancient gene duplications (blue bars). One copy is supposed to be preferentially transposed into the ancestral genome core (pink bar) which then undergoes the major compositional transition (red bars). (From Rayko et al., 2004).

three steps which occurred in duplicated genes during the evolution of the vertebrate genome (see **Fig. 6.21**).

(i) Most of the original gene duplications of interest here seem to have occurred in the fish genomes or earlier in evolution. This is indicated by the compositional divergence which is so strong as not to allow to recognise any correlation between the copies of a given gene (see Jabbari et al., 2003), whereas genes arising from recent duplications show very little compositional divergence, and a good correlation can be observed between orthologous genes from human and *Xenopus* (Cruveiller et al. 1999), or human and zebrafish (Belle et al., 2002).

(ii) While in the original duplication events, genes were tandemly arranged, as suggested by many examples of recent duplications, the preferential transposition of one copy is indicated by the fact that the majority of duplicated genes are found on different human chromosomes. It is interesting to notice that segmental duplications are enriched within pericentromeric and subtelomeric regions in the human genome (Amann et al., 1996; Trask et al., 1998, Eichler, 1999, Horvarth et al., 2000); this bias has been quantitatively tested recently in the working draft of the human reference sequence (Bailey et al., 2001), and enrichment levels were 4.7- to 11.8-fold; this bias appears to be more pronounced for interchromosomal than for intrachromosomal duplications (for a review see Samonte and Eichler, 2002), and for pericentromeric compared to subtelomeric regions.

(iii) In the majority of cases, only one copy underwent the compositional change. Indeed, if a majority of both genes from gene pairs had either undergone the compositional transition, or maintained the original composition, there would be little compositional difference between the two genes of each pair, and **Fig. 6.19B** would not show mirror images. On the other hand, we know that compositional genome changes took place in the "ancestral genome core" of the ancestors of warmblooded vertebrates. The compositional change of one copy occurred, therefore, in the copy located in the "ancestral genome core", whereas the other copy, maintaining the original low GC level remained in the "genome desert", which did not undergo any compositional change.

To sum up, three steps could be identified in the major line of evolution of duplicated genes: (i) the original tandem duplications; (ii) the preferential transposition of one copy of the duplicated genes into the "ancestral genome core"; (iii) the compositional change of this copy.

These steps may suggest (i) that duplications occurred most frequently in the "ancestral genome desert", in agreement with the preferential duplications in GC-poor pericentromeric regions (see Saccone et al., 2002); (ii) that the duplicated copy acquired, in all likelihood, a new function and/or regulation through subfunctionalization (Force et al., 1999; Avaron et al., 2003) or function partitioning (Wagner, 2002) or other mechanisms; and (iii) that transposition of the duplicated copy into the open chromatin of the "ancestral genome core" was generally preferred, as in the case of retroviral integrations (see Tsyba et al., 2004). Interestingly, these latter events led to a further gene enrichment in the "ancestral genome core". Moreover, the expected increased hydrophobicity of the proteins encoded by genes located in the genome core were accompanied by a shortening of introns and by a preferential formation of CpG islands (Rayko et al., 2004).

Part 7
The organization of chromosomes in vertebrates

CHAPTER 1

Isochores and chromosomal bands

Over thirty years ago, fluorescence staining with quinacrine was shown to produce bands on human metaphase chromosomes (Caspersson et al., 1971). This breakthrough, allowed, for the first time, the identification of individual chromosomes through their specific banding pattern and was followed by the development of other banding methods. The most widespread of these relied on treating metaphase chromosomes with dyes after proteolytic digestion or denaturation and produced G bands (Giemsa positive or Giemsa dark bands, which are equivalent to Q bands or Quinacrine bands) and R bands (Reverse bands, which are equivalent to Giemsa negative or Giemsa light bands). Since quinacrine was known to fluoresce with AT-rich DNA (Weisblum and de Haseth, 1972), from the beginning there was an indication that chromosomal bands had something to do with base composition of DNA .

Ten years later, we proposed as a working hypothesis that the G and R bands of the chromosomes from higher vertebrates might be related to isochores (Cuny et al., 1981). The reasons for such a suggestion were the following: (i) as just mentioned, there were indications (summarized by Comings, 1978) that G bands corresponded to AT-rich, late-replicating DNA, and R bands to GC-rich, early replicating DNA; likewise, light (GC-poor) and heavy (GC-rich) isochores were interspersed on chromosomes forming a compositional mosaic; (ii) G banding was very evident in warm-blooded, but not in cold-blooded vertebrates (Bailly et al., 1973; Stock and Mengden, 1975; Schmid, 1978), a feature paralleled by the high and low compositional heterogeneity of the genomes of warm- and cold-blooded vertebrates, respectively; (iii) G banding patterns appeared to be conserved in birds (Stock and Mengden, 1975) and in different orders of mammals (Dutrillaux et al., 1975; Wurster-Hill and Gray, 1979), as are the relative amounts and GC levels of major DNA components in those species; (iv) the amount of DNA per chromosome band, at the highest resolution achieved by Yunis (1976), was compatible with isochore size; (v) gene amplification leads to the formation of homogeneous staining regions in chromosomes (Schimke, 1982), the result expected if the amplified genome segments were smaller in size than isochores, which was the case.

The general suggestion of Cuny et al. (1981) seemed to be confermed by a number of later results (Bernardi, 1989): (i) additional evidence became available for the poor banding (Abe et al., 1988; Medrano et al., 1988; Schmid and Guttenbach, 1988; Yonenaga-Yassuda et al., 1988) and for the limited compositional heterogeneity of the genomes from cold-blooded vertebrates (Bernardi and Bernardi, 1990a), as well as, for the correlation between compositional heterogeneity of DNA and intensity of chromosomal bands in fishes (Medrano et al., 1988; see **Part 4, Section 1.3**); (ii) G bands replicate late, R bands replicate early (Comings, 1978), as do genes previously investigated in replication timing (Furst et al., 1981; Goldman et al., 1984) and localized in GC-poor and GC-rich isochores, respectively (Bernardi et al., 1985b); (iii) genes are preferentially located in R bands

(Korenberg and Engels, 1978; Goldman et al., 1984), as well as in GC-rich isochores (see **Part 5**); (iv) genes located in G and R bands are GC-poor and GC-rich, respectively (Aota and Ikemura, 1986; Ikemura and Aota, 1988), as are genes located in GC-poor and GC-rich isochores (Bernardi et al., 1985b); (v) G bands can be produced by Hae III degradation of metaphase chromosomes (Lima de Faria et al., 1980), as expected from the fact that the GGCC sites split by Hae III are much more frequent in GC-rich compared to GC-poor isochores; in contrast, R bands can be produced by pancreatic DNase degradation of chromosomes protected by GC-specific binding of chromomycin A3 (Schweizer, 1977).

I stressed, however, the fact that the general correlation between isochores and chromosomal bands could only be considered an approximation of the actual situation (Bernardi, 1989), because GC-rich (H1, H2, H3) and GC-poor (L1, L2) DNA components in the human genome are in a 1:2 ratio, whereas R and G bands are in an approximate 1:1 ratio. This discrepancy may also be due, in part, to the fact that DNA density is lower in R bands compared to G bands, an explanation in agreement with the greater DNase accessibility of some chromosomal regions (Garel and Axel, 1976; Weintraub and Groudine, 1976; Kerem et al., 1984) where gene density is higher. An alternative explanation, not exclusive of the former, is that "standard" R bands (*i.e.*, R bands at the resolution of 400 bands), contain more GC-poor isochores (corresponding to the "thin" G bands only revealed by high resolution banding; Yunis, 1976; 1981; Viegas-Pequignot and Dutrillaux, 1978; Sawyer and Hozier, 1986) than "standard" G bands contain GC-rich isochores (corresponding to "thin" R bands). Both explanations appear to be supported by data from Ikemura and Aota (1988), by the compositional mapping of chromosome 21, and by chromosomal compositional mapping (see below).

G and R bands not only differ in their overall isochore make-up, but also in their internal isochore structure as indicated by the following points. G bands are remarkably homogeneous in DNA composition, because they are made up of GC-poor isochores that differ very little from each other in composition. In contrast, R bands are heterogeneous, since the corresponding GC-rich isochores encompass a wide GC-range. This leads to both inter- and intra-band heterogeneity, as shown by the compositional mapping of chromosome 21 (see below). In fact, the interspersion of different GC-rich isochores within individual R bands was already indicated by the finding that genes located in R bands from many chromosomes (Aota and Ikemura, 1986; Ikemura and Aota, 1988) are present in H3 isochores. Since the latter only represent 4–5% of the genome, they are far from accounting for the totality of DNA of R bands (which represent almost 50% of the DNA), and other DNA components must be present. A corollary of this conclusion was that, since gene concentration is highest in the isochores of the H3 family (Bernardi et al., 1985b; Mouchiroud et al., 1991; Zoubak et al., 1996), regions of high and low gene concentration should exist not only in R and G bands, respectively, but also within R bands.

CHAPTER 2

Compositional mapping

2.1. Compositional mapping based on physical maps

The problems discussed in the previous chapter obviously needed an experimental approach to be solved. This approach was **compositional mapping** (Bernardi, 1989), which can be described as follow. Wherever long-range physical maps are available, compositional maps can be constructed by assessing GC levels around landmarks (localized on the physical maps) that can be probed. This simply requires the hybridization of the probes on DNA fractionated according to base composition. If DNA preparations of about 100 kb in size are used, compositional mapping will provide the equivalent of a high-resolution banding, without the uncertanties of cytogenetics. Basically, this approach involved three steps: (i) DNA fractionation (see **Fig. 7.1A**); (ii) hybridization of single-copy sequence probes on the fractions (see **Fig. 7.1B**); and (iii) the construction of the map (see **Fig. 7.1C**).

This experimental approach was originally tried for a set of 50 single-copy sequence probes localized on the long arm of chromosome 21 and provided a direct demonstration for the compositional homogeneity of G bands and for the compositional heterogeneity of R bands, the highest GC levels (corresponding to the H3 DNA component) being in the telomere-proximal regions (Gardiner et al., 1990; see **Fig. 7.1C**). Incidentally, the latter observation agrees with cytogenetical evidence that telomeres almost always correspond to R bands and that the terminal regions of 20 of them (including that of the long arm of chromosome 21) are the regions (the T bands) of human chromosomes that are most resistant to heat denaturation (Dutrillaux, 1973). This suggested that H3 isochores may correspond to these denaturation-resistant telomeric regions and to some similar intercalary regions located on several chromosomes (for example, 11, 19 and 22).

The approach used for the compositional mapping of the long arm of chromosome 21 was extended to (i) the mapping of telomeric probes corresponding to T bands and to R bands, respectively, on compositional DNA fractions (De Sario et al., 1991); in this case, it could be shown that the T band probes hybridized on H3 isochores, the R band probes on H2 and H1 isochores and the telomeric repeat common to all chromosomes on H1, H2 and H3 isochores, as expected; moreover, genes localized in G, R and T bands showed increasing GC levels; (ii) the mapping of the distrophin gene (Bettecken et al., 1992); in this very large (3 Mb) gene only very small (1–2% GC) compositional differences were found to occur; and (iii) the mapping of the Xq26-qter region (Pilia et al., 1993); in this case, the difference between G and R bands first seen on chromosome 21 was also found; moreover, the telomeric regions of the Xq 28band showed very high GC levels compared not only to G bands Xq27.1 and Xq27.3 and to R bands Xq26 and Xq27.1, but also to the centromeric part of Xq28.

This result prompted a more detailed study of band Xq28 (De Sario et al., 1996), which was done by determining (De Sario et al., 1995) the buoyant densities (and GC levels) of

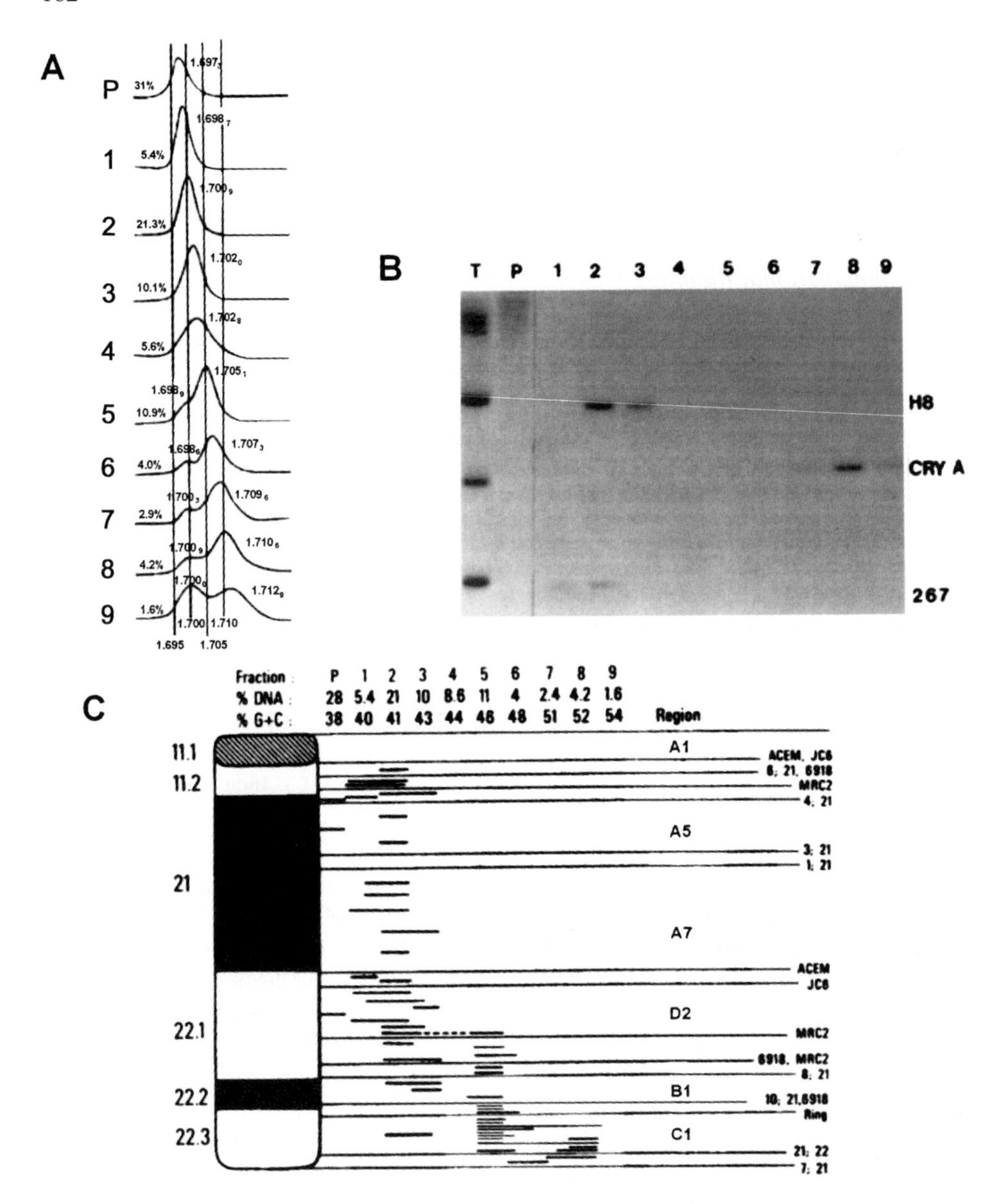

Figure 7.1. **A** Analytical CsCl profiles of human DNA fractions (P stands for Pellet), as obtained after preparative ultracentrifugation in Cs_2SO_4/BAMD at a ligand/nucleotide molar ratio $R_f = 0.14$. Relative amounts and modal buoyant densities are indicated. **B.** Hybridization of probes 267, H8 and CRY A on the DNA fractions from the left panel. **C.** Compositional map of the long arm of human chromosome 21 obtained when hybridisation data are superimposed on the physical map (Gardiner et al., 1988, 1990). Long horizontal lines indicate positions of the breakpoints associated with the rearranged chromosomes listed at the right of the figure. Some regions are not labelled for lack of space. (From Gardiner et al., 1990).

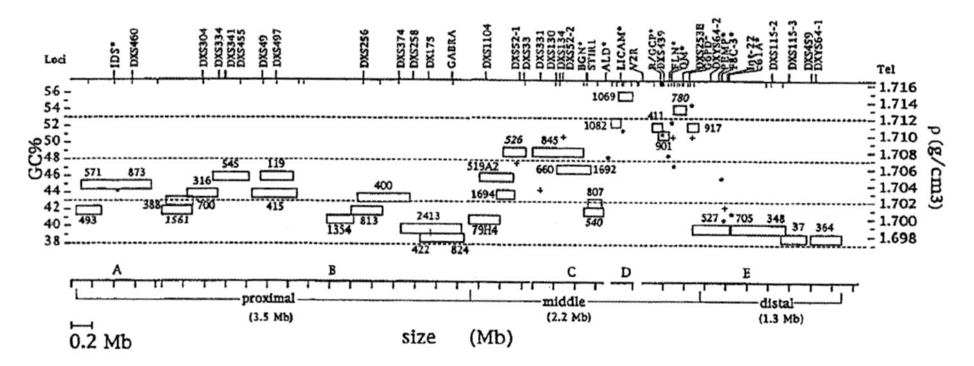

Figure 7.2. Compositional map of human chromosome band Xq28. Yeast artificial chromosomes (YACs), represented as boxes with a height corresponding to 1% GC, are positioned along the physical map on the horizontal scale, and according to their buoyant density and GC level (right and left ordinates respectively) on the vertical scale. Asterisks represent GC levels of isochores calculated from GC_3 values of mapped genes (also indicated by asterisks on gene names on the top part of the figure with the exception, for lack of space, of the EDMD gene). Crosses represent GC levels of isochores as evaluated by hybridization of single copy probes on compositional DNA fraction. The top part of the figure shows a number of mapped genes and loci, as well as the location of CpG islands (thin vertical lines); the region between R/GCP and DXS439 comprises seven CpG islands (unresolved in the Figure). The bottom part of the figure shows YAC contigs A-E, as well as the proximal, middle, and distal regions described in the text. (From De Sario et al., 1996).

YACs (yeast artificial chromosomes) covering this band (see **Fig. 1.9**); this allowed the first construction of an essentially complete compositional map of a chromosomal band, corresponding in this case to 8 Mb of DNA (**Fig. 7.2**). Three regions were observed: (i) a proximal 3.5 Mb region formed by GC-poor L and GC-rich H1 isochores; (ii) a middle 2.2 Mb region essentially formed by a GC-rich H2 isochore and a very GC-rich H3 isochore separated by a GC-poor L isochore; interestingly, the size of the H3 isochore is only 0.2 Mb. Gene and CpG island concentrations increased with the GC levels of the isochores, as expected. Xq28 exemplified a subset of R bands, the H3* bands which are different from the two other subsets, the $H3^{+}$ bands (which are characterized by the predominance of H2 and H3 isochores and by their resistance to heat-denaturation) and the $H3^{-}$ bands (which do not contain H2 and H3 isochores and correspond to the majority of R bands).

A region of human chromosome 21, the 13–14 Mb cen-q21 region, was mapped using the same approach. In this case a compositional map of the centromere and of the subcentromeric region of the long arm (De Sario et al., 1997) was established by determining the GC levels of 11 YACs covering this region which extends from the α-satellite sequences of the C(entromeric) band q11.1, through R band q11.2, to the proximal part of G band q21 (**Fig. 7.3**). The entire region is made up of GC-poor, or L, isochores with only one GC-rich H1 isochore, at least 2 Mb in size, located in band q21. The almost identical GC levels of the centromeric α-satellite repeats (38.5%), of R band q11.2 (39%), and of the proximal part of G band q21 (38–40%) provided a direct demonstration that base composition cannot be the only cause of the cytogenetic differences between C bands, G bands and the majority of R bands, namely the $H3^{-}$ R bands (which do not contain the GC-richest H3 isochores). The results obtained also showed that isochores may be as long as 6 Mb, at least in the GC-poor regions of the genome, and supported previous observations made on

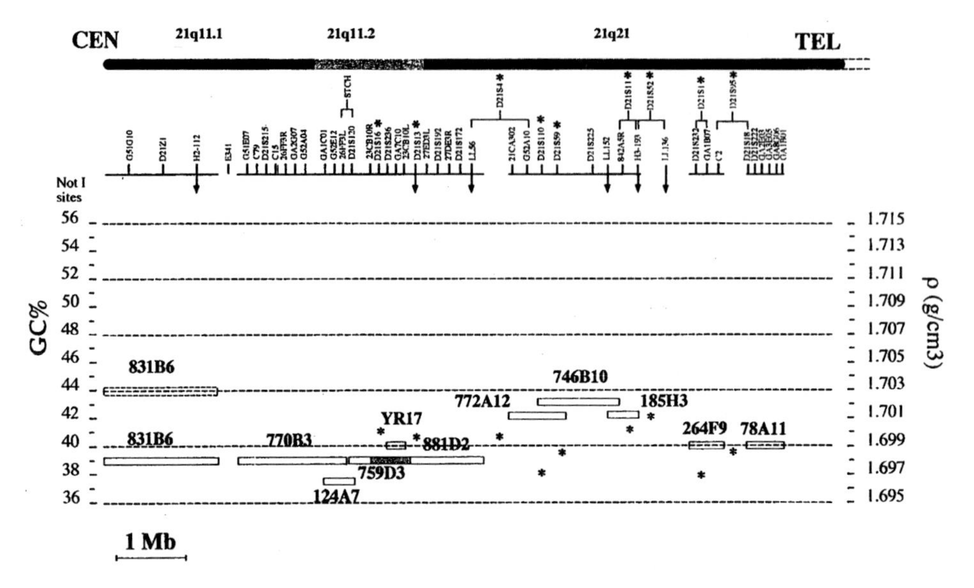

Figure 7.3. Compositional map of the human chromosomal region 21cen-q21. Yeast Artificial Chromosomes (YACs) are represented as boxes (whose height provides the error bar) and are positioned on the physical map according to the STS order (Chumakov et al., 1992) and on the vertical axis according to the measured buoyant densities and the calculated GC levels (right- and left-hand ordinates, respectively). The physical distances were calculated according to the integrated map of Lawrence et al. (1993). Physical distances between adjacent contigs and STSs are not at scale. YAC length are proportional to the measured size. Gaps are represented by breaks in the horizontal line at the top of the figure. The ideogram of chromosome bands in the analyzed region is shown on the top of the figure. YAC 831B6 showed a buoyant density corresponding to 44% GC (broken line box), but its actual GC level is close to 39% GC (solid line box, see text). The overlap region of YACs 759D3 and 881D2 is shaded. Asterisks represent GC levels of isochores as calculated by hybridization of single-copy probes on compositional DNA fractions (Gardiner et al., 1990). The corresponding STSs are also marked with an asterisks. The broken line in YAC 770B3 indicate the absence of STS C79. Arrows represent NotI sites as reported in the literature (Ichikawa et al., 1992; Hattori et al., 1993; Saito et al., 1991). (From De Sario et al., 1997).

Xq28 (De Sario et al., 1996) suggesting that YACs from isochore borders are unstable and/ or difficult to clone. Genes and CpG islands are very rare in the GC-poor region investigated, as expected from the fact that their concentration is proportional to the GC levels of the isochores in which they are contained.

2.2. Chromosomal compositional mapping at a 400-band resolution

Compositional mapping can also be done at the chromosomal level by *in situ* hybridization of compositional DNA fractions corresponding to different isochore families. In the first investigation of this kind (Saccone et al., 1992), it was shown that the hybridization of a DNA fraction derived from H3 isochores produced the highest concentration of signals on two largely coincident subsets of R bands (**Figs. 7.4, 7.5**): (i) the T bands (Dutrillaux, 1973), which are the most heat-denaturation-resistant R bands; and (ii) the chromomycin

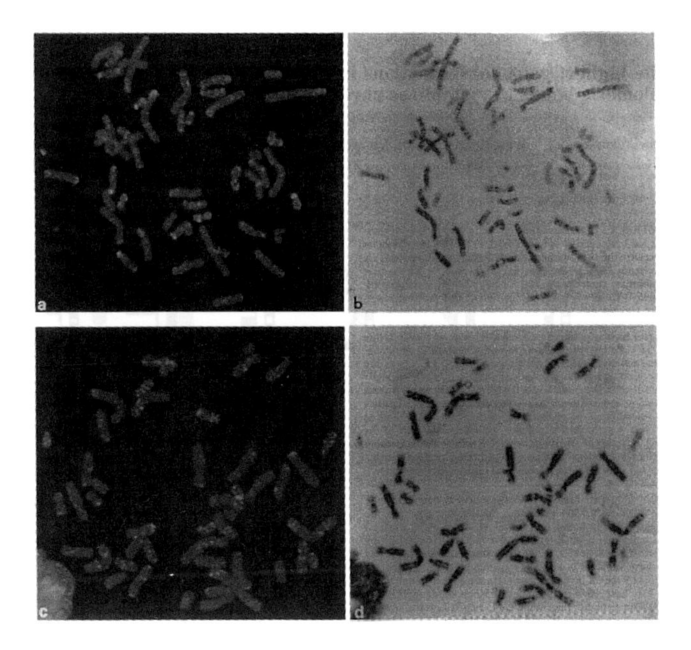

Figure 7.4. The human DNA fraction corresponding to isochore family H3 was hybridized *in situ* to human male metaphase chromosome spreads. **a** and **c** Detection of hybridized biotinylated H3 fraction DNA by fluorescent isothiocyanate-conjugated avidin; chromosomes are counterstained with propidium iodide. **b** and **d** G banding of metaphases shown in **a** and **c**, respectively. In **a**, hybridization was carried out with the H3 DNA probe at a final concentration of 20 ng/μl in the presence of a 1000-fold excess of sonicated unlabeled total human DNA to suppress hybridization of repeated sequences present in the probe. In **c**, a 50-fold excess of a pBR322 plasmid containing the Blur 8 sequence was added to the 1000-fold excess of competitor DNA; the results were the same in both cases. Chromosomes were subsequently G-banded with Wright's stain. (From Saccone et al., 1992).

A3-positive, DAPI-negative bands (Ambros and Sumner, 1987), which are the GC-richest bands of human chromosomes (DAPI is 4,6 diamino-2-phenylindole). *In situ* hybridization of H3 isochore DNA established that T bands comprise GC-rich, gene-rich, single-copy DNA. Indeed the contribution of repetitive DNAs was suppressed by competition with excess unlabeled total human DNA and Alu sequences. The latter are the most abundant type of short interspersed elements, which have their highest density in the GC-richest isochores H2 and H3 (see **Part 6**). This also ruled out the possibility that T bands corresponded to GC-rich satellite DNAs, which was a distinct possibility until then.

In subsequent investigations (Saccone et al., 1993), the chromosomal locations of the other isochore families L1+L2, H1 and H2 were studied by using an improved competition protocol with excess C_0 t 1 DNA, which eliminated the need for a statistical approach. This established (i) that T bands contain not only H3 isochores, but also H2 and some H1 isochores; (ii) that R' bands (*i.e.*, R bands exclusive of T or $H3^+$ bands) are formed, on the average and to almost equal extents, by H1 and L isochores, H2 and H3 isochores being only rarely present; and (iii) that G bands essentially consist of L isochores, H1 isochores being present at low levels.

In further work (Saccone et al., 1996), it was shown that (i) the number of bands

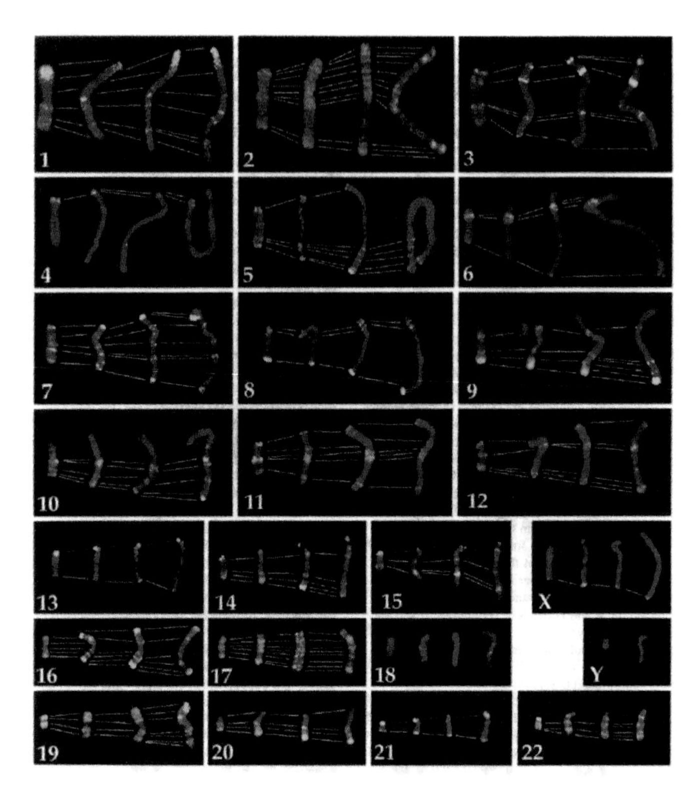

Figure 7.7. Human chromosomes hybridized with the biotin-labeled DNA from the H3 isochore family, at different levels of resolution. The hybridized regions were visualized by fluorescein (yellow signals) and chromosomes were red-stained with propidium iodide. Each panel presents chromosomes with a band resolution ranging from about 300 to about 850. (From Saccone et al., 1999).

Indeed, 23 out of the 28 R_{400} $H3^+$ bands only yielded R_{850} bands containing H3 isochores, whereas only some of the R_{850} bands originating from the other five R_{400} $H3^+$ bands showed H3 hybridization signals. For example (see **Fig. 7.9**), the R_{400} $H3^+$ band 11q13 is resolved into three R_{850} $H3^+$ bands (q13.1, q13.3, and q13.5), and two G_{850} bands (q13.2, and q13.4), whereas the $H3^+$ band 11p15 was one of the five exceptions, with only one (11p15.5) of the three derived R_{850} bands showing hybridization signals. Globally, the coverage of the R_{850} bands derived from the 28 R_{400} $H3^+$ bands was 91% (this calculation was based on band sizes as assessed by Francke, 1994). In no case G_{850} bands derived from the R_{400} $H3^+$ bands showed hybridization signals.

In contrast, only some of the R_{850} bands derived from 23 out of 31 R_{400} H3* bands showed hybridization signals. For example, the H3* band 11q23 (**Fig. 7.9**) yielded only one R_{850} $H3^+$ band (11q23.3; in fact, only the distal part of it was $H3^+$), whereas the other R_{850} band (11q23.1) was $H3^-$. The remaining eight H3* bands showed the features observed in the vast majority of $H3^+$ bands, in that all the derived R_{850} bands were $H3^+$. Globally, 66% (in terms of band sizes; see above) of the R_{850} bands derived from the R_{400} H3* bands contained H3 isochores.

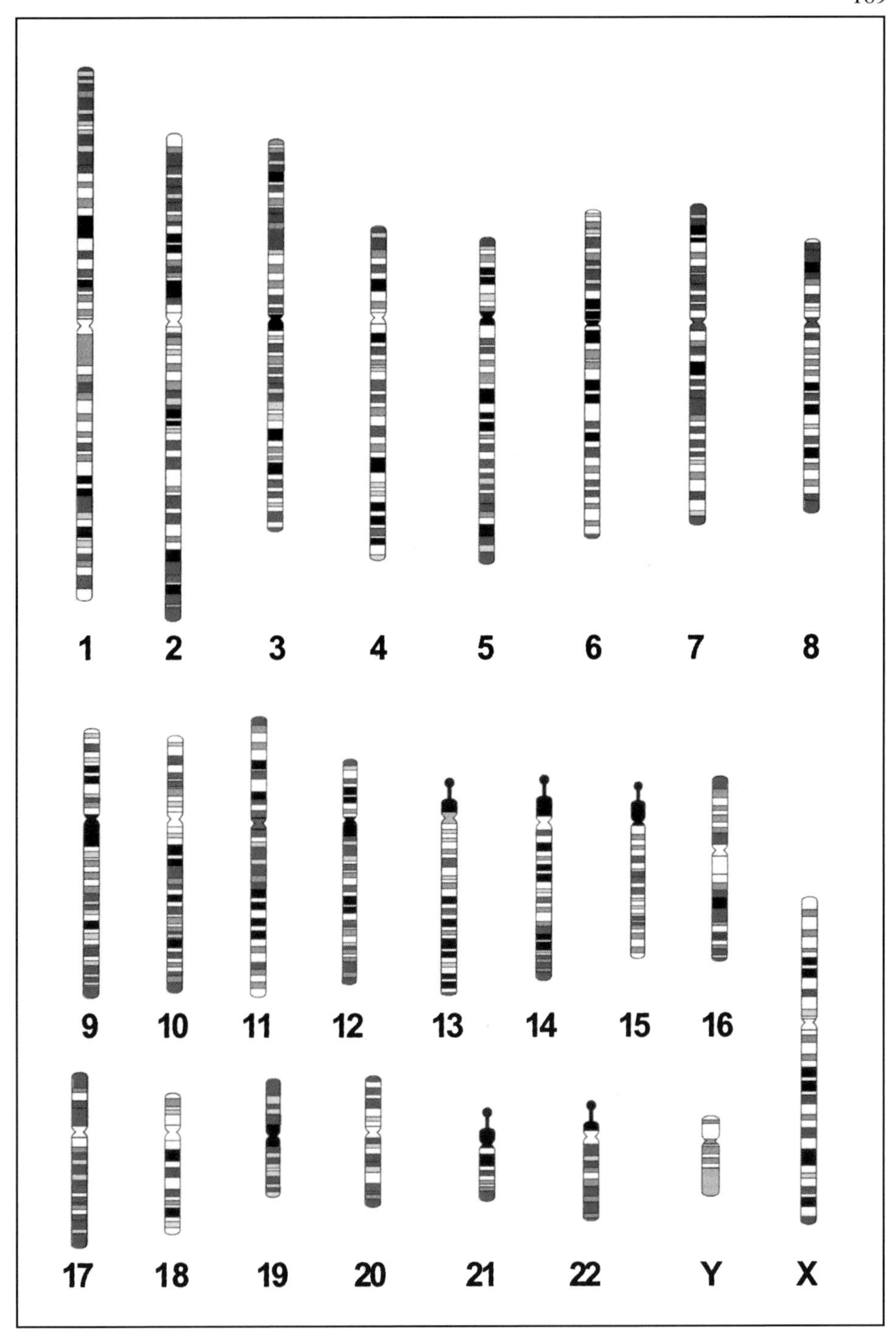

Figure 7.8. Ideogram of human chromosomes at a 850-band resolution (Francke, 1994) showing the $H3^+$ bands as red bands. (From Saccone et al., 1999).

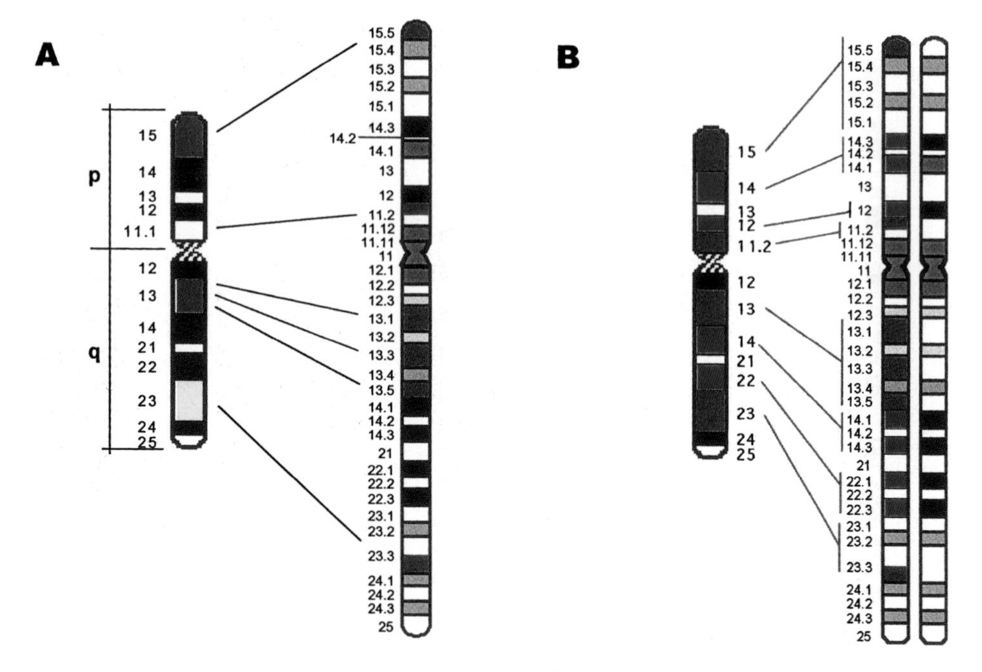

Figure 7.9. **A.** Ideograms of human chromosome 11 at 400 (left) and 850 (right) band resolution showing the chromosomal regions containing H3 isochores. Red, yellow and white bands on the left chromosome indicate the $H3^+$, H3* and $H3^-$ bands. Red bands on the right chromosome indicate the regions that hybridize H3 isochores. (From Bernardi et al., 2000). **B.** Red and blue bands correspond to the GC-richest and GC-poorest bands. The rightmost ideogram is from Francke (1994), with different shades of grey for G bands. (From Bernardi and Saccone, 2002).

As far as the R_{400} $H3^-$ bands are concerned, the higher resolution allowed the identification of 20 bands containing H3 isochores that had not been detected at the lower resolution (Saccone et al., 1996). The majority of these bands were located close to other H3+ or H3* bands (see bands 5q33. 1, 6p21.1, and 12q24.13) and were very thin (see bands 1p13.3, and 7p13). Only some of the R_{850} bands derived from these 20 R_{400} $H3^-$ bands (see **Fig. 7.8**) exhibited hybridization signals (see band 11p11.2 in **Fig. 7.9**). Moreover, in a number of cases, the signal was thinner than the corresponding R_{850} bands (see **Fig. 7.7**), indicating that only part of the R_{850} band contained H3 isochores. Globally, only 9% (in terms of size; see above) of the R_{850} bands derived from the $H3^-$ bands contained H3 isochores. If only the R_{850} bands derived from the 20 $H3^-$ bands were taken into consideration, the coverage was 47%. Incidentally, previous work (Saccone et al., 1996) had shown that H2 and H3 isochores co-localize on metaphase chromosomes, with only four exceptions (the telomeric bands 3q29, 6q27, 13q34, and 20p13) which were $H2^+$ and $H3^-$. Now, these bands were shown to be $H3^+$, indicating also in these cases a co-localization of H3 and H2 isochores.

On the basis of the above results and of the estimated band sizes (Francke, 1994), it can be calculated that about 17% of all bands at a 850-band resolution contain H3 isochores and that about 9%, 6%, and 2% of them derived from R_{400} $H3^+$, H3* and $H3^-$ bands, respectively.

Finally, G bands did not reveal the presence of H3 isochores, the only exceptions being two G_{400} bands, 1p36.2 and 19q13.4, which yielded two R_{850} $H3^+$ bands, 1p36.22, and 19q13.42, respectively (**Fig. 7.8**).

In summary, the results just presented lead to several conclusions. (i) Since the colocalization of H2 and H3 isochores (which represent together 12% of the human genome) in R_{850} $H3^+$ bands appears to be the rule, the fraction of these isochores in those bands (which represent 17% of the total genome) correspond to the majority, 70%, of the DNA contained in them; this value should, however, be higher since DNA is less compact in these bands (see **Chapter 3**). (ii) In some cases, however, the coverage of R_{850} $H3^+$ bands by hybridization signals is overestimated. For example, signals suggest that almost 50% of band Xq28 is $H3^+$, whereas compositional mapping showed that only about 25% is formed by H2/H3 isochores (De Sario et al., 1996). (iii) In a number of R_{850} $H3^+$ bands (see **Fig. 7.8**), H3 hybridization coverage was limited to a fraction of the band (see for example the case of band Xq28); this indicates that the present results provide information concerning a resolution higher than 850 bands; thus, they may correspond, in many cases, to the practical highest resolution that can be reproducibly attained, namely 1,250 bands (Drouin et al., 1994; Scherthan et al., 1994). (iv) 83% of the bands shown in **Fig. 7.8**, namely the R_{850} $H3^-$ and the G_{850} bands, exhibit low or very low gene concentrations; since genome size is remarkably constant in mammals and since such regions are conserved in syntenic regions of chromosomes from mammalian orders that diverged about 100 million years ago (Saccone et al., 1997; Scherthan et al., 1994; Raudsepp et al., 1996; Morescalchi et al., 1997; Chowdhary et al., 1998), the gene-poor majority of the genome should have some functional role (see **Part 12**). (v) Finally, the gene-richest regions correspond to gaps in the first physical map of the human genome (Chumakov et al., 1995), which therefore missed the most interesting regions of the human karyotype (as already stressed by Saccone et al., 1996). The difficulty experienced in cloning these regions into YACs and/or in avoiding high levels of chimerism and deletion is most probably related to their high recombination level, a property which apparently is conserved when these regions are cloned in yeast.

The analysis of the localization of **GC-poor DNA** on metaphase and prometaphase chromosomes will be illustrated in some detail in order to provide an example of the procedure followed. **Fig. 7.10** displays the CsCl profiles, the relative DNA level, and the proportion of the isochore families in the DNA fractions (Saccone et al., 1996) used as probes for the *in situ* hybridizations of GC-poor DNA. L1 isochores are only present in the pellet DNA; L2 isochores are distributed in the pellet and in fractions 1–3; fractions 4 and 5 contain almost exclusively DNA from H1 isochores. In the case of fraction 5 (as well as in the following fractions shown in Fig. 1 of Saccone et al., 1996), a light satellite peak corresponding to DNA from centromeric heterochromatin is present. The pellet DNA, essentially formed by L1 isochores, hybridized on a subset of G bands (**Fig. 7.11**), which were called L1 bands, and corresponded to the G1 and G2 bands, as we will name the two most intensely staining subsets of Francke's R bands (1994). In contrast, the pellet DNA is almost absent in the large majority of the $H3^+_{400}$ and $H3^*_{400}$ bands. This is especially evident in chromosomes 15, 17, 19, and 22 (see **Fig. 7.11**).

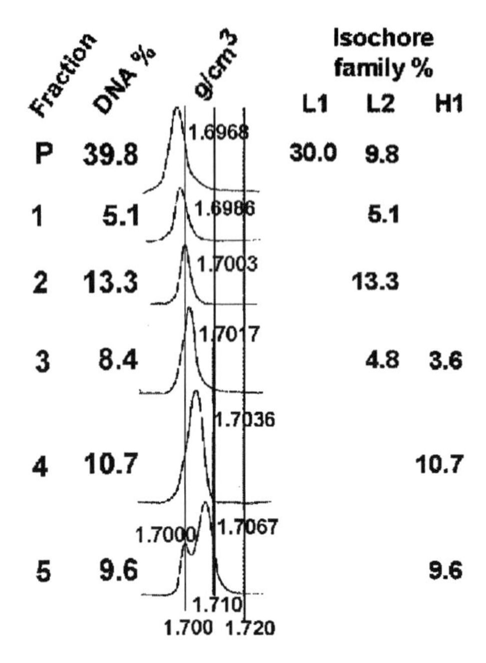

Figure 7.10. CsCl profiles of the human DNA fractions containing the L1, L2 and H1 isochore families (modified from Saccone et al., 1996): fraction numbers (P stands for pellet), relative DNA content, modal buoyant density and relative amounts of isochore families are indicated. (From Federico et al., 2000).

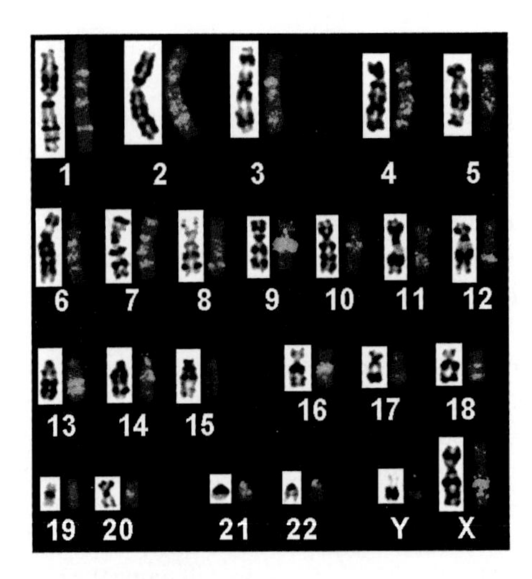

Figure 7.11. L1 isochore banding pattern. The left and right member of each chromosome pair shows the GTG (G bands by trypsin using Giemsa) bands and the L1 isochore hybridization, respectively. Biotinylated L1 isochores present in the pellet DNA shown in Figure 7.10 were detected with avidin-FITC. Chromosomes were stained with propidium iodide. (From Federico et al., 2000).

Interestingly, the pellet DNA represents about 40% of the total genome, most of which belongs to the L1 isochores, whereas the bands on which it is localized correspond to about 26% of all bands (as assessed from the ideogram of Francke, 1994). This indicates that the level of DNA compaction in L1 bands is higher than the average (see below). The high resolution hybridization obtained with pellet DNA gave rise to signals characterized by

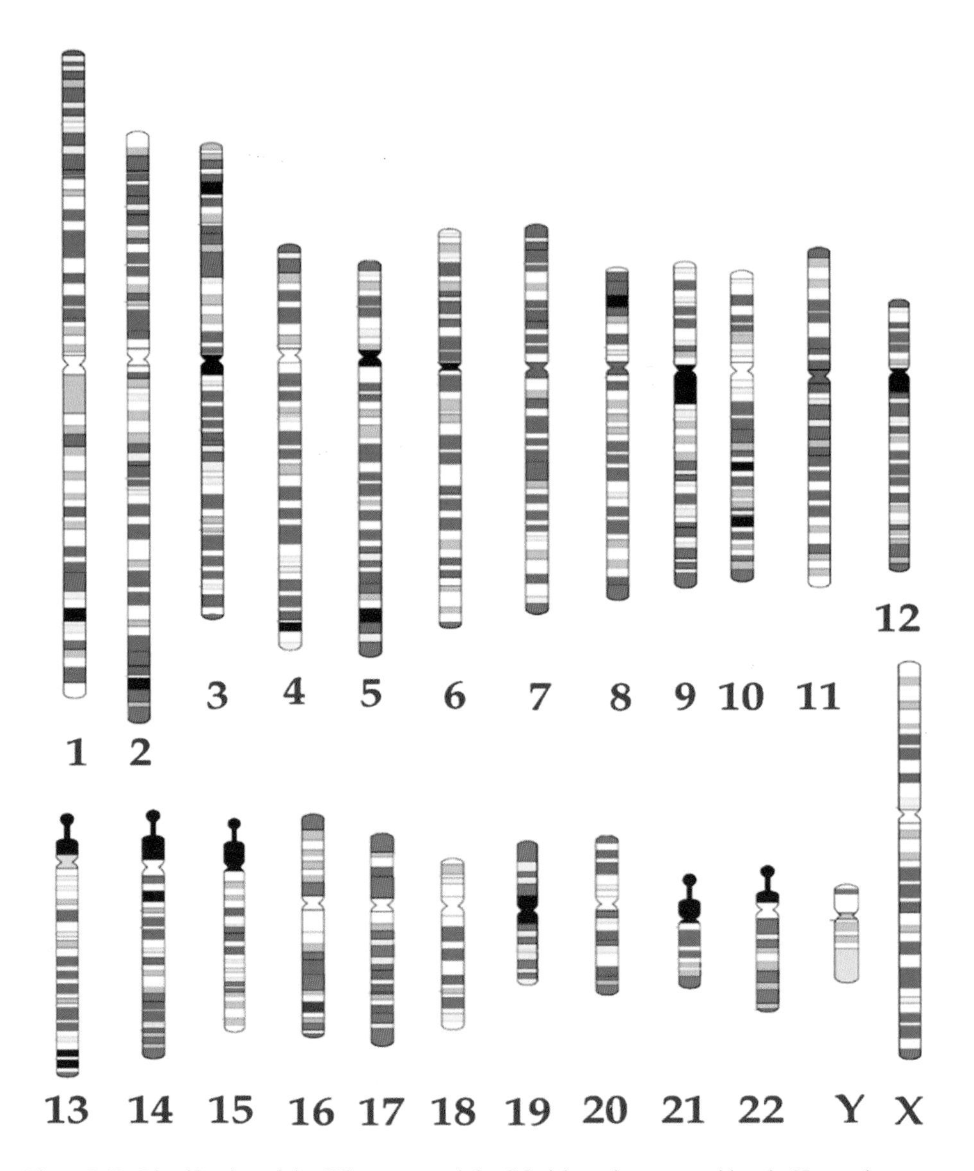

Figure 7.12. Identification of the GC-poorest and the GC-richest chromosomal bands. Human karyotype at a resolution of 850 bands showing the chromosomal bands containing the GC-poorest ($L1^+$, blue bars) and the GC-richest isochores ($H3^+$, red bars). The $H3^-$ R bands are in white. The grey scale of the G bands is according to Francke (1994). (Modified from Federico et al., 2000).

different features. The majority of L1 bands was present in doublets separated by thin non-hybridizing internal bands (see **Fig. 7.12**). The G1 and G2 bands contain almost the totality of the GC-poorest isochores present in the pellet. More precisely, about 85% of the G1 bands and about 40% of the G2 bands showed strong signals, the remaining G2 and G1 bands showing very weak or no signals (see **Fig. 7.12**). Only few (about 4%) G3 bands and no G4 bands seem to contain L1 isochores. These are clear indications that G1+G2 and G3+G4 bands form two compositionally different subsets of G_{850} bands. Finally, some G1 bands do not contain pellet DNA. These bands are 1q41, 2q36.3, 3p24.3, 4q34.3, 5q34, 8p22, 10q23.1, 10q25.1, 13q33.1, 13q33.3, and 14q12. It is possible that L1 isochores are also present in these G1 bands, but in very small amounts. These bands appear to be a small, distinct subset of G1 bands that do not replicate at the end of the S phase of the cell cycle (see **Chapter 6**).

Table 7.1 summarizes the classification of chromosomal bands presented in **Sections 2.2** and **2.3**.

TABLE 7.1

Summarizes the classification of chromosomal bands presented in **Sections 2.2** and **2.3**.

		Bands	Isochores	Reference
At 400-band resolution	R	T bands	H3, H2, H1	Saccone et al., 1993
		R' bands	H1, L	
		G bands	L, H1	
	R	$H3^+$ or T bands	H3, H2	Saccone et al., 1996
		H3* or T' bands	H3, H2	
		H3	H1, L	
		G bands		
At 850-band resolution	R	H3+ bands	H3, H2	Saccone et al., 1999
		H3 bands	H1, L	Federico et al., 2000
	G	L1 bands	H1, L	
		$L1^+$ bands	L1	

Genes, isochores and bands in human chromosomes 21 and 22

Chromosomes 21 and 22 are very different from each other. Together, they exhibit the whole spectrum of chromosomal bands. While chromosome 21 is made of several compositional regions representing all isochore families, chromosome 22 is essentially formed by H2 and H3 isochores, comprises H1 isochores, but contains very few L2 isochores and no

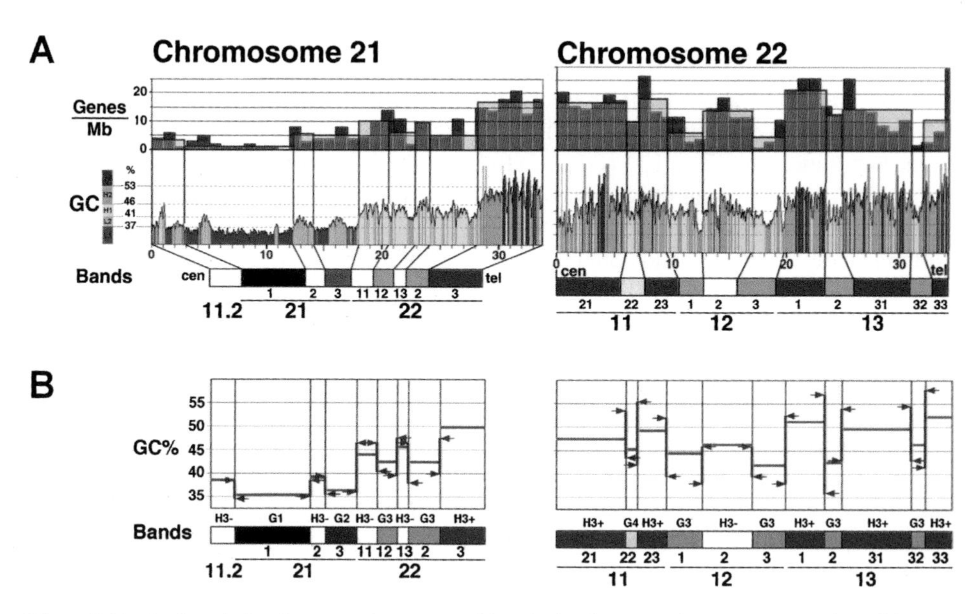

Figure 7.13. **A.** Correlation between chromosomal bands, isochores, and gene concentration of human chromosomes 21 and 22. Bottom to top: 1) **Bands**, ideogram at a resolution of 850 bands showing the four classes of G bands characterized by different staining intensity (Francke, 1994; called here G1 to G4 from black to pale grey; see Fig. 7.14); and the two classes of R bands (Saccone et al., 1999; $H3^+$, red; $H3^-$, white); the two chromosomes are represented according to their cytogenetic size in the karyotype of Francke (1994). 2) **GC**, the GC profiles obtained using a window size of 100 kb; 37%, 41%, 46%, and 53% GC were taken as the upper values for the L1, L2, H1 and H2 isochore families, respectively (Zoubak et al., 1996); the grey boxes indicate DNA sequences not yet available. 3) **Genes/Mb**, gene density calculated as number of genes per Mb; the blue bar plot concerns chromosomal bands, the red plot 1-Mb segments. **B.** Compositional features of chromosomal bands. Bottom to top: 1) **Bands**, band ideograms as in Fig. 7.13. G1-G4, $H3^+$, $H3^-$ bands are indicated. 2) **GC%** average GC level of each chromosomal band (horizontal blue lines), and GC levels observed at band borders (red and blue arrows indicate the GC level on the R and G band side, respectively; vertical red lines indicate the GC difference over 300 kb regions around band borders). Note that all G bands showed lower GC levels than the adjacent R bands, and all R bands showed higher GC levels than the adjacent G bands. These differences were enhanced at band border regions (see above). (From Saccone et al., 2001)

L1 isochores (**Fig. 7.13**). **Compositional fluctuations** are remarkably high in the GC-richest regions compared to the GC-poorest regions. It is possible that the increase in compositional fluctuations may be due to the simultaneous requirement for AT-rich regions to link DNA to the chromosome scaffold and for the very high GC levels of coding sequences and associated CpG islands in H2 and H3 isochores.

Each band type is characterized by a defined isochore content: $H3^+$ and $L1^+$ bands are almost only composed of H2/H3 isochores and of L1/L2 isochores, respectively. The compositionally intermediate $L1^-$ and $H3^-$ bands are more heterogeneous, containing different proportions of L2, H1 and H2 isochores, with a relatively large amount of H1 isochores.

Matching of compositional regions with chromosomal bands (at a 850-band resolution; see **Figs. 7.13A** and **B**) was done using three criteria: the average GC level, the discontinuities in GC level at each band border (see below), and the location of genetic markers assigned to chromosomal bands.

Our identified band borders were in good agreement with the high resolution mapping of chromosome 22 markers of Kirsch et al. (2000), only two exceptions out of 19 markers being observed on contiguous bands but very near the band border. The average GC level of the bands increased from $L1^+$ (G1+G2) to $L1^-$ (G3+G4) among the G bands, and from $H3^-$ to $H3^+$ bands among the R bands. Interestingly, while the GC levels of $H3^+$ bands (47–52% GC) are always higher than those of $L1^+$ bands (35–36% GC), in the case of "intermediate" bands, G bands can be endowed with higher GC levels than those of (non-contiguous) R bands. Indeed, $L1^-$ G bands of chromosome 22 are GC-richer than most $H3^-$ R bands of chromosome 21. This situation was also found in bands from the same chromosome (compare 21q22.12 and 21q22.2 with 21q11.2 and 21q21.2). The average GC level of a G band is, however, always lower than those of the adjacent R bands, and conversely the average GC level of an R band is always higher than those of the contiguous G bands (**Fig. 7.14**). Moreover, the GC levels at each band border (over regions of about 300 kb) are always higher on the R side compared to the G side. Whether the situation just described for chromosomes 21 and 22 is general, remains to be established. This requires, however, a preliminary work to define satisfactory correlations between chromosomal bands and DNA sequences.

Fig. 7.13 shows bar plots of gene densities for chromosomal bands, as well as for 1-Mb regions. These data show gene density ratios as high as 20 between $H3^+$ and $L1^+$ bands, and as high as 30 between some 1-Mb regions. The data were used to illustrate the correlation between gene density and GC level of chromosomal bands. This is best described by two regression lines with different slopes that intersect at 42–43% GC (**Fig. 7.15**), in agreement with previous results obtained on DNA molecules in the 100-kb size range using an independent approach (Zoubak et al., 1996; see **Fig. 5.1**).

Chromosomes 21 and 22 are quite different not only in their composition, but also in their "cytogenetic" size. While the DNA amount of the long arms of both chromosomes is comparable (the difference being less than 3%, as defined by Hattori et al., 2000, and Dunham et al., 1999), the cytogenetic size of chromosome 22 is about 40% greater than that of chromosome 21, as defined at a resolution of 850 bands (Francke, 1994), indicating that DNA is more compact in chromosome 21. While it is possible that the size difference between these two chromosomes was overestimated, similar results were obtained when

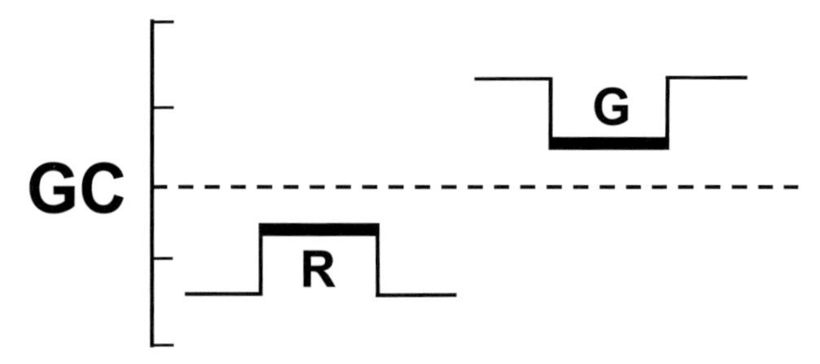

Figure 7.14. In chromosomes 21 and 22, G bands are always flanked by GC-richer bands, whereas R bands are always flanked by GC-poorer G bands. As shown in the figure, G bands may, however, be even GC-richer than other (non-contiguous) R bands.

this analysis was extended to the band level. Indeed, the L1$^+$ band 21q21.1 and the H3$^+$ band 22q13.31 show a similar cytogenetic size and a very different DNA sequence size (about 9 Mb and 6 Mb, respectively), indicating a 50% higher compaction in the L1$^+$ bands compared to H3$^+$ bands. Likewise, we could estimate about 20% more and 14% less DNA than the average in the L1$^+$ band 21q21.1 and H3$^+$ band 21q22.3, respectively. Even if the quantitative aspects of this different compaction are subject to refinement, it is most unlikely that the qualitative conclusion about differential DNA packing is incorrect. These results provide, therefore, independent additional support, that the chromatin of GC-rich

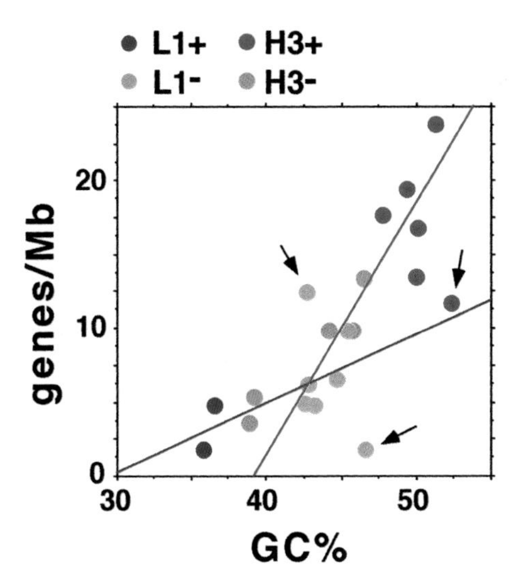

Figure 7.15. The average GC level of each band of chromosomes 21 and 22 was plotted against its gene density. The highest and lowest gene densities were found in H3$^+$ and L1$^+$ bands, respectively, as expected. The remaining G and R bands (the L1$^-$ and H3$^-$ bands) showed gene densities that are correlated with their GC level, independently of their cytogenetic band type (G or R). Three points, indicated by the arrows, represent three outliers (two L1$^-$ and one H3$^+$ bands) not taken into consideration when drawing the regression line. Inclusion of these points does not significantly change the lower slope and changes only slightly the higher slope. (From Saccone et al., 2001.

and gene-rich chromosomal regions has a more "open" structure compared to that of GC-poor and gene-poor regions at prometaphase.

Finally, **Table 7.2** presents the classification, the relative amounts and the average gene densities of bands from chromosomes 21 and 22.

TABLE 7.2
Classification, relative amount, and gene densities of bands from chromosomes 21 and 22[a].

Bands	%[b]	by staining properties[c]		by isochore content[d]		gene density[e]
Giemsa	47%	G1	13.7%	L1+	26.3%	3.0
		G2	12.6%			
		G3	13.1%	$L1^-$	20.7%	6.8
		G4	7.6%			
Reverse	53%			H3–	35.6%	8.6
		T	≈15%	H3+	17.4%	17.3

[a] From Saccone et al. (2001)
[b] Relative amounts of different band types were assessed on the basis of band sizes in the 850-band karyotype of Francke (1994).
[c] Here, we call G1-G4 the G bands characterized by four levels of grey (from black to pale grey) as defined by Francke (1994). T bands are the most heat denaturation-resistant R bands of Dutrillaux (1973). Relative amounts of G bands are estimated from Francke (1994), those of T bands from Dutrillaux (1973).
[d] As defined in Saccone et al. (1996, 1999) and Federico et al. (2000)
[e] This is the average number of genes per Mb found in the bands of chromosomes 21 an 22.

On a different chromosome a very informative analysis of a 17 Mb stretch covering the sequence of human Xq25 and its flanking regions was reported by Ross (2003) and is presented in **Fig. 7.16**. Band Xq25 was selected because it is one of the darkest G bands in the classification of Francke (1994) and because it strongly hybridized to L isochores (Federico et al., 2000). Several features are of interest: (i) the GC increases at both borders of Xq25; (ii) the inversely correlated LINE and SINE distribution; and (iii) the isochore structure of the region obtained from the IsoFinder segmentation results (Oliver et al., 2001). As expected, the gene count shows a great variation, from 0 to 14 genes per Mb.

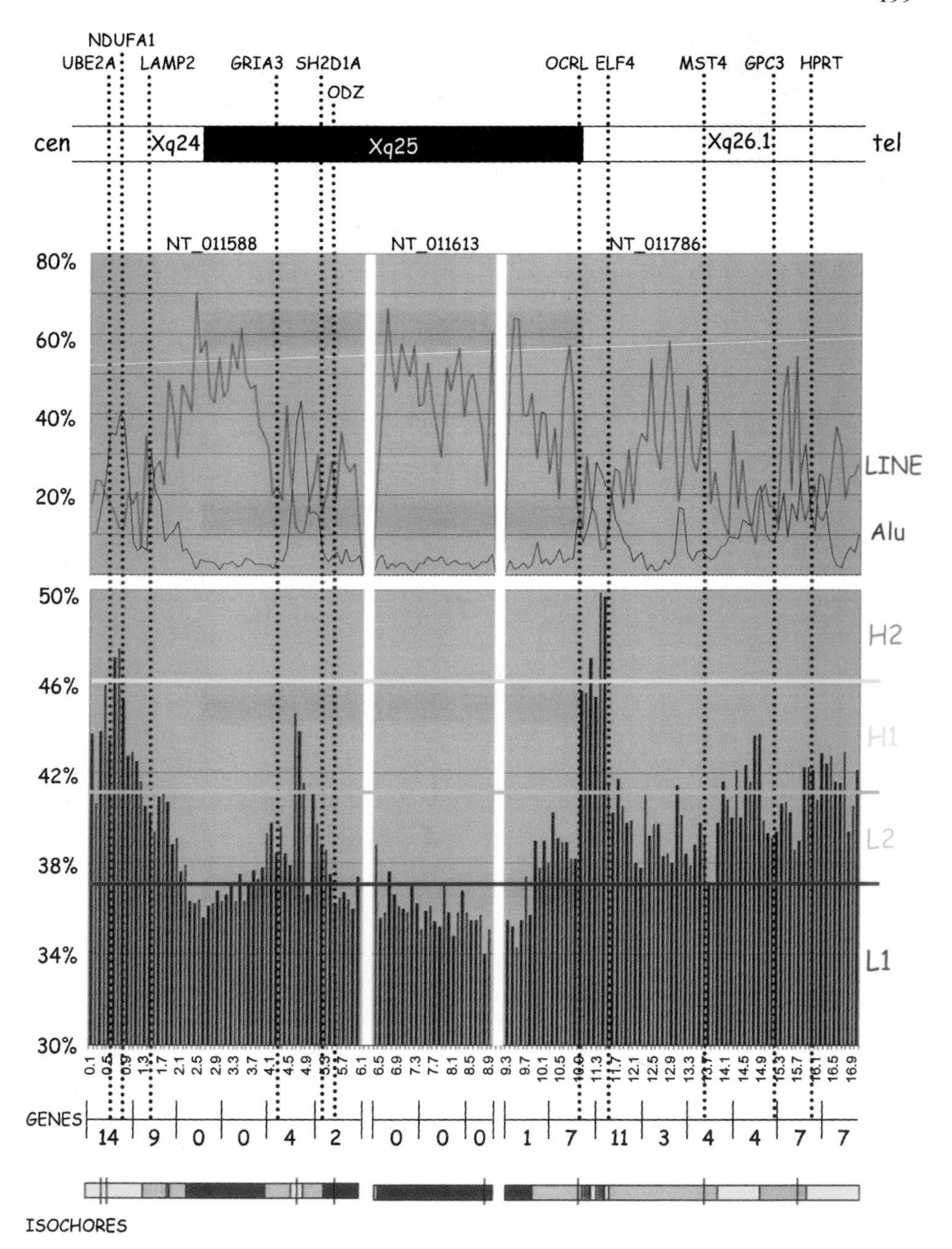

Figure 7.16. Analysis of the sequence of human Xq25 and its flanking regions. The G-banding pattern is represented at the top. A selection of genes is shown on the cytogenetic map. These genes are linked by dotted lines to the 1-Mb sequence intervals in which they lie. The upper histogram shows the percentage of each 100-kb sequence window that is accounted for by LINE or Alu elemensts. The lower histogram shows the G+C content of 100-kb windows (scale in Mb). Horizontal lines show the upper limits designated for L1, L2 and H1 isochores. The two gaps between the three sequances are shown as vertical white lines. Below the G+C plot are the gene counts for each 1-Mb interval. At the base of the figure is the suggested isochore structure of the region taken from the IsoFinder segmentation results. Thin vertical lines indicate predicted boundaries between isochores of the same family.(From Ross, 2003).

Replication timing, recombination and transcription of chromosomal bands

4.1. Replication timing of R and G bands

Chromosome replication is bimodal in all vertebrates. In the case of human chromosomes, the early investigations of Dutrillaux et al. (1976) and Biémont et al. (1978) showed that not only R and G bands differed in replication timing, R bands replicating early, G bands replicating late in the cell cycle, but also that different R bands and different G bands showed different replication timings. In fact, 18 replication timings were identified, at a 300-band resolution. As already mentioned, we identified (at a 400-band resolution) three sets of R bands: H3+, H3* and $H3^-$, endowed with large, moderate and no detectable amount of the gene-richest H3 isochores, respectively. When the replication timings of these three sets of bands were compared with each other, it appeared that the chromosomal bands containing H3 isochores were replicating almost entirely (in the case of H3+ bands) or largely (in the case of H3* bands) at the onset of the S phase, whereas chromosomal bands not containing H3 isochore ($H3^-$ bands) were replicating later (Federico et al., 1998). The existence, at a resolution of 400 bands, of at least three distinct subsets of R band is, therefore, not only supported by their GC and gene concentration, but also by their replication timings (see **Fig. 7. 17**).

As already mentioned, G bands comprise two different subsets of bands, one of which is predominantly composed of L1 isochores, has a higher DNA compaction relative to $H3^+$ bands and corresponds to the darkest G bands of Francke (1994). In contrast, the other subset is composed of L2 and H1 isochores, has less extreme compositional properties and corresponds to the less dark G bands of Francke (see **Fig. 7.18**). When we checked the replication timing of G bands using the data of Dutrillaux et al. (1976) and Biémont et al. (1978), we found that the L1+ bands (or G1+G2; see **Table 7.1**) comprised the bands that replicate at the latest times, whereas the $L1^-$ bands (or G3+G4; see **Table 7.1**) were the earliest replicating among the G bands (Federico et al., 2000). **Fig. 7.19** summarizes all the results just presented.

The best assessment of replication timing in chromosomes 11q and 21q was given by Watanabe et al. (2002; see **Fig. 7.20**). These authors found that replicons are heterogeneous in size (most being 50–450 kb) and that several to 10 (or more) contiguous replicons with origins that fire synchronously at a specific time comprise a megabase-size domain that can be visualized cytogenetically as a band by the replication-banding method. Watanabe et al. (2002) also showed that the frequency of single nucleotide polymorphisms (**SNPs**) was high in the late-replicating, GC-poor regions (see **Fig. 7.20**) and that disease-related genes are concentrated in timing-switch regions, namely at isochore borders. This latter point is not surprising in view of a vast literature indicating that R-bands (which are characterized by frequent compositional discontinuities) and G/R borders are the predo-

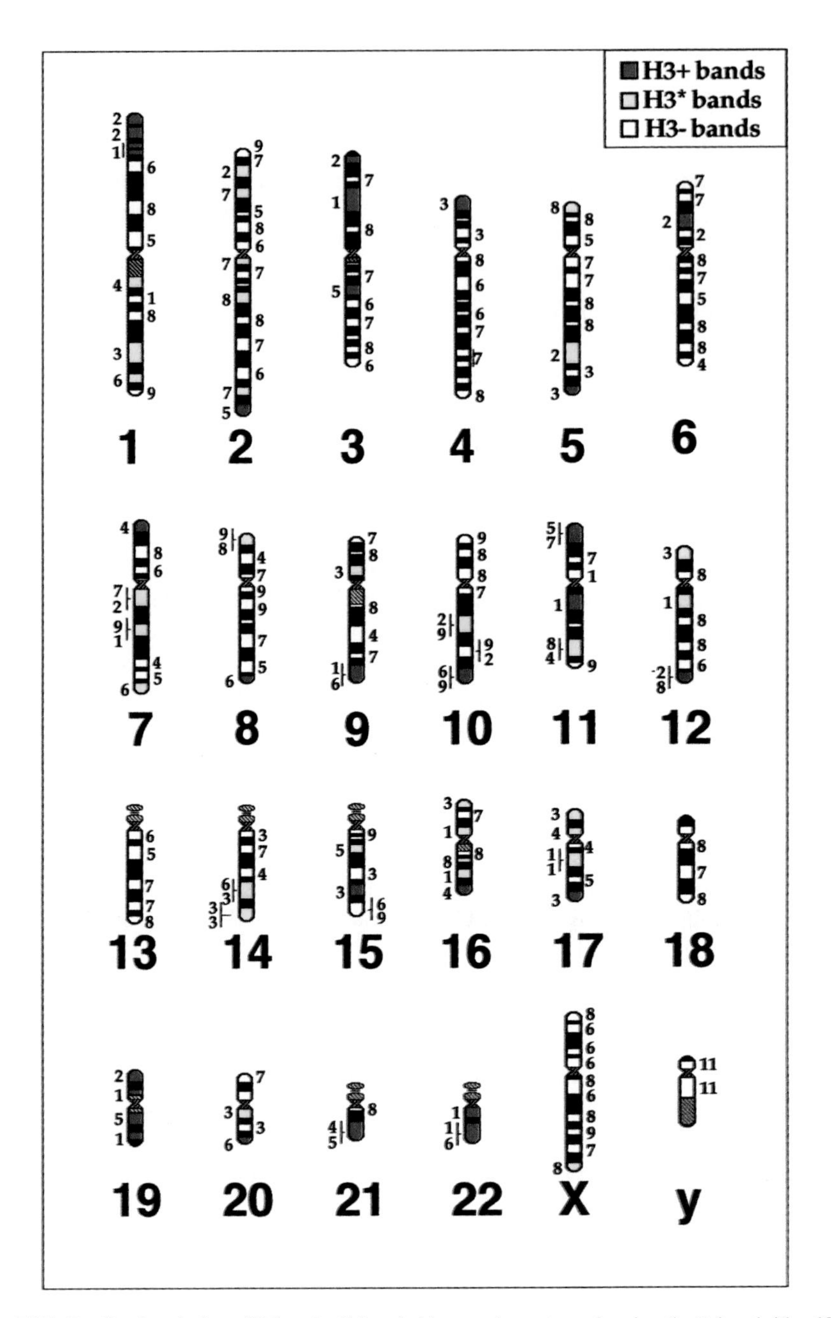

Figure 7.17. Replication timing of R bands. G-banded human karyotype showing the R bands identified by Saccone et al. (1996) as $H3^+$ (red), H3* (yellow) and $H3^-$ (white). G bands are black. The replication class of R bands of Dutrillaux et al. (1976) are indicated (by Arabic numerals) on the left ($H3^+$ and H3* bands) and on the right ($H3^-$ bands) of each chromosome. In some cases, the same R band corresponds to two different replication timings. (From Federico et al., 1998).

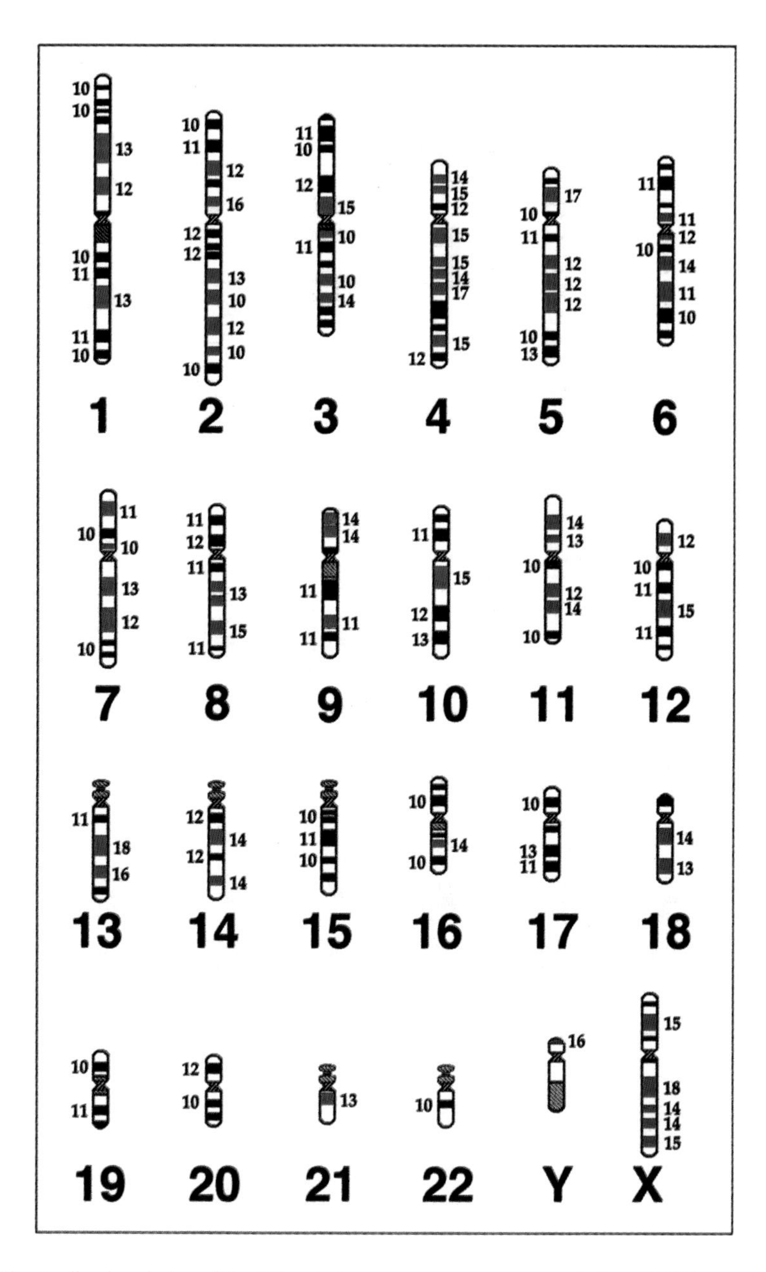

Figure 7.18. Replication timing of the G bands. Human karyotype at a resolution of 400 bands per haploid genome showing the chromosomal bands containing the very GC-poor isochores present in the pellet. Blue, black, and white regions indicate the G bands containing the GC-poorest, L1, isochores, the other G bands, and the R bands, respectively. The numbers on the right and on the left of each chromosome indicate the replication timing of the corresponding regions, as described previously (Dutrillaux et al., 1976; Biémont et al., 1978). (From Federico et al., 2000).

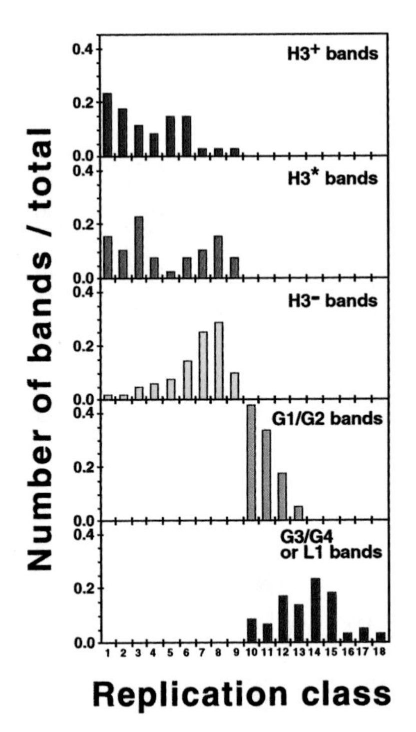

Figure 7.19. Histograms with the replication timing of the chromosomal bands. Histograms showing the human chromosomal bands as distributed in the replication classes described by Dutrillaux et al. (1976) and Biémont et al. (1978). (From Saccone and Bernardi, 2001).

minant sites of exchange processes, including spontaneous translocations, spontaneous and induced sister-chromatid exchanges, the chromosomal abnormalities seen after X-ray and chemical damage, and fragile sites (see Bernardi 1989, 1995, for reviews).

4.2. Recombination in chromosomes

According to a recent review (Nachman, 2002) "*Across the genome, recombination rate varies from low values near 20 to high values close to 5 cM/Mb (centimorgans per megabase). In general, recombination appears to be low near the centromeres of metacentric but not acrocentric chromosomes. Recombination is generally elevated near the ends of chromosomes, regardless of whether or not there is a centromere in proximity*" (see **Fig. 7.21**). The high level of recombination near the ends of chromosomes may be associated with their generally high GC levels, or, as suggested by Bernardi (1989, 1993b, 1995), with the high levels of compositional fluctuation exhibited by GC-rich isochores. Chromosomal rearrangements have two important consequences; the activation of oncogenes by strong promotors that have been put upstream of them by the rearrangement, and the possibility that some chromosome rearrangements lead, in evolutionary time, to reproductive barriers and speciation. It was speculated that the higher incidence of cancer and rate of speciation shown

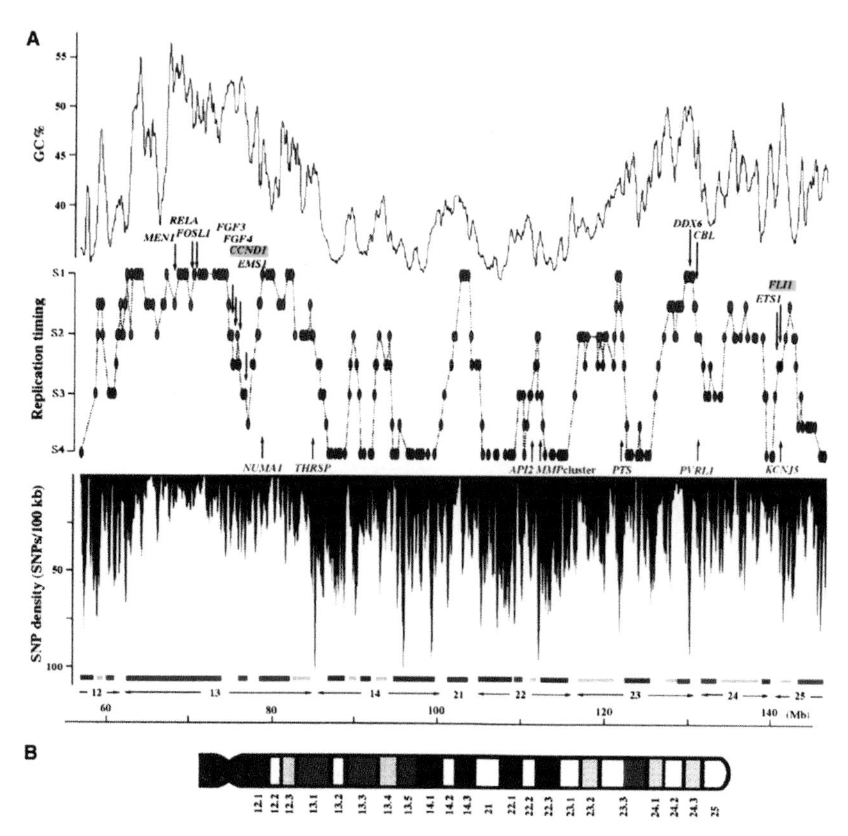

Figure 7.20. Replication timing and GC% and SNP distributions on chromosome 11q. **A.** GC% distribution of the 11q sequence obtained from http://genome.cse.ucsc.edu/ was calculated with a 500 kb window and a 50 kb step. Undetermined bases (Ns) were omitted from the GC% calculation, but the nucleotide positions were not changed. If a total of *N* bases in a window exceeded a half of the window size, the data point was omitted and the neighboring data points were connected by a line. Early and late loci are denoted by red and blue ovals, respectively. Distribution of SNP frequency was calculated with a 200 kb window and a 20 kb step and is indicated by a bar diagram. A megabase-sized zone that is composed primarily of STSs belonging to one timing category is indicated by a horizontal colored line; the *S1* and *S2* zones are indicated by red and pink horizontal lines, respectively, and the *S3* and *S4* zones are indicated by light- and dark-blue lines, respectively. Chromosomal band numbers were assigned by referring to NCBI OMIM, UniSTS, and GENATLAS. Sequences in the subtelomeric region were found to replicate late even though their GC% levels were rather high. **B.** An ideogram of the 11q band pattern reported by Saccone et al. (1999) and Bernardi (2000a). Red indicates bands that are composed mainly of the most GC-rich H3 isochores and thus correspond to T bands. Ordinary R bands are indicated in white. G bands composed primarily of AT-rich L isochores are indicated in black, and those composed of both L and H isochores are indicated in grey. (From Watanabe et al., 2002).

by warm-blooded relative to cold-blooded vertebrates correlate with the higher propensity to chromosome rearrangements of the former due to the larger number of compositional discontinuities in their chromosomes (Bernardi, 1993b).

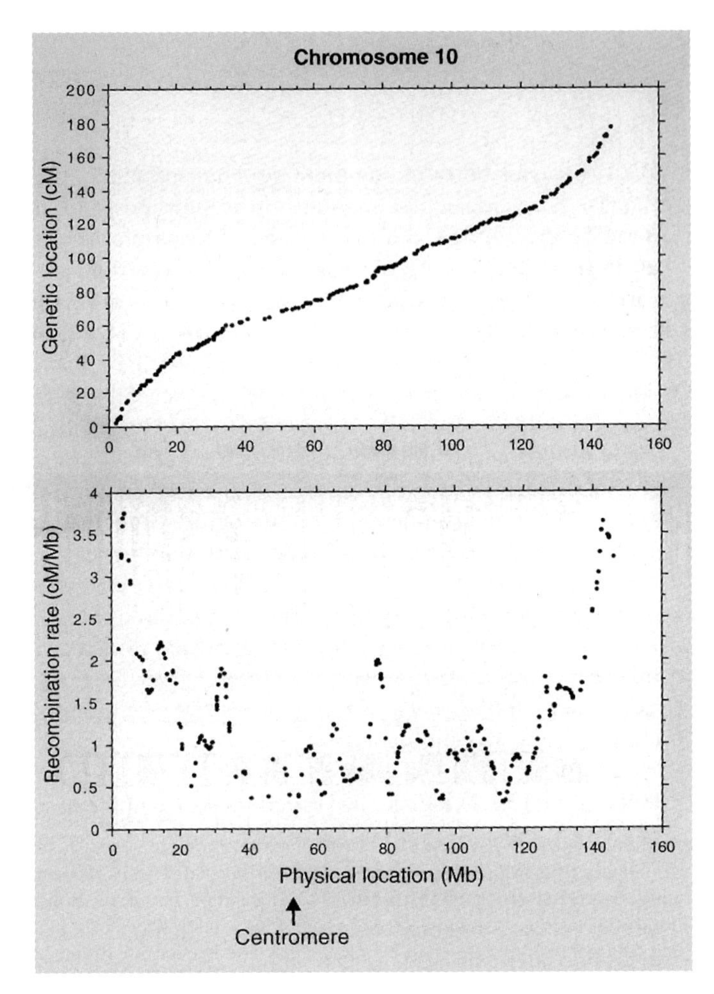

Figure 7.21. Recombination rate variation for chromosome 10 based on data from Kong et al. (2002). The top panel for each chromosome depicts a plot of genetic versus physical position for microsatellite markers; the slope of this curve provides an estimate of recombination rate. The bottom panel gives the recombination rate estimates from Kong et al. (2002) for each marker based on a 3 Mb window. (From Nachman, 2002).

4.3. Transcription of chromosomal bands

As far as transcription is concerned, over 2.45 million SAGE (Serial Analysis of Gene Expression) transcript tags, including 160,000 tags of neuroblastomas, were investigated for 12 tissue types (Caron et al., 2001). SAGE is based on the extraction of a 10-base pair tag from a fixed position in each transcript and the sequencing of thousands of these tags and can quantitatively identify all transcripts expressed in a tissue or cell line (Velcutescu et al., 1995). Algorithms were developed to assign these tags to UniGene clusters and their chromosomal positions. The resulting Human Transcription Map generated gene expression profiles for any chromosomal region in 12 normal and pathologic tissue types (Caron et al., 2001). The map revealed a clustering of highly expressed genes to specific chromosomal regions. Such clusters were called RIDGES (Regions of Increased Gene Expression) by the authors, who found a correlation between gene expression and density of mapped genes for 50–60% of the RIDGES. Since gene density increases with increasing GC, one should, therefore, expect a correlation between gene expression and GC levels of RIDGES, or GC levels of isochores comprising the genes under consideration (as reported by D'Onofrio, 2002). Caron et al. (2001) were apparently not aware of our investigations, because they could have found a correlation between their RIDGES and GC-rich isochores. A surprising result reported by Caron et al. (2001) was that gene density decreases in telomeric regions, since we had shown that most telomeric regions are GC-rich and gene-rich.

Isochores in the interphase nucleus

5.1. Distribution of the GC-richest and GC-poorest isochores in the interphase nucleus of human and chicken.

The *in situ* hybridization of the human GC-richest fraction showed a general distribution of the signals in the central region of interphase nuclei, whereas the GC-poorest fraction was detected prevalently in the peripheral part (**Fig. 7.22A**), all the observed nuclei showing this type of signal distribution (Saccone et al., 2002). Also the hybridization of the chicken GC-richest and GC-poorest DNAs indicates that the former are localized in the interior of the nucleus, whereas the GC-poorest ones are localized in the periphery (**Fig. 7.22B**). The chicken nuclei also showed some large H4 isochore-hybridizing peripheral regions that could be due to the heterochromatin of the Z chromosome (macrosatellite pFN-1; see Hori et al., 1996), which was incompletely competed out by the suppression procedure (see Andreozzi et al., 2001).

5.2. Different compaction of the human GC-richest and GC-poorest chromosomal regions in interphase nuclei

To understand better, at the nuclear level, the chromatin conformation as related to the GC levels of the bands, we investigated in detail (see **Fig. 7.23**) the GC-richest bands 6p21 (probe A) and 9qter (probe B), the GC-intermediate bands 9pter (probe C) and 12q22–23 (probe D), and the GC-poorest band 12q21 (probe E). The GC-profiles shown in **Fig. 7.23** were obtained using the draft human genome sequence (Lander et al., 2001), as redrawn by Pavliček et al. (2002a) and correlated to chromosomal bands using the approach of Saccone et al. (2001). The band DNA probes, together with the corresponding chromosome-specific DNA, were first hybridized on metaphase chromosomes to check the precise localization of the probes on the chromosomes (see **Fig. 7.23**) and to show that no cross-hybridization on other chromosomal sites could be detected with any of the five probes.

Each of the five chromosomal regions (**Fig. 7.23**, probes A to E) were then hybridized on the cell nuclei and the extensions of the hybridization signals in mitotic and interphase cells were compared. The GC-rich 6p21 and 9qter regions showed a wider spreading of the hybridization signals in the nuclei (**Fig. 7.24A** and **B**) relative to that of the chromosome territories. Moreover, a comparison of the same hybridization on metaphase (**Fig. 7.23A** and **B**) and interphase chromosomes (**Fig. 7.24A** and **B**) showed that the extensions of the two band DNAs are clearly more spread out in the nuclei. In contrast, the GC-poor 12q21 region showed a high level of compaction in the nuclei (compare **Figs. 7.23E** and **7.24E**). In fact, the comparison of the hybridizations in the metaphase and interphase chromosomes

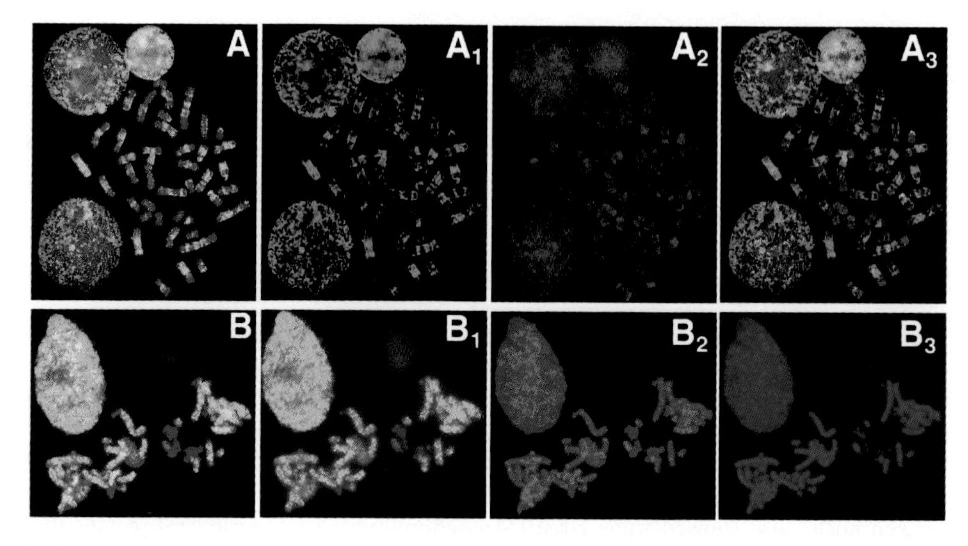

Figure 7.22. Distribution of the GC-poorest and GC-richest isochores in human and chicken nuclei. Human (**A**) and chicken (**B**) chromosomes and nuclei were hybridized with the GC-poorest L1 and the GC-richest H3 isochores. The H3 and the L1 isochores were labeled with biotin and digoxigenin, respectively, and detected with TRITC-avidin (red signals) and FITC-anti-digoxigenin (green signals), respectively. The yellow color is due to overlapping red and green signals. Nuclei and chromosomes were stained with DAPI. The images were obtained with epifluorescence microscopes equipped with appropriate filters, and captured with a CCD camera. MacProbe 4.2.3 and Photoshop 5.0 softwares were used to obtain the images. $\mathbf{A_1}$, $\mathbf{A_2}$, and $\mathbf{A_3}$ are as in **A**, without the DAPI stain, showing the green, the red, and both the green and red signals, respectively. $\mathbf{B_1}$, $\mathbf{B_2}$, and $\mathbf{B_3}$ are as in **B** showing the green signals, the red signals, and the DAPI stain, respectively. (From Saccone et al., 2002).

clearly indicated that hybridization signals of the band-specific DNA are more condensed in the nuclei compared to the mitotic cells.

The compositionally intermediate region 9pter (comprising $H3^-$ and $L1^-$ bands) showed a compaction degree comparable to that observed with the GC-poor band 12q21, the large majority of the nuclei exhibiting the hybridization signal distribution exemplified by **Fig. 7.24C**. Instead, the other GC-intermediate band, 12q22–23 (**Fig. 7.24D**), showed a signal distribution in the nuclei characterized by features intermediate between those of the GC-rich and GC-poor bands, although a number of nuclei (about 25%) were observed with more compact signals, similar to those of the GC-poor band 12q21. In any case, the two GC-intermediate bands tested never showed signals similar to those obtained for the GC-rich bands. It should be noted that the two GC-intermediate regions are located in close contiguity to bands of the GC-poorest class, suggesting that GC-intermediate bands are characterized by compaction features similar to those of the adjacent bands.

The different compaction level of the GC-rich and the GC-poor chromosomal regions is also evident by comparing their hybridizations in the nuclei. In fact, in spite of the fact that the GC-rich 6p21 and the GC-poor 12q21 bands comprise a similar DNA amount (15–20 Mb; see **Fig. 7.24**, probes A and E), the spreading of the signals in the nuclei is much wider

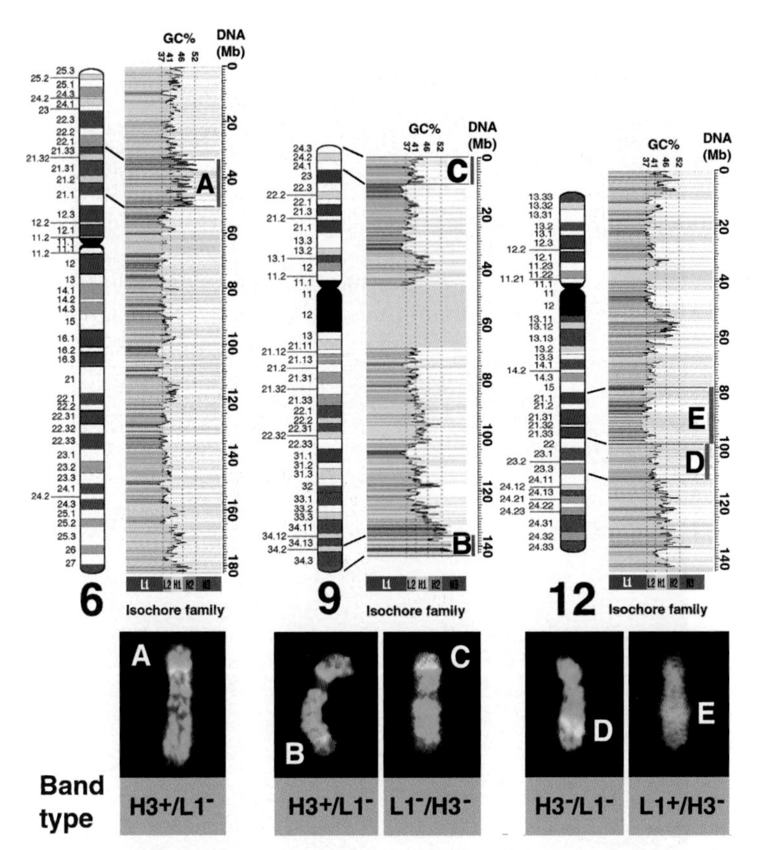

Figure 7.23. Band DNA probes used to demonstrate the differential chromatin compaction. *Top*: Ideograms of human chromosomes 6, 9, and 12 at 850 band resolution, and corresponding GC profiles. The ideograms, show the $H3^+$ bands in red and the $L1^+$ bands in blue. The GC profiles (from Pavlicek et al., 2002a) shown on the right of each chromosome are differentially coloured according to the GC level of the isochore families (L1, blue: $<37\%$, L2, pale blue: $37\% - 41\%$, H1, yellow: $41\% - 46\%$, H2, orange: $46\% - 52\%$, H3, red: $> 52\%$). The chromosomal bands were matched with the GC profiles according to Saccone et al. (2001). **A, B, C, D,** and **E** indicate the probes used for the *in situ* hybridizations. The grey regions in the GC profiles are unsequenced regions. *Bottom*: Chromosomes 6, 9 and 12 showing the hybridization with the biotin-labe led probes **A** to **E**. Chromosomes were painted with digoxigenin-labeled probes. Detection was performed as in Fig. 7.22. Band type indicates the compositional features of the hybridized bands (at a resolution of 850 bands). Probes **A** and **B** detected regions mostly consisting of the GC-richest $H3^+$ bands; probes **C** and **D** detected regions composed by GC intermediate $L1^-$ and $H3^-$ bands, but adjacent to GC-poorest bands; probe **E** detected regions mostly consisting of the GC-poorest $L1^+$ bands. (From Saccone et al., 2002)

for the former compared to the latter (see **Fig. 7.25**). The results obtained with the intermediary band 12q 22–23 are also shown in **Fig. 7.25** for the sake of comparison.

Concerning the nuclear location of the DNA belonging to bands endowed with different compositional features, the results are in agreement with the nuclear distribution of the GC-richest and the GC-poorest isochores (**Fig. 7.24**). In fact, we observed a more peripheral localization of the GC-poor bands compared to a more internal position of the GC-

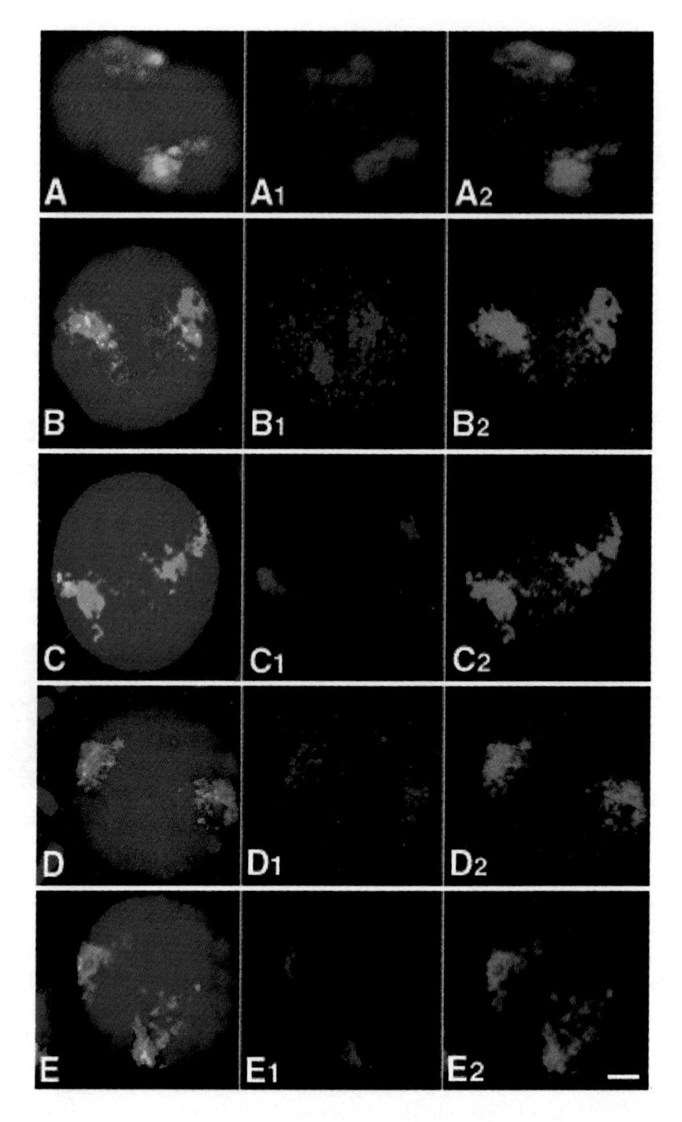

Figure 7.24. Nuclear distribution of chromosomal regions characterized by different GC levels. Chromosomal regions corresponding to bands 6p21 (**A**), 9q34.2-34.3 (**B**), 9p24 (**C**), 12q22-23 (**D**), and 12q21 (**E**) were co-hybridized with the corresponding chromosome probes. Band and chromosome DNAs were biotin- and digoxigenin-labeled, respectively. Detection was done as in Fig. 7.22. Bar: 2 μm. (From Saccone et al., 2002)

rich bands (see **Fig. 7.22**). In the case of chromosome 9, this indicates that different regions of the same chromosome contribute to the DNA located both in the interior and the periphery of the nucleus. Moreover, at a resolution of 850 bands, probes A and B (from the two GC-richest chromosomal regions) and probe E (from one of the GC-poorest regions), detected not only the $H3^+$ and the $L1^+$ bands, respectively, but also the $L1^-$ and the $H3^-$ bands, respectively, that are located between them. This indicates that

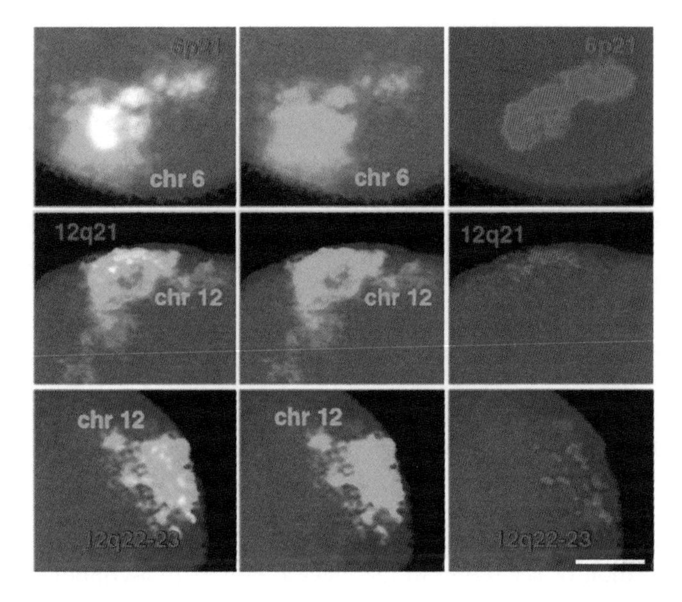

Figure 7.25. Differential compaction of chromatin in human nuclei. Comparison of hybridizations obtained with the GC-richest and the GC-poorest DNA regions (partial nuclei from Fig. 7.24). The band DNA (red signals) and the corresponding chromosome (green signals) are indicated. Nuclei were DAPI stained. The very GC-rich 6p21 band showed a much larger extension compared to the very GC-poor 12q21. Furthermore, the former is more internal compared to the latter. The results obtained with the intermediary band 12q 22-23 are also shown for the sake of comparison. Bar: 2 μm. (From Saccone et al, 2002).

DNA from these very small compositionally intermediate $L1^-$ and $H3^-$ bands also occupy the same nuclear location as the contiguous GC-richest or GC-poorest bands.

5.3. The spatial distribution of genes in interphase nuclei

In conclusion, we have shown that the GC-richest and the GC-poorest DNA of warm-blooded vertebrates (more precisely the DNA characterized by GC levels higher than 50% and lower than 38%), are located in internal and peripheral regions of the nucleus, respectively. The importance of this observation is due to the fact that a number of structural and functional features of the genomes of warm-blooded vertebrates, such as gene density, are correlated with GC levels. Our DNA probes from H3 and L1 isochores detected two subsets of chromosomal bands, namely the $H3^+$ and the $L1^+$ bands, which largely overlap with the T bands of Dutrillaux (1973) and with the two darkest sets of G bands of Francke (1994), respectively. On the other hand, we found that the GC intermediate band DNAs ($L1^-$ and $H3^-$) are characterized by a nuclear location similar to those of the $L1^+$ or $H3^+$ bands that are close to them on mitotic chromosomes. This accounts for the observation (Croft et al., 1999) that chromosome 18 (containing $L1^+$, $L1^-$ and $H3^-$ bands, but no $H3^+$ bands) shows a peripheral nuclear localization, whereas chromosome 19 (which contains $H3^+$, $L1^-$ and $H3^-$ bands, but no $L1^+$ bands) is located in the nuclear interior. In other words, these exceptional situations are due to the fact that these two very small chromosomes lack the GC-richest and the GC-poorest bands, respectively.

In contrast, the large majority of chromosomes comprise both categories of bands (Federico et al., 2000; see also **Fig. 7.26**). It is, therefore, understandable that most chromosomes contribute to both the DNA located in the internal and peripheral parts of the nuclei, according to the cartoon of **Fig. 7.27**. This is possible because the GC-richest and the GC-poorest bands are largely located distally and proximally, respectively (and are, therefore, only rarely adjacent in human chromosomes), and because of the large extension of interphase chromosomes. The observation that many chromosomes, especially the large ones, are spread in the nucleus from the periphery to the interior, even if in different proportions (Boyle et al., 2001), supports a **compositional polarity of the interphase chromosomes.**

We also showed that the two subsets of bands investigated here not only have different locations in the nucleus, but also correspond to different chromatin conformations, the $H3^+$ band DNA being remarkably more relaxed compared to the $L1^+$ band DNA. This is understandable if we consider the special properties of H3 isochores compared to L1 isochores. In fact, the more open chromatin of the $H3^+$ band DNA corresponds to the highest concentration of genes and to the highest level of transcriptional activity. This

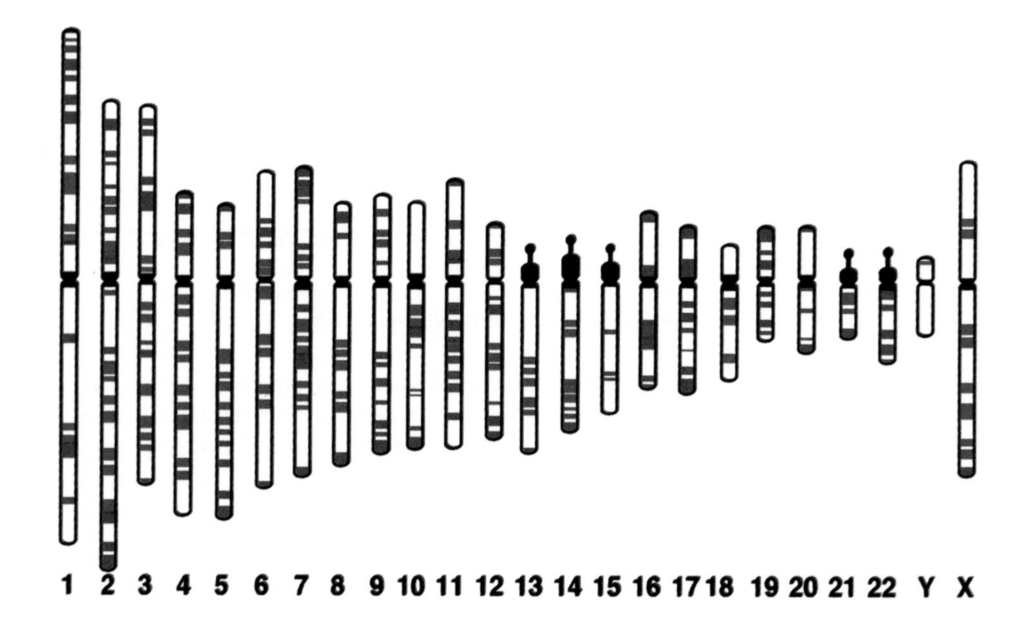

Figure 7.26. Chromosome distribution of the $H3^+$ and $L1^+$ bands. Ideogram of human chromosomes at 850-band resolution showing the position of the $L1^+$ (blue) and the $H3^+$ (red) bands as previously identified on the basis of the concentration of L1 and H3 isochores, respectively. This figure shows that the gene-richest, GC-richest, $H3^+$ bands are prevalently located distally on each chromosome arm and are generally not adjacent to the gene-poorest, GC-poorest, bands, which are prevalently located more proximally. The intermediate bands ($H3^-$, $L1^-$) are left uncolored in order to emphasize this spacing. (From Saccone et al., 2002)

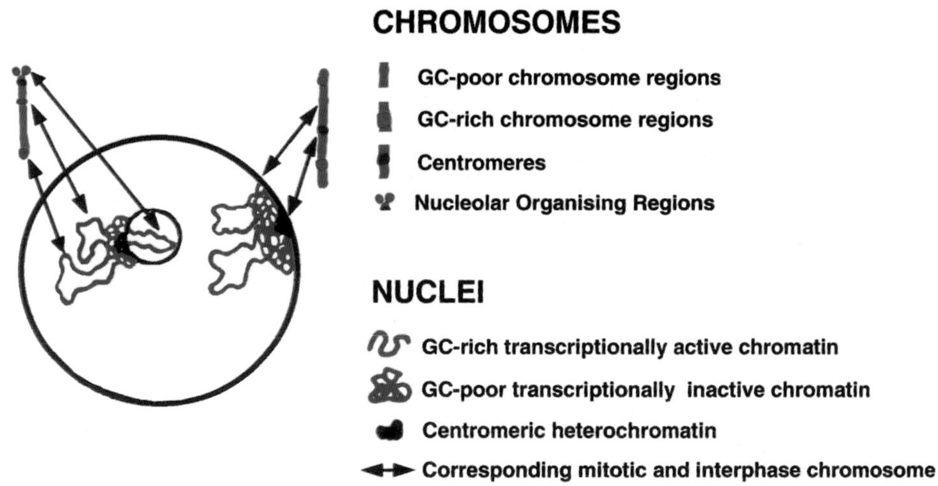

CHROMOSOMES

GC-poor chromosome regions
GC-rich chromosome regions
Centromeres
Nucleolar Organising Regions

NUCLEI

GC-rich transcriptionally active chromatin
GC-poor transcriptionally inactive chromatin
Centromeric heterochromatin
Corresponding mitotic and interphase chromosome

Figure 7.27. A cartoon showing that the centromeric proximal regions of chromosomes tend to be present in compact chromatin structures at the periphery of the interphase nucleus (blue blocks), whereas the distal regions are in an open chromatin structure (red filaments) in the center of the nucleus. Only two chromosomes are depicted. In this simplified drawing entire chromosome arms are represented as completely open. More realistically, the open chromatin loops should concern a number of regions from each chromosome arm.

finding is in general agreement with other data showing that chromatin belonging to G and R bands of mitotic chromosomes is more dense and more open, respectively (Croft et al., 1999; Yokota et al., 1997), and that highly expressed sequences extend outside the chromosome territory (Volpi et al., 2000). Incidentally, the latter observation concerned one of the chromosomal regions, 6p21, which was investigated here.

It should be stressed that the two "gene spaces" previously described, the gene-rich "genome core" and the gene-poor "empty quarter" correspond to the genome compartments located in the interior and at the periphery of the nucleus, respectively. This supports a functional compartmentalization of the nucleus related to nuclear architecture (Strouboulis and Wolffe, 1996). The functional compartmentalization of the cell nucleus was also demonstrated by other findings, such as the different replication timing of DNA located in the internal and peripheral part of the nucleus (Sadoni et al., 1999; Ferreira et al., 1997).

Basically, the chicken genome showed the same compositional properties observed for the human genome, both at the chromosomal and the nuclear level, the only difference being the large number of GC-richest bands that are present on microchromosomes (see Andreozzi et al., 2001). In chicken, the large majority of microchromosome DNA only contribute to the internal part of the nucleus, like the small GC-richest human chromosomes, whereas the large chromosomes, comprising both GC-richest and GC-poorest bands, contribute to both the interior and the periphery of the nucleus.

The different spatial distribution of genes in the nucleus and the different chromatin compaction are apparently general for all vertebrates, as expected from our previous comparative investigations, and as shown by results on *Rana esculenta* (**Fig. 7.28**).

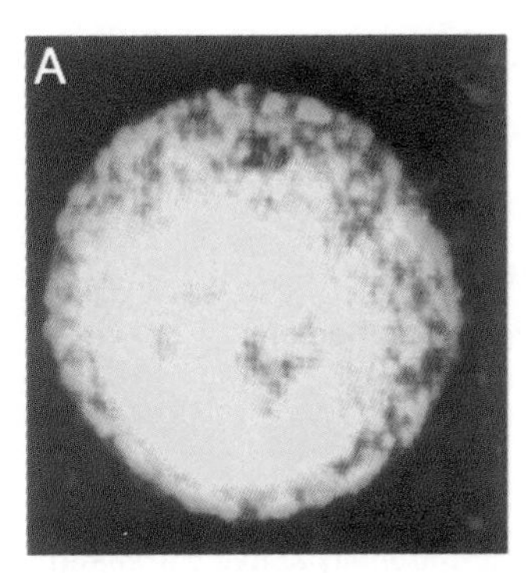

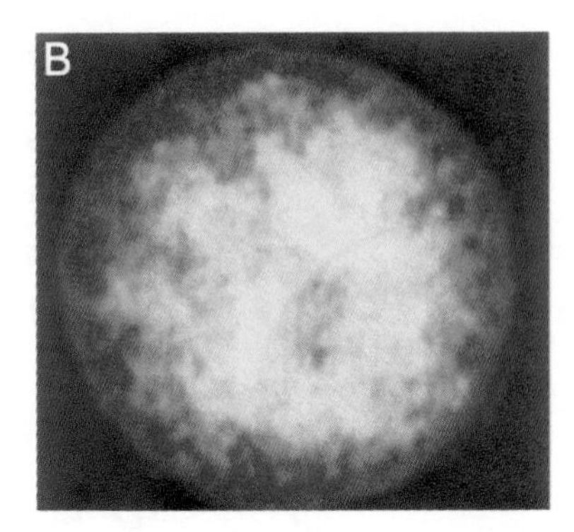

Figure 7.28. Nuclear distribution of chromosomal regions of *Rana esculenta*. GC-poor and GC-rich chicken DNA fractions were labelled with digoxigenin, hybridized on interphase nuclei, and detected with fluorescein-conjugated antibodies (green signals). Nuclei were stained with propidium (red color) **A.** GC-poor isochores are predominant at the nuclear periphery. **B.** GC-rich isochores are predominant at internal locations. The yellow color is due to overlapping red and green signals. (From Federico et al., 2003).

Further studies should now probe in more detail the distribution of H3+ and L1+ isochores in chromosome territories (see Cremer and Cremer, 2001, for a review). The observations of Tajbakhsh et al. (2002) indicated that the former were localized in all subvolumes of the territories at similar frequency, while the latter were more to the interior of the same territories.

Finally, it should be mentioned that a polymer model for the structural organization of chromatin loops in interphase chromosomes was developed by Ostashevsky (1998) on the basis of isochores.

Part 8
The organization of plant genomes

CHAPTER 1

The organization of the nuclear genome of plants

When we started our investigations on the nuclear genomes of plants in 1986, the available information essentially concerned the buoyant densities, methylation and reassociation kinetics for a small number of species (Ingle et al., 1973; Shapiro, 1976; Sorenson, 1984). As far as primary structures were concerned, data were available for less than 200 coding sequences from all plants. So little being known about plant genomes, what we did was to apply to the nuclear genomes of plants the compositional approach that we had previously used to investigate the organization of the genomes of vertebrates.

The earliest results (Salinas et al., 1988) indicated that the nuclear genomes of *Gramineae* exhibited strikingly different compositional patterns compared to those of dicots. Indeed, the compositional distribution of nuclear DNA molecules (in the 50-100 kb size range) from three dicots (pea, sunflower and tobacco) and three *Gramineae* (maize, rice and wheat) were found to be centered around low (41%) and high (45% for rice, 48% for maize and wheat) GC levels, respectively, and to trail towards even higher GC values in maize and wheat.

When the plant sample explored was expanded, it became clear that monocots other than *Gramineae* could exhibit intermediate or even GC-poor compositional patterns, whereas the dicots explored were all GC-poor, with the possible exception of *Oenothera*, which showed, next to a very GC-poor component, several GC-rich DNA components (Matassi et al., 1989; see **Fig. 8.1**).

On the other hand, the compositional distribution of coding sequences from several orders of dicots was found to be narrow, symmetrical and centered around 46% GC, that from monocots (essentially the *Gramineae* barley, maize and wheat) to be broad, asymmetrical and characterized by an upward trend towards high GC values, with the majority of sequences between 60% and 70% GC. Introns exhibited a similar compositional distribution, but remarkably lower GC levels, compared to the exons from the same genes (see below). The compositional differences in coding sequences between *Gramineae* and dicots were confirmed by later work on larger data sets (Carels et al., 1998; see **Fig. 8.2**).

When orthologous coding sequences were compared, points from dicots or from *Gramineae,* respectively, fell on the diagonal line indicating compositional identity, whereas this was not true when comparison were made between coding sequences from *Gramineae* (maize) and dicots (*Arabidopsis*). In this case (**Fig. 8.3**), both the GC and the GC_3 plots showed that, while maize values were close to the diagonal in the low GC or GC_3 range (indicating similarity with *Arabidopsis* values), this was not true for the high GC and GC_3 range in which maize values were much higher, yet linearly correlated, with those of *Arabidopsis*.

In other words, the differences between coding sequences of *Gramineae* and dicots were confirmed by showing that the compositional distribution of homologous coding sequences from several orders of dicots and from *Gramineae* mimick the compositional

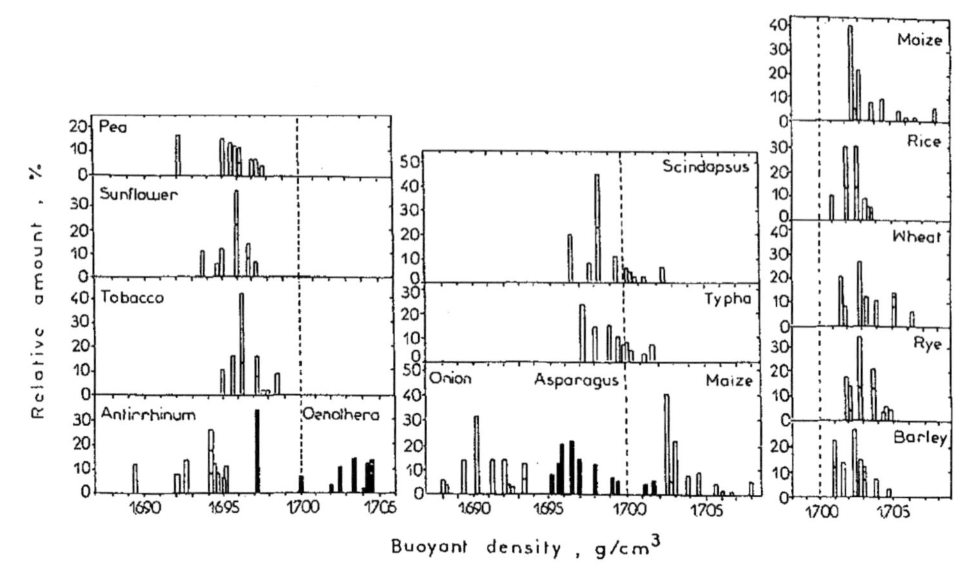

Figure 8.1. Histograms showing the relative amounts and buoyant densities in CsCl of DNA fractions obtained by preparative Cs_2SO_4/BAMD density gradient centrifugation from dicots (left panel) and monocots (middle and right panels). Data for pea, sunflower, tobacco, maize, rice and wheat are from Salinas et al. (1988). The vertical broken line at 1.700 g/cm^3 is shown to provide a reference. Black bars are used whenever required for distinguishing DNA fractions from different plants. Horizontal lines on some bars separate DNA fractions showing the same buoyant densities. (From Matassi et al., 1989).

distributions previously seen for coding sequences in general (Salinas et al., 1988). Indeed, most coding sequences from *Gramineae* were much higher in GC than the homologous sequences of the dicots explored. These differences were even stronger for third codon positions (see above) and led to a striking codon usage for many coding sequences in the case of *Gramineae*.

Introns exhibited a similar compositional distribution, but lower GC levels, compared to exons from the same genes (Salinas et al., 1988; Carels and Bernardi, 2000; see **Fig. 8.4**). The compositional difference between introns and exons was much greater in plants than in vertebrates. Again, an example of this situation from later work (Carels and Bernardi, 2000) is shown in **Fig. 8.5**.

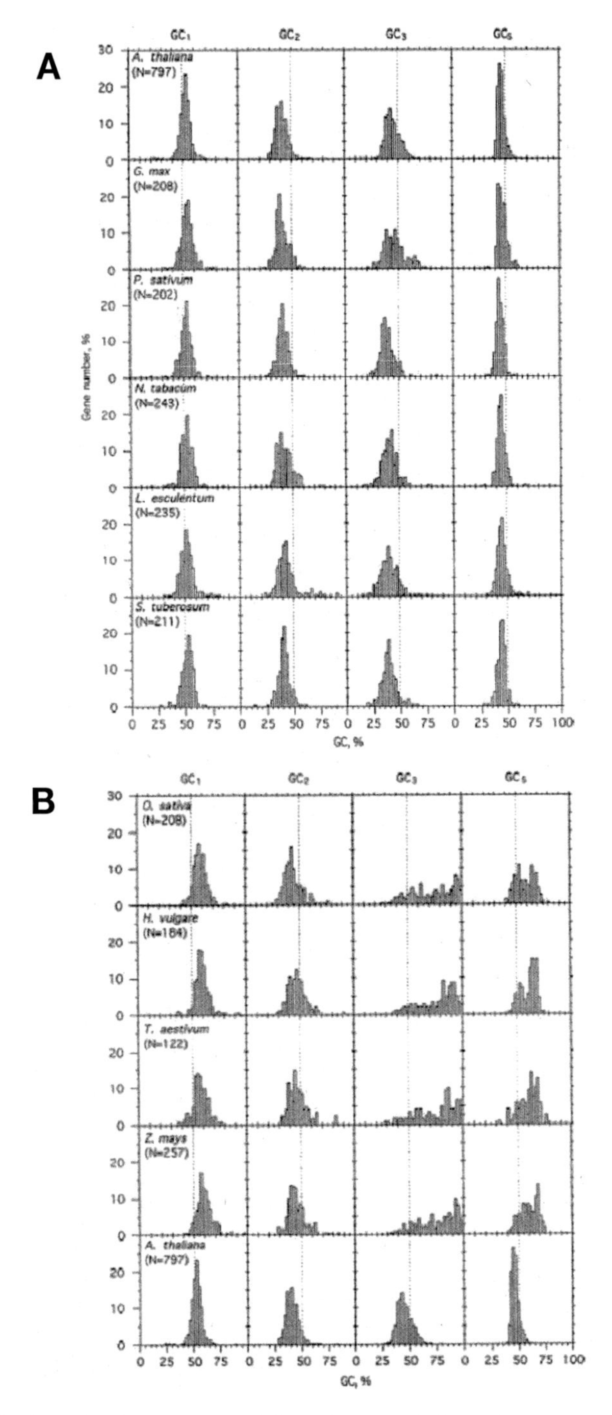

Figure 8.2. Distributions of coding sequences **A** from dicots and **B** from *Gramineae* according to the GC_1, GC_2, GC_3 and GC_s levels, namely the GC levels of first, second and third codon position and of the whole coding sequences, respectively. (From Carels et al., 1998).

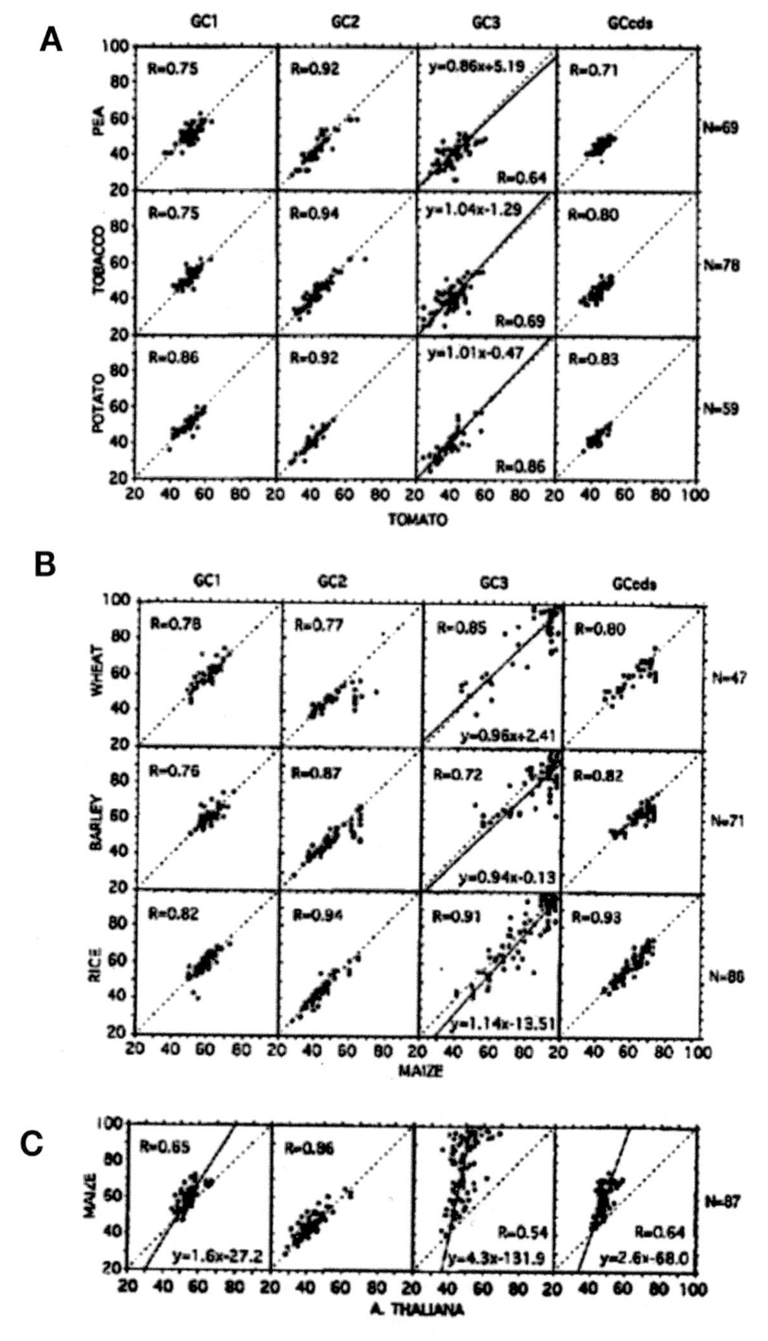

Figure 8.3. GC_1, GC_2, GC_3 and GC_s (GC of coding sequences) of **A** dicots, **B** *Gramineae*, and **C** maize (ordinates) are plotted against the corresponding values of their homologs from **A** tomato, **B** maize, or **C** *Arabidopsis* (abscissas). Orthogonal regression lines (solid), diagonals (dashed), and correlation coefficients (R) are shown. (From Carels et al., 1998).

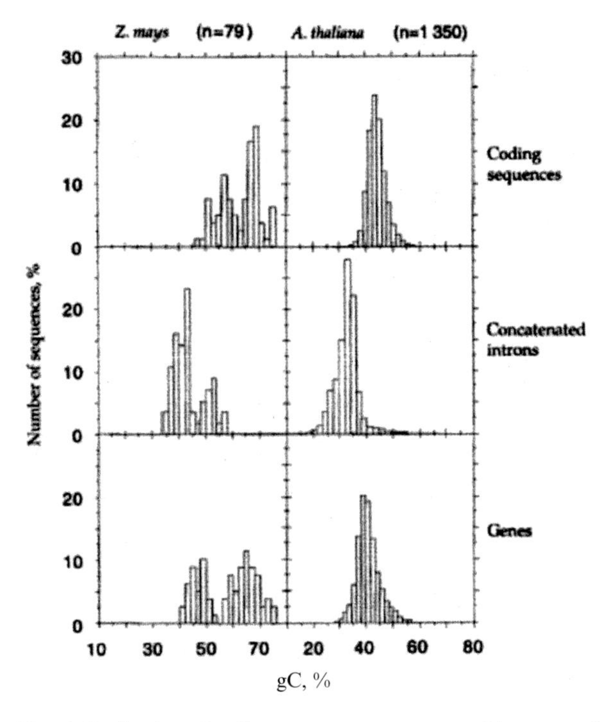

Figure 8.4. Compositional distribution of coding sequences, concatenated introns and genes in maize and *Arabidopsis*. N is the number of genes investigated (From Carels and Bernardi, 2000).

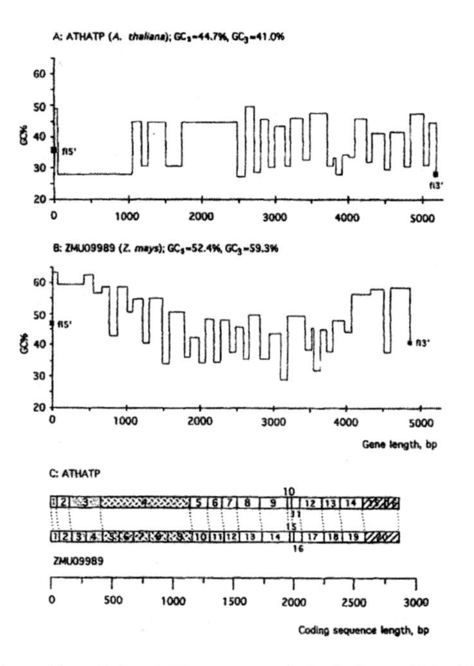

Figure 8.5. Compositional profiles of the ATPase gene of *Arabidopsis* (**A**), ATHATP and of its maize homolog, FMU09989 (**B**). 5' and 3' flanking sequences are 213 and 260 bp long in the case of *Arabidopsis* and 1,445 and 2,995 bp in the case of maize, respectively. Exon correspondence and positions are shown in C. (From Carels and Bernardi, 2000).

CHAPTER 2

Two classes of genes in plants

Two classes of genes were identified in three *Gramineae*, maize, rice, barley, and six dicots, *Arabidopsis*, soybean, pea, tobacco, tomato, potato (Carels and Bernardi, 2000). One class, the GC-rich class, contained genes with no, or few short introns. In contrast, the GC-poor class contained genes with numerous long introns (**Fig. 8.6**). The similarity of the properties of each class, as present in the genomes of maize and *Arabidopsis*, is particularly remarkable in view of the fact that these plants exhibit large differences in genome size (115 Mb for *Arabidopsis*; 2,500 Mb for maize), average intron size, and DNA base com-

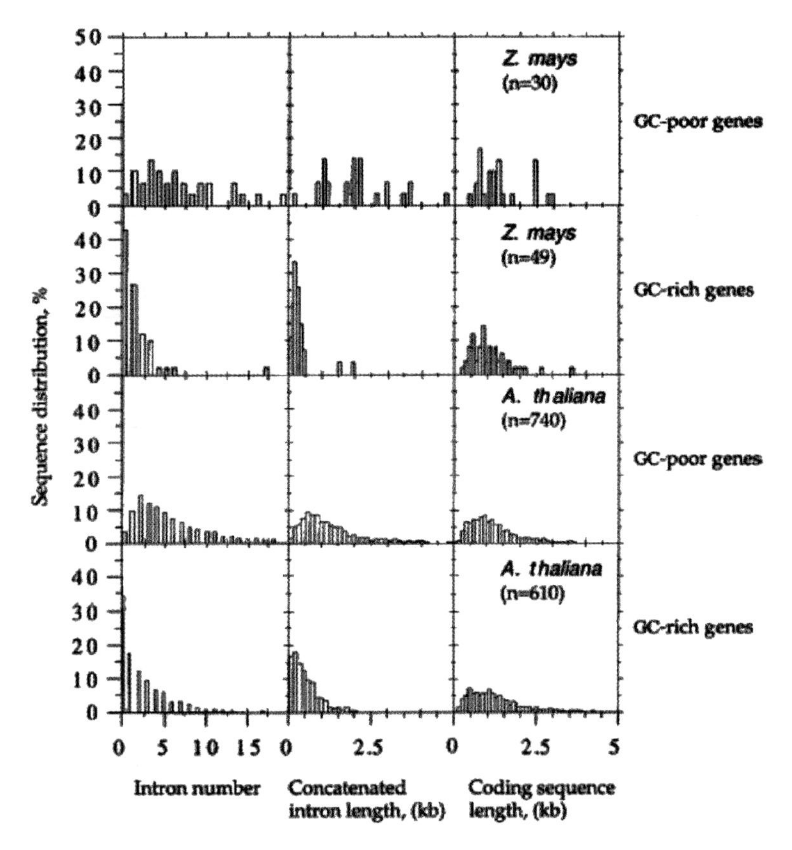

Figure 8.6. Distributions of intron number, length of concatenated introns and coding sequence length in GC-poor and GC-rich genes of maize and *Arabidopsis*. N is the number of genes investigated. An extremely small number of very long concatenated introns in the GC-poor genes from *Arabidopsis* are not represented. (From Carels and Bernardi, 2000).

position. The functional relevance of the two classes of genes is stressed by (i) the conservation in homologous genes from maize and *Arabidopsis* not only of the number of introns and of their positions, but also of the relative size of concatenated introns (**Fig. 8.7**); and (ii) the existence of two similar classes of genes in vertebrates; interestingly, the differences in sizes and numbers of introns in genes from the GC-poor and GC-rich classes are much more striking in plants than in vertebrates. The two classes of genes which differ in intron size have been very recently confirmed (Yu et al., 2002) for *Arabidopsis* and human.

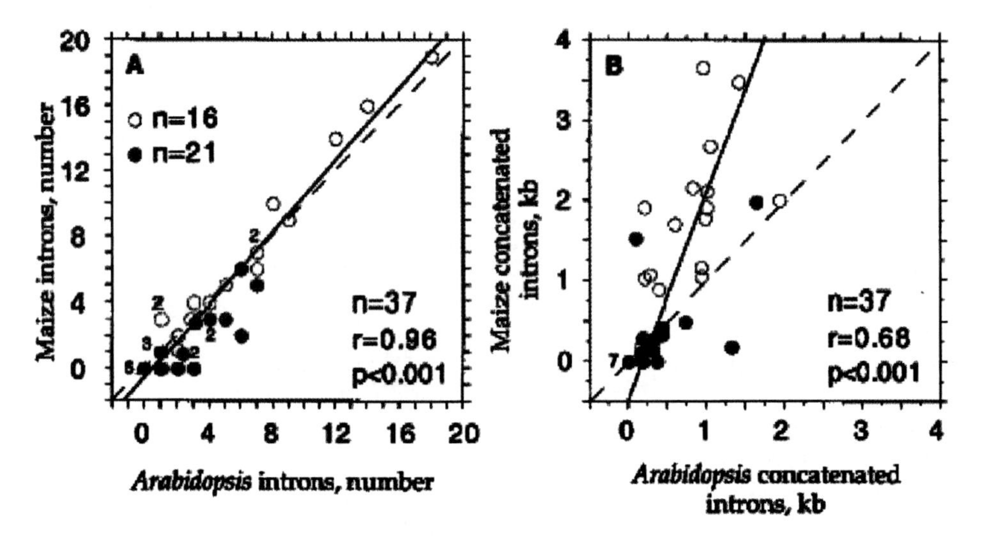

Figure 8.7. Number of introns and length of concatenated introns in homologous genes of maize and *Arabidopsis*. GC-rich genes are represented with solid circles and GC-poor genes with open circles. Number close to circles are the number of genes represented. The solid line is the orthologous regression line through the points, and the broken line is the diagonal. (From Carels and Bernardi, 2000).

An important implication of the results of Carels and Bernardi (2000) derives from the observation that the two classes of genes are conserved in two genomes that widely differ in their GC and GC_3 ranges. Indeed, these ranges are very narrow in *Arabidopsis* (see **Figs. 8.13** and **8.14**) and very broad in maize (see **Fig. 8.8**). Since monocots split from ancestral dicots (about 200 Mya), this suggests that the two classes of genes preceded in evolution the compositional genome transition that led to the increased GC levels of a large percentage of genes from *Gramineae*.

CHAPTER 3

Gene distribution in the genomes of plants

3.1. The gene space in the genomes of Gramineae

The first attempt to investigate the gene distribution in plant genomes was made by localizing 23 nuclear genes from three dicotyledons, pea, sunflower and tobacco, and five monocotyledons of the *Gramineae* family, barley, maize, rice, oat, and wheat, in DNA fractions obtained by preparative centrifugation in Cs_2SO_4/BAMD density gradients (Montero et al., 1990). Each one of these genes (and many other related genes and pseudogenes) was found to be located in DNA fragments (50-100 kb in size) that were less than 1-2% GC apart from each other in any given genome. This was taken as an indication of the existence of isochores in plant genomes, namely of compositionally fairly homogeneous DNA regions at least 100-200 kb in size. Moreover, the GC levels of the 23 coding sequences studied, of their first, second and third codon positions, and of the corresponding introns were found to be linearly correlated with the GC levels of the 'isochores' harbouring those genes. (Coding sequences for seed storage proteins and phytochrome of *Gramineae* deviated from the compositional correlations just described). Finally, CpG doublets of coding sequences were characterized by a shortage that decreased and vanished with increasing GC levels of the sequences.

The general picture which was emerging from investigations on plant genomes suggested a strong similarity with results previously obtained for vertebrate genomes. In apparent agreement with this conclusion, the maize genome (2,500 Mb), on which initially we focused our work, had a relatively similar compositional distribution of DNA molecules and of coding sequences compared to the human genome (3,200 Mb), suggesting a similar gene distribution.

Further investigations led, however, to a completely different picture as far as maize and other *Gramineae* were concerned (for *Arabidopsis* and other plants see **Sections 3.3** and **3.4**). Indeed, when the gene distribution of maize was studied using a larger number of probes, a most striking result was obtained, in that almost all genes were found in DNA fragments (50-150 kb in size) covering an extremely narrow (1-2%) GC range and only representing 10-20% of the genome (Carels et al., 1995). This gene distribution (**Fig. 8.8**), defining what we called a **gene space**, is remarkably different from the gene distribution previously found in the human genome, and deserves several comments. Moreover, storage protein genes were only in part located in the gene space, several of them being comprised in a small **zein space** slightly lower in GC compared to the gene space already mentioned. The correlations originally observed were, in fact, essentially due to the existence of two groups of genes, the GC-poor genes of dicots and the GC-rich genes of *Gramineae*.

(i) The gene space may be visualized as a single family of isochores, since the GC range of the long sequences containing protein-encoding genes (except for some of the zein genes) is

of all the genes tested (histone H3; rbcs, ribulose-1,5-biphosphate carboxylase; Cab, chlorophyll-a/b binding protein; α -amylase) on DNA fractions yielded results essentially identical to those obtained on maize fractions (Montero et al., 1990), Similar conclusions can be drawn from hybridization experiments using maize gene probes. Moreover, probes for genes of storage proteins (high molecular weight glutenin; α , β-gliadin) hybridized to fractions exhibiting slightly lower buoyant densities compared to other genes (Salinas et al., 1988; Montero et al., 1990), as was the case with zein genes in maize.

In the case of rice and barley, we found (**Fig. 8.9**) that the distribution of genes in these genomes is basically similar to that of maize in that all genes, except for ribosomal genes and some storage protein genes, were located in gene spaces that (1) cover GC ranges of 0.8%, 1.0% and 1.6% and represent 12%, 17% and 24% of the genome of barley, maize and rice, respectively (see **Table 8.1**), and largely consist of transposons; (2) have sizes approximately proportional to genome size, suggesting that expansion-contraction phenomena proceed in parallel in the gene space and in the gene-empty regions of the genome; and (3) only hybridize on the gene spaces (and not on the other DNA fractions) of other *Gramineae*.

(vi) The gene space consists of a number of small compositional compartments located on all chromosomes, as judged from the known chromosomal location of maize genes (Ahn et al., 1993). Most of the loci tested in wheat physically map in the distal region of the chromosomes (Flavell et al., 1993), where most recombination events also occur (Flavell et al., 1993; Gill et al., 1993). In contrast, the pericentromeric region of a rye chromosome (Moore et al., 1993) shows a reduced density in unmethylated Not I sites, which are indicative of CpG islands associated with genes (Antequera and Bird, 1988; Gardiner-Garden et al., 1992). Interestingly, the high gene concentrations and the high recombination levels detected in telomeric regions of the wheat genome are reminiscent of similar findings on the human genome (Saccone et al., 1992, 1993) and may reflect a general situation.

(vii) The sequence of the *Adh-1* 280 kb region of maize (**Fig. 8.10**) published by San Miguel et al. (1996) has shown that the composition of this region (47% GC) exactly corresponds to that of the gene space. This region is very largely made up of nested transposons. Indeed, retrotransposition insertions have doubled the size of the maize genome in the last three million years (San Miguel et al., 1998). Presence of species-, genus-

TABLE 8.1

The gene spaces of *Gramineae*[a]

	Compositional range	**GC**	**Relative amount in genome**	**Size**[c]
Plant	**ρ, g/cm^3**[b]	%	%	**Mb**
Maize	1.7020-1.7030	1	17	425
Rice	1.7024-1.7040	1.6	24	100
Barley	1.7017-1.7025	0.8	12	690

[a] From Barakat et al. (1997)

[b] Standard errors on buoyant density values are ± 0.0002

[c] The sizes of the gene spaces were calculated by multiplying their relative amounts by the genome sizes (as obtained from Shields, 1993): 2,500 Mb for maize, 415 Mb for rice, and 5,300 Mb for barley

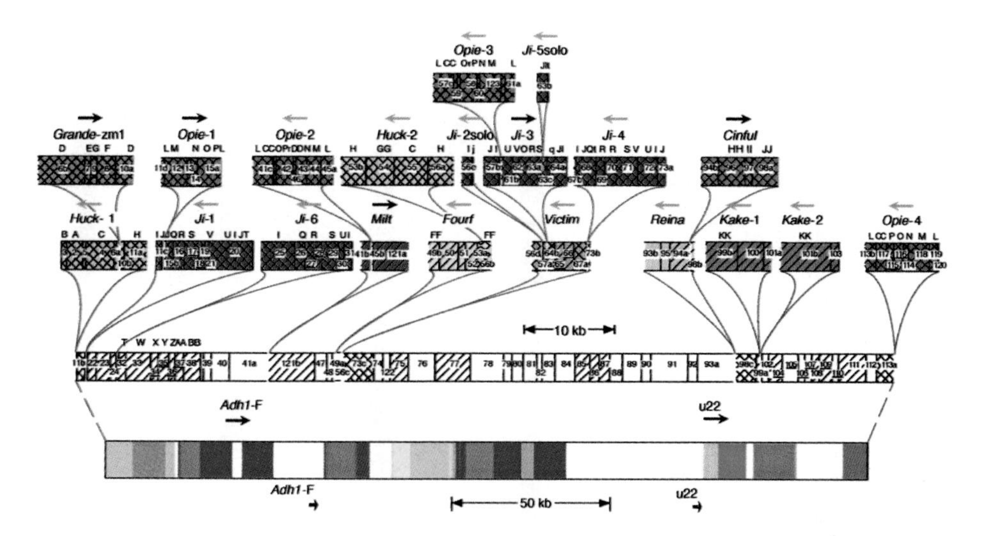

Figure 8.10. Structure of the *Adh1*-F region of maize, showing identified retrotransposons. Bar patterns correspond to the approximate copy numbers: cross-hatched bars, highly repetitive (thousands of copies; hatched bars, repetitive (hundreds of copies); and open bars, low copy number (< 100 copies). Letters above bars indicate the class of repetitive DNA, as defined by hybridization of each highly repetitive fragment to all fragments in the region. Confirmed elements have been positioned above the DNA into which they have inserted. Curved lines below each element converge at the insertion site. The arrow above each element indicates its orientation. Fragments with lower case a, b, or c designations are components of the fragment with the same number. (From San Miguel et al., 1996).

and family-specific repeats strongly suggest repeated invasions by different retrotransposons over time (Sandhu and Gill, 2002, and references therein). Needless to say, this abundance of retrotransposons in the intergenic sequences of the gene space eliminate any compositional correlation between coding sequences and the surrounding non-coding sequences. This is in sharp contrast with the situation found in *Arabidopsis* (see following **Chapter 3.2**).

3.2. Misunderstandings about the gene space of Gramineae

A recent paper (Meyers et al., 2001) claimed that "*An analysis of the GC content of the maize genomic library and that of maize ESTs did not support recently published data that the gene space of maize is found within a narrow GC range*". The claim made by the authors is puzzling in that our actual observations of gene-rich fractions are disputed on the basis of "*the demonstration that maize coding sequences have highly variable GC content*" (something we have known since the work of Salinas et al., 1988), "*that the Huck family of retroelements has a markedly different GC content with respect to the rest of the genome*" (also a well-known point) and "*that Huck is extremely frequent and dispersed in the genome, as well as associated with all other abundant class of repeats. It is therefore, unlikely that its*

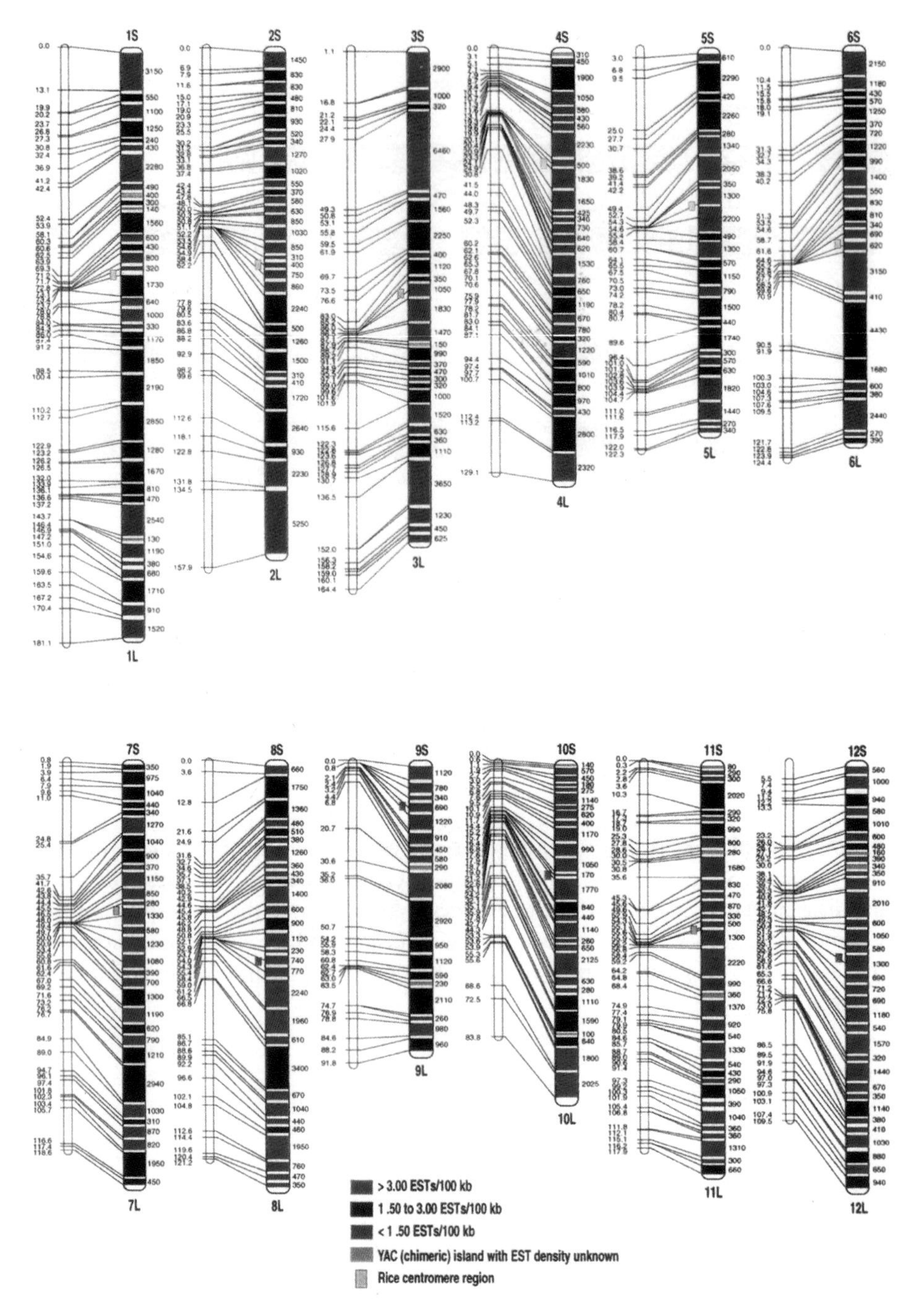

Figure 8.11. Chromosomal distribution of the 6591, rice EST sites. Genetic (left; cM) and YAC-based physical (right; kb) maps are shown for each chromosome. The solid lines across each chromosome show the positions and regions of YAC contigs on the genetic map. Different colors represent different EST densities observed on the individual contigs. (From Wu et al., 2002).

distribution could justify the observations of the Bernardi group... which is still publishing the conflicting claim that nuclear genes in Gramineae are found in a 1-2% GC range (Barakat et al., 2000)". Since we were unable to convince the authors with what they call "*putative gene rich fractions*", we present in **Fig. 8.10** the results of San Miguel et al. (1996) which show the situation of transposons in the 280 kb *Adh* locus of maize and where the GC level is exactly that found (47%) by hybridizing the *Adh* probe on DNA fractions. One should not forget that the majority of maize transposons are in the 42-47% GC range, with Huck being 60% GC, but only representing about 10% of the maize genome and being concentrated around centromeric regions (Ananiev et al., 1998) which represent at least 30% of maize chromosomes. It is understandable, therefore, that the gene space is about 47% GC, and coincides with the modal buoyant density of maize DNA.

Another objection raised for the gene space by Dubcovsky et al. (2001) and San Miguel et al. (2002) was that BACs (Bacterial Artificial Chromosomes) from barley, rice and wheat do not differ significantly in composition from total DNAs. This is, however, not surprising if one considers that the gene spaces of *Gramineae* explored so far practically coincide with the modal buoyant density of DNA (see **Fig. 8.9**).

A similar misunderstanding was apparently due to a superficial reading of our papers. Sasaki et al. (2002) state, for instance, that "*Buoyant density experiments have shown that rice genes are localized in (G+C)-rich islands that occupy 24% of the genome (Barakat et al., 1997). When we plotted the average G+C values against chromosomal position in chromosome 1, however, we did not detected any CpG islands indicating a neutral nucleotide distribution*". As shown in **Fig. 8.9**, the "*islands*" (to use the authors' nomenclature) forming 21% of the genome (see **Table 8.1**) are very close to the modal GC level of rice DNA and are not GC-rich. It is, therefore, not surprising that G+C-rich islands (mistakenly called CpG islands by the authors) were not found. Incidentally, the assessment of genes by Sasaki et al. (2002) presents the same problem as that of Yu et al. (2002; see **Fig. 10.13**), namely an extraordinary abundance of imaginary genes.

In fact, our original observation of a gene space in maize (Carels et al., 1995) was followed by other findings pointing towards the existence of gene-rich regions in cereals. Studies on the genome organization of wheat suggested that more than 85% of the genes are present in less than 10% of the chromosomal regions (Gill et al., 1996a,b; Sandhu and Gill, 2002). Results by other authors (Panstruga et al., 1998; Feuillet and Keller, 1999) point in the same direction. The most recent and comprehensive evidence comes from the mapping of over 6500 Expressed Sequence Tag (EST) sites in rice (Wu et al., 2002; see **Fig. 8.11**).

3.3. The gene space of other plants

In order to decide whether the genome organization and gene distribution found in *Gramineae* are specific to this family of plants or related to genome size, we have studied the gene distribution in two dicots, pea and tomato, and one non-graminaceous monocot, date-palm (Barakat et al., 1999). The genome of these plants are 5,000, 1,000 and 250 Mb in size, respectively. While in pea most genes are localized in a gene space formed by long gene clusters that cover a 1% GC range and represent about 20% of the genome, in tomato

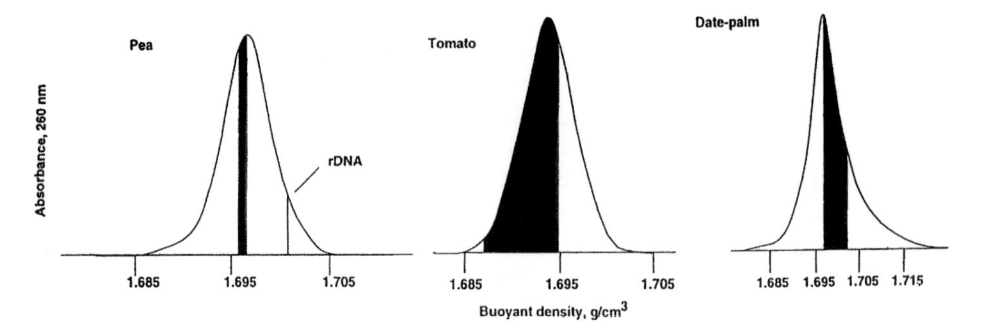

Figure 8.12. CsCl profile of nuclear DNAs from pea, tomato and date palm is obtained by centrifugation in an analytical density gradient. The black areas correspond to the gene space. (From Barakat et al., 1999).

the vast majority of genes are spread over the genome, covering a 5% GC range which represent 60% of the genome. In date-palm, genes are distributed in DNA fractions covering a 5% GC range and representing 40% of the genome (**Fig. 8.12**). We conclude that the concentration of genes in a narrow gene space is not an exclusive property of *Gramineae,* because it is also found in a dicot, pea, whereas tomato and date-palm show gene distributions which are closer to that of *Arabidopsis* (see following chapter) than to that of maize.

3.4. Distribution of genes in the genome of Arabidopsis

When we analyzed (Barakat et al., 1998) the small (*ca.* 115 Mb) nuclear genome of *Arabidopsis thaliana*, which comprises more than 25,000 genes, we could show that its organization is drastically different from that of the genome of *Gramineae* (see **Fig. 8.13**). Indeed, (i) genes are distributed over about 85% of the main band of DNA in CsCl and cover an 8% GC range; (ii) ORFs are fairly evenly distributed in long (> 50 kb) sequences from GenBank that amount to about 10 Mb; and (iii) the GC levels of protein-coding sequences (and of their third codon positions) are correlated with the GC levels of their flanking sequences (**Fig. 8.13**).

The recent genome sequence of *Arabidopsis* (Arabidopsis Genome Initiative, 2000) allowed us to construct the corresponding compositional map (**Fig. 8.14**) which is characterized by only a few moderately GC-rich regions in an overall GC-poor genome. This dicot contains only 20-30% of repeated sequences, half of which are highly repetitive, the other half being moderately repetitive (Meyerowitz, 1992). Moreover, the *Arabidopsis* genome contains very few retrotransposons (the *Ta* and *Athila* families occur in 15 and 150 copies, respectively; Thompson et al., 1996) and is characterized by a very low methylation level (less than 6% of all C are methylated; Leutwiler et al., 1984), the majority of methylation being in centromere-associated tandem repeats, rDNA arrays, and transposons or retrotransposon-derived elements.

Finally, the fact that other members of the *Brassicaceae* family (which diverged very recently from *Arabidopsis*) are endowed with large genomes supports the view that the small genome of *Arabidopsis* is the result of a marked contraction, in which all (or most) of

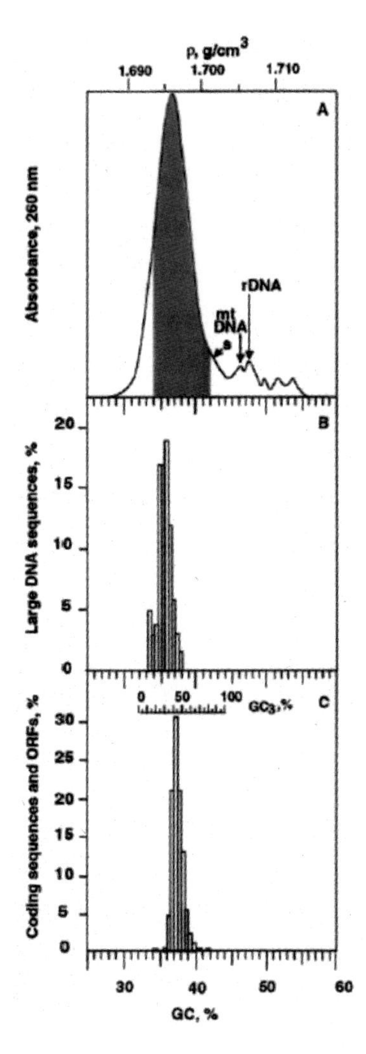

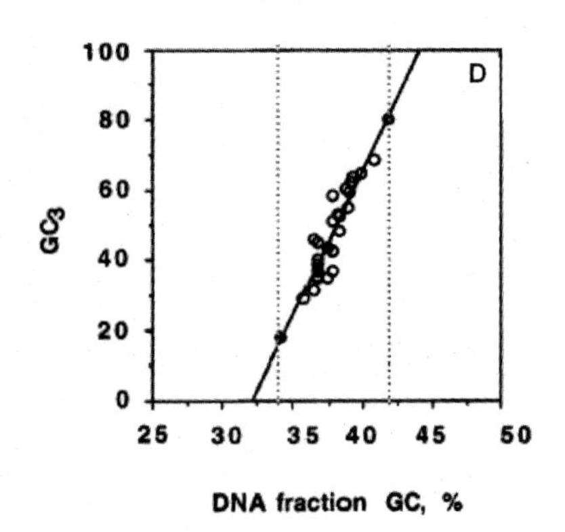

Figure 8.13. **A.** Absorbance profile of *Arabidopsis* nuclear DNA as obtained by centrifugation in a CsCl analytical density gradient. The shoulder may correspond to contaminating chloroplast DNA, the following small peaks to contaminating mitochondrial DNA (ρ = 1.706 g/cm^3), rDNA (ρ = 1.707 g/cm^3), and to three satellite DNAs. The shaded area corresponds to the DNA fractions containing nuclear protein-encoding genes. **B.** Compositional distribution of large (> 50 kb) GenBank DNA sequences from *Arabidopsis*. **C.** Gene distribution obtained by plotting the relative number of *Arabidopsis* genes against their GC_3 values (top scale); 2,490 sequences from GenBank (release 103; October 15, 1997) were used to construct the histogram. In **C**, the common GC abscissa of the three plots represents the GC values of the DNA fractions containing the genes. **D.** Plot of GC_3 of genes (circles) versus GC values of DNA fractions corresponding to the hybridization peaks. The solid circles represent the two extreme GC_3 values of *Arabidopsis* genes as found in GenBank. The vertical broken lines indicate the GC range of the DNA fractions containing the genes. This was used to define in **A** the DNA (shaded area) range in which genes are located. (From Barakat et al., 1998).

the large gene-empty regions separating the gene clusters disappeared, as also did most transposons located in intergenic sequences. The contraction undergone by the *Arabidopsis* genome is reminiscent of that shown by the genomes of *Tetraodontiformes*. In this case, highly and moderately repeated sequences were greatly reduced (Pizon et al., 1984) compared with the genomes of its ancestral order *Perciformes* and introns also underwent a contraction. Similar phenomena have been observed in the *Arabidopsis* introns (Carels and Bernardi, 2000).

Chromosome 1

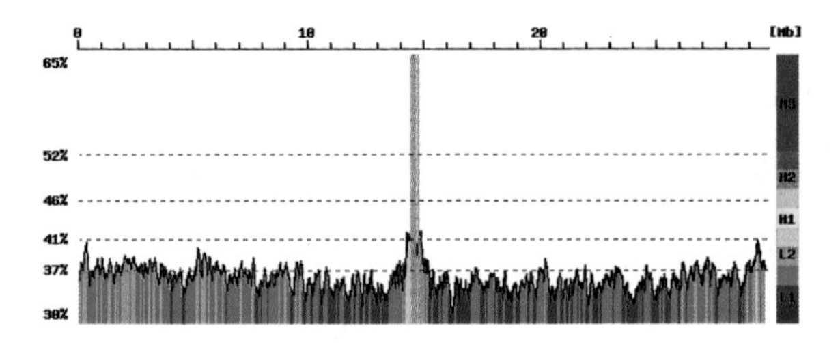

Chromosome 2

Chromosome 3

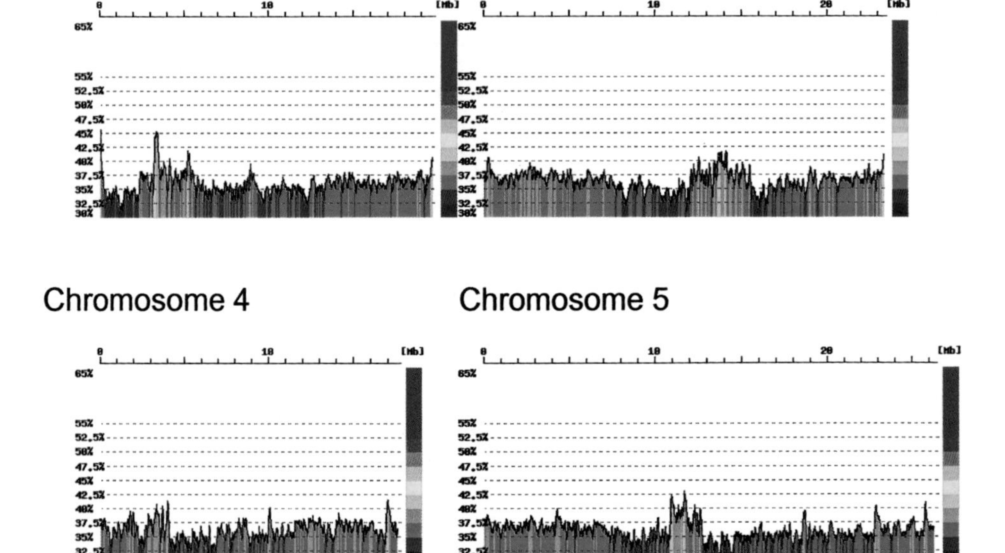

Figure 8.14. Compositional profiles of *Arabidopsis thaliana* chromosomes. (Modified from Pačes et al., 2004).

3.5. *A comparison of the genomes of* Arabidopsis *and* Gramineae

Two major differences distinguish the genome of *Arabidopsis* from the genomes of *Gramineae*. The first one concerns the nature of intergenic sequences. In the genomes of *Gramineae*, intergenic sequences within the gene clusters that form the gene space (see **Fig. 8.16**) are compositionally rather homogeneous, and largely formed by transposons (San Miguel et al., 1996). Indeed, all transposons investigated so far, which include *Mu*, *Ac* and the majority of *Cin4* elements, are exclusively located within the gene space. This is

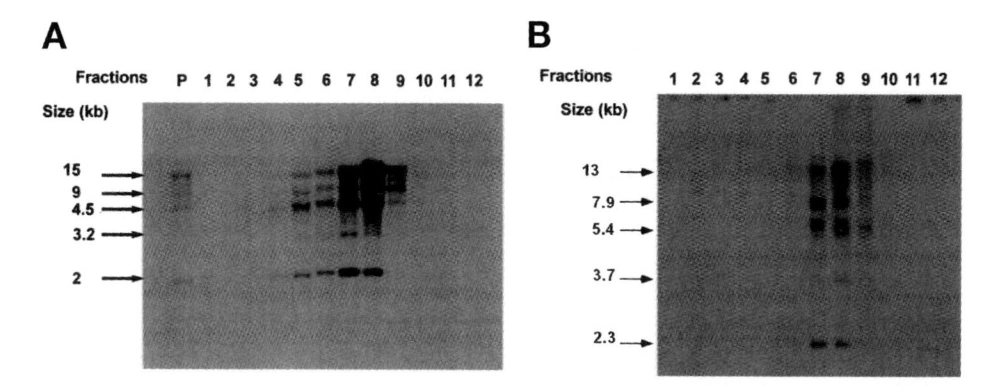

Figure 8.15. **A.** Localization of T-DNA integration in the genome of transformed *Arabidopsis*. Transgenic plants were obtained by transformation with *Agrobacterium* plasmids, which contained a *uidA* (GUS) reporter gene. A GUS probe was hybridized on Cs_2SO_4/BAMD DNA fractions restricted with *Hin*dIII, an enzyme not cutting the GUS sequence. *Arabidopsis* DNA was pooled from 30 plants. P, pellet; 1-12, DNA fractions. **B.** Hybridization of GUS sequence on Cs_2SO_4/BAMD fractions from rice. 1-12, DNA fractions. (From Barakat et al., 2000).

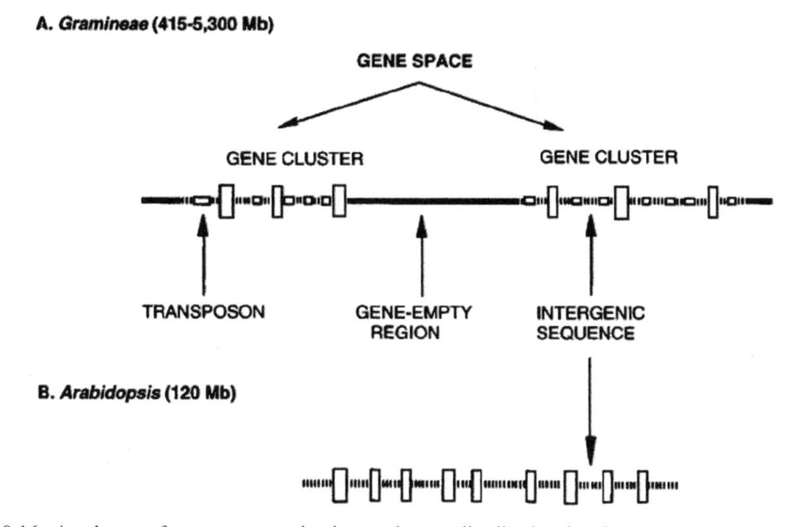

Figure 8.16. A scheme of genome organization and gene distribution in plant genomes. **A.** In the large genomes of *Gramineae*, genes (large vertical boxes) are present in long gene clusters, which are separated from each other by gene-empty regions formed by repeated sequences (thick solid line). The ensemble of gene clusters forms the gene space. The intergenic sequences are compositionally very homogeneous because largely formed by transposons (small horizontal boxes in the intergenic sequences). **B.** The small genome of *Arabidopsis* essentially differs from the genomes of *Gramineae* because of (i) the disappearance (or very strong reduction) of gene-empty regions; (ii) the practical absence of transposons in intergenic sequences; and (iii) the higher gene density. (From Barakat et al., 1998).

also true for the other transposons localized near the *Adh-1* gene by San Miguel et al. (1996). Maize transposable elements appear, therefore, to be located in gene-rich, transcriptionally active regions. As a consequence of the massive integration of transposons into the gene space of *Gramineae*, compositional correlations between coding and flanking sequences are barely detectable. In contrast, in the *Arabidopsis* genome, which comprises a negligible amount of transposons (Thompson et al., 1996), intergenic sequences are compositionally well correlated with coding sequences (see **Fig. 8.13**).

Interestingly, when we investigated the integration of a T-DNA (transferred DNA from *Agrobacterium* plasmids) in the genomes of *Arabidopsis* and rice, we found different patterns of integration, which are correlated with the different gene distributions (**Fig. 8.15**). While T-DNA integrates essentially everywhere in the *Arabidopsis* genome, integration in the rice genome was detected only in the gene space, namely in the gene-rich, transcriptionally active, regions (Barakat et al., 2000). This suggests a correlation between integration and transcriptional activity, in agreement with previous observations on transcribed proviral sequences that also integrate into transcriptionally active regions of mammalian genomes (see **Part 6**).

The second difference concerns gene distribution. As shown in **Fig. 8.16** in the genomes of *Gramineae*, genes are present in long gene clusters separated by long gene-empty regions. The ensemble of long gene clusters forms the gene space which corresponds to 12-24% of the large genomes studied so far. Interestingly, this organization is basically the same over a more than 10-fold range of genome sizes, adding one more basic property to those shared by the genomes of *Gramineae*, like the collinearity of genetic maps and the compositional correlation among orthologous genes. In contrast, in *Arabidopsis*, genes are fairly evenly distributed over regions amounting to about 85% of the genome, whereas

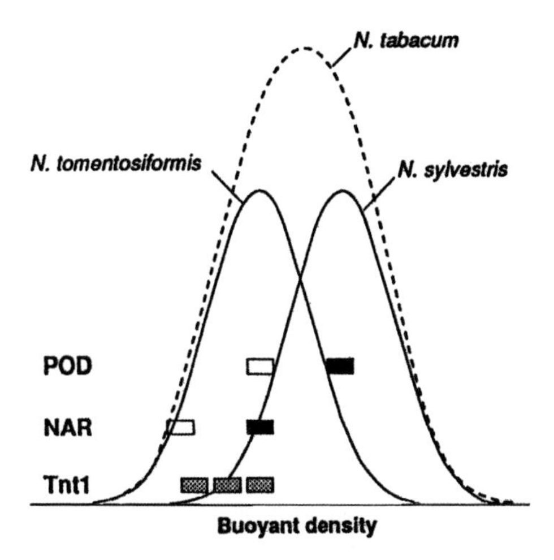

Figure 8.17. Scheme of the DNA and gene distribution in the amphidiploid genome of tobacco. The two Gaussian curves (shifted apart form each other for the sake of clarity) represent the DNA distributions of *N. tomentosiformis* and *N. sylvestris*, respectively; the envelope of the two curves (broken line) represents that of *N. tabacum*. Black boxes represent genes (NAR=nitrate reductase; POD=lignin forming peroxidase) localized in the GC-rich and GC-poor isochores of *N. tomentosiformis*. Gray boxes represent the distribution of Tnt1, which may be accounted for by its mobility. (From Matassi et al., 1991).

gene-empty regions are greatly reduced and may largely consist of the repeated sequences localized in centromeres and telomeres.

3.6. The bimodal gene distribution in the tobacco genome

When we studied the compositional distribution of six genes (or small multigene families) and one family of transposable elements, *Tnt1*, in DNA fractions from tobacco (*Nicotiana tabacum*) separated according to base composition, we could show that gene distribution is bimodal and that such bimodality is due to the different base composition of the two parental genomes of tobacco (*N. sylvestris* and *N. tomentosiformis*) and to the different parental origin of the genes tested (Matassi et al., 1991).

These results indicate a conservation of the compositional patterns of the two parental genomes, as well as a conservation of the gene localization. (**Fig. 8.17**) Moreover, they suggest the absence of any extensive recombination between two genomes which have been in the same nucleus for a time which has been estimated to be less than 6 million years, whereas the two parental genomes have an estimated life of about 75 million years (Okamuro and Goldberg, 1985).

In conclusion, gene distribution is bimodal in the tobacco genome and this bimodality is due to the amphidiploid nature of this genome. The lack of recombination seen in the tobacco genome fits with the concept of **nuclear architecture**, namely with the concept of spatial ordering of chromosomes such as that seen in sexual hybrids between barley and rye (see Heslop-Harrison and Bennet, 1990, for a review). Interestingly, a spatial chromosome distribution was also reported in the case of hybrids between *N. plumbaginifolia* and *N. sylvestris*. Telocentric chromosomes of *N. plumbaginifolia* were positioned predominantly at the periphery of metaphase plates, whereas *N. sylvestris* chromosomes occupied the center (Gleba et al., 1987). Another consideration derived from the tobacco results is that the mixing of two genomes is less obvious than proposed by some authors (see for example, Margulis and Sagan, 2002).

3.7. Methylation patterns in the nuclear genomes of plants

Methylation was investigated in compositional fractions of nuclear DNA preparations (50-100 kb in size) from five plants (onion, maize, rye, pea and tobacco), and was found to increase from GC-poor to GC-rich fractions. This methylation gradient showed different patterns in different plants and appears, therefore, to represent a novel, characteristic genome feaure which concerns the noncoding, intergenic sequences that make up the bulk of the plant genomes investigated and mainly consist of repetitive sequences. Our data, as well as data from literature, indicate that both mC and mC/mC+C increase with increasing GC of plant genomes (**Fig. 8.18**).

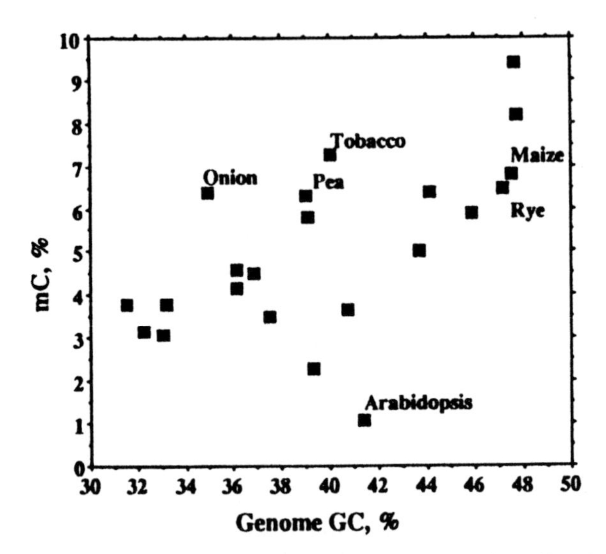

Figure 8.18. Plot of mC against the GC levels of plant DNAs. Plants investigated by Matassi et al. (1992) as well as *Arabidopsis* are indicated. Data for other plants are from references given in the original paper. The correlation coefficient was 0.61 (p=0.0028). (From Matassi et al., 1992).

Part 9
The compositional patterns of the genomes of invertebrates, unicellular eukaryotes and prokaryotes

CHAPTER 1

The genome of a Urochordate, *Ciona intestinalis*

The results obtained on the genome of vertebrates prompted investigations on the genome of other eukaryotic genomes. Among those, the genomes of the immediate ancestors of vertebrates, **Urochordates** and **Cephalochordates**, were of special interest to us in view of our work having been focused on the organization and evolution of the vertebrate genome. Our knowledge of the genome of *Amphioxus*, a Cephalochordate, is still too limited to be discussed here. In contrast, we have recently obtained some information about the genome of a Urochordate, *Ciona intestinalis*, which will be presented in this Chapter.

The CsCl analytical profile of DNA from *Ciona* is characterized by a main peak with a modal buoyant density of 1.6945 g/cm^3, corresponding to 35.16% GC, and by two minor peaks characterized by a lower (1.6707 g/cm^3) and a higher (1.7096 g/cm^3) buoyant density (**Fig. 9.1A**). The lighter peak probably is mitochondrial DNA, the heavier peak was demonstrated to correspond to ribosomal DNA by hybridization of an appropriate probe on a preparative CsCl gradient of *Ciona* DNA. The analytical CsCl profile of the main peak of **Fig. 9.1A** covers a narrow range. The standard deviation of the CsCl profile (which is the sum of standard deviations due to compositional heterogeneity and Brownian diffusion), was estimated to be 2.36 by fitting a gaussian curve (**Fig. 9.1B**).

Sixteen single-copy coding sequences were hybridized on DNA fractions from shallow CsCl gradients. GC levels of the fractions corresponding to hybridization peaks are indicated on the analytical CsCl profile in **Fig. 9.2A.** The GC and GC$_3$ levels of the coding sequences are shown in **Fig. 9.2B**. They cover most of the compositional spectrum of the

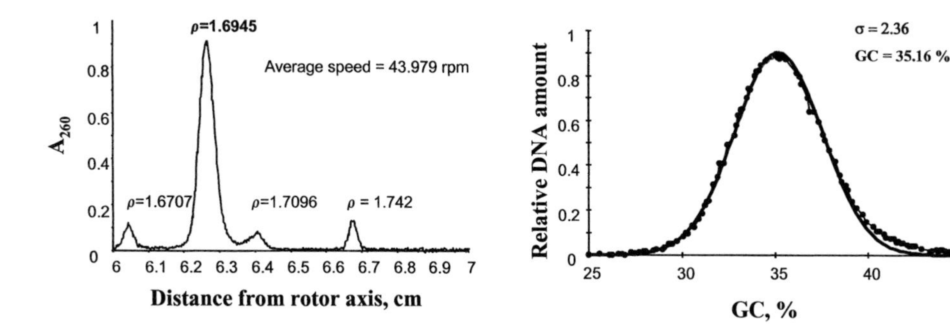

Figure 9.1. **A.** Profile of *Ciona* DNA (molecular size > 100 kb), as obtained by analytical ultracentrifugation to sedimentation equilibrium in a CsCl gradient. The 1.742 g/cm^3 peak is the peak of the marker DNA from phage 2C. **B.** Fit of analytical profile of *Ciona* DNA (curve) with a gaussian curve (dotted line). This analysis was performed using the MacCurveFit program. The GC scale was obtained from the buoyant density values using the relationship of Schildkraut et al. (1962). (From De Luca et al., 2002).

available coding sequences (see below) from *Ciona*.

Fig. 9.3A shows a plot of GC_3 levels of the sixteen coding sequences analyzed against the GC levels of the corresponding long sequences (GC fractions) in which the coding sequences were localized. The regression equation was used to position the GC_3 distribution of the coding sequences available in data bank (about 100 sequences) relative to the CsCl profile (**Fig. 9.3B**).

The local gene concentration was then calculated by dividing each bar value of the GC_3 histogram (bins of 5% width) by the corresponding value on the CsCl profile. This approach (Zoubak et al., 1996) estimates gene concentration across the compositional range of the genome and shows that gene density in *Ciona* is only shifted by 1% GC towards higher values from the DNA peak obtained by analytical ultracentrifugation (**Fig. 9.3B**). A very similar result was obtained by hybridizing a cDNA preparation from a mobile larval stage of *Ciona* on DNA fractions from a shallow gradient (**Fig. 9.4**).

In conclusion, the genome of *Ciona* is characterized by a very symmetrical compositional distribution of DNA molecules and by a gene distribution which is barely shifted to higher GC values compared to the distribution of DNA. This is a situation which is remarkably different from that found in fishes where both distributions trail towards high GC values. (see **Part 4**). A speculation (modified from De Luca et al., 2002) on the implication of these differences as far as the formation of two gene spaces of vertebrates are concerned follows.

The very first consideration here is to take into account that genome size is 160 Mb in *Ciona* (Satoh et al., 2002) and about 400 Mb in the fishes endowed with the smallest genomes, like *Arothron diadematus* and other fishes of the order *Tetraodontiformes* (*e.g., Fugu rubripes, Tetraodon viridiformis*; see **Table 4.2**). In the case of *A. diadematus*, the amounts of rapidly and intermediate reassociating repeated sequences are very small (6% and 7%, respectively), in fact the smallest reported so far for a vertebrate genome (Pizon et al., 1984). In other words, even when in its most compact form, a fish genome is more than twice as large as the genome of a urochordate, like *Ciona*. This is a strong indication of a

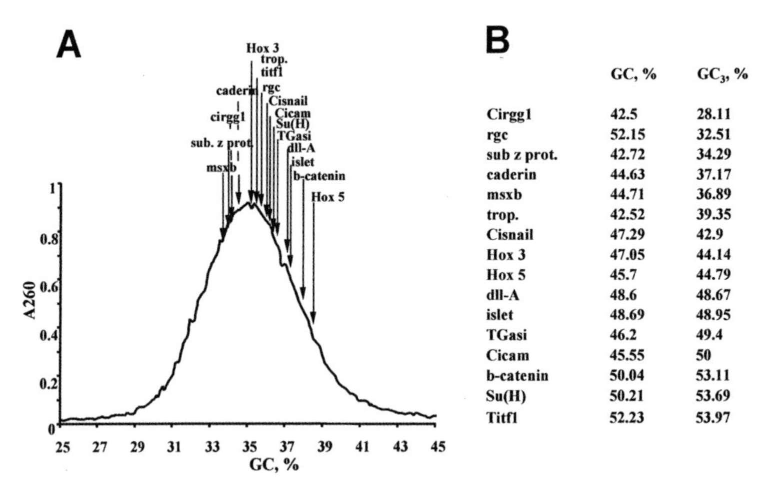

	GC, %	GC_3, %
Cirggl	42.5	28.11
rgc	52.15	32.51
sub z prot.	42.72	34.29
caderin	44.63	37.17
msxb	44.71	36.89
trop.	42.52	39.35
Cisnail	47.29	42.9
Hox 3	47.05	44.14
Hox 5	45.7	44.79
dll-A	48.6	48.67
islet	48.69	48.95
TGasi	46.2	49.4
Cicam	45.55	50
b-catenin	50.04	53.11
Su(H)	50.21	53.69
Titfl	52.23	53.97

Figure 9.2. **A.** Localization of the coding sequences used in hybridisation experiments on the DNA fractions from *Ciona*. **B.** GC and GC_3 levels of the coding sequences are listed. (From De Luca et al., 2002).

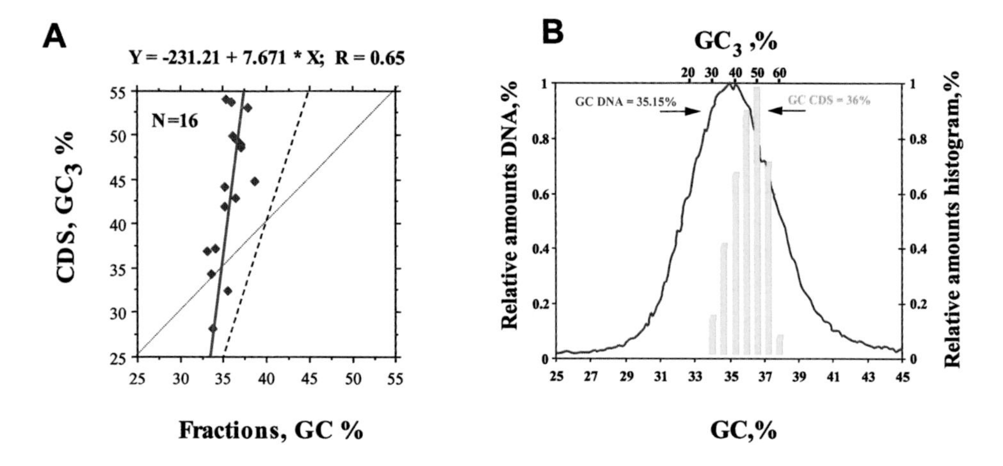

Figure 9.3. **A.** Correlation between GC_3 levels of 16 coding sequences (CDS) from *Ciona* used for hybridization experiments and the GC levels of the DNA fractions in which genes were localized. The range of angles corresponding to a 5% confidence interval, as calculated according to Jolicoeur (1990), is from 80.4° to 84.7°. The slope corresponds to an angle of 82.6°. The corresponding regression line for human coding sequences and the diagonal are also shown by a thick and a thin broken line, respectively. **B.** Profile of gene concentration in the *Ciona* genome, as obtained by dividing the relative number of genes in each 5% GC_3 interval of the histogram of gene distribution (yellow bars) by the corresponding relative amount of DNA deduced from the CsCl profile (blue line). The positioning of the GC_3 histogram relative to the CsCl profile is based on the correlation shown in **A.** (From De Luca et al., 2002).

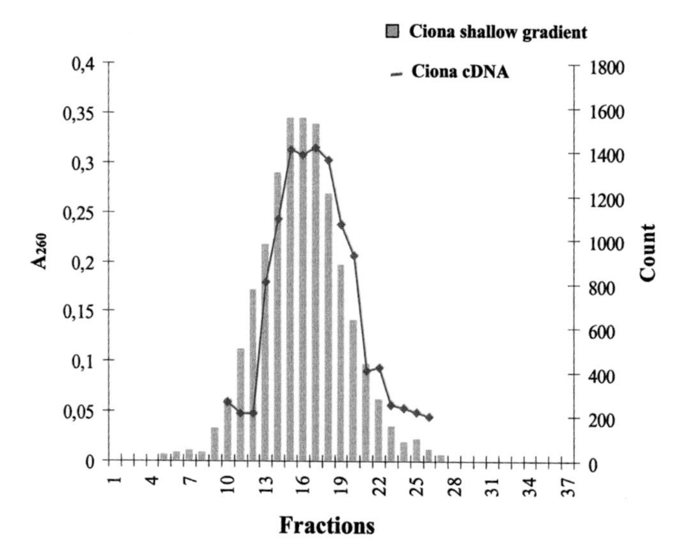

Figure 9.4. Profile of gene distribution obtained using *Ciona* cDNA at mobile larval stage as a probe for hybridization (red curve) The histogram shows the DNA distribution obtained from shallow gradient method. 100 μg of DNA were loaded on the gradient. (From De Luca et al., 2002)

genome duplication in the ancestral line leading to fishes, in agreement with the original proposal by Ohno (1970) and some current ideas. In fact, this genome duplication was accompanied by an increase of intergenic sequences, as indicated by the fact that, neglecting cases of later polyploidizations, as in Salmonids, the "average" genome size of fishes is around 1,000 Mb (Bernardi and Bernardi, 1990a,b; see **Part 4**). This suggests that one possible mechanism for the formation of the two gene spaces in the vertebrate genome was (i) the existence of two compartments characterized by different levels of gene expression in urochordates, a point that can be tested when the annotated sequence of the *Ciona* genome will become available; (ii) a sequence expansion of intergenic sequences (and introns) in the low-expression compartment which became gene-poor compared to the rest; (iii) the preferential integration in the ancestral genome core of copies of duplicated genes (see **Part 6**). This proposal is presented in the scheme of **Fig. 9.5**, which also displays a phylogenetic tree of bilaterian animals.

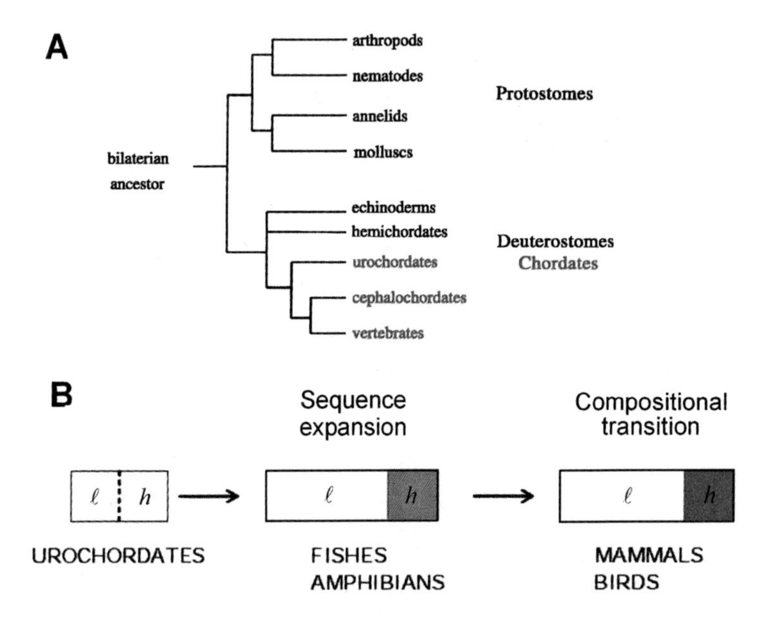

Figure 9.5. **A.** Phylogeny of bilaterian animals. *Ciona intestinalis* is a member of the urochordates, the most primitively branching clade of chordates. Chordates in turn are deuterostomes, one of two great divisions (along with protostomes) of bilaterian animals. (Modified from Delhal et al., 2002). **B.** A hypothetical scheme of genome evolution from urochordates to fishes, to mammals and birds. In urochordates, DNA and gene distribution are essentially uniform. Two "ancestral gene spaces", characterized by a low (*l*) and a high (*h*) level of gene expression, are however, supposed to exist. The genome duplication event leading to fishes was accompanied by a sequence expansion of *l* space, with a consequent decrease of gene concentration in that gene space, and by a preferential insertion of duplicated genes in the "ancestral genome core" which was characterized by a slightly higher GC level (indicated by the pink color). This "ancestral genome core" underwent a further stronger increase in GC indicated by the red color to become the 'genome core' of mammals and birds. (Modified from De Luca et al., 2002).

CHAPTER 2

The genome of *Drosophila melanogaster*

Since our investigations on the *Drosophila melanogaster* genome (Jabbari and Bernardi, 2000) were done before the complete sequence became available (Adams et al., 2000), it was necessary for us to confirm that the sequence sample that we used (1,095 DNA fragments > 50 kb in size) was a good representative of the whole genome. It was shown, therefore, in a study of the compositional distribution of large DNA segments, that the GC histogram of *Drosophila* DNA sequences larger than 50 kb had a good fit to the CsCl profile (**Fig. 9.6**). The latter was obtained using *Drosophila* DNA having approximately the same molecular weight as the large sequences. **Fig. 9.6** also shows the same comparison for the human genome (5,456 DNA fragments of > 50 kb).

The compositional homogeneity of large DNA segments was investigated by plotting the GC levels of 50 and 100 kb segments from both *Drosophila* and human DNAs against those of the following (contiguous) segments (**Fig. 9.7**). Slopes close to unity were found in all cases. Correlation coefficients for the two sizes were 0.71 and 0.85 for *Drosophila*, 0.83 and 0.88 for human. Since correlation coefficients for 10 kb and 20 kb segments in *Drosophila* were 0.58 and 0.67, respectively, whereas in human the corresponding values were 0.79 and 0.83, it appears that correlation coefficients for *Drosophila* reach levels close to the

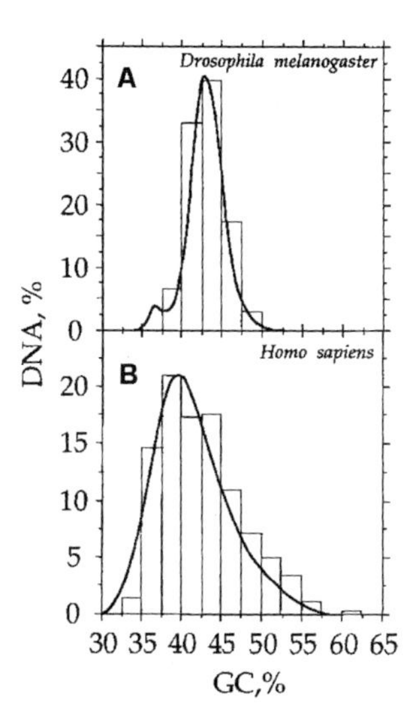

Figure 9.6. The GC histogram of large genomic sequences (> 50 kb) from *Drosophila* (**A**) and human (**B**) are superimposed on the CsCl profiles of the corresponding DNAs, as obtained by analytical ultracentrifugation (from Thiery et al., 1976, and Zoubak et al., 1996, respectively). Buoyant densities were converted to GC levels according to Schildkraut et al. (1962). (From Jabbari and Bernardi, 2000).

human ones only for the larger window sizes (100 kb), whereas the short-range compositional heterogeneity is higher in *Drosophila*.

As far as the plots of **Fig. 9.7** are concerned, it should be noted that a perfectly homogeneous genome would produce a diagram consisting of a single point located on the diagonal, whereas a correlation coefficient of unity (and a unity slope) would correspond to a set of perfectly homogeneous tracts, *i.e.*, to the compositional identity of all pairs of adjacent segments (points outside the diagonal corresponding to compositional differences), and randomly generated sequences would not exhibit any correlation (not shown). Although data available at the time of the work did not allow using window sizes larger than 100 kb, one could still conclude that DNA segments with compositional homogeneity may reach sizes of at least 200 kb, because of the closeness of GC levels of a number of adjacent 100 kb segments.

The compositional correlations of third codon positions with long (> 50 kb) DNA segments in which the coding sequences were present **(Fig. 9.8A)** showed a slope of the orthogonal regression line equal to 5.5 (the range corresponding to a 5% error being 5.1–5.9) and a correlation coefficient r of 0.42 ($p < 0.0001$). These values are consistent with those of the correlations between GC_3 levels and the GC levels of 5' and 3' flanking sequences 20 kb in size. The slope is higher than that observed for human, 4.09 (**Fig. 9.8B**), yet the difference in angles is small, only 3° (see Clay et al., 1996).

The distribution of GC_3 values of *Drosophila* coding sequences (**Fig. 9.8**) covers a range

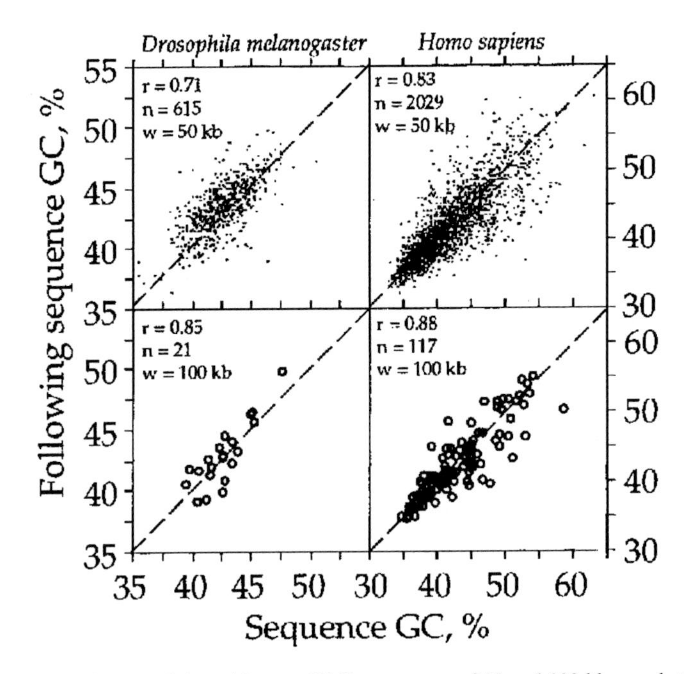

Figure 9.7. GC levels of *Drosophila* and human DNA sequences of 50 and 100 kb are plotted against the GC levels of the following contiguous sequences of the same size. r is the correlation coefficient, n the sample size and w the window size. The diagonal is shown as a broken line. (From Jabbari and Bernardi, 2000).

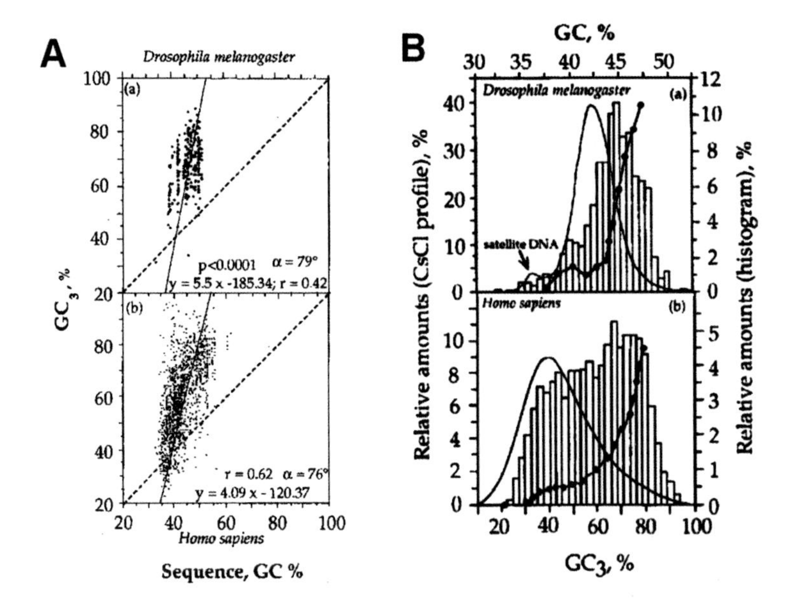

Figure 9.8. **A.** Plot of GC_3 levels of *Drosophila* and human coding sequences against GC levels of large (> 50 kb) DNA sequences embedding the coding sequences. The equations (and slope angles, α) of the orthogonal regression line are shown. **B.** Histogram of gene numbers in 2.5% GC_3 bins from 958 *Drosophila* coding sequences. The regression equation of **A** (upper panel) was used to position the coding sequences histogram on the CsCl profile. Points show the ratios of corresponding values of histogram and CsCl profile, *i.e.*, the relative gene concentration. The lower panel shows the corresponding results for the human genome. (From Jabbari and Bernardi, 2000).

that is comparable to that covered by human coding sequences, but the distribution profile is much narrower and centered on higher values, the mean GC_3 value being 66.2 ± 11.6.

The regression equation of **Fig. 9.8A** (upper panel) was used to position the GC_3 distribution relative to the CsCl profile (**Fig. 9.8B,** upper panel). The local gene concentration was then calculated by dividing each bar value of the GC_3 histogram by the corresponding value on the CsCl profile. This approach (Zoubak et al., 1996) estimates gene concentrations across the compositional range of the genome, and shows that gene density in *Drosophila* is characterized by low levels in the GC-poorest part of the genome (and, expectedly, by no genes in the GC-poor satellite DNA) and by a linear increase over the rest of the genome. **Fig. 9.8B** shows that the profiles of gene distribution are comparable in the human and *Drosophila* genomes, in that gene concentration parallels GC concentration in both cases. However, gene concentration showed a 7-fold difference between the GC-poorest and the GC-richest regions of the *Drosophila* genome *versus* a 20-fold factor in the case of the human genome. A still lower, 3-fold, difference was found in the genome of *Anopheles gambiae*, which, incidentally, is 45% GC (and not 35% GC, as claimed by Holt et al., 2002) on the average (Jabbari and Bernardi 2003b). A factor that could conceivably limit the accuracy of gene distribution estimates in the *Drosophila* genome is the lower correlation coefficient of the plot of GC_3 *vs.* GC of large sequences, compared to the corresponding human plot. However, the error in the slope of the regression lines is small.

Fig. 9.9 shows that the compositional patterns of chromosomes from *Drosophila* are characterized by a predominance of relatively GC-rich structures (corresponding to H1 isochores) with very few GC-richer tracts but a relative abundance of GC-poor stretches (corresponding to L2 isochores).

In conclusion, the *Drosophila* genome shows a similarity to the human genome in its gene distribution, although the gradient of gene concentration is less steep in *Drosophila* compared with human. The evolutionary reasons for such similarities will be discussed in **Part 12, section 1.2**. Moreover, the short range heterogeneity of the *Drosophila* genome is higher than that of human.

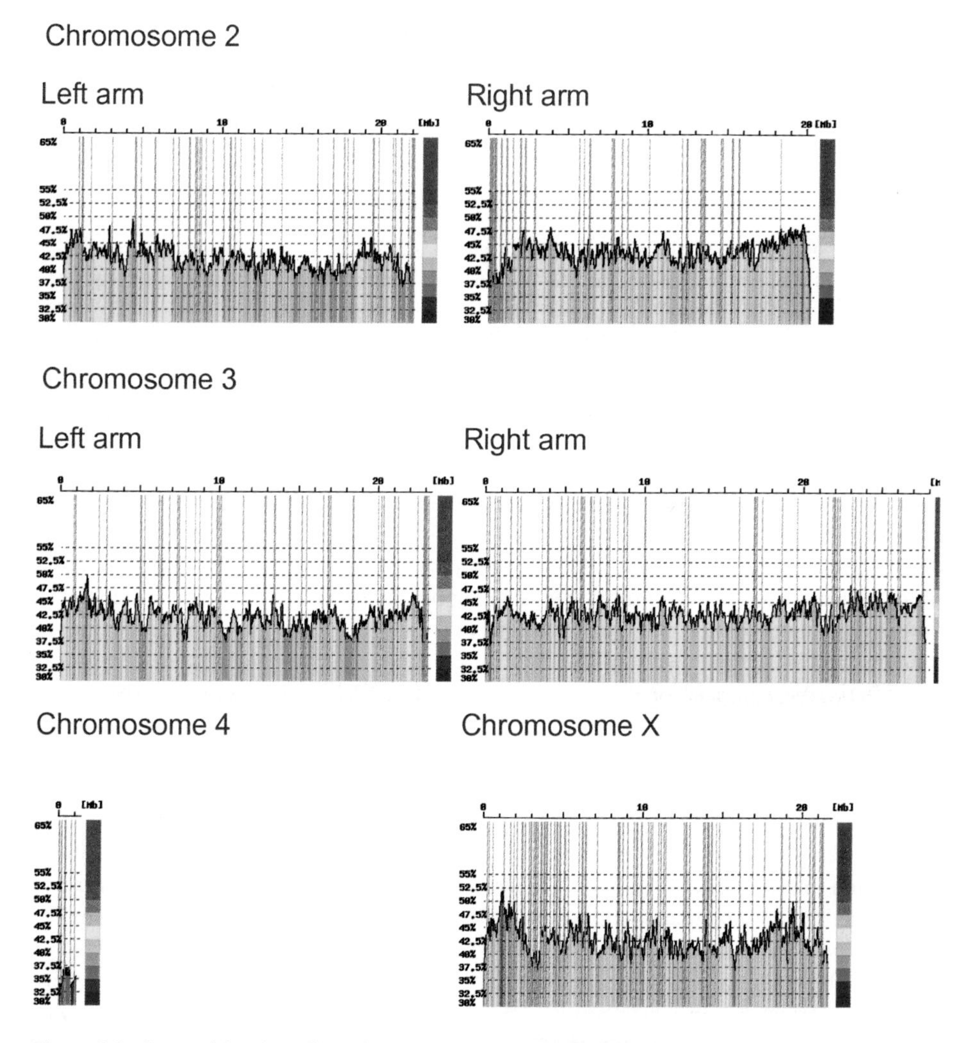

Figure 9.9. Compositional profiles of *Drosophila melanogaster* chromosomes. Vertical grey lines correspond to gaps in the sequence. (Modified from Pačes et al., 2004)

The genome of *Caenorhabditis elegans*

The nuclear genome of *Caenorhabditis elegans* is 97 Mb in size and comprises over 19,000 genes rather uniformly distributed over six chromosomes (*Caenorhabditis elegans* Sequencing Consortium, 1998). Compositionally, its DNA is only 36% GC. The compositional patterns of *C. elegans* chromosomes are presented in **Fig. 9.10** to show the difference between the autosomes, which are GC-poor and relatively heterogeneous, and the X chromosome, which is GC-poorer and more homogeneous. On the X chromosome, genes appear at a lower density and are more evenly distributed compared to autosomes. In contrast, the frequency of EST matches varies according to their position along the auto-

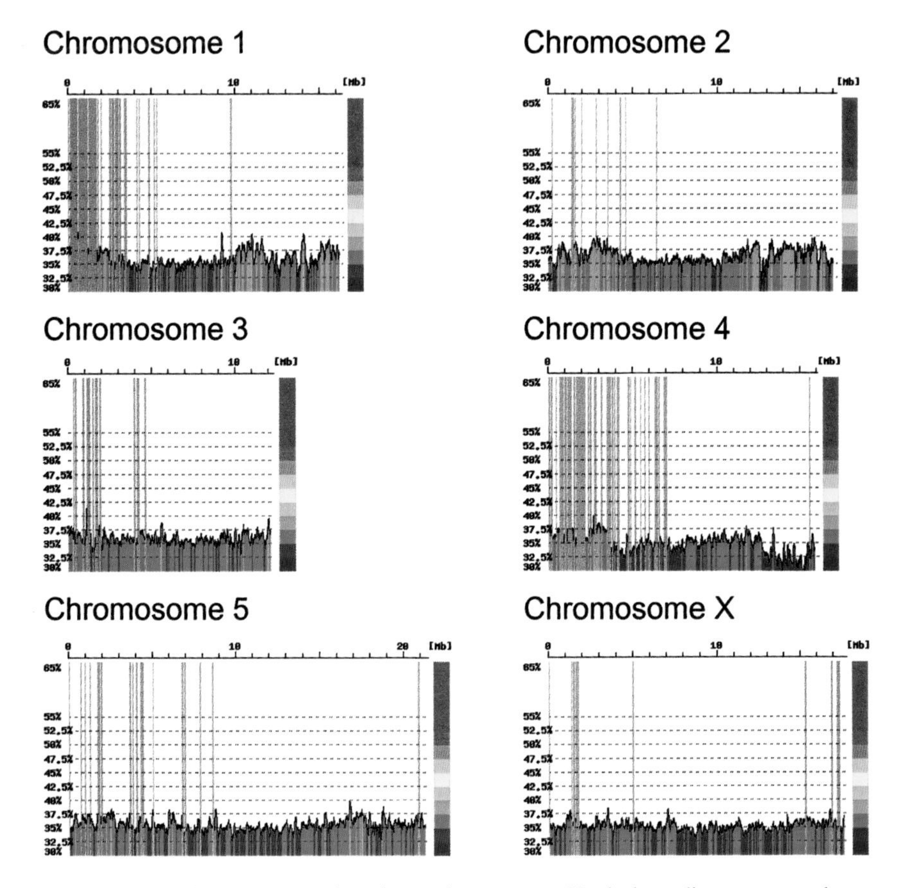

Figure 9.10. Compositional patterns of *C. elegans* chromosomes. Vertical grey lines correspond to gaps in the sequence. (Modified from Pačes et al., 2004)

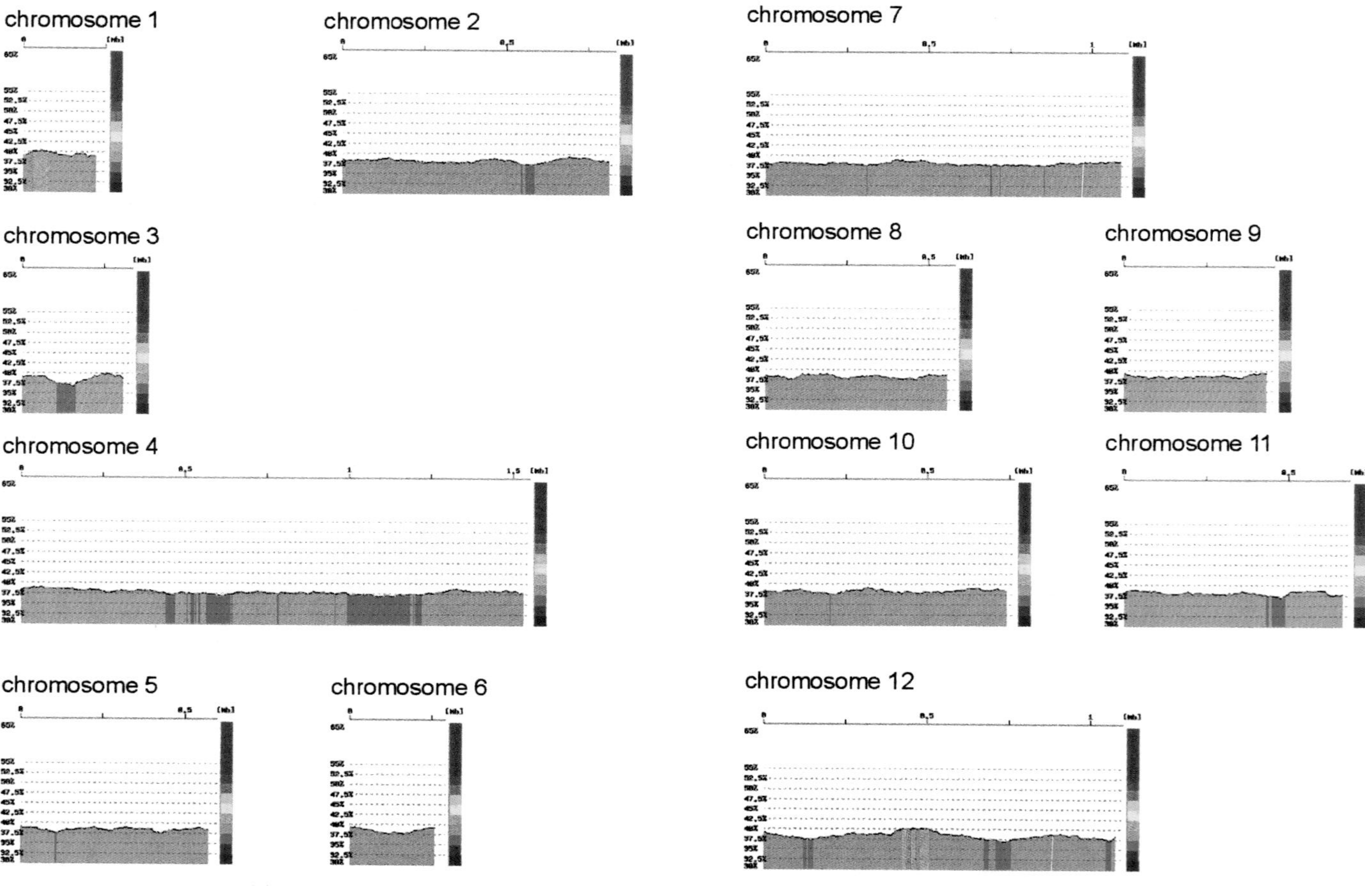

Figure 9.11. Compositional patterns of yeast chromosomes. (Modified from Pačes et al., 2004).

standard deviation (10%) and values from 20% to 97%, and the ESAG genes have a lower GC content compared to housekeeping genes (Michels, 1987; Musto et al., 1995); (3) satellite and middle repetitive DNA are scarce in the *T. brucei* genome.

We have also investigated the compositional distributions of exons and their codon positions, as well as the codon usage and amino acid composition of the nuclear genomes of the African and American trypanosomes *T. brucei* and *T. cruzi*. Very large differences between the two species were found in all the properties investigated. The most striking differences concern the compositional distributions of third codon positions and the extremely large nucleotide divergence of third codon positions for homologous genes encoding proteins that are highly conserved in their amino acid sequences. Moreover, when coding sequences from each species were divided into two groups according to the GC levels in third codon positions, very different codon usages and amino acid compositions were found (Musto et al., 1994).

In conclusion, the compositional patterns of the Trypanosome genomes indicate that **compartmentalization is a phylogenetically very widespread situation in eukaryotes**, as already predicted (Bernardi et al., 1985b).

(iii) **Plasmodia**. A compositional compartmentalization was demonstrated in the nuclear DNA of *Plasmodium cynomolgi*, which consists of isochores likely to average 100 kb (McCutchan et al., 1988).

A striking feature of *P. falciparum*, the unicellular parasite responsible for the most virulent and widespread form of human malaria, is that it hosts the GC-poorest (22% GC) nuclear genome known so far (Pollak et al., 1982, McCutchan et al., 1984). This genome, which only comprises 3 Mb of DNA (Weber, 1988) organized in 14 chromosomes (Kemp et al., 1987; Wellems et al., 1987), is, therefore, an excellent model for studying compositional constraints and their effects.

The analysis of sequence data has provided useful information about the genes encoded in the nuclear genome of *P. falciparum*. The most relevant features are the following: (i) the coding strand is purine-rich; (ii) A is predominant in all codon positions; (iii) the third codon positions are extremely GC-poor and, as a consequence, codon usage is strongly biased (Weber, 1988; Hyde and Sims, 1987).

In an analysis of the nuclear coding sequences from *P. falciparum*, comprising 175 kb (Musto et al., 1995), we found that the trends described previously with more limited data sets are still valid. We then tried to understand whether the biases already noted are species-specific or determined by the composition of the genome, and whether they are different for sequences that display different expression levels and certainly are under different evolutionary pressures, like housekeeping and antigen genes (see also **Part 11**).

In the case of the *P. falciparum* genes that we analyzed here, we noted that the biases in base composition of different codon positions, amino acid frequencies and codon preferences are almost identical in housekeeping and antigen sequences. If we take into account that both the level of expression and the evolutionary constraints over these sequences are most unlikely to be the same, it is reasonable to conclude (i) that the extremely biased composition of the genome (Pollak et al., 1982; McCutchan et al., 1984) is the major factor in determining codon preferences and amino acid frequencies; and (ii) that the compositional constraints (Bernardi and Bernardi, 1986a) operate in the same direction over all the translated sequences and their codon positions (Musto et al., 1995, 1997, 1999).

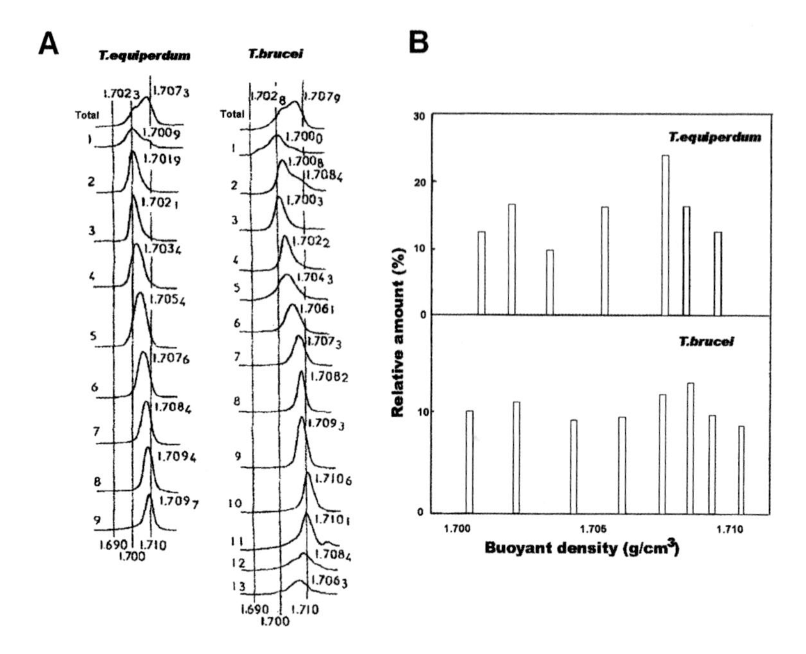

Figure 9.12. **A.** Analytical CsCl profiles of unfractionated (total) nuclear DNAs from *T. brucei* and *T. equiperdum* and of the compositional fractions obtained by preparative centrifugation in a Cs_2SO_4/BAMD gradient. **B.** Histograms displaying the relative amounts of the major DNA components of *T. brucei* and *T. equiperdum,* as estimated from the data of **A.** In *T. equiperdum,* fractions 2 and 3 and fractions 8 and 9, respectively, were pooled because of their very close modal buoyant densities. Likewise, in *T. brucei,* fractions 1, 2 (except for the 1.708 g/cm^3 component) and 3, and fractions 10 and 11, respectively, were pooled. (From Isacchi et al., 1993).

Our results further support the hypothesis that among the various factors that certainly influence the architecture of the coding (and non-coding) sequences, the compositional constraints on the genome (or on isochores) are most relevant, as already proposed (Bernardi et al., 1985b).

CHAPTER 5

Compositional heterogeneity in prokaryotic genomes

5.1. CsCl gradient ultracentrifugation and traditional fixed-length window analyses

Almost fifty years ago, Lee et al. (1956) reported that bacterial DNAs cover a very wide spectrum of base compositions, a result confirmed by Belozerski and Spirin (1958). Shortly after Meselson et al. (1957) introduced density gradient centrifugation of DNA as a new experimental approach in molecular biology, Rolfe and Meselson (1959) and Sueoka et al. (1959) discovered that the modal buoyant density of bacterial DNAs was correlated with their average base composition, GC, the molar ratio of guanine+cytosine. At the same time, they realized that bacterial DNAs were characterized by very narrow band profiles compared to mammalian DNA. The difference in heterogeneity was overestimated by the fact that the standard mammalian DNA used at that time, calf thymus DNA, comprises 23% GC-rich satellites (as we showed 20 years later; Macaya et al., 1978). The idea of remarkably homogenous genomes stuck, however, to bacteria until present. A detailed analysis by Bernaola-Galvan et al. (2004), outlined below, refutes a widely accepted idea that bacteria and archaea normally exhibit low GC contrasts. This study extended an early compilation of heterogeneity data from ultracentrifuge work on *Clostridium, Haemophilus, Micrococcus, Streptococcus, Bacillus subtilis* and *Escherichia coli* (Cuny et al., 1981).

Fig. 9.13 shows intragenomic GC heterogeneity at different scales for human and four bacterial species, as assessed using the traditional fixed-window approach. The difference between a homogeneous (*Chlamydia trachomatis*) and a heterogeneous example (*Xylella fastidiosa*) is striking. Whereas *C. trachomatis* exhibits only a slightly higher heterogeneity than a random sequence, at all scales from 100 bp to 100 kb, *X. fastidiosa* approaches the heterogeneity of human DNA at scales around 5 kb.

5.2. Generalized fixed-length window approaches

Measuring compositional heterogeneity *via* the standard deviation of GC levels among fixed-length segments has two important advantages. First, the behavior of this measure, including anomalies and biases, is well understood and has been extensively studied for more than a century. Second, in the absence of sequence information, CsCl density gradient ultracentrifugation can be used to obtain good estimates of the fixed-length fragment distribution of a genome's GC levels, and of its standard deviation.

Every measure of genomic heterogeneity has, however, characteristic drawbacks. It is therefore advisable to confirm results obtained *via* one measure by other measures, and by checking that heterogeneities are not simply due to a few short "outlier" regions. For example, the standard deviation of fixed-length fragments is sensitive to outlier DNA

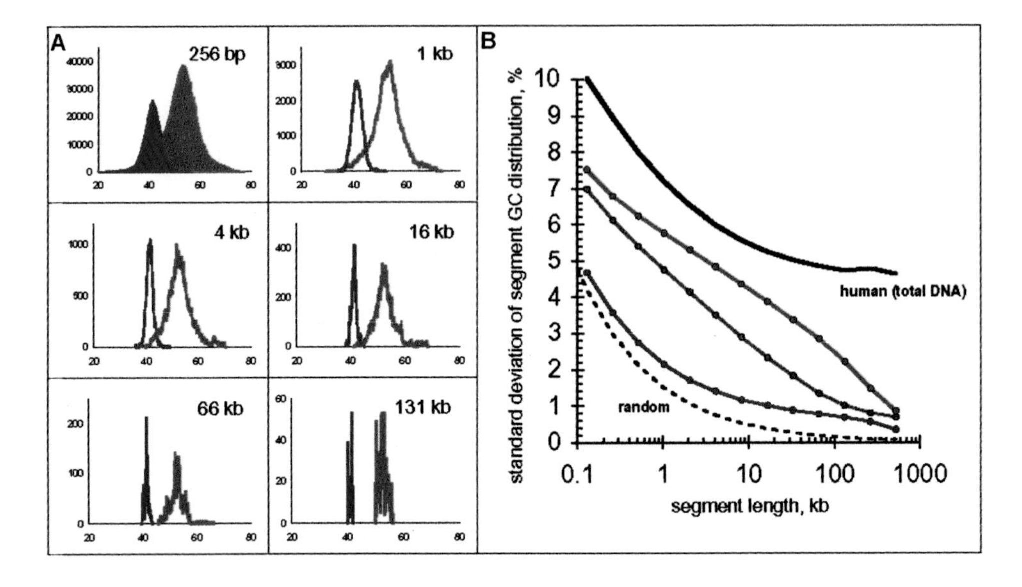

Figure 9.13. Fixed-length window GC level histograms (**A**) and their standard deviations (**B**) for different species. Window sizes chosen for the histograms are successive powers of 2 bp, from 256 bp to 131,072 bp. A homogeneous genome (*Chlamydia trachomatis*; blue curves) and a heterogeneous genome (*Xylella fastidiosa*; red curves) are shown in both representations. The standard deviations are shown also for the human genome (highest, black curve with plateau due to isochore structure) and for a random DNA sequence consisting of independent, identically distributed nucleotides (lowest, dashed black curve; valid for GC levels between 30% and 70%). An historically important bacterium, *Escherichia coli* (grey middle curve) is also included for comparison. (From Bernaola-Galvan et al., 2004).

having GC levels that are far from the mean (*e.g.*, satellites or rRNA genes). Another drawback of this measure is seen when genomes are organized into relatively homogeneous regions of very different sizes. A plot of standard deviation *vs.* window length, as in **Fig. 9.13**, then only gives an indirect picture. We therefore applied several methods, which allowed us not only to explore different kinds of heterogeneity, but also to confirm the robustness of some conclusions.

The usual fixed-length window approach can be generalized in at least three ways. A first way is to keep fixed windows, but to weight the nucleotides inside each window according to their positions within the window. A second way is to partition the entire sequence, not into fixed-length segments, but into variable-length segments whose lengths are chosen by the GC variation itself. A third way is to find a representation of heterogeneity that separates it into contributions from codon position differences and from heterogeneities above the scale of genes.

Traditional moving-window plots use a box filter: each GC base pair within a fixed length window is counted once. From such windows' GC levels one can recover their standard deviation, the traditional measure of heterogeneity. Alternative filters can be designed, which weight the GC counts according to their position in the window. The weighting can, for example, be made to follow a Gaussian form, in order to emphasize the center of the window (**Fig. 9.14**). When window counts are represented by color or brightness rather than by the height of a line plot, a single plot can accommodate the full range of possible window sizes. From such plots one can quickly locate the homogeneous and heterogeneous regions present in a genome at any scale, and assess their stability when the window size changes. Such plots allow quick comparisons between the overall contrast levels present in different genomes, as can be seen in **Fig. 9.14**. They also localize the regions that cause the contrasts, and indicate if they are confined to a single locus or ubiquitous. *X. fastidiosa* contains one striking, very GC-rich (blue) locus, but also shorter, similarly GC-rich regions throughout its genome. Even after excluding these regions, the overall heterogeneity remains distinctly higher than in *C. trachomatis*. The same conclusion is reached via segmentation analysis (see below).

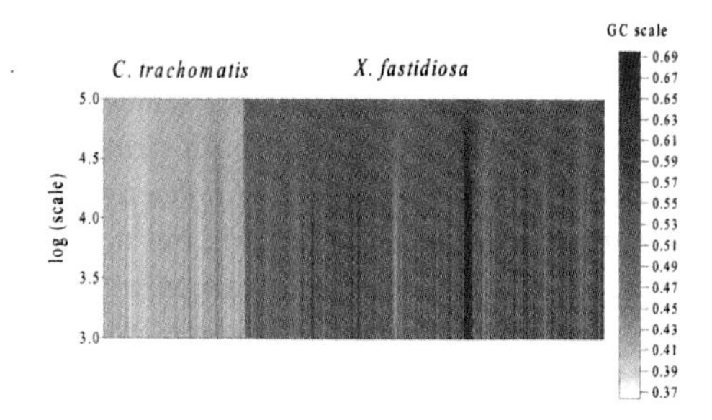

Figure 9.14. Fixed-length moving window plots of the GC heterogeneity of two completely sequenced bacterial genomes, *C. trachomatis* and *X. fastidiosa*, using a modification of the traditional box filter. GC counts are given smaller weights when they are located farther from the window's midpoint, according to a Gaussian weight curve. GC content (expressed as a fraction of 1) is shown as color (dark blue in GC-rich regions). Each horizontal transect is a moving-window plot for a different window size; along the vertical axis, window sizes increase logarithmically from 1 kb to 100 kb. (From Bernaola-Galvan et al., 2004).

5.3. Intrinsic segmentation methods

A second alternative to traditional fixed-length window scans is a rigorous partitioning, using criteria that involve the GC variation within the sequence, into relatively homogeneous segments of different sizes. No restrictions are placed on the lengths of the segments. A genome can be partitioned in this way by recursive top-down segmentation (Bernaola-Galvan et al., 1996). First, the entire sequence is searched to find a partition into two parts that would maximize their GC contrast. A chosen significance threshold or stopping criterion then decides if the contrast is large enough to justify the proposed partitioning. The process is iterated until a contrast between two adjacent (sub)segments becomes weaker than the chosen threshold. We applied a recently developed segmentation variant (Oliver et al., 2001, 2002) which incorporates two further key improvements, making it particularly suitable for locating boundaries between fairly homogeneous, isochore-like regions: (i) the cuts on the sequence are made one by one, in a hierarchical way, which ensures the most statistically significant cut at each scale; and (ii) short-scale sequence heterogeneity below 3 kb is filtered out. Such coarse graining of the sequence corresponds

two or more of these measures, we consistently found the following genomes among the most heterogeneous species: *Xylella fastidiosa, Aeropyrum pernix, Halobacterium sp., Mycoplasma genitalium, Neisseria meningitidis*, the Sakai strain of *Escherichia coli*, and *Vibrio cholerae* (both chromosomes). Among the most homogeneous species we found *Agrobacterium tumefaciens, Chlamydia trachomatis, Rickettsia prowazekii*, and *Campylobacter jejuni*.

Although some genomes' heterogeneities may be partly due to horizontal transfers of DNA from species having different GC levels, such transfers are unlikely to be the only source of heterogeneity in the genomes we have analyzed. For example, the robustly heterogeneous *Mycoplasma genitalium* is often considered to contain almost no recent horizontal acquisitions (Ochman et al., 2000; Kerr et al., 1997). In at least some genomes, therefore, alternative explanations must be considered, such as region-dependent selection on base composition.

5.4. Does intragenomic heterogeneity in E. coli *arise from exogenous or endogenous DNA?*

Prokaryotic genomes show a wide range of intragenomic compositional heterogeneities, which can reach values comparable to those of regional heterogeneities in mammalian genomes (see **Chapter 1**). In fact, we reported, more than 20 years ago, that fish genomes were very often more compositionally homogenous than bacterial genomes in the same size and composition range (Hudson et al., 1980; **Fig. 3.10**)

The presence of considerable GC contrasts within the apparently endogenous DNA of some genomes, and its absence in others, remains enigmatic. The similarities between the GC distributions of prokaryotes and vertebrates in shape and variety (albeit not in width) raise a question that has received relatively little attention until now: could some of the factors shaping GC contrasts in prokaryotes be similar to those shaping the much more pronounced GC contrasts in vertebrates?

The extent to which lateral transfers can be recognized by compositional contrast is still an unsolved issue. Some authors have, in essence, suggested that it is the recently acquired foreign DNA that occupies most of the GC-poor tail of the *E. coli* genome's asymmetric GC distribution, for example, so that without such foreign DNA the GC distribution would be thinner and almost symmetric. Large GC contrasts in *E. coli* are thus interpreted as likely footprints of relatively recent lateral transfers, and are thought to erode with time. This line of reasoning was partly motivated by early comparisons of sequenced genes in *E. coli* and *Salmonella enterica*, which suggested that genes present in one lineage but absent in the other, *i.e.*, genes that could have arrived by lateral transfer, were frequently aberrant in base composition and/or codon usage. Such results suggested that other compositional anomalies might also indicate a foreign origin of the DNA (see, *e.g.*, Groisman et al., 1993; Lawrence and Ochman, 1997; Ochman et al., 2000).

It is still difficult to evaluate the utility of this working hypothesis as a general principle for prokaryotes, since the currently used compositional criteria often use cutoffs that are obtained on the basis of particular assumptions, rather than on a dataset of lateral transfer genes that have been independently verified for the species studied. An alternative to the lateral transfer explanation may be selection on base composition in certain regions of prokaryotic genomes, as was pointed out already by Syvanen (1994). More generally, Sueoka (1992) has offered the following perspective: "*The (relatively) small intragenomic heterogeneity of G+C contents in unicellular organisms such as E. coli and yeast may be regarded as a simple form of isochores (domains of unique G+C contents). It is a likely possibility that the intragenomic heterogeneity of DNA G+C content is a ubiquitous phenomenon from bacteria to mammals and that the only difference is the extent of heterogeneity between unicellular and multicellular organisms.*"

Perhaps the best-known examples of genes that experience compositional constraints in prokaryotes are the ribosomal RNA (rRNA) genes: hyperthermophiles apparently experience strong selective pressure to keep their rRNA genes at high GC levels. The high GC levels ensure the correct secondary structures for the rRNA at high temperatures, and are sometimes maintained in spite of a quite different GC level of the rest of the genome: GC levels of rRNA genes (and tRNA genes) in hyperthermophiles appear to correlate preferentially with temperature, rather than with the overall genomic GC content (Galtier and Lobry, 1997; Hurst and Merchant, 2001); see, however, note added in proof in **Part 12, Chapter 5**). Exceptional compositions of the region surrounding the origin or terminus of replication might also reflect constraints, *e.g.* temperature-dependent constraints for structures that are needed for efficient functioning of the *ori* sequence. Constraints of this type have been shown experimentally for *ori* sequences in bacteriophage G4 and in the mitochondrial genome of yeast (Goursot et al., 1988, and refs. therein). There appears to be no reason, *a priori*, why GC constraints should not also conserve the GC content of some regions containing protein-coding genes.

5.5. Inter- and intra-genomic GC distributions

Datasets collected during the past 50 years consistently show a wide range of genomic GC contents among prokaryotic species and genera, although they do not yet suggest a particular shape of the intergenomic GC distribution. In view of the many and diverse prokaryotic taxa that are still being discovered and the strong biases in currently available samples (Hugenholtz, 2002), it seems that it is still too early for even taxonomically filtered subsamples of the available genomic GC levels to yield a representative distribution of bacteria or archeaea.

Well before the issue of frequent lateral transfers arose, ultracentrifugation results on bacterial GC heterogeneity (Sueoka, 1959; Guild, 1963; Yamagishi, 1974) had indicated that long regions of distinct base composition should exist within the genomes of some bacteria. The existence of such regions has been confirmed by segmentation analyses of GC contrasts in sequenced prokaryotic genomes (Bernaola-Galvan et al., 2004). In some prokaryotic genomes, one finds compositional mosaicism, while in others, such as *M. genitalium*, the variation around the genome resembles a wave rather than a mosaic, but the reasons for the diverse types of systematic, large-scale variation remain an open question. The relative contributions of horizontal transfer from chromosomes of distantly related species, of phages, of recombinogenicity and of selection on base composition, for example, are still largely unresolved.

Part 10
Gene composition and protein structure

CHAPTER 1

The universal correlations

GC_1, GC_2 and GC_3, the GC levels of first, second and third codon positions, are correlated (with very high correlation coefficients; see **Fig. 3.21**) with the GC levels of the corresponding prokaryotic and viral genomes as well as with the GC levels of the corresponding coding sequences from vertebrates (Bernardi and Bernardi, 1985, 1986a), so indicating the existence of compositional correlations (both inter- and intra-genomic) among the three codon positions of prokaryotic and eukaryotic genes. A similar point was also made for prokaryotic genes by Wada and Suyama (1985) and by Sueoka (1988). When large data sets became available for human genes, compositional correlations between first, second and third positions were defined (D'Onofrio et al., 1991; see **Fig. 10.1**). Since these correlations were essentially identical from human to prokaryotes, they were called **universal correlations** (D'Onofrio and Bernardi, 1992).

In fact, the partial overlap of points in plots like that of **Fig. 10.1** obscures their actual distribution in the diagram. This distribution is shown in **Fig. 10.2** which presents histograms of compositional classes corresponding to 2.5% GC intervals in first + second codon positions and 5% GC intervals in third codon positions. The data of **Fig. 10.2** stress the fact that the positive correlations seen in **Fig. 10.1** correspond, as expected, to lines passing through the most abundant classes of genes. It should be noticed, however, that the abundant classes at 75–80% GC in third codon positions and at 55-60% in first + second positions suggest that the linear relationship tends to bend to the right at such high values; and that only very few extreme genes lie outside lines that are parallel to the linear relationship, but are shifted by ±5% GC along the abscissa axis (note that the set of strongly deviating genes of **Fig. 10.1**, not represented in **Fig. 10.2** because outside the frame, would also belong to this class).

The scatter of points in the plots of **Fig. 10.1** deserved to be analyzed further. Indeed, this scatter was obviously not due to any experimental error (because the points derive from primary structures), but to some intrinsic factor. In fact, it can be shown that the scatter of points along the horizontal axis in **Fig. 10.1** is due to particular frequencies of amino acids in the encoded proteins. Moreover, in coding sequences with very GC-rich first and second codon positions (~70% GC; see **Fig. 10.1**, bottom frame), third codon positions tend to be GC-poor (30–50% GC), as if a **compensation** was needed to keep the overall GC level of the coding sequence within certain limits. This point had been made earlier by Wada and Suyama (1985, 1986).

Because of the considerations just made, it is of interest to divide amino acids (see **Fig. 10.3**) into three classes: (i) those that only contain G and/or C in the first and second positions of their codons; (ii) those that only contain A and/or T; and (iii) those that contain G and/or C as well as A and/or T. The GC class comprises four amino acids, alanine, arginine (quartet codons), glycine, and proline; the AT class comprises seven amino acids, asparagine, isoleucine, leucine (duet codons), lysine, methionine, phenylala-

Interestingly, the intragenomic correlations of **Fig. 10.1** also hold intergenomically (D'Onofrio and Bernardi, 1992; D'Onofrio et al., 1999a; **Fig. 10.4**).

The universal correlations and the hydrophobicity of proteins

(i) Two implications of the positive correlations between GC_3 and GC_1, or between GC_3 and GC_2, are that increases in GC_3 should be accompanied (1) by increases of aminoacids encoded by codons of the GC class (the codons having G and/or C in first and second positions; see **Fig. 10.3**) and by decreases of aminoacids encoded by codons of the AT class (the codons having A and/or T in first and second positions); and (2) by increases in quartet codons and decreases in duet codons, simply because the former are GC-richer than the latter in first and second positions. Indeed, the first point was verified (D'Onofrio et al., 1991) and confirmed by recent data (D'Onofrio et al., 1999a,b); and the second point was also found to be true (**Fig. 10.5**).

Fig. 10.6 displays the Grantham (1980) representation of the genetic code. Incidentally, the terms quartets, duets etc. introduced by Grantham (1980) should be preferred to "four-fold degenerate" and "two-fold degenerate". Indeed, the term "**degeneracy of the code**" was introduced at a time when the only role of codons was visualized in terms of encoding amino acids, and the role of third codon positions in the accuracy and speed of translation (Akashi, 1998) was not even imagined.

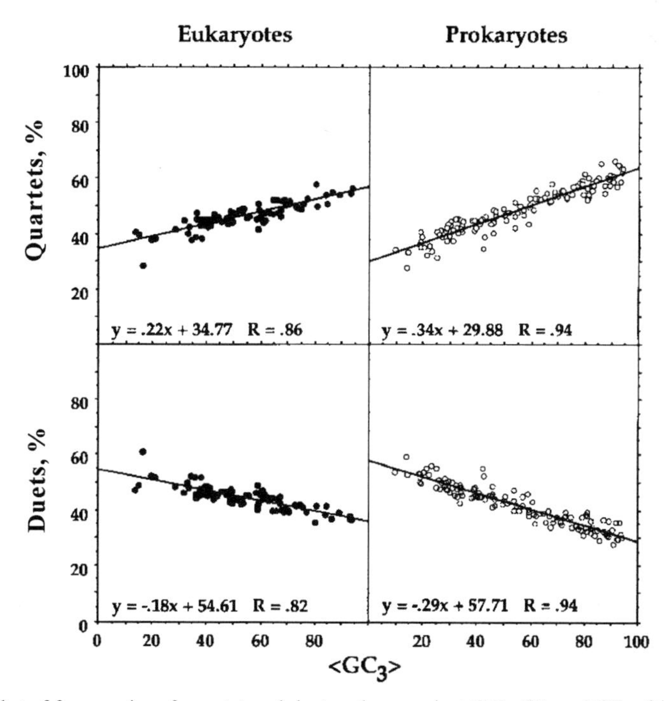

Figure 10.5. Plot of frequencies of quartet and duet codons against GC_3. (From D'Onofrio et al., 1999a).

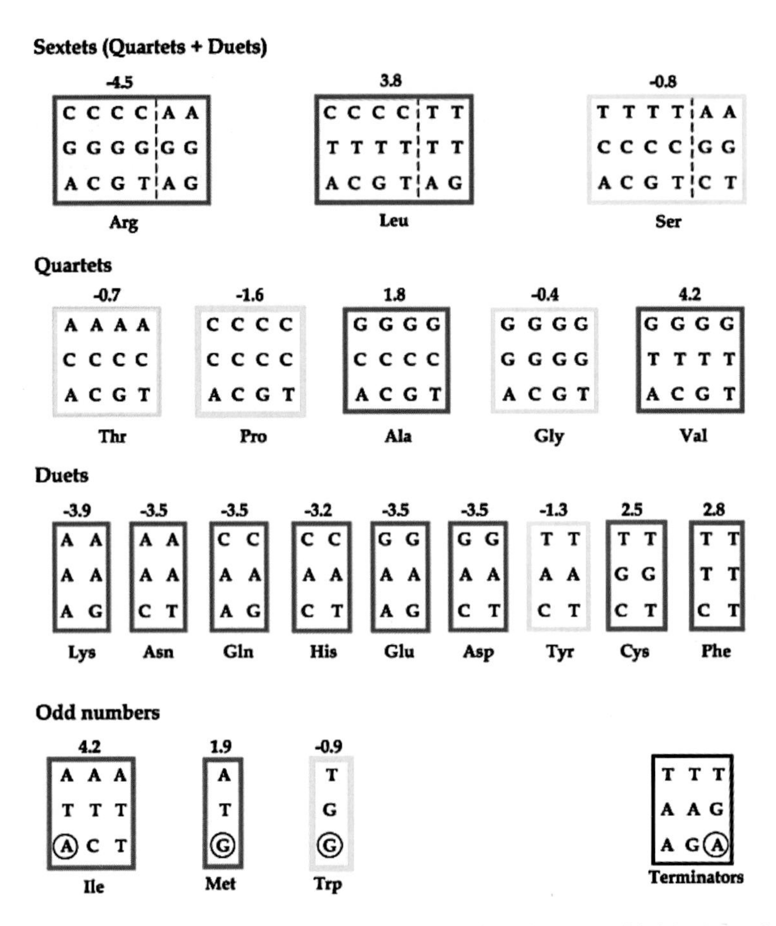

Figure 10.6. The Grantham (1980) representation of the genetic code was modified in that codons rather than anticodons are displayed, and hydropathy values for aminoacids are shown using the scale of Kyte and Doolittle (1982). Red, blue and yellow boxes indicate the most hydrophobic, the most hydrophilic and the amphipathic aminoacids, respectively. The black box comprises terminator codons. (From D'Onofrio et al., 1999a,b).

(ii) In both bacteria and vertebrates, GC_3 is positively correlated not only with the frequency of aminoacids of the GC class, but also with the hydrophobicity of aminoacids (D'Onofrio et al., 1999a,b; **Fig. 10.7**). Interestingly, although the slopes of the two regression lines of **Fig. 10.7** are identical, prokaryotic values are higher on average than vertebrate values. This difference is accompanied by another remarkable feature, namely by a cysteine level which is half as high in prokaryotes as in vertebrates. This suggests that the higher hydrophobicities of prokaryotic proteins were replaced, as a stabilizing factor, by the higher levels of disulphide bridges present in eukaryotic proteins.

The results of **Fig. 10.7** are in agreement with Naylor et al. (1995), but in striking disagreement with Gu et al. (1998). The latter authors studied the *dna*A protein from *E. coli* and 14 other bacteria, plus 10 to 14 other prokaryotic proteins, and claimed that "*both strongly hydrophobic and strongly hydrophilic amino acids tend to change to ambivalent* *amino acids, suggesting that the majority of these amino acid substitutions are not caused by positive Darwinian selection"* and that their *"results can be easily explained under the neutralist view by assuming that either most of these amino acid changes are nearly neutral or their selective disadvantage are not large, so that substitutions can still occur when mutation pressure is strong*". We will try to explain, first why the findings reported by Gu et al. (1998) are different from those presented here and, second why their interpretation is different.

Gu et al. (1998) divided amino acids into three classes, according to the GC levels of first and second positions of the corresponding codons (essentially as previously done by Jukes and Bhushan, 1986, and by D'Onofrio et al., 1991), and plotted the frequencies of amino acids against genomic GC and found that the GC-rich class increased, the intermediate class remained constant, and the GC-poor class decreased. These results could be predicted from the plots of the frequencies of amino acids *vs.* GC_{1+2} of D'Onofrio et al. (1991). More interestingly, they plotted the frequencies of external (Asp, Glu, Lys, Arg, His, Asn and Gln), internal (Phe, Leu, Ile, Met, Val, Tyr and Trp) and ambivalent (Ala, Pro, Gly, Ser, Thr, Cys) aminoacids (the classification of amino acids was from Dickerson and Geis, 1983) against genomic GC and found an increase in ambivalent amino acids and a parallel decrease in both external and internal aminoacids. Hence, their conclusion.

The authors correctly noted that the classification of external (hydrophilic), internal (hydrophobic) and ambivalent (amphypathic) amino acids which they used "*is not unambiguous*". Indeed, Ala and Cys (classified as ambivalent by Dickerson and Geis, 1983) are

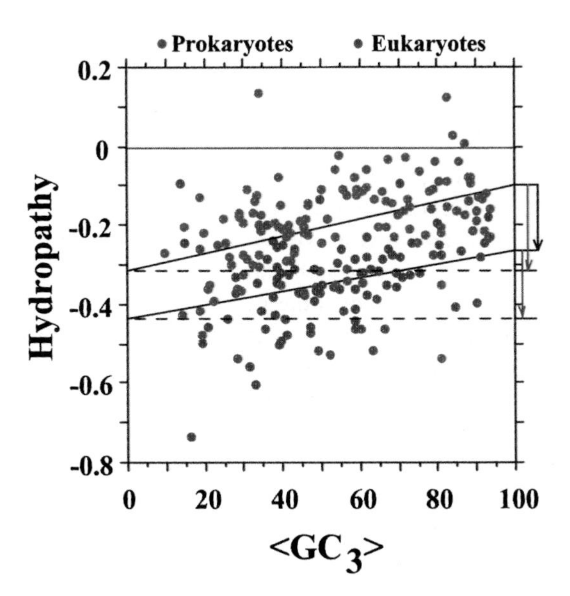

Figure 10.7. Hydropathy values of proteins from prokaryotes (red circles) [y=0.0021x –0.30; R=0.44] and eukaryotes (blue circles) [y=0.0017x –0.44; R=0.27] plotted against $\langle GC_3 \rangle$. An increase in hydropathy corresponds to an increase in hydrophobicity. Arrows indicate that the difference in hydrophobicity of prokaryotic and eukaryotic proteins has about the same magnitude as that between proteins encoded by GC-rich and GC-poor proteins. (From D'Onofrio et al., 1999a,b).

hydrophobic, and Tyr and Trp (classified as internal by Dickerson and Geis, 1983) are "intermediate" amino acids according not only to Kyte and Doolittle (1982), but also to other authors (see *e.g.* Rose et al., 1986; Engelman et al., 1986; Fiser et al., 1996). The problem, in the case of Gu et al. (1998), is however, not simply the assignment of a given amino acid to a given class, but the fact that they considered all amino acids in a given class as equivalent in hydropathy. This is incorrect. If the hydropathy values for each amino acid are used, following Kyte and Doolittle (1982), it is clear that only hydrophylic amino acids decrease with increasing <GC> (or <GC_3>), whereas both hydrophobic and amphypathic amino acids increase. Needless to say, this conclusion is in sharp contrast with that of Gu et al. (1998). Indeed, <GC_3> increases in coding sequences are accompanied by increases in the hydrophobicity of the encoded proteins (see **Fig. 10.7**), and, therefore, according to a widely accepted viewpoint, by an increase in protein stability and by a faster folding, namely by structurally and functionally meaningful changes.

(iii) If one considers the existence of universal correlations and of the correlation between GC_3 and the hydrophobicity of amino acids, one would be tempted to conclude with Eyre-Walker and Hurst (2001) that "*genes in the (G+C)-rich isochores yield proteins with different amino-acid compositions (D'Onofrio et al., 1991) and hydropathies (D'Onofrio et al., 1999a) to those in the (G+C)-poor isochores; both features seem to be a consequence of the correlation between isochore G+C content and* GC_{12}". This conclusion, is, however, wrong because of the following points (see Jabbari et al., 2003a, for further details).

First of all, the authors' use of "*isochore GC*" invokes GC_3, the GC level of third codon positions, which is strongly correlated with isochore GC (Bernardi et al., 1985b; Bernardi and Bernardi, 1986a), R being 0.82 for human genes (Clay et al., 1996). Since what is under consideration here are amino acid composition and hydropathy, the problem must concern primarily the encoded proteins, and therefore the correlation between GC_3 and GC_{1+2} (the average GC levels of first + second codon positions; D'Onofrio et al., 1991). Incidentally, the correlation between isochore GC and GC_{1+2} also holds, as expected from the data of Bernardi and Bernardi, 1986a; see **Fig. 10.8**).

Second, the authors' corrected statement concerns two existing correlations, which are shown in **Fig. 10.9**: (1) the correlation between GC_3 and GC_1/GC_2, and (2) the correlation between GC_3 and hydropathy, which is also valid for intergenomic comparisons (D'Onofrio et al., 1999a), as well as (3) a tacit hypothesis, namely that the correlation between GC_3 and hydropathy holds for GC_{1+2}. The authors' argument indicates an implicit assumption, that the correlations under consideration are transitive (see **Fig. 10.10**).

From a purely statistical point of view, transitivity of correlations is valid only if correlation coefficients are large enough. For example, if A and B are correlated with a positive correlation coefficient R, and if B and C are correlated with the same positive correlation coefficient R, then A is guaranteed to correlate positively with C only if $2R^2 > 1$ (eq. 1), *i.e.*, if $R > \sqrt{2}/2 = 0.707$. As has been emphasized by Kendall and Stuart (1976, pp. 424–428), such a relation "*is by no means a trivial result*". If R_{AB}, R_{BC} and R_{AC} describe the three correlations concerned, then the best general relation that can be derived is $1+2R_{AB}R_{BC}R_{AC}- R_{AB}^2-R_{BC}^2-R_{AC}^2 \geq 0$. In the case of two equal, positive correlations $R_{AB} = R_{BC}$, we obtain $R_{AC} \geq 2R_{AB}^2 - 1$, from which eq. 1 follows. In the more general case of two arbitrary positive correlations R_{AB}, R_{BC}, transitivity can be invoked only if $R_{AB}^2 + R_{BC}^2 > 1$ (eq. 2; see also Appendix of Jabbari et al., 2003a). However, the correlation

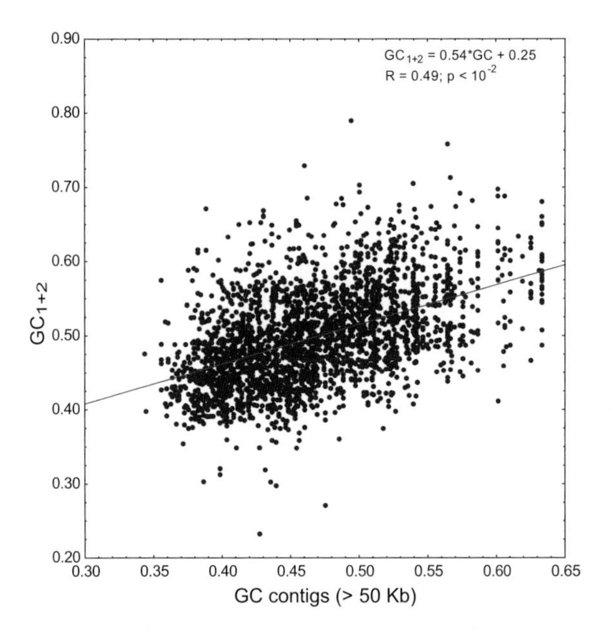

Figure 10.8. Correlation between GC level of large (> 50kb) DNA sequences from human and GC_{1+2} of the embedded coding sequences. (From Jabbari et al., 2003a).

coefficient of GC_3 *vs.* GC_{1+2} is 0.42 ($p < 10^{-4}$; Clay et al., 1996), that of GC_3 *vs.* hydropathy is 0.17, and that of GC_{1+2} *vs.* hydropathy is – 0.09 (**Fig. 10.9**). Thus transitivity cannot be used in the case under consideration.

Third, **Fig. 10.10** shows that while GC_3 is positively correlated with hydropathy, both GC_1 and GC_2 show slightly negative correlations with hydropathy, despite the fact that the correlations among the three codon positions are positive. Therefore, the statement that the positive correlation observed between GC_3 and hydropathy should be the result of the positive correlation between GC_3 and GC_{1+2} is incorrect.

As just shown, the correlation between GC_3 and hydrophobicity does not depend upon the correlation between GC_3 and GC_2 or GC_{1+2}. On the other hand, obviously the hydrophobicity of proteins depends upon the hydrophobicity of amino acids. The question then should be asked about the reason(s) for the correlation between GC_3 and hydrophobicity. The reason is that quartets and duets are positively and negatively correlated with GC_3, respectively (**Fig. 10.5**; D'Onofrio et al., 1999a), and quartets comprise three abundant hydrophobic amino acids, duets six abundant hydrophilic amino acids. Expectedly, a plot of hydrophobic and amphipathic amino acids increase with increasing GC_3, whereas hydrophilic amino acids decrease (**Fig. 10.11**). The resulting positive correlation between hydrophobicity and GC_3 is reduced in magnitude only by the opposing effect of the strongly hydrophobic triplet, isoleucine, and a strongly hydrophilic sextet, arginine.

An independent argument against the conclusions of Eyre-Walker and Hurst (2001) comes from the observation (D'Onofrio et al., 2002) that gravy scores of the secondary

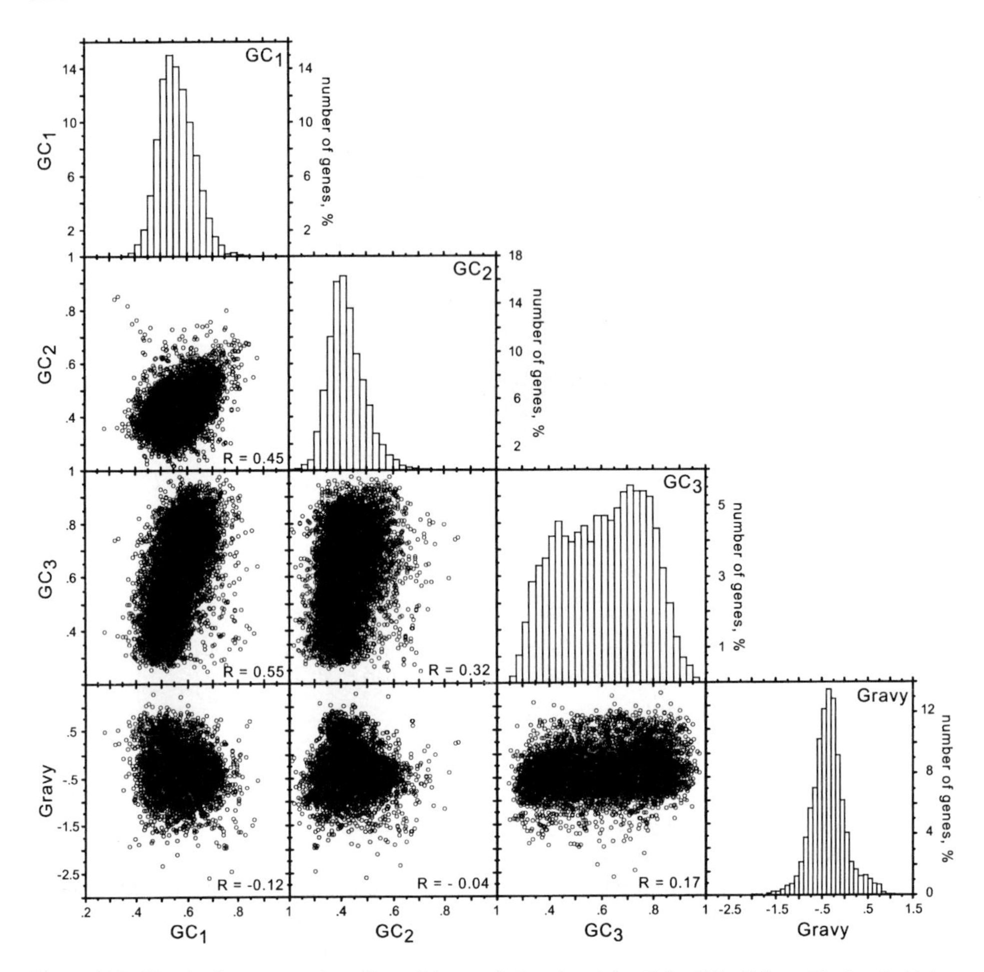

Figure 10.9. Matrix plot representing all possible correlations involving GC_1, GC_2, GC_3 and hydrophobicity. Correlation coefficients (R) are indicated. The range of each variable (lowest value, highest value) is given in the histograms. Sample size is N=20,148. Well-annotated human genes were extracted from GenBank (Release 127, 15 December 2001) using the ACNUC retrieval system (Gouy et al., 1985). Hydropathy values of each encoded protein (gravy) were calculated (Lobry and Gautier, 1994) as $\Sigma(n_i/N)H_i$, where n_i is the number of occurrence of a given amino acid, N is the total number of amino acids in the protein and H_i is the assigned value of hydrophobicity according to the scale of Kyte and Doolittle (1982). (From Jabbari et al., 2003a).

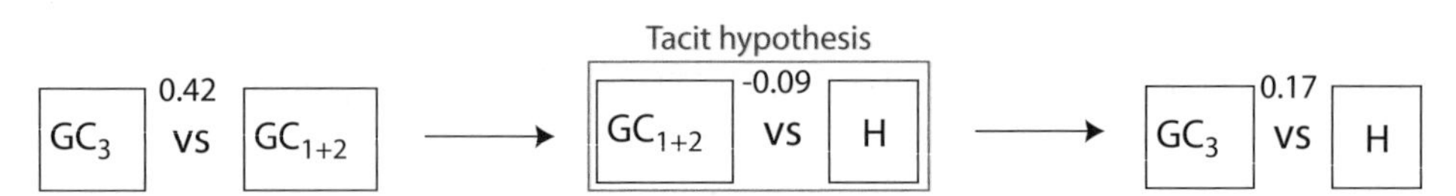

Figure 10.10. Scheme showing the essence of the argument in Eyre-Walker and Hurst (2001): the transitivity of correlations (R values are indicated in bold face), and the assumption that a correlation exists between GC_{1+2} and hydropathy (red box). (From Jabbari et al., 2003a).

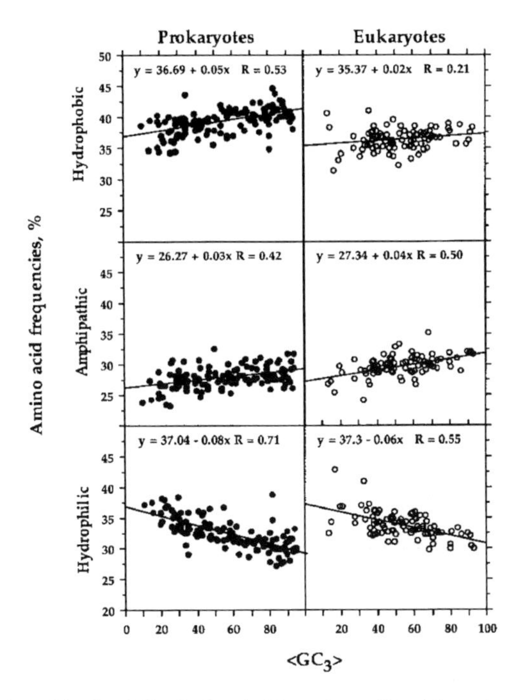

Figure 10.11. Frequencies of hydrophilic, hydrophobic and amphipathic amino acids from prokaryotes (closed circles) and eukaryotes (open circles) plotted against GC_3. (From D'Onofrio et al., 1999a).

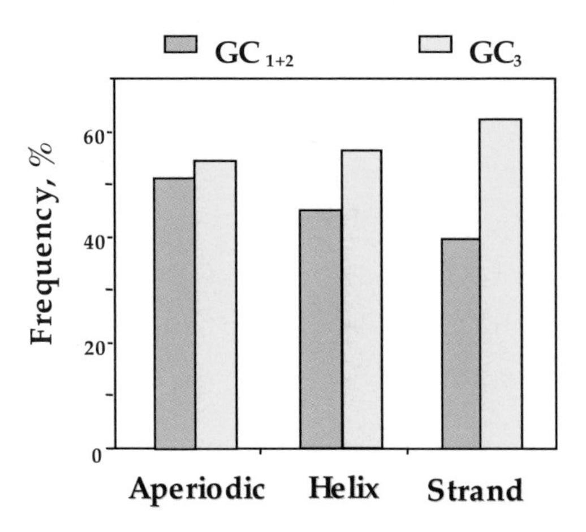

Figure 10.12. Histogram of the frequencies of GC_3 and GC_{1+2} in the three secondary structures of the proteins, which were ordered according to increasing hydrophobicity. (From D'Onofrio et al., 2002).

structures of proteins showed a positive trend with GC_3, as was previously found at the protein level (D'Onofrio et al., 1999a), and a negative trend with GC_{1+2} (see **Fig. 10.12**).

The universal correlation and imaginary genes

When the first draft sequences of entire rice genomes were published (Yu et al., 2002; Goff et al., 2002), the most striking discovery was the unusual gene set that Yu et al. (2002) proposed largely on the basis of predictions. Indeed, roughly half of the protein-coding genes in rice had, apparently, no significant homologs in the previously sequenced genome of *Arabidopsis*, nor (in most cases) in any other known taxon's sequenced DNA.

Two interpretations are possible for these findings. The first one is that most true rice genes do have orthologs (or close paralogs) in other species, *i.e.*, that most of the unmatched *de novo* predictions are incorrect. The second interpretation is that about half of all true rice genes indeed have no homologs in any sequenced species, including rice. This interpretation would imply, however, that about 20,000 genes were formed (or altered beyond recognition) at most since the split of monocotyledons from dicotyledons 200 Mya. This becomes difficult to envisage as soon as we examine the striking differences between the matched and unmatched genes of rice.

First, as Yu et al. (2002) themselves noted, the unmatched genes (*i.e.* the genes with no detectable homologs in *Arabidopsis*) have coding regions that, on average, are only about half as long as those from matched genes (the genes with homologs). Second, they have twice as many introns as normal in the 200 bp - 2 kb range. Thus associated splicing constraints would have to be met, in addition to the already quite heavy requirements for a *de novo* gene to evolve to functionality (or even translatability). Third, the distribution of GC levels from the putative gene set contains a hump of unusually GC-rich coding sequences (Yu et al., 2002). Further inspection (**Fig. 10.13A**) reveals that this hump is mainly due to unmatched genes that fail to show differences in GC among the three codon positions, and that have GC_2 (second position levels) soaring to heights that are virtually unknown either in prokaryotic or eukaryotic coding sequences (Cruveiller et al., 2003b). In particular, the unmatched genes violate a linear relationship between GC_2 and GC_3 that is largely conserved (given enough intragenomic variability) from human to *E. coli* (see D'Onofrio and Bernardi, 1992; Cruveiller et al., 2003b and references therein). These facts make it even more unlikely that accidental sequestering of such extreme noncoding DNA would result in a functional gene that is translated *in vivo*. Finally, both transposon/repetitive DNA and lateral transfer explanations for the predicted (yet anomalous) rice genes can be essentially ruled out by the copy-number and homology searches conducted by Yu et al. (2002).

Similar problems were found for chromosome 1 (Sasaki et al., 2002; **Fig. 10.13B**) and for chromosome 10 (The Rice Chromosome 10 Sequencing Consortium; **Fig. 10.13C**).

To help decide if the apparent presence of such large numbers of compositionally anomalous genes is a mere prediction artifact, one can characterize the likely structural and functional properties of the encoded proteins. We compared a set of obvious deviants (roughly a thousand genes with $GC_2 > 65\%$, a level that is extremely rare among experi-

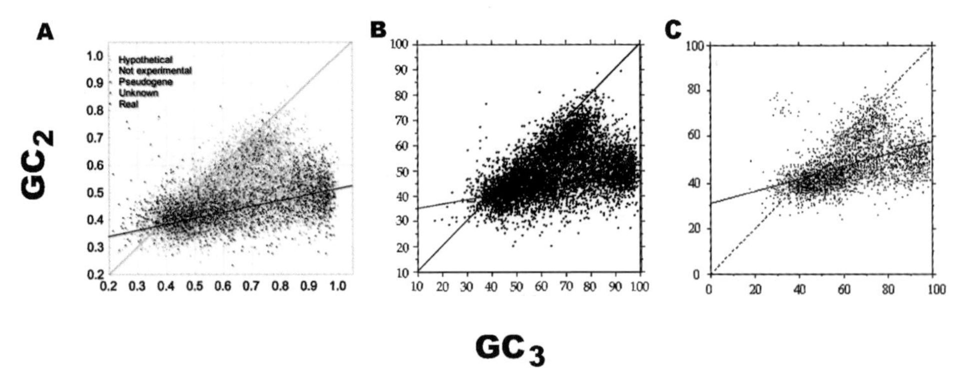

Figure 10.13. Scatterplots of GC_2 *vs.* GC_3 levels in predicted and experimentally indentified rice genes. The diagonal (GC_2=GC_3) is indicated. **A.** The gene set (N=10,087) from Yu et al. (2002) was partitioned into 5 classes according to the annotations (real genes, not experimental, unknown, pseudogenes and hypothetical). (From Cruveiller et al., 2003a). **B.** The predicted set of protein-coding sequences of rice chromosome 1 obtained by Sasaki et al. (2002). (From Cruveiller et al., 2003b). **C.** The gene set from Chromosome 10 (The Rice Chromosome 10 Sequencing Consortium, 2003).

mentally determined genes) and a similarly sized set of obvious conformers ($GC_2 < 37\%$) at the protein level. The GC_2-rich deviants, almost all of unmatched type, had essentially twice the percentage of coil structures, and half the percentage of helices, compared to the compositionally normal group; only the turn and sheet proportions were similar in both protein sets.

In summary, the set of protein-coding genes that are now proposed for rice remains largely enigmatic. Clearly, we may be on the verge of discovering a new type of anomalous but functional proteins, specific to *Gramineae*, with highly unusual compositional and structural features, and containing few or none of the domains that are known so far. Alternatively, if it turns out that 40-50% of the published *de novo* predictions for rice were incorrect, as we believe, this should point the way for improvements and future caution when using sparsely validated prediction algorithms.

CHAPTER 4

Compositional gene landscapes

4.1. Large-scale-features of the human gene landscapes

GC levels in third positions (GC_3) are almost free of constraints at the protein level, while those in second positions (GC_2) are almost completely determined by the gene product. Therefore, their joint distribution, or **the compositional gene landscape displays relations between DNA/RNA and the proteins**.

Among taxa that are well-represented in sequence databases, GC_2 and GC_3 levels exhibit a tendency to cluster along a widely conserved, straight line in the landscape, delineated by the major axis or orthogonal regression line. The correlation to which this linearity corresponds is found in species as distant as human and *E. coli*, and the major axis is consistently close to the line $GC_3 = 6\ GC_2 - 2$. In other words, a 1% change in GC_2 corresponds roughly to a 6% change in GC_3, and the two codon positions have similar GC levels when either of them is around 40%. This intragenomic correlation, and the correlation between first/second and third codon positions (GC_{1+2} *vs.* GC_3), are well conserved among vertebrates. Similar intergenomic correlations are found for eukaryotes or prokaryotes, when each genome is represented by its genome-wide gene averages (see Bernardi and Bernardi, 1985, 1986; Wada and Suyama, 1985; Sueoka, 1988; D'Onofrio et al., 1991,1999a; D'Onofrio and Bernardi, 1992). Such conservation of the major axis can be used to detect incorrectly predicted genes, even in previously uncharacterized genomes (Cruveiller et al., 2003a,b; see **Chapter 3**).

Very large numbers of non-redundant, reliable gene sequences, which are now available for several vertebrate species, allow higher levels of resolution than were previously possible, and prompted us to analyze their landscapes (two-dimensional distributions) in more detail. We focussed on human, where a nonredundant set of 10,218 genes was available from a well-curated human sequence database. We also studied the landscapes for three other vertebrates that are represented by large sequence databases, namely mouse, chicken and *Xenopus* (Cruveiller et al., 2003c). The warm-blooded vertebrate landscapes are adequately described by a structure of peaks and valleys along the landscapes' crests, whereas the *Xenopus* landscape only shows a short crest with two peaks.

Fig. 10.14 shows three 'aerial' views of the (GC_2, GC_3) landscape of human coding sequences: a scatterplot (panel A), a 2-dimensional histogram (panel B; 37 × 37 bins) and a smooth contour plot of the same region, as obtained by interpolation (panel C; 21 × 21 bins). Where points are dense in the scatterplot, or there are many genes per bin, the altitude of the GC_2/GC_3 landscape is high. It is difficult to see the structure by eye from the raw scatterplot, but the histogram and especially the contour representation reveal the high, narrow, nearly linear crest of the landscape, punctuated by peaks and valleys situated along or close to it. We have called such contour plots 'Sorrento plots', for the 2002 Symposium on Molecular Evolution in which they were first presented and served as the logo.

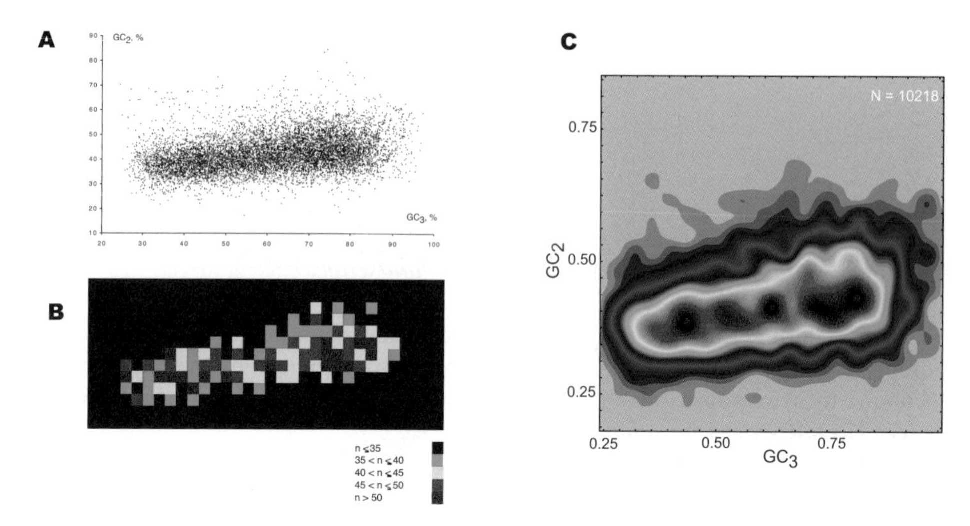

Figure 10.14. 2-dimensional representations of the landscape of GC levels in second and third positions (GC_2, GC_3) of 10218 curated human genes: **A**, scatterplot; **B**, bivariate histogram and **C**, smoothed contour plot. (From Cruveiller et al., 2003c).

The rise to the crest region is steep: few bins contain intermediate gene numbers between the high frequencies in the narrow crest region and the low frequencies in the surrounding plain. At least five elevations can be identified along the crest. The five peaks or components have similar heights, so that each of them represents about 20% of the gene count (the difference between the basal height of the crest region and the further elevations corresponding, however, to about 5% of the gene count in the data set). Analysis of the genes in each peak region showed that none are explained by the overrepresentation of a few gene families or superfamilies such as immunoglobulins, T cell receptors, major histocompatibility complex, histones or kinases. The multiple peaks in human are superimposed on a broad background feature of the crest, which is present also in cow and chicken. This background feature is the presence of two smooth, broad hills of approximately the same size, separated by a wide pass or saddle (at 56% GC_3 in human). The fine-scale features of the crest involve the number, positions, sizes and shapes of the peaks along the crest, which rise above the two broad hills.

In contrast, the cold-blooded vertebrate, *Xenopus* (**Figure 10.15**), differs from the warm-blooded vertebrates, having only a short crest with two peaks.

The human sequences shown in **Fig. 10.14** should already cover at least 15–30% of all human genes as they are commonly defined (Fields et al., 1994; Venter et al., 2001). Unless there turn out to be major, unsuspected biases in identifying human genes, it seems reasonable to expect a similar landscape for all of the genes, including those that have not yet been reliably identified. Furthermore, the available set of *bona fide* genic sequences, *i.e.*, those that are experimentally verified or risk-free, is unlikely to grow much in the next few years. The results presented here for human may be as comprehensive as any that will be obtained from larger gene sets in the near future.

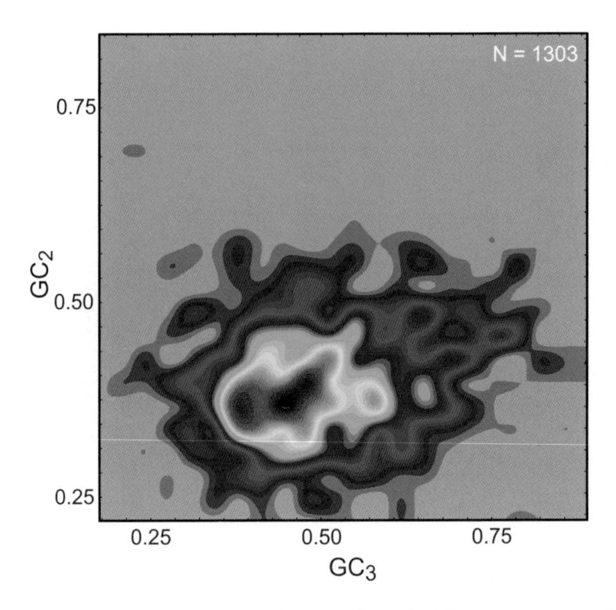

Figure 10.15. Frequency distributions of GC_2 and GC_3 values of 1303 *Xenopus* genes. (From Cruveiller et al., 2003c).

4.2. Gene landscapes correspond to protein landscapes

We tried to partially reconstruct the human landscape, *i.e.*, the altitudes along the landscape's crest, using combinations from the 8 amino acids having G or C in their second position. **Fig. 10.16** shows an example of a contour plot in which GC_2 has been replaced by the frequency of the 4 residues having C in second position: Ala, Pro, Ser and Thr. The general peak structure of this 4-amino acid landscape resembles that of the corresponding GC_2 landscape (**Figure 10.14**), although there are still differences. Further studies along such lines may help to understand the factors contributing to the peaks, and which base compositions of DNA (reflected in GC_3) and/or amino acid compositions (reflected in GC_2) are preferentially represented in vertebrate genomes.

4.3. Gene landscapes correspond to experimentally determined DNA landscapes

The number of peaks found in the human (five), mouse (four), chicken (six) and *Xenopus* (two) landscapes correspond to those that one might expect from the major components of bulk DNA. These DNA components were determined experimentally (Macaya et al., 1976; Cuny et al., 1981) by silver- and BAMD-assisted density gradient ultracentrifugation. BAMD is a sequence-specific DNA ligand that binds preferentially to AT-rich DNA, and that may also exhibit a preference for certain AT-rich oligonucleotides. The positions and extents of the two wide hills that define the main crest of the landscape correspond very well to the low GC and high GC DNA components in these species (see, *e.g.*, Bernardi, 2001). The individual peaks in the GC_2, GC_3 landscape show, in general, a good concor-

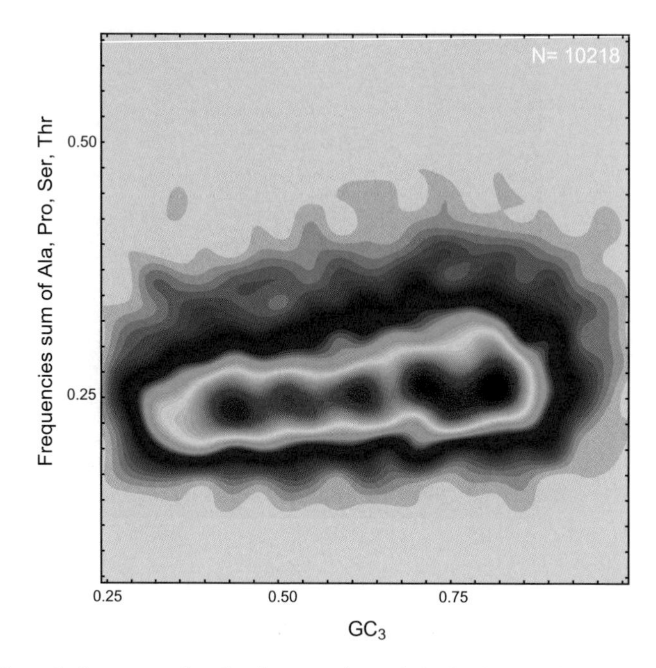

Figure 10.16. Smoothed contour plot showing a variant of the landscape of 10,218 human genes: the vertical axis is the summed frequencies of alanine, proline, serine, and threonine instead of GC_2, the horizontal axis is GC_3. The four amino acids used to recreate this landscape are frequent and all have cytosine in second position in four codons. (From Cruveiller et al., 2003c).

dance with the experimental peaks deduced from bulk DNA, albeit with some minor differences. Although the expected and observed values almost coincide for the third ($\approx$ 60% GC_3) and fourth ($\approx$ 73% GC_3) peaks, the spacing of the other landscape maxima is slightly narrower than expected. Thus the GC-poorest peak, named L1, would have been expected at a GC_3 value just below 39%, corresponding to chromosomal regions of about 39% GC. Instead it is observed, both in the contour plot and in the GC_3 histogram, around 43–44% GC_3, corresponding to chromosomal regions of about 40% GC. The small error at the bulk DNA level (1% GC) is however within expected error margins.

In conclusion, an analysis not of bulk DNA, but of genes yielded multiple peaks in histograms of codon position GC levels. These peaks appear to correspond to the experimentally observed components of bulk human DNA, and may have a biological meaning, although we have not yet been able to identify it. Thus, the human landscape is formed by a small number of underlying, discrete distributions of GC_3 and GC_2 corresponding to particular, well-defined types of genes, genic/protein regions and/or chromosomal environments, and provides another evidence for compositional discontinuities in the human genome.

CHAPTER 5

Nucleotide substitutions and composition in coding sequences. Correlations with protein structure

5.1. Synonymous and nonsynonymous substitution rates in mammalian genes are correlated with each other

It is well known that the rate of nonsynonymous substitutions is extremely variable among genes, as revealed a long time ago by investigations on the rates of amino acid substitution (Dickerson, 1971; Dayhoff, 1972). Indeed, rates range, in a human/murid comparison (Li, 1997), from zero (in the genes coding for histones 3 and 4) to 3.06×10^{-9} substitutions per site per year (in the gene for γ-interferon). The average for the small set of genes taken into consideration by Li (1997) is $0.74\ (\pm 0.67) \times 10^{-9}$ substitutions per site per year.

The more than 300-fold range in nonsynonymous substitution rates reported by Li (1997) in different genes (neglecting those showing no substitution) may be ascribed to differences in the rate of mutation and/or to the intensity of selection. As for the first possibility, the difference in mutation rate is not nearly large enough to account by itself for the phenomenon under consideration. This conclusion can be drawn from the fact that synonymous mutation rates for mammalian genes cover a range which is much smaller than that of nonsynonymous substitutions (see below). The most important factor in determining the rate of nonsynonymous substitution appears, therefore, to be the **intensity of selection**. The higher the deleterious effects of amino acid replacement, the higher the chances of the mutation being eliminated by negative selection (advantageous mutations being very rare compared to deleterious mutations, they are neglected here). Indeed, negative selection is indicated by the actual amino acid changes observed, functionally crucial amino acids not showing replacement and conservative substitutions being predominant over nonconservative ones.

As far as the rate of synonymous substitutions is concerned, the average value of 3.51 $(\pm 1.01) \times 10^{-9}$ per site per year reported for 47 genes from human and rodents (Li, 1997) is misleading in that it conveys the idea of an average rate not very different from that reported for pseudogenes, thus reinforcing the idea (see Wolfe et al., 1989) that the synonymous substitution rate is close to the mutation rate. In fact, a larger set of over 300 human/murid genes has shown a 20-fold range of synonymous rates (Bernardi et al., 1993; Wolfe and Sharp, 1993). This finding immediately raises a question: If the fastest rate observed is equal to the mutation rate (a point which remains to be demonstrated), what is the cause for the 20-fold slower rate exhibited by other genes?

An answer to this question came from investigations on the **frequency and compositional patterns of synonymous substitutions in orthologous genes** from four mammalian orders (Cacciò et al., 1995; Zoubak et al., 1995). These studies showed (i) that the frequencies of conserved, intermediate, and variable positions (defined as the positions showing no change, one change, or more than one change, respectively) of quartet and duet codons are

structure of the encoded protein, since all divergent segments are located on the surface of the molecule, facing one side (almost parallel to the cell membrane) on the exposed surface of the organism (see **Figs. 10.17** and **10.18**). Although the observation that divergent residues preferably lie on surface areas of proteins is not new, the finding that **both synonymous and nonsynonymous rates are correlated with protein structure** is indeed a strong indication that structural constraints could be responsible for the correlation.

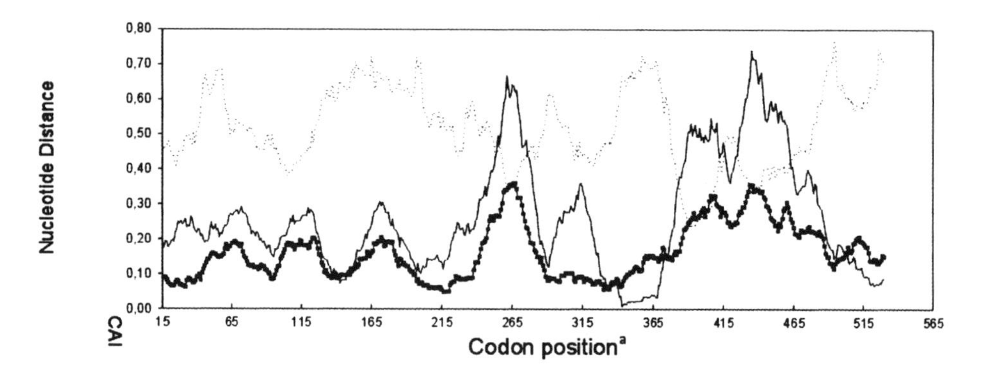

Figure 10.17. Profiles of synonymous distance (thin line), nonsynonymous distance (thick line) and Codon Adaptation Index (dotted line) for the GP63 gene of *Leishmania*. The window size used was 30 codons, shifting one codon at a time. Each window was labeled according to the codon falling in the middle so that the first window is assigned to codon 15. The correlation coefficients were calculated using non-overlapping windows to insure the independence of sampling points. (From Alvarez-Valin et al., 2000).

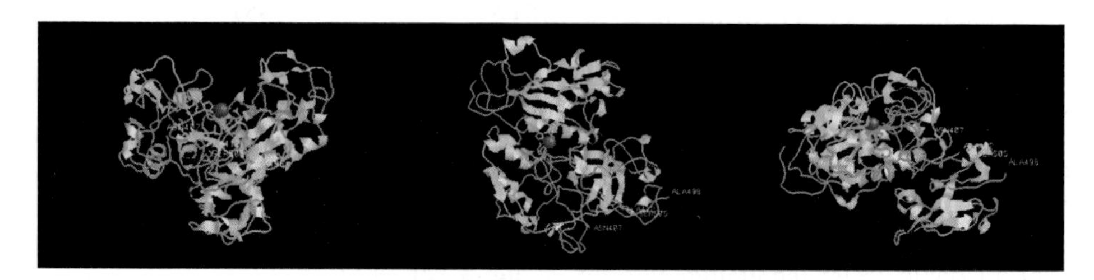

Figure 10.18. Three different views of the three-dimensional structure of the GP63 protein from *Leishmania*. Divergent regions are represented by cyan color. The only variable region close to the anchoring site is shown in green. The amino acids composing the catalytic site are represented in green. The Zn++ atom appears in red. The membrane anchoring tail is also represented in red. The amino acids that indicate the beginning of the non-resolved segments of the protein (residues Asn407, Ala412, Ala498, and Ser505) are labeled to show their location. (From Alvarez-Valin et al., 2000).

5.4. Base compositions at nonsynonymous positions are correlated with protein structure and with the genetic code

The nucleotide frequencies in the second codon positions of genes are remarkably different for the coding regions that correspond to different secondary structures in the encoded proteins, namely, helix, β-strand and aperiodic structures (**Fig. 10.19**). Indeed, hydrophobic and hydrophilic amino acids are encoded by codons having U or A, respectively, in their second position. Moreover, the β-strand structure is strongly hydrophobic, while aperiodic structures contain more hydrophilic amino acids.

The relationship between nucleotide frequencies and protein secondary structures is associated not only with the physico-chemical properties of these structures, but also with the organisation of the genetic code. In fact, this organisation seems to have evolved so as to preserve the secondary structures of proteins by preventing deleterious amino acid substitutions that could modify the physico-chemical properties required for an optimal structure (Chiusano et al., 2000).

Indeed, the physico-chemical properties of protein structures, which are strongly dependent on amino acid composition, are correlated with well-defined choices in the second codon position of the corresponding coding sequences. This link between protein structure and the second position of the codons suggests that the organisation of the genetic code reflects somehow the secondary structures of proteins. One of the principal adaptive forces driving the organisation of the code could have been, therefore, protein structure. Even though this could be in part expected, there are still no data that clearly indicate whether the link between the amino acid properties and the organisation of the code evolved because these properties are structural determinants, or because these properties are involved in the codon-anticodon interactions that could have promoted the organisation of the genetic code according to the stereochemical hypothesis (Woese, 1967; Shimizu, 1982; Yarus, 1991; see Di Giulio, 1997, for a review). Indeed, earlier attempts to link protein secondary

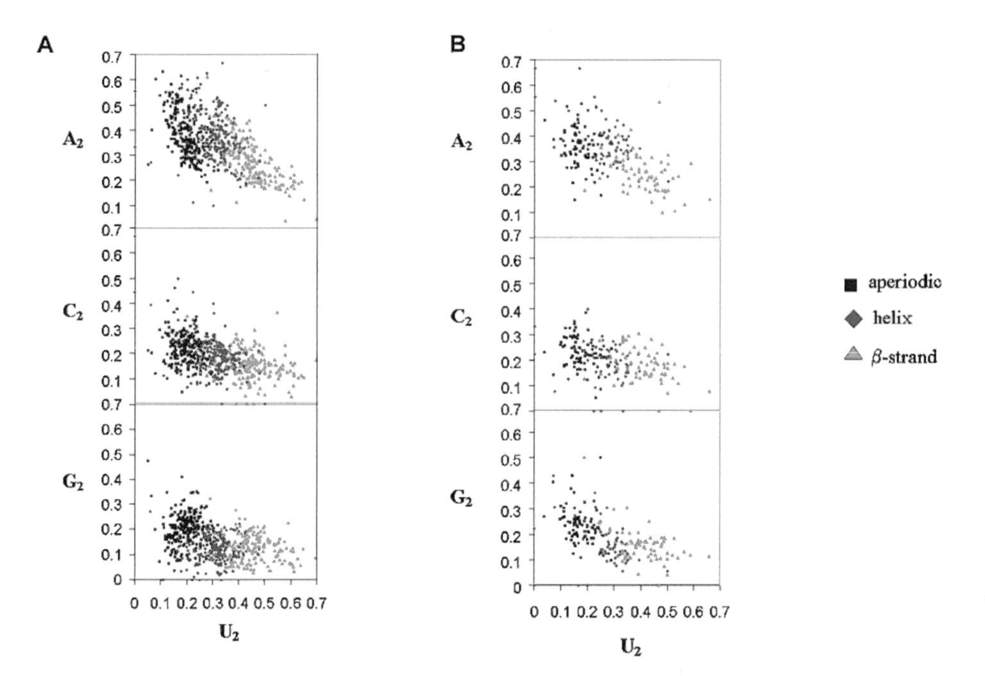

Figure 10.19. Scatterplots of A_2, C_2 and G_2 versus U_2 in (**A**) human and (**B**) prokaryotic data. (From Chiusano et al., 2000).

structures to the genetic code have been unsuccessful (Salemme et al., 1977; Goodman and Moore, 1977). More recent investigations that analyze the relationship between the genetic code and the putative primitive structures have led to the conclusions that β-turn (Jurka and Smith, 1987) and β-strand structures (Di Giulio, 1996) are linked to the structure of the genetic code, suggesting that these structures could have moulded the code. However, none of these previous approaches defined a link between the organisation of the genetic code and protein structure so clearly. The stereochemical hypothesis suggests that the genetic code originated from the interactions between codons or anticodons and amino acids. These interactions depend on the physico-chemical properties of amino acids and anticodons. Remarkably, the same physico-chemical properties of amino acids are also determinants for the correct folding of proteins. In other words, the genetic code would have been organised *via* forces related to the interactions between amino acids and their codons/anticodons, the same amino acid determinants of this interaction being also linked to properties involved in the organisation of the secondary structures of the proteins.

In conclusion, protein secondary structures are reflected in the genetic code because secondary structures were the three-dimensional elements of the proteins that had to be preserved at the time the genetic code was organised. Thus, in order to reduce dramatic translation errors while assembling the amino acid sequence required for the determination of a given structure, amino acids with similar physico-chemical properties were grouped by identical second positions in their codons, leading to their present organisation in the columns of the standard representation of the code (**Table 10.2**).

Based on this hypothesis, further details can be discussed, such as the clear separation of aperiodic structure and β-strand structure, shown by the second codon position analysis.

TABLE 10.2
The standard genetic code.

1st base in codon	2nd base in codon: U	C	A	G	3rd base in codon
U	Phe	Ser	Tyr	Cys	U
	Phe	Ser	Tyr	Cys	C
	Leu	Ser	STOP	STOP	A
	Leu	Ser	STOP	Trp	G
C	Leu	Pro	His	Arg	U
	Leu	Pro	His	Arg	C
	Leu	Pro	Gln	Arg	A
	Leu	Pro	Gln	Arg	G
A	Ile	Thr	Asn	Ser	U
	Ile	Thr	Asn	Ser	C
	Ile	Thr	Lys	Arg	A
	Met	Thr	Lys	Arg	G
G	Val	Ala	Asp	Gly	U
	Val	Ala	Asp	Gly	C
	Val	Ala	Glu	Gly	A
	Val	Ala	Glu	Gly	G

The separation suggests that these two structures were fundamental when the genetic code was organised. The intermediary features of the helix structure demonstrate its flexibility in terms of structural requirements, but also stress that the helix structure is not clearly distinguished by the genetic code. Actually the most frequent amino acids in helix (like leucine and glutamic acid) have opposite physico-chemical properties. As a consequence, the most preferred amino acids in the helix structure cannot be grouped according to common physico-chemical properties of the lateral chains, and thus cannot be selected through common choices in the second codon positions. Alternatively, the poorer imprinting of helix in the genetic code, can be interpreted as supporting the idea that this is a late appearing structure, whose emergence could have taken place after the rules of the code were already established. Moreover, the analysis of the preferred second codon position usage in turn structure has shown that this structure is better separated from the β-strand than the remaining aperiodic structures (Chiusano et al., 2000). This sharp separation could be viewed as supporting the hypothesis that the primitive structures could have been the β-turn (Jurka and Smith, 1987) and/or the β-strand (Brack and Orgel, 1975; Di Giulio, 1996).

5.5. Base compositions at synonymous positions are correlated with protein structure

The base composition of different codon positions was investigated on human sequences encoding 62 proteins with known crystallographic structures (D'Onofrio et al., 2002). This analysis (which superseded a previous one by Chiusano et al., 1999, performed on 34 predicted structures) concerned not only the inter- but also the intra-structure differences, as seen not only in the total set of coding sequences, but also in three subsets characterized by different GC_3 levels. As far as the whole set is concerned, both inter- and intra-structure differences were significant, with the only exceptions of A_3 and T_3 that were not different between helix and aperiodic, and between strand and helix, respectively, and of $A_3 - T_3$ that were not significantly different in helix (see **Fig. 10.20**).

When the three subsets of sequences characterized by different GC levels were analyzed, the number of significant inter- and intra-structure differences increased from the GC_3-poor to the GC_3-rich subset. In the latter set, almost all differences were significant, the exceptions concerning intra-structure differences in helix, and inter-structure differences of G_3 (strand/helix; strand/aperiodic), A_3 (helix/aperiodic) and T_3 (strand/helix). This is an important result because it shows that the compositional transition led to an increase rather than to a decrease of the significant compositional differences among the sequences corresponding to different secondary structures of proteins. The observed differences were not due to a different codon usage, but were affected by the biased occurrences of the amino acids among the secondary structures of the proteins.

Therefore, the physico-chemical properties of the secondary structures of the proteins are the main factor affecting the base composition not only at the second (Chiusano et al., 2000; see preceding section), but also at the third codon positions. The observation that DNA segments encoding different secondary structures exhibit different synonymous and nonsynonymous rates, as well as the fact that these segments also have different "silent" base composition suggest that protein structure is very likely a significant factor that affects

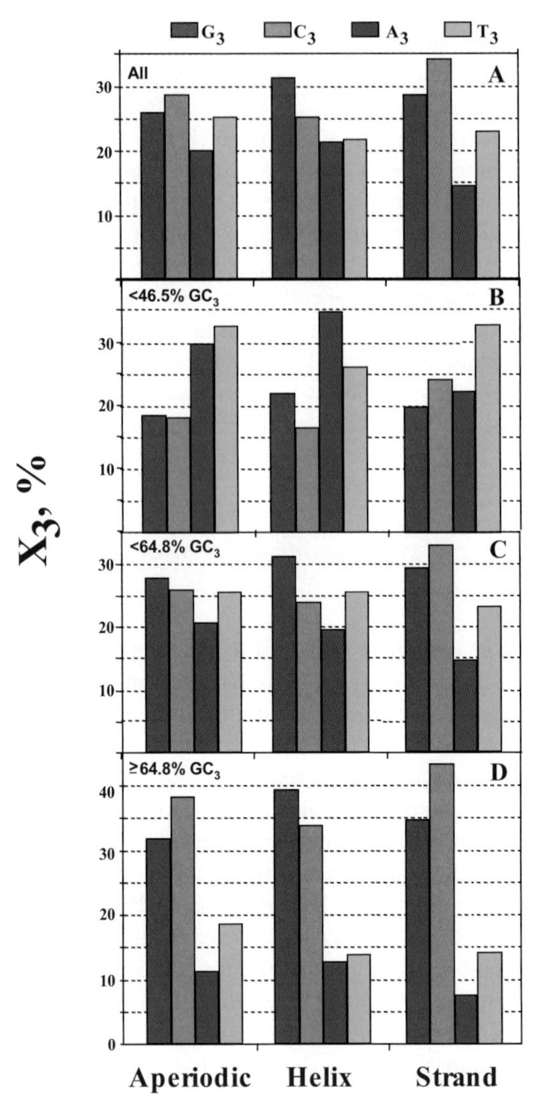

Figure 10.20. Average base composition in third codon positions of genes encoding 62 human proteins with known crystallographic structures. **A** concerns all data, **B**, **C** and **D** three subsets characterized by the GC_3 values indicated. Structures were ordered according to increasing hydrophobicity. (From D'Onofrio et al., 2002).

in a similar way the synonymous and nonsynonymous rates. This notion is further reinforced by the observation that in the gene encoding the GP63 protein from *Leishmania* the relationship between protein structure and substitution rates is straightforward (see **Section 5.3**).

Part 11
The compositional evolution of vertebrate genomes

CsCl profiles. In the transitional, or shifting mode, one part of the DNA molecules changes its composition. Two **intragenomic shifts** have been recognized. The **major shifts (Fig. 11.1A)** concern the changes undergone by the gene-dense part of the vertebrate genome (as well as by the coding sequences embedded in them) at the transitions between reptiles and mammals or between reptiles and birds. The **minor shift (Fig. 11.1B)** concerns the compositional changes undergone by the GC-poorest and the GC-richest isochores at the emergence of murids. In this case, the two extreme compositions converge towards average values. In contrast, the **whole genome** or **horizontal shifts (Fig. 11.1C)** concern the totality of the genome which becomes GC-poorer or GC-richer, without any change in its compositional pattern. Finally, in the conservative mode (which precedes and follows the shifts), no change takes place (**Fig. 11.1D**).

One should recall here that the compositional patterns of DNA molecules are correlated with the compositional patterns of coding sequences (see **Part 3**, **Chapter 2**, **Figs. 3.17** and **3.19**). One could, therefore, schematically represent the compositional modes of evolution using coding sequence patterns instead of the patterns of DNA molecules.

CHAPTER 2

The maintenance of compositional patterns

2.1. The maintenance of the compositional patterns of warm-blooded vertebrates

The existence of a compositionally conservative mode of evolution of warm-blooded vertebrates was initially suggested by the similarity of their isochore patterns (which may, however, undergo very limited "whole-genome shifts", to be discussed in **Chapter 5**), and by the similarity of their coding sequence patterns. Another line of evidence was provided by **GC_3, G_3 and C_3 plots** concerning orthologous genes of species from different mammalian orders. **Fig. 11.2** shows G_3 and C_3 plots for human and calf genes (artiodactyls share with primates the "**general mammalian pattern**", the most widespread mammalian pattern, as defined by the CsCl profiles; see **Part 4**).

Fig. 11.2 shows that the regression lines of the human/calf plots pass through the origin and are characterized by unity slopes and very high correlation coefficients, 0.88–0.89. In other words, average G_3 and C_3 values of orthologous genes from human and calf are very close to each other over the entire GC_3 range, including GC_3 values up to 90–100%. Now, extant placental mammalian orders diverged some 100 Mya (million years ago; a very approximate figure based on molecular data; Murphy et al., 2001; Eizirik et al., 2001; Hedges and Kumar, 1999) from a common ancestor according to a **star-like phylogeny** (see **Fig. 11.3**). In other words, they evolved independently of each other for about 100 Myrs. One should conclude, therefore, that the very many mutations that took place during this long time interval led to no change or only to small non-directional changes in the com-

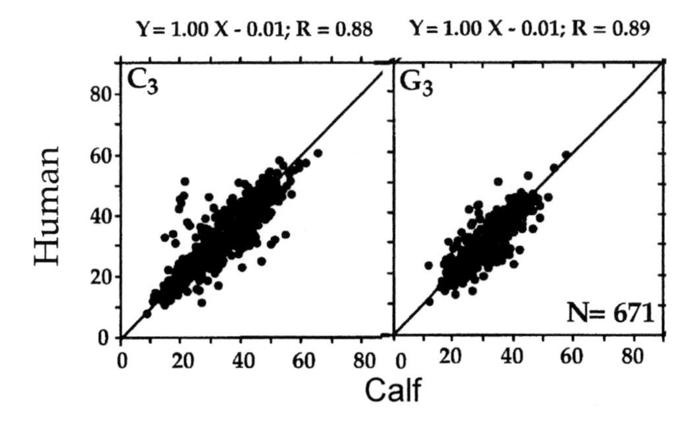

Figure 11.2. Correlation between G_3 and C_3 values of orthologous genes from human and calf. The orthogonal regression lines coincide with the diagonals. The equations of the regression lines and the correlation coefficients are shown. N is the number of gene pairs explored. (From Bernardi, 2000b).

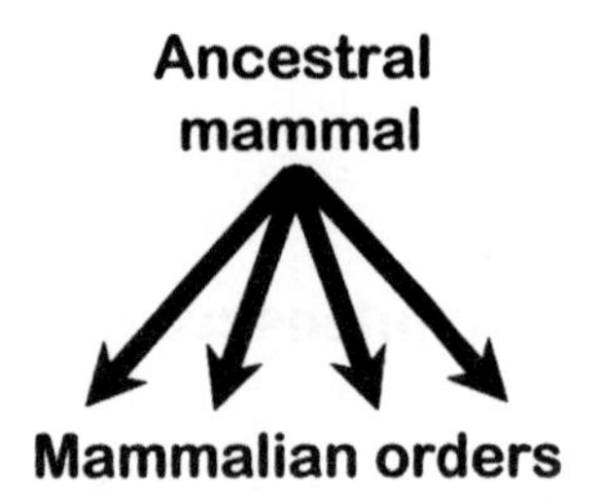

Figure 11.3. Star-like phylogeny of mammalian orders. The figure shows that mammalian orders evolved from a common ancestor independently from each other.

positional patterns of both DNA and coding sequences. Because of the linear correlations between GC_3 and isochore GC, these plots confirm the conservation of isochore patterns during this time interval. A very similar result was obtained when comparing orthologous genes from birds belonging to different avian orders (Kadi et al., 1993; Mouchiroud and Bernardi, 1993).

The correlations between GC_3 values of mammals (and birds) from different orders are very important because GC_3 values are, in turn, correlated with the GC levels of surrounding DNA segments about 200 kb in size. Since in both cases correlation coefficients are very high (0.88 in **Fig. 11.2**, 0.82 in the GC_3 *vs.* isochore GC correlations of human genes; see Clay et al., 1996), eq. 2 of **Part 10, Chapter 2**, applies. Transitivity of correlations then leads to the conclusion that long DNA stretches, at least 200 kb in size, around orthologous genes are compositionally correlated with each other in mammals and birds. In other words, the conservative mode of mammalian and avian evolution also concerns compositional patterns, as already realized many years ago (Bernardi et al., 1988). A detailed example of this situation is given in **Chapter 4.2**.

2.2. The conservative mode of evolution and codon usage

Another way to investigate the conservative mode of evolution is by plotting the **average frequency of codons** from orthologous genes. **Fig. 11.4** presents the results obtained for the human/calf comparison. For reasons which will become clear in the following chapter, the gene data set was split into a GC-poor ($GC_3 < 60\%$) and a GC-rich ($GC_3 > 60\%$) subset. The two **difference histograms** (human values less calf values) so obtained are almost indistinguishable from the baseline. In other words, the average frequency of codons (and, therefore, also of aminoacids) from all orthologous genes that were compared is the same in the two species.

On the other hand, as we first noted (Bernardi et al., 1985b), "*a different codon strategy is used for different genes located in the same genome*", and such different strategy "*is mainly determined by the location of genes in heavy or light isochores*". This was shown by demon-

⟶

Figure 11.4. Histograms comparing average codon frequencies of orthologous genes from human and calf (Cruveiller and Bernardi, unpublished results).

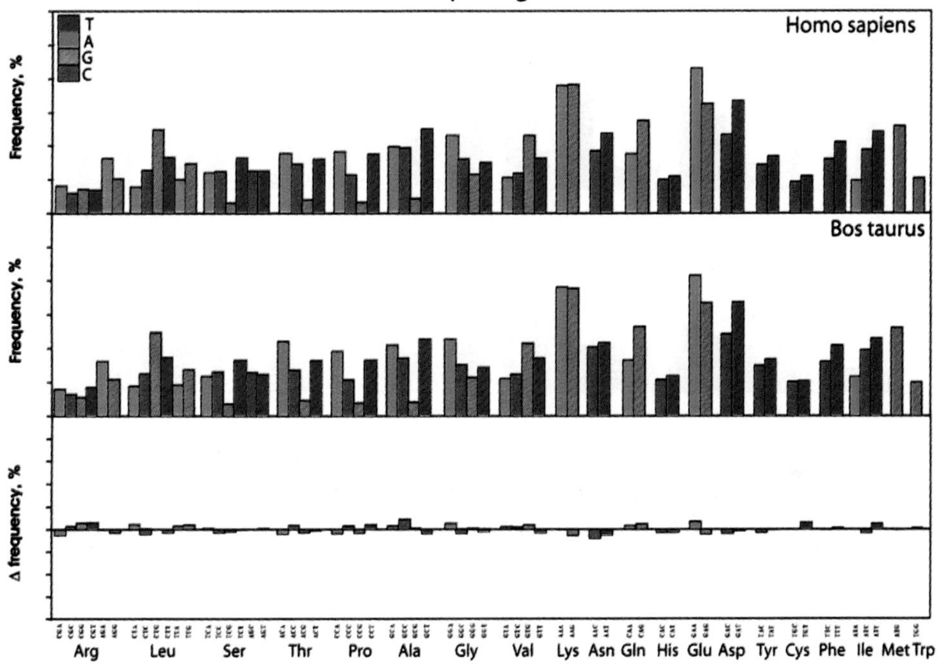

GC-rich genes

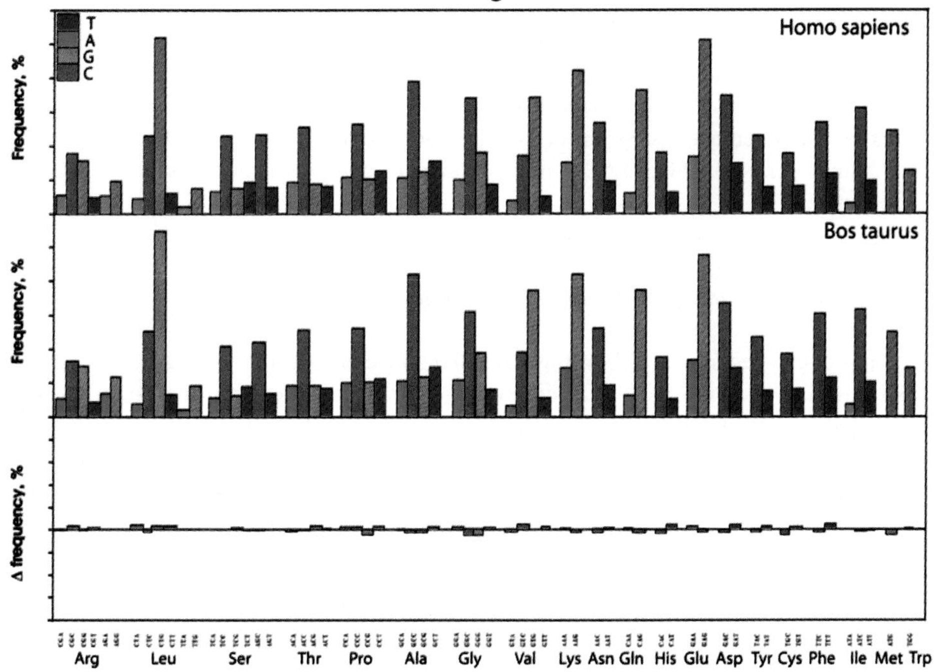

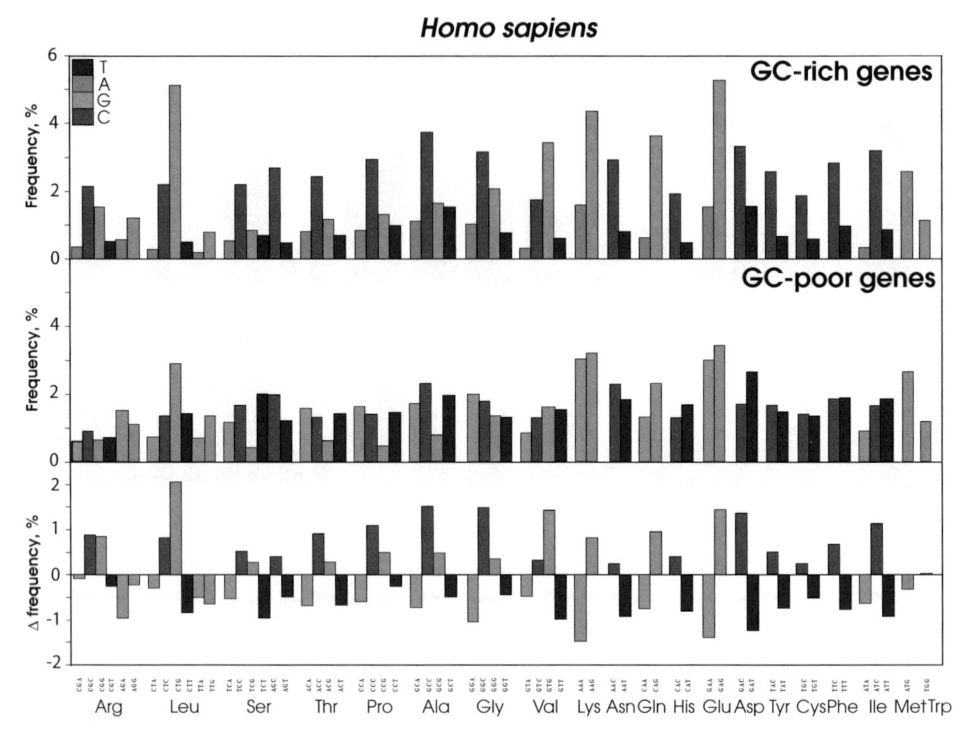

Figure 11.5. Histograms comparing average codon frequencies of GC-rich and GC-poor human genes (Cruveiller and Bernardi, unpublished results).

strating a correlation between GC_3 and the GC levels of the isochores in which the genes were embedded (see **Part 3**, **Chapter 3**). Another way to demonstrate this point is to construct the difference histogram of average codon frequencies from GC-rich and GC-poor human genes (see **Fig. 11.5**). A more detailed discussion of codon usage is presented in **Section 3.2**.

2.3. Mutational biases in the human genome

The approach used in our laboratory to infer and compare the underlying pattern of mutations among genes located in different isochores consisted in using mutation data bases that are available for many genes associated with human genetic diseases. The assumption of this approach is that such mutations reflect the actual frequency distribution of mutations as they arise in populations. It should be noted that these mutations were not yet subjected to the sieving effect of selection, in contrast to the pattern of nucleotide substitutions that depends upon two processes, mutation and fixation, the latter being influenced by selection. Inferring the mutational pattern from the observed pattern of mutations responsible of genetic alterations is not new. In fact, previous investigations have used this kind of information for analysing the mutational dynamics of human genes (see Krawczak and Cooper, 1996).

In order to investigate whether the pattern of mutations differs among genes having different GC_3 levels, we analyzed the mutational spectrum of genes with different average GC_3 levels. **Fig. 11.6** shows that there is very little if any variation in the expected GC content of a neutral sequence evolving only under the effect of mutations. Interestingly enough, the results show that, independently of whether CpGs, and/or repeated elements are included, the variations of the expected GC content for sequences evolving only under

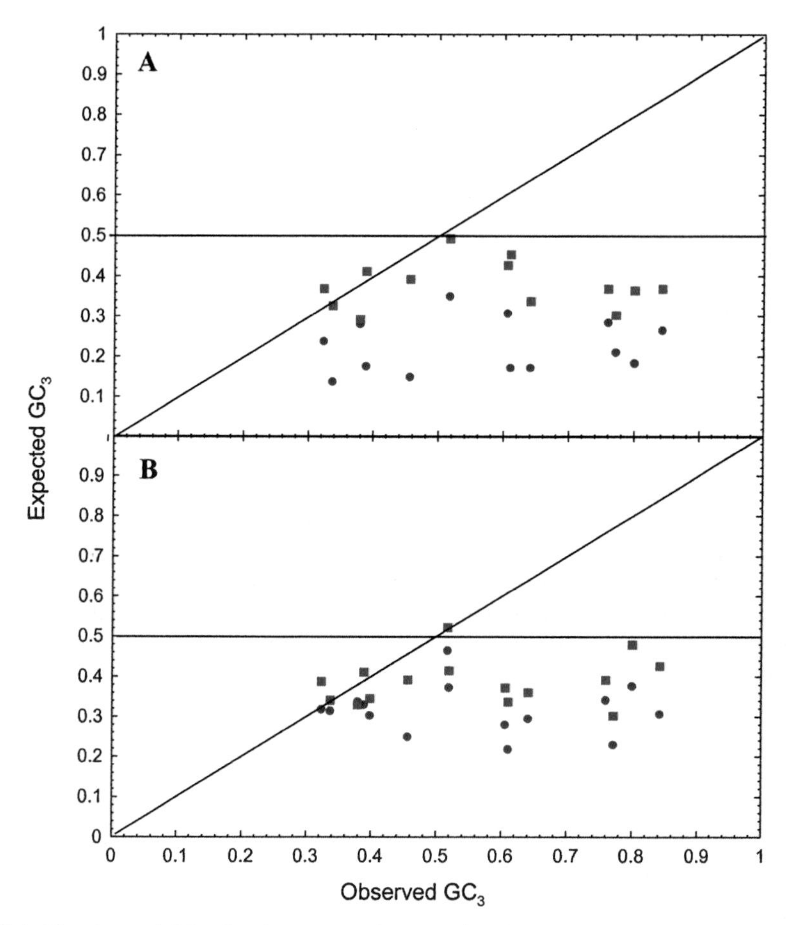

Figure 11.6. The observed GC_3 of each gene was plotted against the GC_3 level that would be expected if it depended solely on the mutational pattern. There are four different expectations for each gene, **A** when repeated elements are considered, with CpG included (blue circles), or excluded (red squares); **B** when repeated elements are not considered, with CpG sites taken into account or not (symbols as for **A**). The diagonal shows the line where points should be located if the GC_3 level was determined by the mutational input, so that the vertical distance between each point and this line indicates how far away is the observed GC_3 from the mutational equilibrium. The horizontal lines at GC = 0.5 indicate the borders between AT bias and GC bias. (Modified from Alvarez-Valin et al., 2002).

the effect of mutations are always in the region of AT biases (with the exception of one of the results obtained from PAX6).

Fig. 11.6 also indicates that the mutational patterns in GC_3-rich genes do not appear to explain their GC_3-richness. Indeed, GC_3-rich and very GC_3-rich genes exhibit patterns of mutations that yield expectations of neutral GC level that are much lower than the actual GC_3. In fact, **the actual GC_3 values are between two and three times higher than the expectations obtained from mutational data**. Therefore, **Fig. 11.6** strongly suggests that the variability in GC_3 cannot be attributed to variations in the mutational input along the genome but to fixational biases. In other words, even if GC_3-rich genes are subjected to AT biases these genes remain GC_3-rich, indicating that the majority of new mutations that are introduced in the populations, which are GC $\rightarrow$ AT, are eliminated. This conclusion also suggests that **the compositional heterogeneity of the isochore pattern is under selective constraints** since, whatever factor is responsible for maintaining high GC_3 levels, if affects the whole isochore, as indicated by the strong correlation between the GC level of the isochore and GC_3.

The results of **Fig. 11.6** are in agreement with recent analyses on single nucleotide polymorphisms (Eyre-Walker, 1999; Smith and Eyre-Walker, 2001). These authors found that, contrary to what would be expected if the GC content was solely determined by mutational biases (and assuming the equilibrium condition holds), the number of segregating GC $\rightarrow$ AT polymorphisms is significantly higher than that of AT $\rightarrow$ GC polymorphisms. Obviously, if the sequences were in mutational equilibrium, the two kinds of polymorphisms should be present in equal amounts.

The problem under consideration here has been investigated also through other approaches. The analysis of substitution patterns in pseudogenes can be used to infer the underlying pattern of mutations since pseudogenes are supposed to be quite free of selectional constraints. While some analyses suggest that all pseudogenes (from both GC-poor and GC-rich isochores) are under AT biases (Gojobori et al., 1982), a more recent paper on two globin pseudogenes located in different isochores (Francino and Ochman, 1999) proposed that the patterns of substitutions are not the same. According to these authors, the expected GC level that is obtained from the β globin gene (located in a GC-poor isochore) would be 0.4, whereas the expected GC level inferred from the α globin pseudogene (located in a GC-rich isochore) would be 0.57, suggesting a GC bias in GC-rich genes and an AT bias in GC-poor genes. This conclusion is contradicted by the results on the AT bias obtained on genes, pseudogenes, repeated sequences and single nucleotide polymorphisms and, therefore, the general conclusion by Francino and Ochman (1999), that isochores result from mutation and not from selection, is certainly not supported.

The two major compositional shifts in vertebrate genomes

3.1. The major shifts

Two **major, intragenomic, compositional shifts** took place in the genomes of the ancestors of present-day mammals and birds, respectively (see **Fig. 11.1**). These shifts, originally observed at the DNA level, consist in the appearance in the genomes of warm-blooded vertebrates of a relatively small percentage (10–15%) of GC-rich DNA molecules (the "**genome core**") that are GC-poorer in the genomes of most cold-blooded vertebrates (**Fig. 3.17**).

The GC-richest 10–15% of the genomes of cold-blooded vertebrates hybridize, however, single-copy DNA from human H3 isochores (Bernardi, 1995; see **Fig. 5.6**). This, as well as other lines of evidence (see **Part 5, Section 1.3**), indicate that in cold-blooded vertebrates there is an "**ancestral genome core**" located in the GC-richest fractions of the genome. These GC-richest fractions are, however, not as GC-rich as the corresponding fractions of the genomes of warm-blooded vertebrates. Since the genomes of mammals and birds derive from those of two different lines of ancestral reptiles (see **Section 3.3**), these findings indicate that two major compositional changes independently occurred in the distinct ancestral lines leading to warm-blooded vertebrates, and that they concerned a small part of the genome, which is, interestingly, the gene-richest part of it. In warm-blooded vertebrates, the majority of the genome which did not undergo the compositional transition was called the **genome desert** (or the empty quarter) or the **paleogenome**, the minority which did, the "**genome core**", or the **neogenome** (Bernardi, 1989; see **Fig. 11.7**). The paleogenome/neogenome terminology should, however, be abandoned, because it may

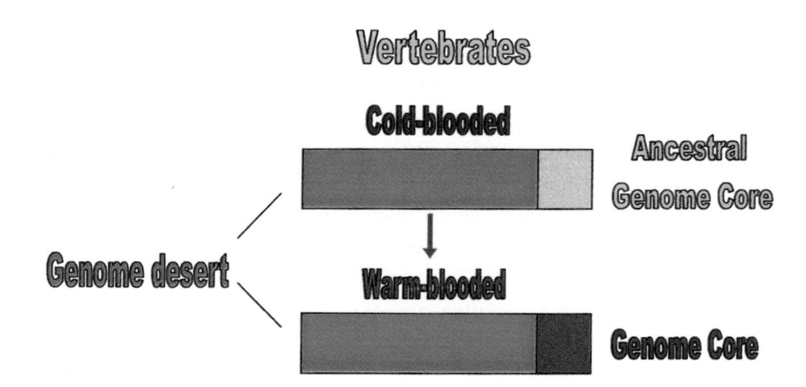

Figure 11.7. Scheme of the compositional transition shown by the genomes of cold- to warm-vertebrates. The "empty quarter" (or "paleogenome") is GC-poor and gene-poor (blue box) and did not undergo any compositional change. The gene-dense, moderately GC-rich "ancestral genome core" (pink box) underwent a compositional change into a gene-dense, GC-rich "genome core" (or "neogenome"; red box).

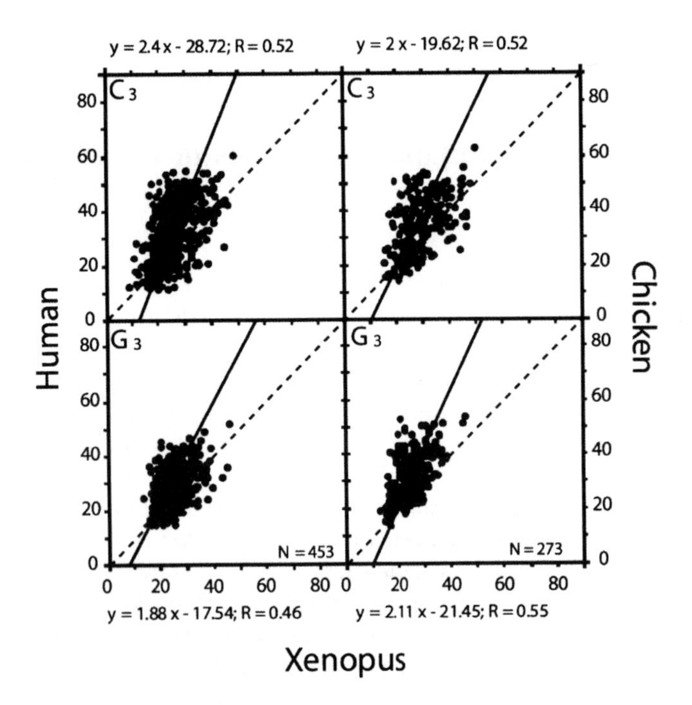

Figure 11.8. Correlation between G_3 and C_3 values of orthologous genes from human, or chicken, and *Xenopus*. The orthogonal regression lines are shown together with the diagonals (dashed lines). The equations of the regression lines and the correlation coefficients are indicated. N is the number of gene pairs explored. (From Bernardi, 2000b).

convey the idea that the neogenome is a novel compartment of the genome instead of a compositionally modified compartment.

The two major shifts were then detected at the coding sequence level (Bernardi et al., 1985b, 1988; Perrin and Bernardi, 1987; Mouchiroud et al., 1987, 1988; Bernardi and Bernardi, 1991). Indeed, coding sequences from cold-blooded vertebrates are relatively homogeneous from a compositional viewpoint and generally characterized by low GC levels, whereas coding sequences from warm-blooded vertebrates are compositionally much more heterogeneous and reach very high GC levels (**Fig. 3.17**), up to 100% GC in the third codon positions of avian genes.

Another evidence for the major shifts was obtained by comparing the nucleotides in third codon positions of orthologous genes from human and *Xenopus* (as initially shown by Perrin and Bernardi, 1987; Mouchiroud et al., 1987, 1988; and Bernardi and Bernardi, 1991). When the orthologous genes of human/*Xenopus* were investigated in their GC_3, or in their G_3 and C_3 values (**Fig. 11.8**), points were scattered about the diagonal in the low GC

→

Figure 11.9. **A.** Codon frequencies of human (top) and *Xenopus* (middle) GC-poor (0-60% GC_3) orthologous genes. The bottom histogram shows the differences in codon frequencies. **B.** Codon frequencies of human (top) and *Xenopus* (middle) GC-rich (60-100% GC_3) orthologous genes. The bottom histogram shows the differences in codon frequencies. (From Cruveiller et al., 2000).

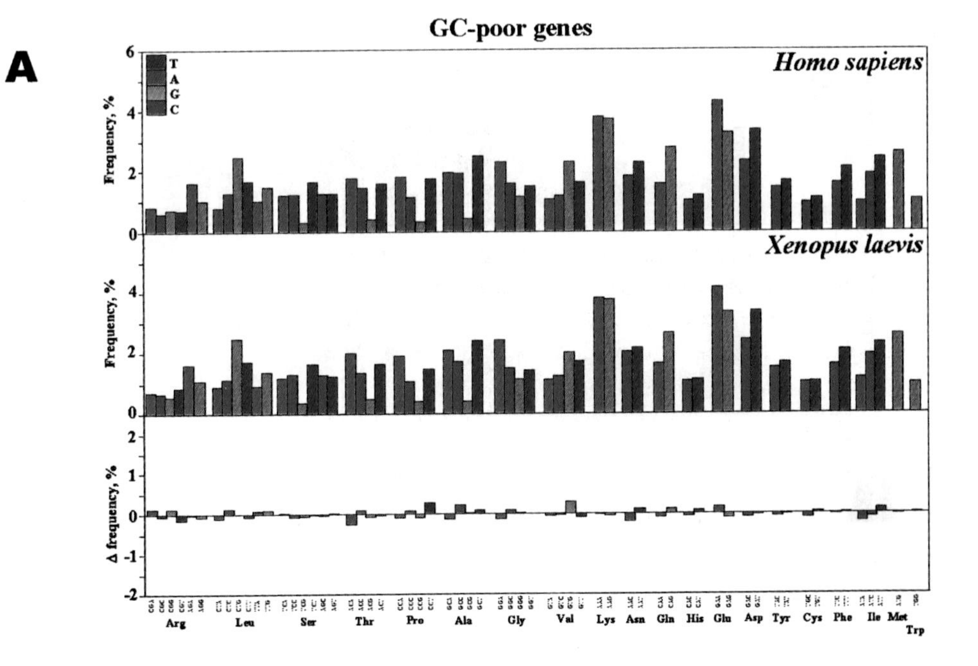
A
GC-poor genes
Homo sapiens
Xenopus laevis
T
A
G
C
Frequency, %
Δ frequency, %
Arg
Leu
Ser
Thr
Pro
Ala
Gly
Val
Lys
Asn
Gln
His
Glu
Asp
Tyr
Cys
Phe
Ile
Met
Trp

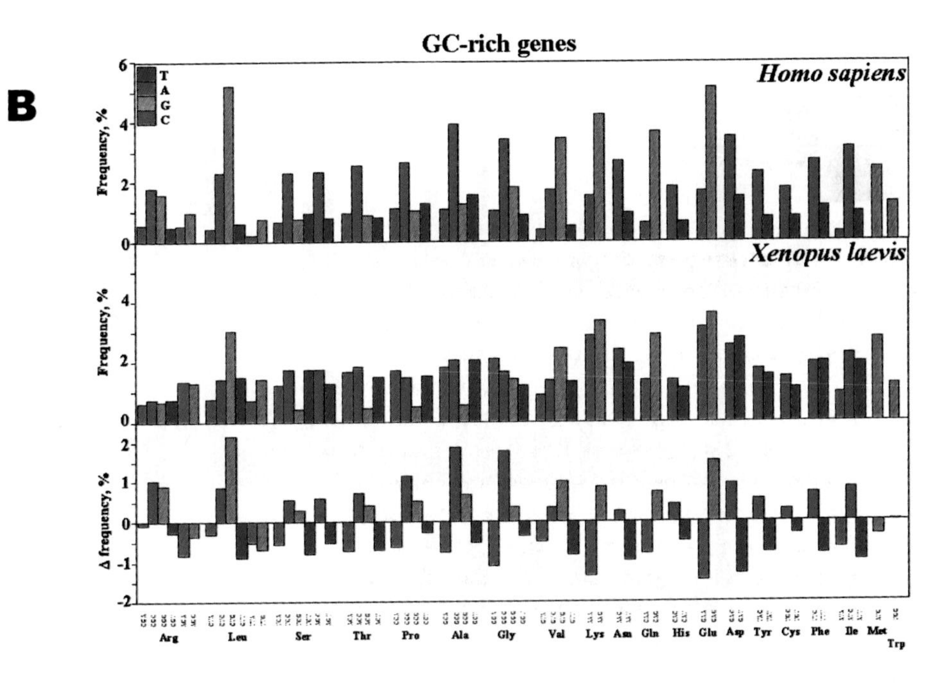
B
GC-rich genes
Homo sapiens
Xenopus laevis
T
A
G
C
Frequency, %
Δ frequency, %
Arg
Leu
Ser
Thr
Pro
Ala
Gly
Val
Lys
Asn
Gln
His
Glu
Asp
Tyr
Cys
Phe
Ile
Met
Trp

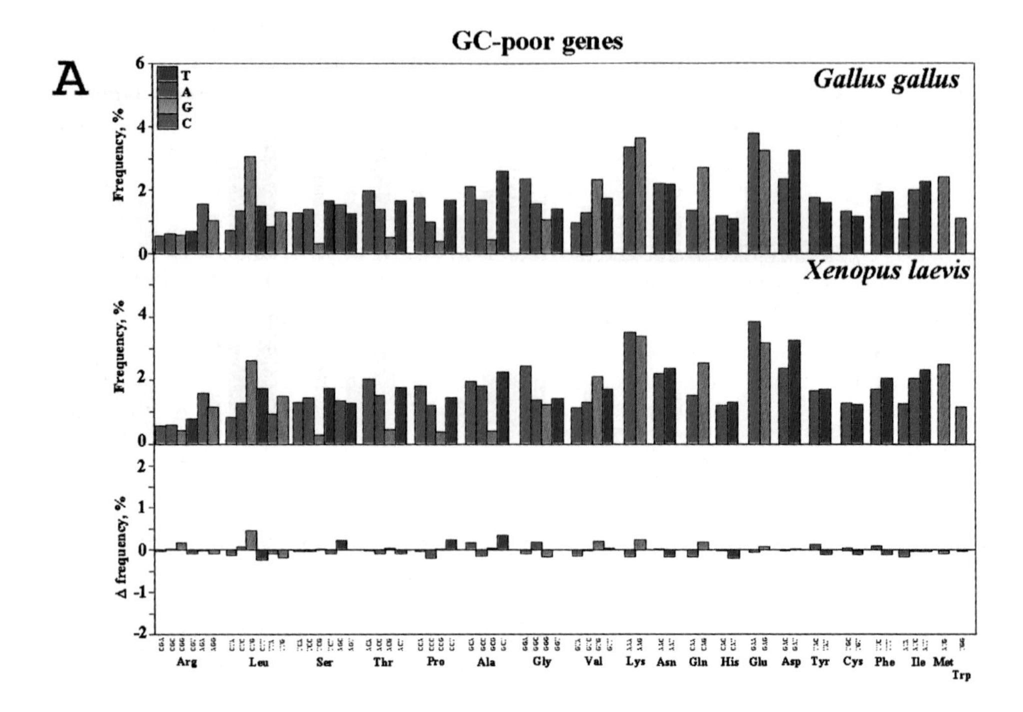
GC-poor genes
A
T
A
G
C
Gallus gallus
Xenopus laevis
Frequency, %
Δ frequency, %
Arg
Leu
Ser
Thr
Pro
Ala
Gly
Val
Lys
Asn
Gln
His
Glu
Asp
Tyr
Cys
Phe
Ile
Met
Trp

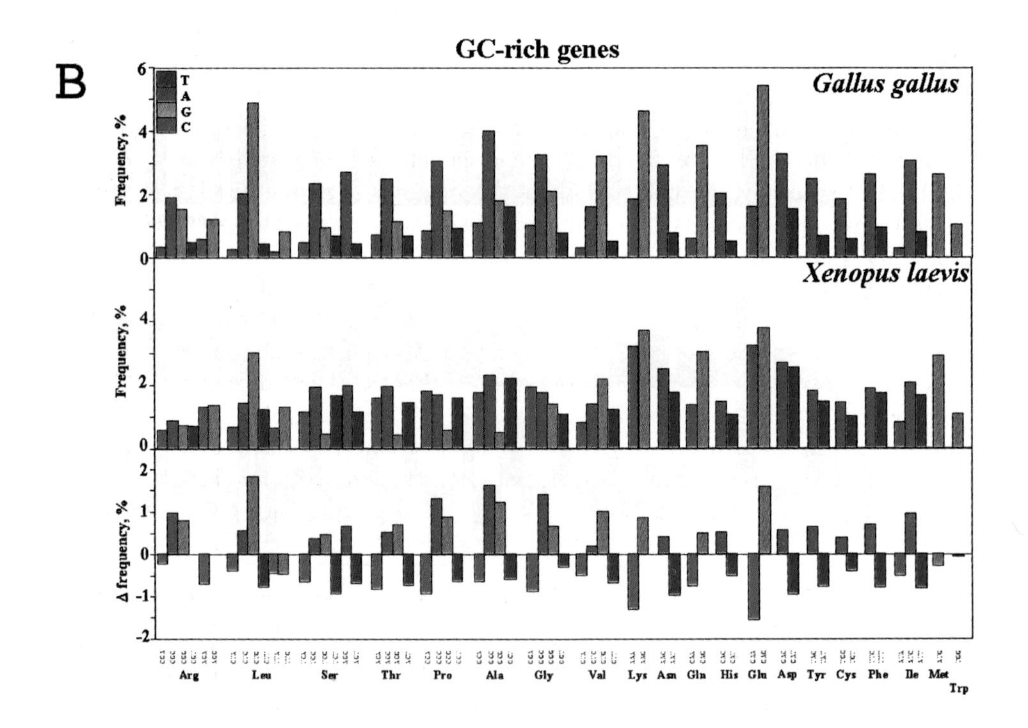
GC-rich genes
B
T
A
G
C
Gallus gallus
Xenopus laevis
Frequency, %
Δ frequency, %
Arg
Leu
Ser
Thr
Pro
Ala
Gly
Val
Lys
Asn
Gln
His
Glu
Asp
Tyr
Cys
Phe
Ile
Met
Trp

range, showing no directional changes between the two species, whereas human gene values were increasingly higher, on the average, compared to the corresponding *Xenopus* gene values, as GC_3 values increased.

This produced regression lines having slopes that were close to 2. Very similar results were obtained when comparing orthologous genes from chicken and *Xenopus* (**Fig. 11.10**). Since very good correlations exist between GC_3 and GC levels of extended DNA regions flanking the coding sequences (Bernardi et al., 1985b; Clay et al., 1996), the results of **Fig. 11.10** also confirm that large regions surrounding orthologous genes underwent (or, alternatively, did not undergo) the compositional transition (Bernardi, 1995).

Two additional observations were reported concerning the compositional transitions under consideration. First, codon frequencies and codon usage were essentially unchanged for the orthologous genes that had not been affected by the major shift, whereas they were drastically changed for those that had been affected (Cruveiller et al., 2000). In this analysis, the sets of orthologous gene pairs from human/*Xenopus* and chicken/*Xenopus* were split into two sub-sets according to the GC_3 levels of the GC-richer organism: the "GC-poor genes" (0%-60% GC_3), and the "GC-rich genes" (60%-100% GC_3). As far as GC_3-poor genes are concerned, the results are very clear-cut in that differences in amino acids and codon usage between *Xenopus* and human or chicken are extremely small, confirming the data of Cruveiller et al. (1999). In other words, the average levels of amino acids and the average codon frequencies of the two sets of orthologous data are very similar and the hydropathy of the proteins was the same (**Figs. 11.9, 11.10**). In contrast, GC_3-rich genes of human and chicken show remarkable differences in the levels of encoded amino acids and codon frequencies relative to *Xenopus* and an increase in the hydrophobicity of proteins (**Figs. 11.9, 11.10**).

The second observation was that the regression lines between GC_3, G_3 and C_3 values of orthologous genes from human and chicken showed a high correlation coefficient and coincided with the diagonal, indicating that the genes that had not undergone the transition and those that had, were, to a large extent, the same sets of genes in the two species (**Fig. 11.11**). Expectedly, the scatter about the diagonal was larger than in the corresponding calf/human plot of **Fig. 11.2**, but difference histograms are close to the baseline when chicken and human data are compared (**Fig. 11.12**).

In the last case, it is remarkable that codon choices are the same even for quartet codons. One might imagine that the similarity in codon preferences between human and chicken for each set of genes is nothing but the predictable result of the fact that genes with similar GC_3 levels are compared. One might then expect for the case of GC_3 data set, that G- and C-ending codons exhibit higher values in both species. However, **Fig. 11.12** shows that codon frequencies in quartets is the same for both species. For instance in the leucine codon group, CTG was by far the most frequent codon in both species, while CTC was the second in the order of frequency, whereas in the glycine codon group, GGC is the most frequent codon in both species. The same comparison for the remaining codon groups and

←

Figure 11.10. **A.** Codon frequencies of chicken (top) and *Xenopus* (middle) GC-poor (0-60% GC_3) orthologous genes. The bottom histogram shows the differences in codon frequencies. **B.** Codon frequencies of chicken (top) and *Xenopus* (middle) GC-rich (60-100% GC_3) orthologous genes. The bottom histogram shows the differences in codon frequencies. (From Cruveiller et al., 2000).

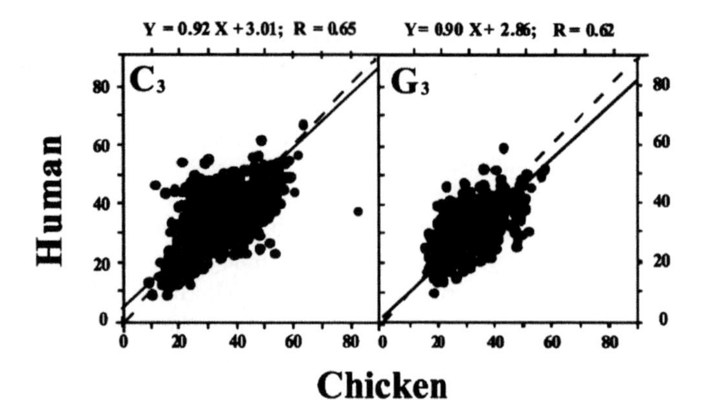

Figure 11.11. Correlation between G_3 and C_3 values of orthologous genes of human and chicken. Other indications are as in Fig. 11.6. (From Bernardi, 2000b).

also for the case of GC_3-poor data set clearly indicates that the similarity in codon frequencies go much beyond what would be expected from the simpler similarity in GC_3 levels.

It should be stressed that CsCl profiles and GC_3 plots of avian genomes indicate an extremely high level of similarity among different orders (Kadi et al., 1993; Mouchiroud and Bernardi, 1993). This similarity also applies to the mammalian genomes belonging to the "general pattern" (Sabeur et al., 1993; Mouchiroud and Bernardi, 1993), and to the "murid pattern" (Douady et al., 2000), respectively. In other words, what has been shown here for human and chicken genes is generally valid for mammalian and avian genomes.

To summarize, (i) the compositional transitions concerned the genes and the intergenic sequences that are located in the GC-richest isochores of the genomes of warm-blooded vertebrates, namely in the isochores that correspond to the gene-dense regions of the vertebrate genome, the "genome core"; in contrast, they did not concern the genes and the intergenic sequences from the gene-poor "genome desert" (see Bernardi, 2000a), which remained compositionally stable, on the average, since the early vertebrates; (ii) the transitions occurred (and were similar) in the independent ancestral lines of mammals and birds, but in almost no cold-blooded vertebrate (the heterogeneous genomes of reptiles will be discussed later; only very few cold-blooded vertebrates have relatively GC-rich genomes, but these genomes are compositionally much more homogeneous than the genomes of warm-blooded vertebrates; see Bernardi and Bernardi, 1990a,b and **Part 4**); (iii) the transitions stopped before the appearance of present-day mammals and birds, as indicated by the very similar patterns found in different mammalian and avian orders, respectively (see the human/calf comparisons of **Fig. 11.2** for an example).

These findings indicate that the compositional transitions affecting the "genome core" of

→

Figure 11.12. **A.** Codon frequencies of human (top) and chicken (middle) GC-poor (0-60% GC_3) orthologous genes. The bottom histogram shows the differences in codon frequencies. **B.** Codon frequencies of human (top) and chicken (middle) GC-rich (60-100% GC_3) orthologous genes. The bottom histogram shows the differences in codon frequencies. (From Cruveiller et al., 2000).

the ancestors of mammals and birds, had already reached a **compositional equilibrium** at least at the times of appearance of present-day mammals and birds, and that, from those times on, the compositional patterns resulting from the cold- to warm-blooded transitions

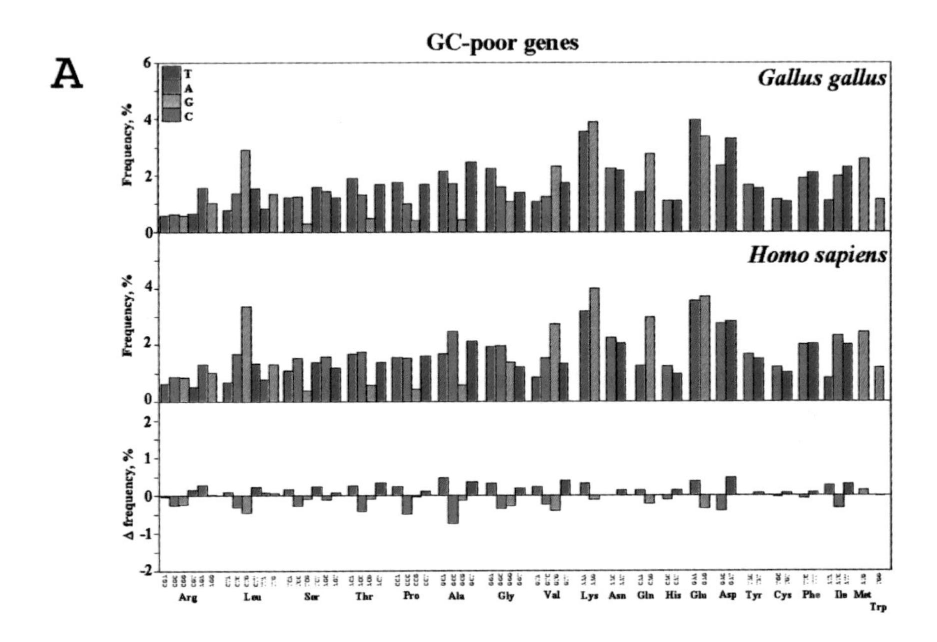

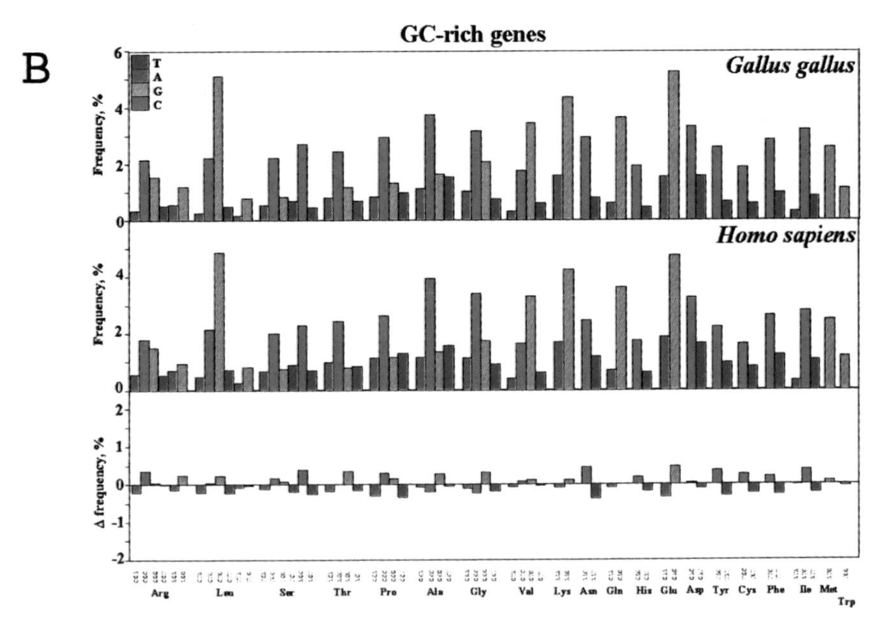

TABLE 11.1

Relative Synonymous Codon Usage (RSCU) values of *P. falciparum* (*Pf*) and *S. aureus* (*Sa*)[a]

aa	Codon	*Pf*	*Sa*	aa	Codon	*Pf*	*Sa*	aa	Codon	*Pf*	*Sa*	aa	Codon	*Pf*	*Sa*
Phe	TTT	1.6	1.4	Ser	TCT	1.4	1.4	Tyr	TAT	1.8	1.6	Cys	TGT	1.8	1.3
Phe	TTC	0.4	0.6	Ser	TCC	0.4	0.2	Tyr	TAC	0.2	0.4	Cys	TGC	0.2	0.7
Leu	TTA	4.1	3.3	Ser	TCA	1.9	1.8	End	TAA	*	*	End	TGA	*	*
Leu	TTG	0.7	0.9	Ser	TCG	0.2	0.3	End	TAG	*	*	Trp	TGG	1.0	1.0
Leu	CTT	0.7	0.8	Pro	CCT	1.3	1.5	His	CAT	1.6	1.5	Arg	CGT	0.8	1.8
Leu	CTC	0.1	0.2	Pro	CCC	0.3	0.2	His	CAC	0.4	0.5	Arg	CGC	0.0	0.6
Leu	CTA	0.3	0.6	Pro	CCA	2.4	1.8	Gln	CAA	1.8	1.7	Arg	CGA	0.4	0.6
Leu	CTG	0.1	0.2	Pro	CCG	0.1	0.5	Gln	CAG	0.2	0.3	Arg	CGG	0.0	0.2
Ile	ATT	1.4	1.6	Thr	ACT	1.3	1.2	Asn	AAT	1.7	1.4	Ser	AGT	1.7	1.5
Ile	ATC	0.2	0.5	Thr	ACC	0.5	0.2	Asn	AAC	0.3	0.6	Ser	AGC	0.4	0.8
Ile	ATA	1.4	0.9	Thr	ACA	2.0	2.1	Lys	AAA	1.7	1.6	Arg	AGA	4.2	2.5
Met	ATG	1.0	1.0	Thr	ACG	0.2	0.5	Lys	AAG	0.3	0.4	Arg	AGG	0.6	0.4
Val	GTT	1.8	1.6	Ala	GCT	1.9	1.4	Asp	GAT	1.8	1.6	Gly	GGT	1.9	1.8
Val	GTC	0.2	0.4	Ala	GCC	0.4	0.3	Asp	GAC	0.2	0.5	Gly	GGC	0.1	0.7
Val	GTA	1.7	1.5	Ala	GCA	1.6	1.9	Glu	GAA	1.8	1.6	Gly	GGA	1.8	1.2
Val	GTG	0.3	0.5	Ala	GCG	0.1	0.4	Glu	GAG	0.2	0.4	Gly	GGG	0.2	0.4

[a] From Musto et al. (1995). *Pf* and *Sa* are *P. falciparum* and *S. aureus*, respectively. RSCU were calculated according to Sharp et al., 1986. Asterisks indicate stop codons; these were not taken into account.

aureus (see **Part 9** and **Table 11.1**), although some translational selection is still detectable (Musto et al., 1999).

To sum up, codon usage is due to two main selective factors: 1) compositional constraints (which we visualize as the result of natural selection shaping up the compositional patterns of genomes), **and 2) translational selection**. When the former are very biased, the latter becomes barely detectable (one should remember that at 100% GC, 50% of codons are not used) and, *vice versa*, when the compositional constraints are not biased (and genomes are in middle GC range), translational selection becomes quite visible.

The importance of compositional constraints on codon usage may be demonstrated in another way. As shown in **Figs. 11.9A** and **11.10A**, there is **no difference in average codon frequency** between orthologous GC-poor genes of *Xenopus* and human or chicken. In other words, the average codon frequency has been conserved in GC-poor genes since at least the common ancestor of tetrapods. (In fact, data by Romero et al., 2003, indicate a conservation dating back to Cyprinids, a fish family belonging to the superorder *Ostariophysi*; see **Part 4**). In sharp contrast, **Figs. 11.9B** and **11.10B** show large differences in average codon frequency in GC-rich genes of *Xenopus* and human/chicken. Again, however, average codon frequencies of GC-rich genes are very close to each other in warm-blooded vertebrates (human and chicken; **Fig. 11.12B**).

3.3. Other changes accompanying the major shifts

The compositional transitions that took place in the genomes of the ancestors of mammals and birds involved more than the compositional changes just described. Indeed, (i) **CpG islands** were formed (See **Part 5, Chapter 2**); these regulatory sequences, about 1 kb in size, mainly located 5' of GC-rich genes and characterized by high levels of GC and unmethylated CpG doublets, do not exist in fishes and amphibians; interestingly they do not exist even in reptiles, like crocodile and turtle, that exhibit some compositional heterogeneity (Aïssani and Bernardi, 1991a); these results rule out the proposal (Bird, 1987) that unmethylated CpG islands are remnants of the less methylated genomes of invertebrates ancestral to vertebrates; (ii) **nucleotide context changes** towards G and C took place around the AUG initiator codon of GC-rich human genes relative to GC-poor human genes and genes from cold-blooded vertebrates (Pesole et al., 1999; see **Part 5. 1.2**); like (i) above, this point indicates that the major shifts also caused **compositional changes in regulatory sequences**; (iii) **H3$^+$ bands** appeared in metaphase chromosomes (see **Part 7**) as a consequence of the GC increase corresponding to the compositional transition; (iv) **karyotype changes** and **speciation** increased (Bernardi, 1993b), probably in association with a higher level of recombination in distal parts of chromosomes of warm-blooded vertebrates compared to cold-blooded vertebrates; it has been argued (Bernardi, 1993b) that the increased genome instability associated with the higher recombinogenicity is a disadvantage, which is compensated by a higher genetic variability, an advantage in terms of coping with environmental changes and/or exploration of new ecological niches; a similar reasoning was already made for the mitochondrial genome of yeast (see **Part 2, Chapter 3.5**). (v) **DNA methylation** and **CpG doublet** concentration decreased by a factor of two in mammals and birds, compared to fishes and amphibians (Jabbari et al., 1997; see **Part 5, Chapter 3**, and the following section).

This methylation/CpG transition deserves a more detailed discussion. An analysis of the data presented in **Part 5** indicates that (i) two positive correlations hold between the 5mC and GC levels of the genome of fishes/amphibians and mammals/birds, respectively (see **Fig. 5.13**); (ii) the higher methylation of fishes and amphibians is not related to the higher amounts of repetitive DNA sequences (see **Fig. 5.14**); and (iii) the 5mC and CpG observed/expected values show no overlap between the two groups of vertebrates and suggest the existence of **two equilibria in 5mC and CpG levels** (see **Fig. 11.14**). Several important questions then arise concerning (i) the two equilibria in methylation and CpG shortage; (ii) the transition between the two equilibria; and (iii) the causes of the methylation/CpG transition.

The **existence of two equilibria** of DNA methylation and CpG in vertebrate genomes is stressed by the very similar methylation levels of mammals belonging to orders separated from each other by 100 million years. As already mentioned in **Part 5**, because of the star-like phylogeny of mammalian orders (see **Fig. 11.3**), this situation indicates that the ancestral mammalian genome also showed low CpG (o/e) and 5mC levels. This conclusion contradicts the hypothesis (Cooper and Krawczak, 1989) that the vertebrate genome originally was strongly methylated (and GC-rich) and that the methylated CpG doublets (as well as the GC levels) subsequently underwent a monotonous decay (see **Fig. 11.15**). An additional difficulty of that model was that the ancestors of vertebrates, the cephalochor-

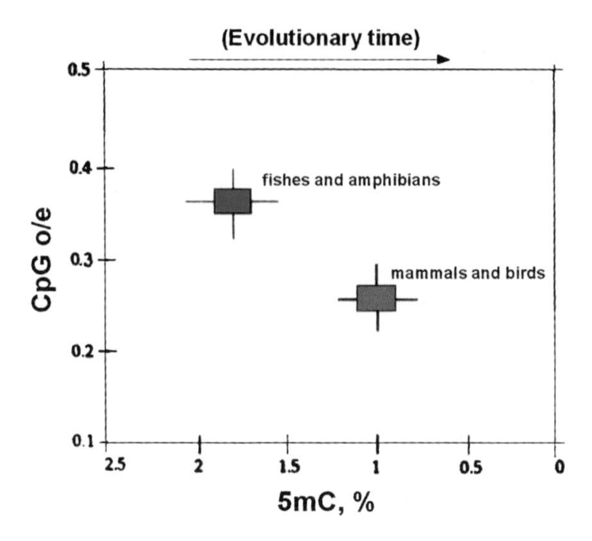

Figure 11.14. Plot of CpG observed/expected ratio (CpG o/e) against 5mC level in the genomes of cold-blooded and warm-blooded vertebrates. CpG o/e values were taken from the available literature for four cold-blooded vertebrates (three fishes and one amphibian) and nine warm-blooded vertebrates (eight mammals and one bird). All available methylation data were used, namely 32 cold-blooded vertebrate and 46 warm-blooded vertebrate species, respectively. Horizontal and vertical lines crossing the boxes correspond to standard deviations. Note the reverse scale on the 5mC axis. This was done to put fishes/amphibians and mammals/birds in an essentially time perspective. (From Jabbari et al., 1997).

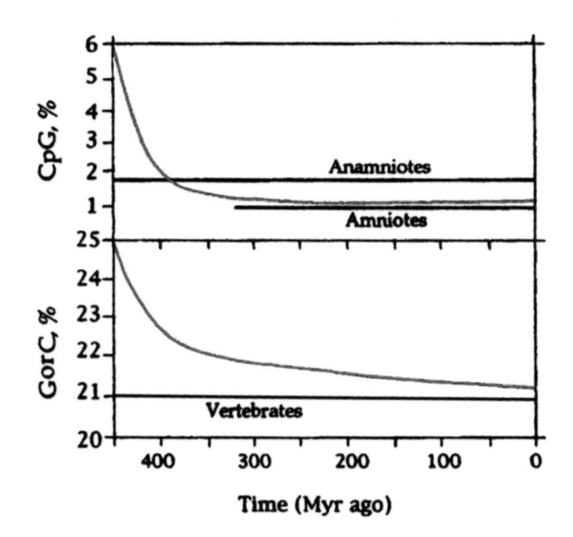

Figure 11.15. The results of Jabbari et al. (1997) on DNA methylation of fishes/amphibians (upper horizontal line) and of reptiles/mammals/birds (lower horizontal line) and the average GC level (horizontal line in the bottom frame) are compared with the monotonous decrease of CpG (red line) and GC (blue line) hypothesized by Cooper and Krawczak (1989). (Modified from Jabbari et al., 1997).

dates and urochordates, are characterized by genomes that are neither GC-rich nor strongly methylated, making it difficult to understand how and when the hypothesized original high GC and methylation of vertebrates had originated. The considerations just presented also contradict the hypothesis of an equilibrium between CpG depletion and CpG formation (Sved and Bird, 1990) which also predicts a strong decrease in CpG during vertebrate evolution (this decrease being, however, less rapid than that predicted by Cooper and Krawczak, 1989).

Concerning the **methylation/CpG levels changes**, the results available so far suggest that reptiles are characterized by a lower methylation level, compared to fishes and amphibians, close to those of mammals and birds placing the methylation transition at the appearance of the common ancestor of reptiles and mammals, namely at the appearance of amniotes.

One should, however, keep in mind that the number of reptilian DNAs analysed so far is small, that rather different levels were found in the few reptiles investigated and that no estimate is available for the CpG shortage. Under these circumstances, the question as to whether the low methylation level is present in all or only in some reptiles is still open. Therefore, two possibilities should be considered, namely, that either the common reptilian ancestor of mammals and birds had already undergone a decrease in methylation which was then transmitted by descent to extant reptiles, mammals and birds; or that the methylation change occurred independently in reptiles (possibly, only in some of them), in mammals and in birds.

As far as the **causes** for the methylation/CpG transition are concerned, they will be discussed in **Part 12**, along with the causes for the major shifts.

CHAPTER 4

The minor shift of murids

4.1. Differences in the compositional patterns of murids and other mammals

Although characteristic differences between the CsCl profiles of murids and other mammals were first observed by Thiery et al. (1976; see **Table 3.2**) only further more detailed investigations definitely proved that the mouse genome did not exhibit the GC-richest DNA components that were present in the human genome (Salinas et al., 1986; Zerial et al., 1986a). These findings were confirmed by investigations at the coding sequence level, which demonstrated in murid genes characterized by extreme base compositions (highest and lowest GC levels) a **minor, intragenomic, shift** (see **Fig. 11.16**) relative to orthologous genes from mammals exhibiting the general pattern (Mouchiroud et al., 1988). Indeed, the GC-richest and GC-poorest genes of the murids are less GC-rich and less GC-poor, respectively, than their orthologs from other mammals, whereas the genes with intermediate GC values are unchanged. The very high correlation coefficient of the human/mouse plot of **Fig. 11.16** indicates a **conservation of the order of GC levels in orthologous genes from human and murids** (Mouchiroud et al., 1988). Now, it is known that murids exhibit (i) rates of synonymous substitutions that are higher, by a factor of 5 to 10, relative to human coding sequences (Wu and Li, 1985; Gu and Li, 1992); (ii) a defective repair system (see Holliday, 1995); and (iii) a compositional pattern that is derived from the general mammalian pattern (Galtier and Mouchiroud, 1998).

As far as the last point is concerned, if murids arose relatively late in mammalian

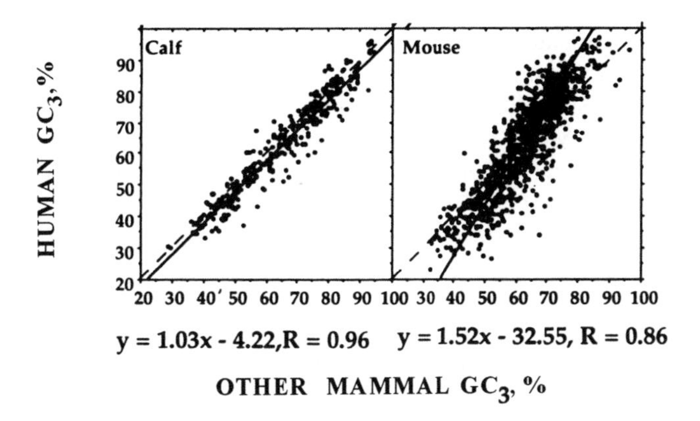

Figure 11.16. Correlations between GC_3 values of orthologous genes from human and calf and human and mouse. The orthogonal regression lines practically coincide with the diagonal (broken line) in the case of calf, but diverge in the case of mouse. The equations of the regression lines and the correlation coefficients are shown. (From Bernardi, 2000a).

evolution, as indicated by the paleontological record (Carroll, 1988), it was difficult to escape the conclusion that their genome underwent a narrowing of its compositional distribution compared to the mammals from which they were derived (Mouchiroud et al., 1988). Molecular data were interpreted, however, to suggest (Easteal, 1990; Li et al., 1990; Bulmer et al., 1991) that *Myomorpha* branched off before the divergence (at 100 Myrs) among carnivores, lagomorphs, artiodactyls and primates, a divergence of mammalian orders which all share the general pattern. The absence of a fossil record for over 50 Myrs obviously outweigh the molecular data and favour a derived compared to a primitive murid pattern (Sabeur et al., 1993).

Under these circumstances, it is interesting to observe that the higher mutational input "randomizes" the composition of synonymous positions by reducing the difference between the extreme GC values found in the general mammalian pattern. Indeed, under a mutational bias model *à la* Francino and Ochman (1999), one would expect instead an increase of the differences between extreme values. It is also remarkable that this **"randomization"** process also concerns non-coding sequences, and even regulatory sequences, causing for instance, an **"erosion" of CpG islands** (Aïssani and Bernardi, 1991a,b; Matsuo et al., 1993). In the case of α globin, for example, the human gene is very GC-rich and is associated with a CpG island, whereas the mouse gene is less GC-rich and has lost the island.

4.2. Isochore conservation in MHC loci of human and mouse

An important question concerns the effect of the minor shift on the isochore pattern. This problem has been explored on all syntenic regions shared by human and mouse, that were large enough to comprise more than one isochore. The results of this investigation (Pavlíček et al., 2002b) indicate that the human isochore pattern is still recognizable in the mouse chromosomes after the minor shift. This will be illustrated here by considering the major histocompatibility (MHC) locus.

The MHC region of mammalian genomes embeds a series of highly polymorphic genes that are expressed on the surface of several cell types, including the T and B lymphocytes that form an integral part of the adaptive immune system. In human, the MHC (or HLA) locus covers roughly 4 Mb of chromosome 6, while in mouse the MHC (or H2) locus is located in a largely syntenic region on chromosome 17. The locus has long been of intrinsic interest in view of its central role in conferring immunity, of its polymorphism, and of its paralogous loci on human chromosomes 1, 9 and 19 (Endo et al., 1997; MHC Sequencing Consortium, 1999). More recently, it has provided a paradigmatic example of the sharp boundaries that can exist between mammalian isochores. The MHC region in human has also provided the first examples of a precise link between isochore boundaries and replication timing switchpoints at the sequence level. The central, GC-poor isochore (**Fig. 11.17**, blue and yellow landscape for human) is replicated late, the two GC-rich isochores that flank it (orange and red) are replicated early, and the sharp boundaries of the central isochore both coincide with the switches in replication timing (Tenzen et al., 1997; the MHC Sequencing Consortium, 1999; and references therein). This region therefore provides a particularly clear illustration of the genome-wide tendency of GC-rich regions (and

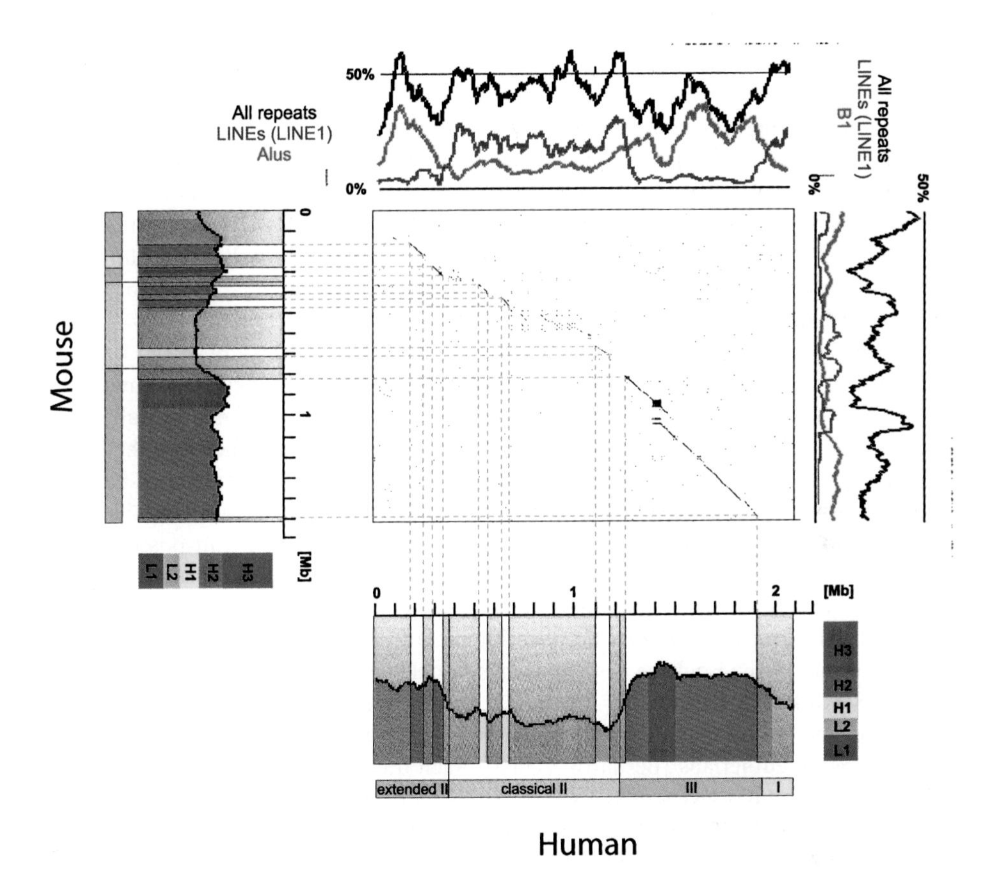

Figure 11.17. Color-coded GC scans, sequence similarity dot plot and repetitive element contents for homologous parts of human and mouse MHC loci. In both species the scale increases from the proximal (centromeric, upper/left) beginning of the sequence to its distal (telomeric) end. GC levels were plotted using overlapping 100 kb sliding windows and partitioned into five color-coded intervals, representing the five isochore families in human: their limits, indicated by horizontal dashed lines, are as in Bernardi (2001). Bright colors correspond to dot plot diagonals and show extended sequence homology, dull colors show regions with no or only local homology. In both species, the locus spans three distinct isochores, characterized in human by red and orange for the two GC-rich isochores, and by blue and yellow for the central GC-poor isochore. The MHC regions are indicated by bars below the GC scans. These regions correspond to the classes of MHC genes (MHC Sequencing Consortium, 1999; Stephens et al., 1999), and are blockwise syntenic between human (HLA) and mouse (H2), with the exception of the H2-K subregion in mouse (short yellow bar near the proximal end). Opposite each GC scan are the corresponding contributions of repetitive elements, which closely follow expectations from genome-wide analyses: *e.g.* in human the LINE1 elements are more frequent in L1 and L2 DNA, while Alus are most frequent in H2 DNA (Bernardi, 2001, and references therein). (From Pavlicek et al., 2002b).

the bands in which they are located) to replicate earlier than GC-poor regions (Federico et al., 2000).

The recent sequencing of the mouse MHC locus offers a first opportunity to compare a contiguously sequenced, nearly syntenic region spanning three clearly delimited isochores in human and mouse. Such a comparison is of particular interest because, as already mentioned, human and mouse represent the two principal patterns of base compositional organization in eutherians: mouse and rat belong to a subset of rodents (coinciding essentially with the myomorphs) that differ from other eutherians in having less dramatic contrasts in GC level among and within their isochores (reviewed in Bernardi, 2000; see **Part 11**). This phenomenon also affects the compositionally prominent, typically unmethylated CpG islands, which surround the promoters of nearly all housekeeping and many tissue-specific genes: in mouse they are lower, shorter and less abundant than in human (see **Part 5, Chapter 2**) .

As can be seen in **Fig. 11.17**, the isochore structure of the locus is conserved between human and mouse, not only in the regions showing sequence homology (as estimated by standard similarity criteria), but also in the intervening regions (represented by dull colors and the absence of diagonals in the dot plot). The intervening regions between the bright homology stripes include the H2-K class I sub-region in mouse, which is located between the two most proximal homology regions shown in **Fig. 11.17** (~0.25 Mb from the proximal beginning of the sequence), and which is thought to have been transposed from a distant class I region near the distal end of the sequence. The H2-K insertion locally disrupts the synteny of genes, the otherwise strict partitioning of the classes into different regions and the sequence similarity between human and mouse, but it does not interrupt the homogeneity of the isochore or the compositional synteny of the locus.

The conservation of the number and extent of the isochores is not altogether trivial, since human and mouse span the largest differences in compositional genome organization that exist among eutherians (Bernardi, 2000a). Such a conservation could, however, be expected from the GC_3 levels of the genes in this locus, from the strong compositional correlation that exists between the GC_3 of genes and the GC levels of the much longer regions of DNA that embed them, and from the correlation between genic GC_3 levels of orthologous genes in human and mouse. Furthermore, it is of interest that the general correlation between the GC levels of corresponding regions in human and mouse loci remains clearly visible (from the **similar landscapes** of the GC scans) even where local sequence similarity is weak or insignificant.

The GC-rich proximal parts (left for human, upper for mouse) correspond to the extended class II MHC region, the central GC-poor parts in both species correspond to the 'classical' (immunological) class II region, and the distal parts represent the class III region and the beginning of the class I region in human (Stephens et al., 1999). Although class III and extended class II are both GC-rich, they do not have the same degree of polymorphism: the polymorphism of the extended class II in human, although not as high as that of the classical class II, is closer to it than to the low polymorphism in class III. In line with the observations on polymorphism in human, homology between human and mouse is again typically well conserved in MHC class III, while the level of conservation is lower for the extended and especially for the very GC-poor classical MHC class II, where homology is limited to a few short regions. The MHC locus thus shows on the whole **a higher level of evolutionary conservation of the GC-rich, gene-rich regions**, and less conservation of the long GC-poor segment. Moreover, the lengths of the GC-rich class III and extended class II regions are similar for both human and mouse, while the GC-poor classical class II region is significantly shorter in mouse than in human. This difference, which corresponds to a loss of GC-poor DNA in an ancestor of mouse, is in agreement with a general propensity of more compact vertebrate genomes to preferential DNA contraction in the GC-poorer, and typically gene-poorer, regions (Bernardi, 1995; see **Fig. 5.14C**).

The **general reduction of compositional contrast in mouse** is visible also in the MHC region: the difference in GC level between the central and flanking isochores is much less pronounced than in human. The precise localization of isochore boundaries in mouse, and the prediction of corresponding replication timing switchpoints, could thus profit from comparisons with orthologous human sequences. Indeed, combined GC scan/dot plot alignments such as the one shown in **Fig. 11.17** could help reveal isochore boundaries: for example, the proximal boundary of the GC-poor isochore in mouse is compositionally less sharp than the distal boundary in mouse, or than the two boundaries in human. In the case of this central MHC isochore in mouse, all comparisons (homology of sequences, homology of corresponding syntenic genes, corresponding changes in GC level) consistently point to the same location for its proximal boundary, namely within a short region ($\ll$ 100 kb) delimited by two homology stripes, located approximately 350 kb into the mouse sequence, and indicated by a division in **Fig. 11.17**. More detailed analyses by experiments (*cf.* Tenzen et al., 1997) and segmentation algorithms (*cf.* Oliver et al., 2001) should help to confirm its precise position.

The repeat content of the human MHC sequence in **Fig. 11.17** is 45.5%, Alus comprising 16.1% and LINE1 elements 15.6% of the sequence. The mouse sequence contains 6% B1 sequences (the most recent mouse SINE class, analogous to Alus in human) and 3.1% LINE1 elements. These numbers are very close to previous genome-wide estimates, at corresponding GC levels, of repeat densities in mouse and human. It is interesting that, in spite of independent insertions of these classes of retroelements, their isochore distribution is similar in both species. Human and mouse LINE1 elements are, as expected (see **Part 6**), more dense in the GC-poor part, namely classical class II, while SINE repeats, human Alus and mouse B1s, are more frequent in the GC-rich parts of the locus (extended class II and class III). We have proposed that negative selection on compositionally non-matching elements is the main factor influencing the distribution of LINES and SINES (see **Part 6**).

The functional meaning of the strict regional partitioning of MHC gene classes into different isochores, and of the conservation of this partitioning despite high recombinogenicity during the evolution of the locus, speak for a strong selection that maintains the precise isochore structure of this region.

Recent observation showed the isochore conservation between all human, mouse and feline classical and extended class II MHC regions (Yuhki et al., 2003). Despite many rearrangements and extensive gain and loss of genes during long separate evolution, the human and cat classical and extended class II regions show nearly identical GC content. The cat extended and classical class II regions maintain 49.1% and 41.2% GC content, whereas the same human regions have 49.9% and 41.3 % of GC (Yuhki et al., 2003). Given the weak similarity in the majority of DNA sequences without coding capacity, the selection on the global structure and composition of large genomic regions rather than conservation of individual bases or genes, seems to be the main factor retaining the similar GC content in large genomic segments of mammals.

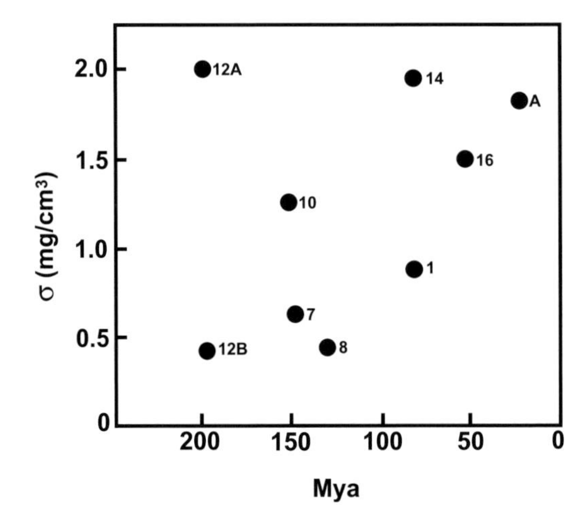

Figure 11.18. Standard deviations of average modal buoyant densities in CsCl, (, of DNAs from fish species belonging to the same order, family, or genus (in the only case of *Aphyosemion*) are plotted against the time of the appearance of these groups. Groups are: *Lamniformes*, 1; *Cypriniformes*, 7; *Salmonidae*, 8; *Gadiformes*, 10; *Aplocheilidae*, 12A; *Cyprinodontidae*, 12B; *Aphyosemion*, A; *Perciformes*, 14; *Tetraodontiformes*, 16. Numbers correspond to those of Table 2 of the original paper. (From Bernardi and Bernardi, 1990b).

Part 12
Natural selection and genetic drift in genome evolution: the neo-selectionist model

CHAPTER 1

Molecular evolution theories and vertebrate genomics

1.1. Molecular evolution theories

Molecular evolution theories have been discussed in detail in a number of textbooks (*e.g.*, Nei, 1987; Gillespie, 1991; Hartl and Clark, 1997; Graur and Li, 2000; Ridley, 2004). Here we will, therefore, only sketch out the selection theory, the neutral theory and the nearly neutral theory, in order to ask whether they can account for, or at least be compatible with, our observations on the structural and evolutionary genomics of vertebrates.

According to the **classical selection theory**, derived from neo-Darwinian views, natural selection, "*the preservation of favourable variations and the rejection of injurious variations*" (Darwin, 1859), or the differential multiplication of mutant types, essentially occurs through the elimination of the progeny of individual organisms carrying deleterious mutations (**negative, or purifying, selection**), and only very rarely *via* the preferential propagation of organisms with advantageous mutations (**positive selection**; **Fig. 12.1A**; see also Gillespie, 1991). In this view, it is assumed that changes in frequency of a substitution depend essentially on selection. A problem with this hypothesis is that selection cannot apply to the vast majority of the noncoding sequences that represent 97–98% of a typical mammalian genome, such as the human genome. Indeed, if only a few percent of noncoding sequences are conserved in evolution (see below), it is difficult to see how a single substitution can be selected for or against out of three billion nucleotides (in the case of mammalian genomes), because its selection coefficient can only be extremely small.

Thirty five years ago, Kimura (1968) launched the idea (supported from a biochemical point of view by King and Jukes, 1969, who called it "*non-darwinian evolution*") that a large number of mutations must be neutral. Kimura's paper of 1968 was the first step in the development of the well-known **neutral theory**, more precisely called the **neutral mutation-random drift theory** (Kimura, 1983), in which "*the main cause of evolutionary change at the molecular level - changes in the genetic material itself - is random fixation of selectively neutral or nearly neutral mutants rather than positive Darwinian selection*". Kimura stressed, however, that "*The neutral theory is not antagonistic to the cherished view that evolution of form and function is guided by Darwinian selection, but brings out another facet of the evolutionary process by emphasizing the much greater role of mutation pressure and random drift at the molecular level*", in which "*increases and decreases in the mutant frequencies are due mainly to chance*". As shown in **Fig. 12.1B**, the neutral theory drastically changed the view proposed by the classical selection theory. The problem with the neutral theory is the increasing evidence for selection not only in coding sequences (including third codon positions, as shown in **Parts 10** and **11**) and regulatory sequences (promoters, enhancers), but also in noncoding "conserved sequences", whose amount is estimated to be around 10% in the human genome (see Hardison, 2000; Shabalina et al., 2001; Dermitzakis et al., 2002).

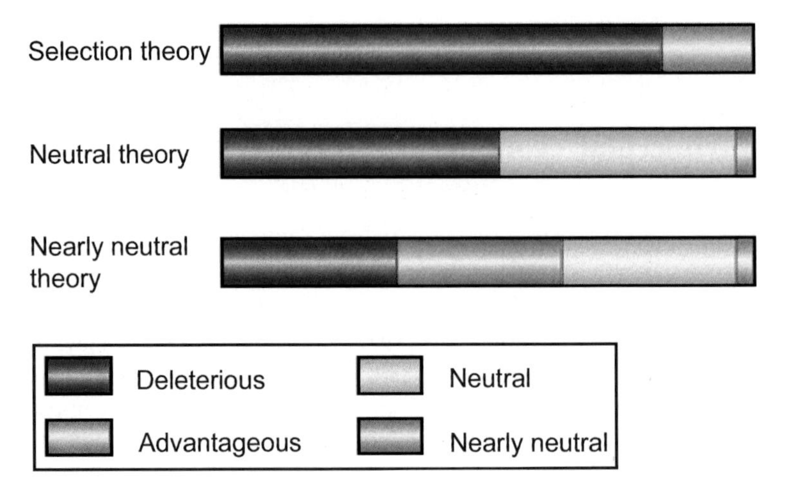

Figure 12.1. Selectionist, neutral and nearly neutral theories. **A.** Selectionist theory: early neo-Darwinian theories assumed that all mutations would affect fitness and, therefore, would be advantageous or deleterious, but not neutral. **B.** Neutral theory: the neutral theory considered that, for most proteins, neutral mutations exceeded those that were advantageous, but that differences in the relative proportions of neutral sites would influence the rate of molecular evolution (that is, more neutral sites would produce a faster overall rate of change). **C.** Nearly neutral theory: the fate of mutations with only slight positive or negative effect on fitness will depend on how population size affects the outcome. Figure modified with permission from Ohta (2002). (From Bromham and Penny, 2003).

The nearly neutral theory of Ohta (1971, 1987, 1992) **(Fig. 12.1C)** "*described how the rate of molecular evolution could vary not only with changes in the mutation rate* (as postulated by the neutral theory), *but also through the changing balance between selection and drift*". As still phrased by Bromham and Penny (2003), "*the nearly neutral theory considered three categories of mutations: mutations for which selection is the predominant force* ($4N_e\sigma_s > 3$; *where* N_e *is the effective population size and* (σ_s *is the variance of the selection coefficient s*), *nearly neutral mutations which are governed by both selection and drift* ($3 \geq 4N_e\sigma_s \geq 0.2$) *and effectively neutral mutations, the fate of which is determined only by drift* ($4N_e\sigma_s < 0.2$). *The fate of mutations with only slight positive or negative effect on fitness will depend on how population size affects the outcome*". The problem of the nearly neutral theory in the case of the vertebrates is the small population sizes of the ancestral reptilian lines that underwent the compositional changes at the emergence of mammals and birds, and the small population sizes of the independent lines that maintained those changes. These population sizes essentially preclude selection and equate, in such cases, the neutral and nearly neutral theories.

The relative roles of selection and neutrality in genome evolution were discussed for a long time by population geneticists without reaching a solution (see Hey, 1999). It was even said that "*for almost two decades the field of molecular evolution was almost paralyzed by the selectionist-neutralist debate*" (Kondrashov, 2000). As **a first approximation** solution to the problem, we would like to propose that the selection theory applies to "conserved sequences" (including coding and regulatory sequences), but not to the vast majority of noncoding sequences, whereas the neutral theory would apply to the vast majority of noncoding sequences, but only in part to "conserved sequences" (as defined above). This latter view has been called "***the weak form of the neutral theory****, the strong one asserting that almost all DNA changes are effectively neutral*" (Crow, 2003).

In spite of their differences, both the selection and the neutral theories shared a common point, namely that they took into consideration only the **fixation of single-nucleotide mutations**. This is understandable not only because they were formulated by population geneticists, who concentrated on coding sequences and proteins, but also because they were developed when our understanding of eukaryotic genome organization was very poor. For example, nobody could imagine, at that time, that coding sequences might represent as little as 2–3% of the human genome, or that vertebrate genomes are mosaics of isochores.

1.2. Structural genomics of vertebrates

We will now focus on the specific questions mentioned at the beginning of this chapter, namely whether the selection and the neutral theories can account for our results on the structural and evolutionary genomics of vertebrates, or, failing that, whether they can, at least, be reconciled with our results.

The unexpected discoveries that we made when investigating the genomes of vertebrates (summarized in **Table 12.1** and below) went against the prevailing views on genome organization (see **Parts 1, 3, 5**). More importantly for this discussion, our results could not be accounted for by any explanation of eukaryotic genome evolution that was only based on selection, or on random fixation, of single-nucleotide mutations and did not consider **compositional changes** and **regional effects** (see **Chapter 2**). We will now outline our main conclusions on structural genomics of vertebrates following the points of **Table 12.1**, and argue why they cannot be explained by the selection theory or by the neutral theory.

TABLE 12.1
Structural genomics of vertebrates

(i)	**Genome compartmentalization** (*discontinuous compositional heterogeneity, isochores*)
(ii)	**Genome phenotype** (*compositional patterns of isochores and coding sequences*)
(iii)	**Genomic code** (*compositional correlations between coding and non-coding sequences and between codon positions; functional compatibility between coding and noncoding sequences*)
(iv)	**Gene distribution in the genome and in chromosomes** (*gene spaces: genome core, genome desert*)
(v)	**Correlations of gene distribution with:** *structural properties (intron and UTR size, chromatin structure) and functional properties (gene expression, replication timing, recombination levels*

(i) The **compositional compartmentalization of the genome**, namely its discontinuous compositional heterogeneity and the existence of isochores, was not only antagonistic to models of continuous heterogeneity (see **Part 3**, **Section 1.4**), but also could not be explained by either the selection or the neutral theories. Indeed, it is impossible to form and maintain an isochore structure simply by selection of single nucleotides, or random fixation of neutral mutations, as any simulation can show. The recently discovered AT bias in the mutation process (see **Part 11**) makes the formation and maintenance of GC-rich isochores of warm-blooded vertebrates even more difficult to explain.

(ii) Likewise, the maintenance of different **genome phenotypes**, namely the different compositional patterns of isochores and coding sequences exhibited by different classes, and even orders, of vertebrates, cannot be explained by the classical selection or the neutral theories. For instance, if only single-nucleotide mutations were fixed, how could this process lead to the different genome phenotypes of amphibians/fishes and of warm-blooded vertebrates? And how could this process account for the maintenance of the very similar genome phenotypes of mammals, on the one hand, and birds, on the other, over geological times?

(iii) The **genomic code**, the compositional correlations between coding and noncoding sequences (which provide the strongest support for the concept of the genome as a **coordinated ensemble**; see **Part 1**) and the universal compositional correlations between different codon positions, cannot be accounted for by evolutionary processes involving the random fixation of neutral mutations or selection on single nucleotides. An implication of the compositional correlation between coding and noncoding sequences concerns the functional compatibility of these sequences, exemplified by the effect of compositionally matching or non-matching contiguous noncoding sequences on the transcription of integrated viral sequences (see **Part 6**). This **genomic fitness,** as we will call the functional compatibility, cannot be understood in the light of the classical selection or the neutral theories.

(iv) The totally unexpected **bimodal distribution of genes** into two compositionally different gene spaces, the gene-rich genome core and the gene-poor genome desert, implies the existence of evolutionary processes (*e.g.*, expansion-contraction phenomena, insertion and deletion of transposons) indicating that **the order and the complexity of genome organization is far beyond the explanatory power of any stochastic mutation process alone or of selection on single nucleotides**.

(v) The **correlations of the bimodal gene distribution with a) structural properties**, such as the sizes of introns or of untranslated sequences and open/closed chromatin structures; **and b) functional properties**, such as the level of gene expression, the bimodal (early-late) replication timing, and the differences in recombination levels, point to a complexity in genome organization that was not (and could not be) even imagined when the selection and neutral theories were proposed.

At this point, it should be stressed that while the classical selection and the neutral theories cannot account for our findings, they are compatible with them, as it will be shown in the following **Chapter 2**.

1.3. Our previous conclusions

Since the classical selection theory was known to be untenable for the vast majority of noncoding sequences in eukaryotic genomes (see above), we originally concentrated on why our findings could not be explained by the neutral theory. In 1986, we summarized our views on compositional constraints (*i.e.*, the factors that are responsible for the compositional properties of the genomes) and genome evolution as follows: "*Nucleotide sequences of all genomes are subject to compositional constraints that (1) affect, to about the same extent, both coding and non-coding sequences; (2) influence not only the structure and function of the genome, but also those of transcripts and proteins; (3) are the result of environmental pressures; and (4) largely control the fixation of mutations. These findings indicate (1) that non-coding sequences are associated with biological functions; (2) that the organismal phenotype comprises two components, the "classical phenotype", corresponding to the "gene products", and a "genome phenotype", which is defined by the "compositional constraints"; and (3) that natural selection plays a more important role in genome evolution than do random events*" (Bernardi and Bernardi, 1986a). Seventeen years later, we find that nothing substantial needs to be changed in those statements.

While the basic tenet of the neutral theory, that the "*main cause of evolutionary change at the molecular level is random fixation of selectively neutral or nearly neutral mutants*", could not explain our findings, the initial proposal by Kimura (1968) and by King and Jukes (1969) that **the majority of mutations could only be neutral or nearly neutral**, still held. Obviously, this was so not so because a number of changes in coding and regulatory sequences may be neutral, but mainly because up to 97–98% of many vertebrate genomes are made up by noncoding sequences. Indeed, most changes could occur only in those sequences, and they could not be, as a rule, under any significant selection. There was, then, a need to reconcile this conclusion with our explanation based on natural selection, which also seemed to be inescapable, especially since many years of intense scrutiny, criticisms and search for alternative explanations had, if anything, strengthened it (see **Chapters 5** and **6**). This reconciliation required, however, a **paradigm shift**.

Originally, we stated that "*intragenomic GC-level changes clearly indicate that most mutations, in both coding and non-coding sequences, are fixed not at random, but under the influence of compositional constraints, in compliance with a general compositional strategy involving, in all likelihood, both negative and positive selection. Random fixation of neutral mutations (Kimura, 1968, 1983, 1986; King and Jukes, 1969) certainly also occurs, but only to an extent such that the general compositional strategy and relationships are not blurred*" (Bernardi and Bernardi, 1986a). Even if these statements can still be considered as valid, they were very general, and did not make clear how the compositional constraints influenced the fixation of mutations.

The first step towards such a paradigm shift was made fifteen years ago when we proposed a "**regional model**" for the selection process, in which isochores and the corresponding chromatin domains were visualized as selection units (Bernardi et al., 1988; this model was well understood by Gillespie, 1991; p. 95). At that time, we noted that "*the regional negative and positive selection at the isochore level are strongly reminiscent of a process postulated to diminish the genetic load associated with "standard" selection, the "forward creep – back leap"* of Zuckerkandl (1975). This process was later formulated by Zuckerkandl (1986) as follows: "*In certain parts of the genome, notably in zones of highly repetitive sequences, mutations may be freely accepted as neutral until the sequence adulteration of a large segment passes a certain threshold. At that time the adulterated sequence may be eliminated by negative selection and may be substituted by a better sequence, one regenerated by amplification from an appropriate master sequence. This process obviously would reduce radically the proportion of events of positive or negative selection necessary to maintain the sequence motifs.*"

Our regional selection model was presented in some detail ten years ago (Bernardi, 1993a) and is summarized in **Fig. 12.2.** In the conservative mode of evolution of vertebrate genomes, its key feature is the existence of compositional thresholds, within which mutations could be neutral or nearly neutral, but beyond which the sequence under consideration is selected against in the case of the conservative mode of evolution. In the shifting or transitional mode, the compositional changes (GC increases) of the genomes of warm-blooded vertebrates were visualized as due to positive selection, again working at the regional level.

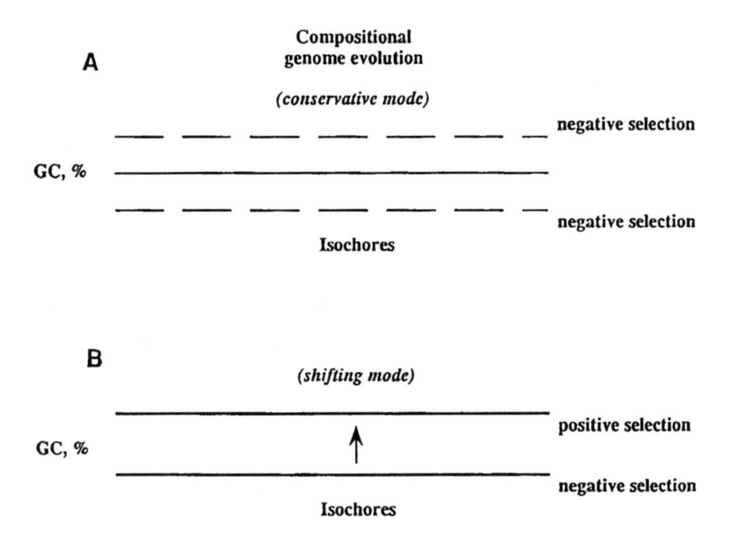

Figure 12.2. **A.** Scheme of negative selection in the conservative mode of evolution. Isochores (solid lines) that drift beyond the GC thresholds indicated by the broken lines are counterselected. **B.** Scheme of negative and positive selection in the transitional or shifting mode of genomes evolution. Isochores (solid lines) with decreasing GC levels are counterselected, whereas those with increasing GC levels are selected for. (From Bernardi 1993a).

Natural selection in the maintenance of compositional patterns of vertebrate genomes: the neo-selectionist model

The compositional conservation of third codon positions of orthologous genes from different mammalian orders (belonging to the general compositional pattern), or from different avian orders, over times close to 100 million years (see **Part 11**), implies the elimination of changes that would lower the high and very high GC_3 values (up to 100%) that are abundant in both classes. The same requirement must be satisfied by the noncoding intergenic and intragenic sequences that are compositionally correlated with the GC_3 values of the corresponding genes. In fact, there is a need to counteract not only the random mutations that accumulate during geological times, and that would lower the GC_3 levels to 50%, but also the mutational AT bias (including the deamination process that changes mCpG into TpG/CpA doublets), as well as compositional changes resulting from recombination and from the insertion of transposons endowed with base compositions different from those of the target isochores.

In the case of coding and regulatory sequences, this process can be visualized within the classical framework of negative selection acting on single-nucleotide changes at many *loci*, because of their effect on the thermal stability of DNA and RNA (see **Chapter 4**), on the hydrophobicity (and, as a consequence, on the rate of folding; see **Chapter 4**), and on the secondary structure of the encoded proteins. In the case of third codon positions, an additional selection factor, translational selection, was demonstrated in amphibians (Musto et al., 2001) and fishes (Romero et al., 2003), and is likely to operate in all vertebrates (see **Part 11**). It is impossible, however, to invoke this single-nucleotide process for intergenic sequences, or for intragenic noncoding sequences (*e.g.*, for the long introns of GC-poor genes), because in these cases selection on individual changes cannot be strong enough (a point already made in the preceding **Chapter 1**).

In the case of intergenic sequences, and noncoding sequences in general, a possibility is that selection will operate on "**critical changes**". As shown in **Fig. 12.3**, one can assume that the high average GC level of the GC-rich isochore under consideration (set in **Fig 12.3** at 55% GC; see also **Fig. 3.20A**) corresponds to a compositional optimum, because of its compositional conservation over geological times in different mammalian or avian orders. We also know, in addition, that there is a strong mutational AT bias, at least in the human genome, and a deamination of mCpG, hence the higher density of changes (blue arrows) pointing towards AT, compared to the changes (red arrows) pointing towards GC. One can further assume that changes keeping the average GC levels within the thresholds (indicated by the horizontal broken lines) are tolerated and do not change the structure of DNA and chromatin nor the fitness of the individual. When enough changes to AT have accumulated in a given region, a **few additional "critical changes"** (shown as pointing towards AT because of the strong mutational AT bias) may push the **average regional GC level beyond a threshold and cause a "phase transition"**, *i.e.* a sudden structural change in DNA and chromatin. It should be clear that the "critical changes" are simply the last changes in the overall process that pushes the average regional GC level beyond a threshold. (There is no point, at this time, to be more specific on such phase transition, but obviously cooperative effects play a role). Changes of this kind will affect, in turn, the expression of genes located within the isochore under consideration, as indicated by the demonstrated effect of base composition of flanking, noncoding sequences on the expression of proviral sequences (see **Part 6**). It is likely that, in most cases, such effects are only mildly deleterious. An accumulation of such compositionally altered isochores in the genome may, however, lead to a decreased fitness of the individual and to the elimination of the progeny carrying such regions by negative selection.

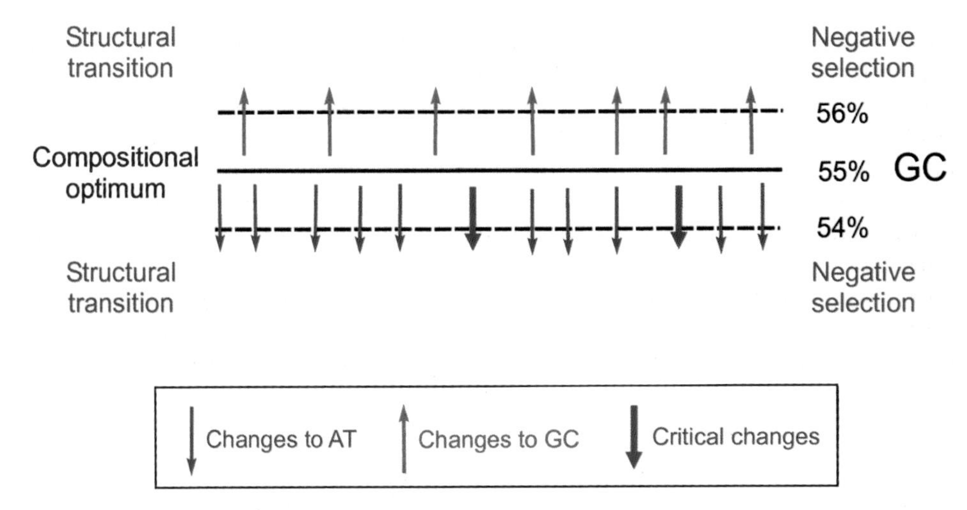

Figure 12.3. A scheme of the conservative mode of compositional evolution of a GC-rich noncoding intergenic region from a warm-blooded vertebrate. The compositional optimum of the region is set at 55% GC. The arbitrary thresholds of the tolerated range are indicated by horizontal broken lines. Red and blue arrows refer to changes towards GC or AT, respectively. Thick arrows refer to critical changes (see text).

Interestingly, this model predicts that **the fitness of an individual is not only a genetic property, but also a genomic property**, as already proposed on the basis of the results on transcription of integrated viral sequences (see **Part 6**). Needless to say, this model sees the individual as the target of selection, in agreement with much earlier ideas summarized by Mayr (1988). This model can be tested. Indeed, it predicts that syntenic noncoding sequences (perhaps even from different human populations) might reveal compositional differences. If such situations will be found by future investigations, one could speculate about the possibility of **genomic diseases**, in which genes and regulatory sequences do not carry single deleterious changes as in **genetic diseases**, but the neighbouring noncoding sequences show an altered composition. Another prediction of the model is the existence of two dynamics of negative selection, one for the individual changes and one for the regional changes.

Some points of the proposed model need to be discussed further.

(i) What has just been presented is an improved version of the model first published ten years ago (Bernardi, 1993a; see **Fig. 12.2**), from which it essentially differs in proposing the existence of critical changes, that are responsible for the phase transition, leading to a defective DNA/chromatin structure. It should be stressed that while "critical changes" obviously are single-nucleotide changes, they differ from the single-nucleotide changes taken into consideration by the selection and neutral theories in that they cause "regional effects", structural DNA/chromatin changes that may be deleterious for the function of genes located in or near that region. In functional terms, such deleterious changes in noncoding sequences are comparable to deleterious changes in coding or in regulatory sequences, as already mentioned.

(ii) A major point of the model is that most changes that accumulate in the 97–98% of the genome formed by noncoding sequences, and that slowly lead to pushing the sequence beyond the thresholds, **while potentially deleterious, *per se* are neutral or nearly neutral**. Obviously, this model eliminates the objection of the impossibly high number of selective deaths required by selection acting on individual mutations. The model also accounts for the fact that standard comparisons of orthologous mammalian coding sequences (which are compositionally very close) will not reveal that the sequences are themselves the result of the maintenance of, say, sequences that became GC-rich at the reptilian-mammalian transition.

(iii) Until the structural changes in DNA/chromatin have occurred, and while regional selection has not acted, only neutral and nearly neutral changes will be detected. In other words, the features of molecular evolution that take place within certain compositional

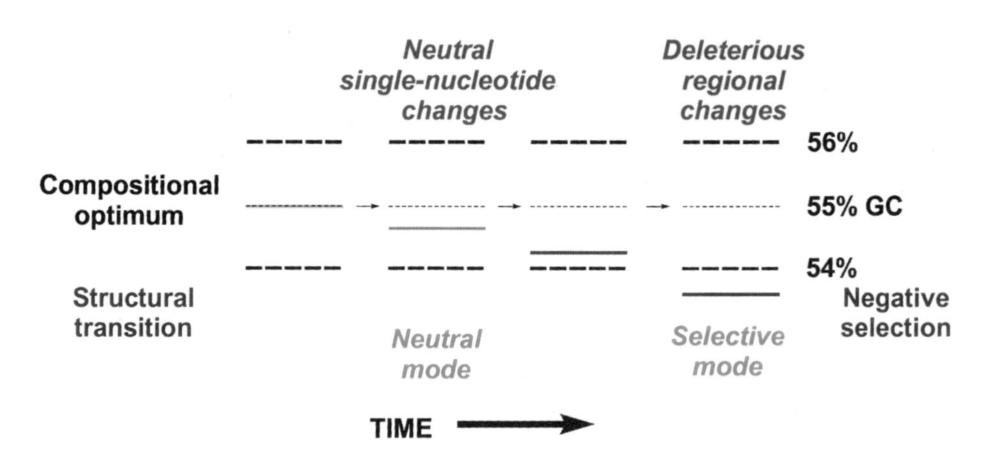

Figure 12.4. Time course of typical compositional changes of a GC-rich noncoding intergenic region from a warm-blooded vertebrate in the conservative mode. It is assumed that the overall compositional trend is dominated by a mutational AT bias. In an early phase, the average GC level of the region, initially visualized at its compositional optimum (set at 55% GC), is decreasing but remains within a tolerated range (whose arbitrary thresholds are indicated by the heavy broken lines). In a late phase, the average GC trespasses the lower threshold, the region undergoes a structural change which is deleterious for gene expression and is potentially subject to negative selection. Until then, the changes are neutral or nearly neutral (see also the text).

thresholds can be accounted by the neutral theory. They represent, however, only one mode of genome evolution, the "neutral mode". The other mode, the "selective mode", keeps the compositional pattern, the genome phenotype, under its dominant, pervasive control (**Fig. 12.4**).

It should be stressed that the drift pushing a region beyond the acceptable thresholds is not only influenced by mutational biases and mC deamination, but also by recombination, biased gene conversion, and transposon insertion. While these factors may slow down, speed up, or alter the direction of, the compositional drift, the final control is still exerted by natural selection. In the case of transposon insertions, it is conceivable that situations may arise in which there is no need for critical changes to alter the DNA/chromatin structure. For instance, the insertion of a large LINE element in a short GC-rich intergenic sequence might, by itself, lead to deleterious compositional regional changes that will be selected against (see also **Part 6**).

(iv) The picture presented so far is in agreement with our previous conclusions (Bernardi and Bernardi, 1986a) that "*our data do not support the view that non-coding sequences can be equated with functionless "junk DNA" (Ohno, 1972). Rather, it suggests that non-coding sequences do play a physiological role, one that may concern the modulation of basic genome functions (see below). This suggestion, although not a new one (Britten and Davidson, 1969; Zuckerkandl, 1976, 1986; Davidson and Britten, 1979) no longer rests on "adaptive stories", which can rightly be criticized (Gould and Lewontin, 1979; Doolittle and Sapienza, 1980; Orgel and Crick, 1980), but rather on the newly demonstrated compositional constraints. Interestingly, identical conclusions have been reached, on the basis of different evidence, for the non-coding sequences of the mitochondrial genome of yeast (Bernardi, 1982a, 1983; de Zamaroczy and Bernardi, 1985, 1986a,b, 1987).*" In fact, the structure of the eukaryotic genome resembles that of a protein, where the essential functional features of the active site (the exons in the case of the genome) are linked and maintained in three-dimensional space by the intervening polypeptides (the intergenic sequences and introns in the case of the genome).

(v) While in this chapter the emphasis has been placed on negative selection, it is obvious that **positive selection** also plays a role in coding sequences.

In conclusion, if the **classical selection** acts on the **classical phenotype** (essentially the proteins and their expression), the **compositional selection** proposed here would act on the **genome phenotype** (the compositional properties of the genome and their functional implications). This compositional selection would not only act at the individual nucleotide level (like the classical selection; *e.g.*, at third codon positions), but also at the **regional level** through **critical changes**. What is essentially proposed for the bulk of DNA forming the vast expanses of noncoding sequences is an input of changes which *per se* are neutral or nearly neutral or slightly deleterious, but which are followed by a **negative (purifying) selection at the regional level (Fig. 12.4)**. We propose to call this new model of genome evolution the **neo-selectionist model. This model incorporates the neutral changes into a selectionist frame**, very much as the neutral theory incorporated the classical selection changes as far as the classical phenotype was concerned.

CHAPTER 3

Natural selection in the major shifts

In the case of the major compositional genome shifts, some degree of positive selection is certainly operational (and helpful) on changes in coding sequences, introns, regulatory sequences, and possibly even on "critical changes" (see **Chapter 2**). There is, however, no strict need for it. Indeed, the progressive increase in body temperature that accompanied the emergence of warm-blooded vertebrates (see **Part 11**), and that we visualize as the main selective factor responsible for the major compositional transitions (see **Chapter 4**), may (i) shift the compositional optimum to higher GC levels, (ii) shift upwards the lower threshold of the accepted compositional range, and (iii) eliminate increasingly higher average GC levels by negative (purifying) selection **(Fig. 12.5)**.

The present body temperature of mammals was probably attained over a considerable lapse of time. In principle, this process could have taken of the order of 100 Myrs (the time difference between the appearance of mammalian–like reptiles, 320 Mya, and that of early mammals, 220 Mya; time estimates are from Carroll, 1987). Some recent data suggest, however, time spans that are much shorter than the maximal ones quoted above. Indeed, the discovery of feathered dinosaurs (Norell et al., 2002), while providing additional evidence for the independent emergence of homeothermy in mammals and birds, helps in defining the time of appearance of homeothermy in birds and suggests a relatively short time span for its acquisition in the ancestors of birds.

Figure 12.5. A scheme of the shift of compositional optimum caused by a gradual increase in body temperature. At the same time, thresholds within which changes are tolerated (broken lines) are shifted upwards. GC values are only indicative.

CHAPTER 4

The causes of the major shifts

4.1. Compositional changes and natural selection

The first explanation (Bernardi and Bernardi, 1986a) that we proposed for the "major shifts" experienced by the genomes of the ancestors of mammals and birds, and for the maintenance of the new patterns, was that they were due to **natural selection** acting on the **genome phenotype**, namely on the compositional pattern of the genomes. As a working hypothesis, we suggested that the selective advantages provided by the compositional genome transitions accompanying the emergence of warm- from cold-blooded vertebrates were the higher thermodynamic stabilities of DNA, of RNA, and of the proteins encoded by the newly formed GC-rich coding sequences, all these advantages being achieved simultaneously (Bernardi and Bernardi, 1986a). The main reasons for this **thermodynamic stability hypothesis** were that (i) vertebrates are a very small taxon sharing most genetic and genomic properties; in other words, the vast majority of inputs influencing genome composition are, in all likelihood, very similar; (ii) a major difference between cold- and warm-blooded vertebrates is body temperature; (iii) the "major shifts" were never observed in fishes and amphibians (the majority of vertebrates), involved both coding and noncoding sequences, and only concerned 10–15% of the genome, interestingly the gene-richest part of it. One more general reason was the expectation from basic thermodynamics that the structures of DNA, RNA and proteins should be affected by an increase in body temperature.

The following sections will consider results that support the thermodynamic stability hypothesis. Objections and alternative explanations will be critically presented in **Chapters 5** and **6**.

4.2. The thermodynamic stability hypothesis: DNA results

Several lines of evidence support the thermodynamic stability hypothesis at the DNA level. While their strength varies from weak to strong, their cumulative weight in favor of the hypothesis is overwhelming.

The rationale for a first set of approaches ([i, ii] and [iii] below) to test the hypothesis is the following. If the difference in compositional patterns of cold- and warm-blooded vertebrates is body temperature, *T*b, one should be able to find cold-blooded vertebrates that have been living at high temperature for a long enough time to exhibit compositional features in their genomes that approach those of warm-blooded vertebrates (indeed, short exposures to high temperatures may be taken care of by physiological responses). Unfortunately, this approach is not without problems, 1) because body temperature is far from constant in cold-blooded vertebrates, especially in reptiles, the most interesting class; 2) because the parameter of real interest, the maximal *T*b sustained for extended periods of time, *T*max, is not generally known; 3) because "control" species may not be available.

(i) **The compositional heterogeneity of genomes of cold-blooded vertebrates living at high temperatures.** We looked (Bernardi and Bernardi, 1986a; 1990a) for differences between the genomes of fishes living at high temperatures (about 40°C), such as *Alcolapia alcalicus grahami* (formerly *Tilapia grahami* and later *Oreochromis grahami*) from Lake Magadi and Lake Natron, Kenya (**Fig. 12.6A**), or *Cyprinodon salinus* from Death Valley, CA, USA (**Fig. 12.6B**) and their close relatives living at lower temperatures (about 20°C) by investigating Cs_2SO_4/BAMD fractions in analytical CsCl density gradients. We found, in both cases, that fishes living at high temperatures showed DNA fractions reaching higher GC values compared to those derived from the related species living at lower temperatures (**Fig. 12.6**). While these changes were going in the expected direction, no further characterization of the GC-rich fractions was made, except for the demonstration that the GC-rich fractions did not correspond to satellite DNAs. Another pair of congeneric fishes currently

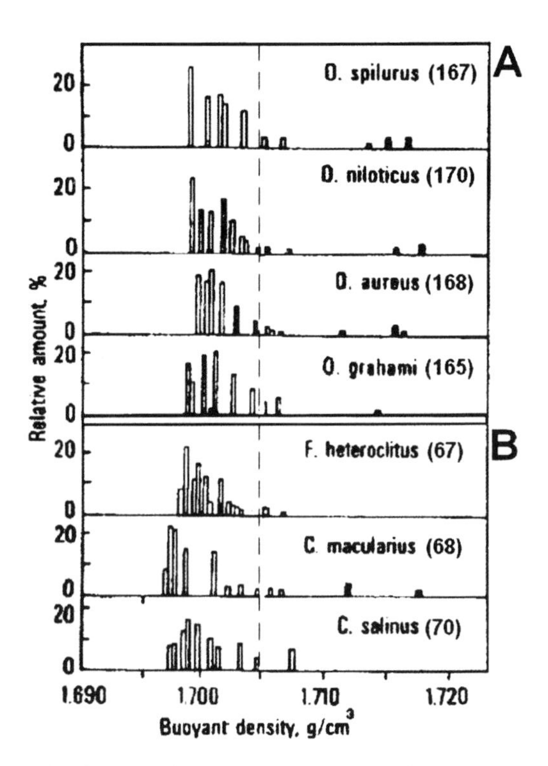

Figure 12.6. Histograms showing the relative amounts and modal buoyant densities in CsCl of DNA fractions obtained by preparative Cs_2SO_4/BAMD density gradient centrifugation from **A** *A.a. grahami*, **B** *C. salinus* and their close relatives. Black bars correspond to satellite components. Fish nomenclature was not updated for the *Oreochromis* genus presented in **A** (see text). For the numbering of fishes see Table 4.2. (Modified from Bernardi and Bernardi, 1990a).

under study is *Gillichthys seta* (living at a high temperature) and *G. mirabilis* (living at a moderate temperature). In this case, a shift of the DNA fractions hybridizing cDNA (from both species) from lower to higher GC levels was observed in *G. seta* compared to *G. mirabilis* (G. Bucciarelli, unpublished observations). All the above findings can, however, only be considered as preliminary. If they are mentioned here, it is because the DNA fractions obtained from the species pairs mentioned may be investigated further at the sequence level. Along a similar line, a comparison of two fish species belonging to the same sub-family (*Tetraodontinae*) *Takifugu rubripes*, a fish living in temperate sea water, and *Tetraodon viridiformis*, a tropical fish (Froese and Pauly, 2002), show that the latter exhibit a higher GC level (49% *vs.* 45%) and a stronger positive asymmetry for long DNA stretches (see **Fig. 4.5A, B**).

(ii) **GC_3 and body temperature in vertebrates.** In apparent contrast with the thermal stability hypothesis, Belle et al. (2002) reported no correlation between either the mean GC_3, or the standard deviation of GC_3 (two ways of evaluating the spreading of the GC_3 histogram towards higher values), and body temperature of 18 cold-blooded and two warm-blooded vertebrates. These authors, therefore, concluded that "*the thermal stability hypothesis does not appear to explain general patterns of composition*".

In contradiction with this conclusion, a plot of the maximal (or mean) body temperatures of cold-blooded vertebrates against the standard deviation of their GC_3 values (data of Belle et al., 2003) showed a significant (yet, admittedly, borderline) correlation, provided that the data from alligator and turtle were neglected. This elimination is justified by the fact that these species were not only represented by a very low number of genes (16 and 17), but also showed the largest and smallest standard deviations, respectively (see **Fig. 12.7**), in

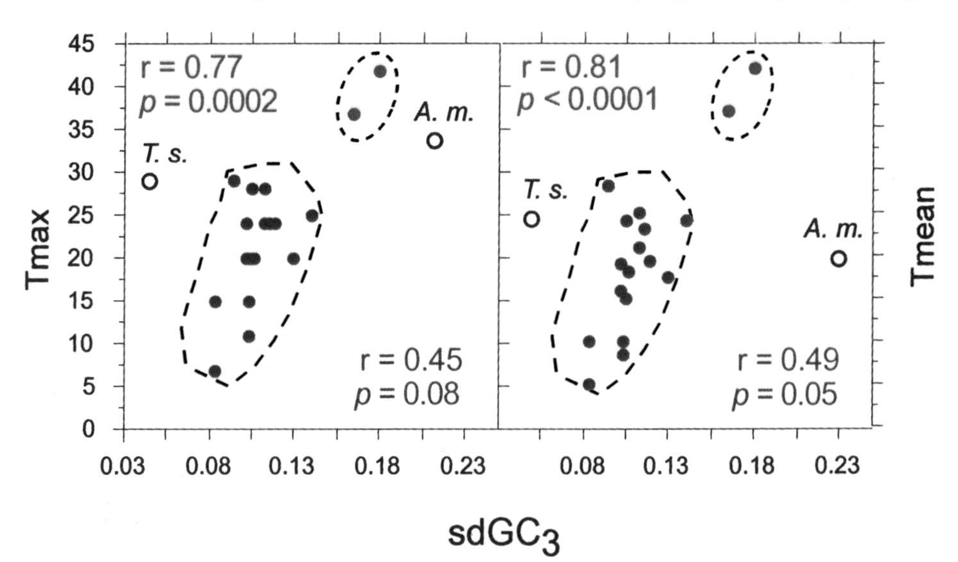

Figure 12.7. The standard deviations of GC_3 of genes from cold- (blue points) and warm-blooded (red points) vertebrates are plotted against their mean body temperature. Data are from Belle et al. (2002). Correlation coefficients and *p* values neglecting the two outliers (*Alligator* and *Trachemys*) are shown. Red figures concern all the points, blue figures cold-blooded vertebrates only. (From Jabbari et al., 2003c).

spite of the closeness of CsCl profile asymmetry and heterogeneity of the two species under consideration (see **Table 4.4B**). This was an indication of a gene sample problem.

A conclusion suggested by the results of **Fig. 12.7** is that the **compositional heterogeneity** of third codon positions of vertebrates (which essentially depends upon the relative amounts of GC-rich genes) **may well be correlated with body temperature**. While this approach can, and will, be pursued further, comparisons of aligned orthologous genes are needed to provide a final proof for the differences.

(iii) **CpG frequency and DNA methylation.** This different approach relies on the evaluation of two parameters, CpG and mC, that are related to body temperature and show very marked differences between cold- and warm-blooded vertebrates. Concerning the transition in methylation and CpG levels of vertebrates, we suggested that the two-fold lower 5mC and CpG levels of warm-blooded vertebrates compared to cold-blooded vertebrates might have been caused by a higher deamination rate related to their higher body temperature (Jabbari et al., 1997). Indeed, deamination of 5mC residues in double-stranded DNA has a strong temperature dependence (Shen et al., 1994). The fact that the genomes of at least some reptiles are similar to those of warm-blooded vertebrates in their methylation might be accounted for by the relatively high maximal body temperature reached by them (see **Part 5**). It should also be stressed that the genomes of reptiles show a higher degree of variability in parameters like compositional heterogeneity, CsCl profile asymmetry, and mC, compared to the genomes from other vertebrate classes (Bernardi and Bernardi, 1990a; Hughes et al., 2002). This variability (see **Fig. 4.9** for asymmetry values and **Fig. 5.13** for mC values) might be related to differences in body temperature, which is known to cover a broad range in reptiles (see **Chapter 5**).

It should be stressed that both mC and CpG frequency can be measured in a very precise way, especially if assessed on compositional fractions or on sequenced DNA segments. Indeed, this approach can take advantage of the excellent correlations of both CpG and mC with isochore GC to ensure measurements which can be compared among genomes covering different GC ranges and which can eliminate anomalous data points related to satellite DNAs (see **Fig. 5.15**). A comparison of CpG values from 40-kb DNA segments of two fish genomes (*B. rerio* and *T. rubripes*, zebrafish and fugu) with 40-kb DNA segments from a warm-blooded-vertebrate (*H. sapiens*) shows the difference expected on the basis of the results of **Part 5**. A plot of comparable data from the genome of platypus (*Ornithorhynchus anatinus*), which has a body temperature of 30°-32°, shows an intermediate behaviour (**Fig. 12.8**).

A suggestion derived from the above results and considerations is that the compositional heterogeneity of both DNA and third codon positions could be compared with CpG and methylation levels instead of being compared with body temperature. Indeed, non only can CpG and methylation be precisely measured, but these parameters are likely to integrate time and temperature effects, thus solving the major problem encountered when using only direct temperature measurement in this kind of experimental approach.

(iv) **The interphase nucleus of warm-blooded vertebrates.** As already mentioned, any explanation of DNA thermodynamic stability should answer three crucial questions concerning the major compositional transitions, namely 1) why only a small part, 10–15%, of the genomes of the ancestors of mammals and birds underwent the changes; 2) why the changes concerned both coding and noncoding sequences; and 3) why the changes did not

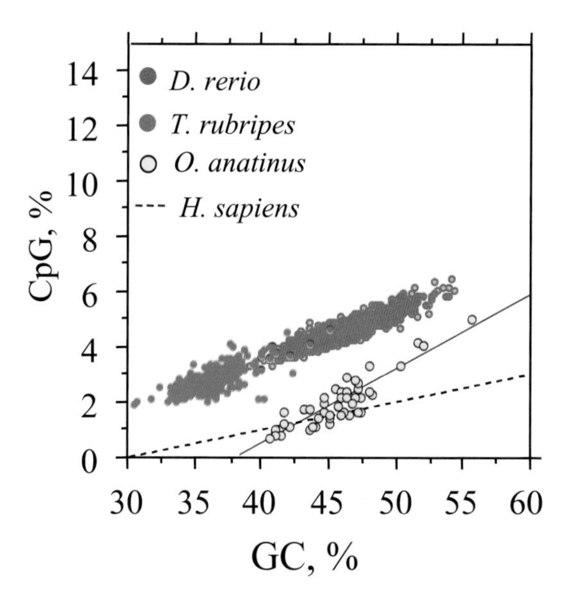

Figure 12.8. CpG levels are plotted against GC levels for 40-kb genome segments of *Drosophila*, zebrafish, pufferfish, platypus, and human. (From Jabbari and Bernardi, 2004b).

take place in fishes and amphibians. Some findings on the interphase nucleus of warm-blooded vertebrates (Saccone et al., 2002) provide answers to these questions.

Let us recall that the genome core, the GC- and gene-richest H2/H3 isochores 1) have their highest concentration in a set of R bands of metaphase chromosomes, the $H3^+$ bands (Saccone et al., 1992; 1993; 1996), that largely coincide with the T bands, previously identified as particularly resistant to thermal denaturation (Dutrillaux, 1973); 2) tend to be located in telomeric-proximal regions of prometaphase chromosomes, whereas the GC- and gene-poorest $L1^+$ isochores tend to occupy centromeric-proximal positions (see **Fig. 7.25**); and 3) occupy very large volumes in the center of the nucleus at interphase, whereas $L1^+$ isochores are packed against the nuclear membrane (Saccone et al., 2002; see **Part 7, Section 5.3**).

Since in cold-blooded vertebrates the gene-dense regions of the ancestral genome core are also centrally located in the nucleus, whereas the gene-poor regions of the empty quarter are also packed at the periphery (see **Fig. 7.28**), then **a very plausible explanation** for the differential compositional transition of the ancestral genome core is that, as body temperature increased with the appearance of homeothermy, the DNA of the open chromatin of the genome core needed to be stabilized by an increasing GC level. This requirement was made more serious by the fact that DNA of warm-blooded vertebrates is characterized by a low level of DNA methylation, a destabilizing factor for DNA (see **Part 5**). In contrast, this GC increase was conceivably not needed by the DNA of the "genome desert", which was already stabilized by its dense chromatin packing at the nuclear periphery (**Fig. 12.9**). This stabilization would also not have been required in cold-blooded vertebrates, because of their lower body temperature and higher DNA methylation.

(v) **The genomes of mammals and birds.** The independent yet parallel compositional

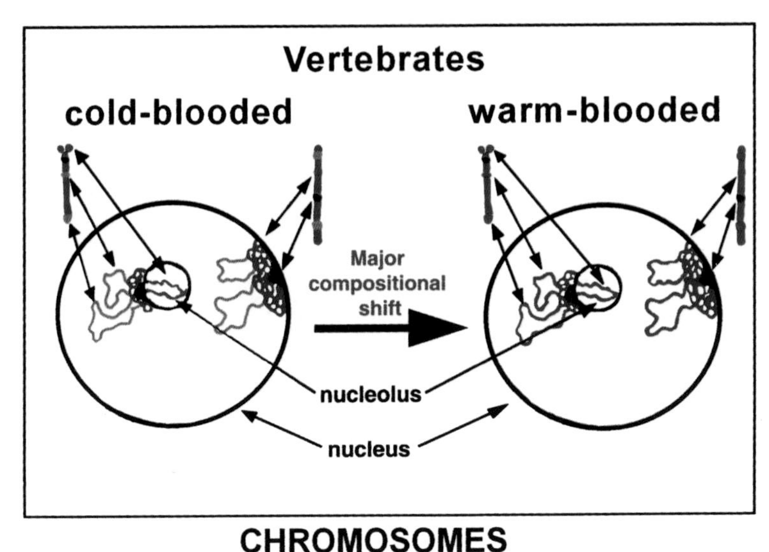

Figure 12.9. A model for the major compositional transition and the formation of GC-rich isochores. Two chromosomes are represented in their metaphase and interphase configurations, one chromosome carrying the GC-rich nucleolar organizer. In warm-blooded vertebrates, the chromosomal regions with the highest concentration of genes (the $H3^+$ bands, or "genome core") are coloured in red. In the nucleus, these regions are much more "open" compared to the remaining gene-poor regions (the "genome desert") coloured in blue (see Part 7, Chapter 5). A similar situation is present in cold-blooded vertebrates, where the "ancestral genome core" (which is less GC-rich than the "genome core" of warm-blooded vertebrates, and is coloured in pink) is also centrally located in the nucleus (Fig. 7.28) and, in all likelihood, also characterized by a decondensed chromatin. The "ancestral genome core" underwent a compositional transition that increased its thermodynamical stability at the higher body temperature of warm-blooded vertebrates. This stabilization was not needed in cold-blooded vertebrates, because of their lower body temperature and higher DNA methylation, nor in the "genome desert" of warm-blooded vertebrates, where the stability was provided by the compact chromatin structure itself. This could explain why the compositional changes were regional, instead of concerning the totality of the genome. (From Saccone et al., 2002).

changes that occurred in the genomes of mammals and birds (see **Figs. 11.11, 11.12**) provide one of the strongest arguments in favor of the thermostability hypothesis. Indeed, the tree of **Fig. 12.10** supports **two independent, convergent, compositional transitions lead-**

ing to the genomes of mammals and birds, which postulate a common cause that can be visualized as a permanent high body temperature. This view is also supported by the different origins and times of appearance of mammals (from Therapsids, about 220 Mya, million years ago) and birds (from Dinosaurs, about 150 Mya) and by the recent discovery of feathered Dinosaurs (Norell et al., 2002), an indication of an emerging homeothermy in avian ancestors (see **Chapter 5**). It should be noted that birds, which have a higher body temperature than mammals (41° *vs.* 37°, on the average), also reach higher GC

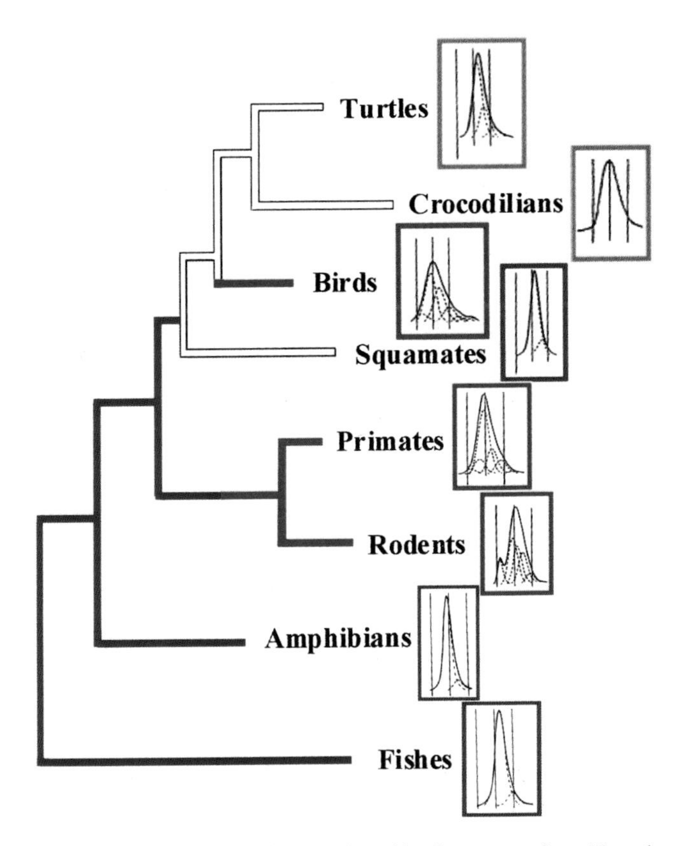

Figure 12.10. The maximum likelihood phylogeny of combined sequences from 11 nuclear proteins (1943 aminoacids; from Hedges and Poling, 1999) is shown together with the branches leading to Amphibians and Fishes, and the CsCl profiles (from Thiery et al., 1976) of representative species, *S. irideus, X. laevis, T. graeca, C. niloticus, I. iguana, G. gallus, M. musculus* and *H. sapiens* (the three vertical bars correspond to buoyant densities of 1.690, 1.700 and 1.710 g/cm^3). Red, yellow and blue frames indicate strongly heterogeneous, moderately heterogeneous and fairly homogeneous genomes (a resolved profile for crocodile was not available; see however Part 4). Red, blue and white branches refer to warm-blooded, cold-blooded and "intermediate" vertebrates, respectively (see text; modified from Cruveiller et al., 2000).

values in their H4 isochores, which do not exist in mammalian genomes.

It should be stressed here that mammals and birds on the one hand, and amphibians and fishes on the other, exhibit very similar compositional patterns within each class (some exceptions concern fishes living at high temperatures), whereas the compositional patterns of reptiles are "intermediate", in the sense that they cover a broad range of heterogeneity.

(vi) **The genomes of plants and insects.** Another line of (admittedly circumstantial) evidence for the role of temperature in compositional shifts is provided by observations made on other organisms which are related to each other, while being phylogenetically very distant from mammals and birds. Indeed, most genes from *Gramineae* exhibit a much higher GC_3 level relative to orthologous genes from dicots (Salinas et al., 1988; Matassi et al., 1989; Carels et al., 1998; see **Fig. 12.11A**). It should be noted that, while this compositional transition took place, in all likelihood, between an ancestral monocot and *Gramineae*, it can only be investigated at present by comparing orthologous genes from *Gramineae* and dicots (which are ancestral to monocots) due to the lack of appropriate gene samples from monocots ancestral to *Gramineae* (see **Part 8**).

On the other hand, many genes from *Drosophila melanogaster* (and *Anopheles gambiae*; not shown), which exhibit a wide compositional range (with the GC-rich regions enriched in genes; Jabbari and Bernardi, 2000; see **Part 9** and Myers et al., 2000), were systematically GC-richer (**Fig. 12.11B**) than their orthologs from the GC-poorer, compositionally homogeneous (Wobus, 1975), genome of *Chironomus thummi*.

The situations found in *Gramineae* relative to dicots, and in *Drosophila* relative to *Chironomus*, are very reminiscent of the human/*Xenopus* or chicken/*Xenopus* comparisons (**Figs. 11.8, 11.9, 11.10**). At this point, these intragenomic compositional transitions may either be similar by sheer coincidence, or because they are due to the same factors. We argued that *Gramineae*, in contrast to the reference dicots, originated from hot, arid regions and had to stand very high maximal temperatures for geological times (Bernardi et al., 1988). Likewise, *Drosophila* larvae are exposed to temperatures as high as 45° (Feder, 1996) and *Anopheles* is a tropical insect, whereas *Chironomus thummi* is a dipteran from

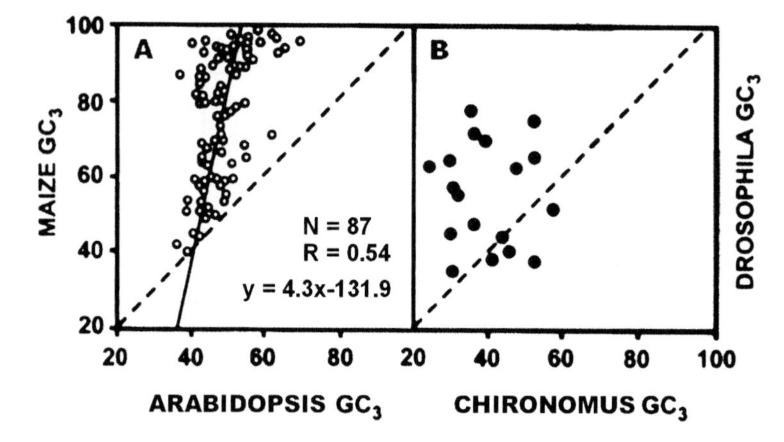

Figure 12.11. **A.** GC_3 values of maize genes are plotted against the GC_3 values of their homologs from *Arabidopsis* genes (Carels et al., 1998). **B.** GC_3 values of *Drosophila* genes are plotted against the GC_3 values of their homologs from Chironomus. (From Bernardi, 2000b).

sub-arctic regions. It is, therefore, possible that a common factor, selection for an increased thermal stability, is responsible for these similar compositional genome transitions which again concern only one part, the gene-rich part, of the genome.

(vii) **The genomes of prokaryotes.** Scattered observations correlating optimal growth temperature of thermophiles and genome GC level can be found in the literature (see for example Kagawa et al., 1984). The most significant ones concern closely related eubacteria (Cleans and Berkeley, 1986; Whitman et al., 1992). Very recently, a comparison of the complete genomes of *Corynebacterium efficiens* with *C. glutamicum* became available (Nishio et al., 2003). These close relative strains have different growth temperature optima. While *C. efficiens* can grow and produce glutamate above 40°, the optimal temperature for *C. glutamicum* is around 30°. The GC levels are 63.4% in *C. efficiens* but only 53.8% in *C. glutamicum*, showing an excellent agreement with the idea that differences in optimal growth temperature are well correlated with GC difference in closely related bacteria (see also **Section 4.4, Chapter 5** and note added in proof in **Chapter 5**).

4.3. The thermodynamic stability hypothesis: RNA results

As for **RNA stability**, abundant evidence indicates that high GC stabilizes RNA by the formation of stem-and-loop structures (Hasegawa et al., 1979; Wada and Suyama, 1986). As an example, the GC level of ribosomal RNA of bacteria is well correlated with their optimum growth temperature (Galtier and Lobry, 1997).

These results are similar to the *in vivo* data on the effect of growth temperature on the structure and function of *ori* sequences in the mitochondrial genome of yeast (Goursot et al., 1988; see **Part 2**). In this case too, increasing temperature from 28°C to 33°C led to the melting of AT stems in the *ori* sequence of some petite mutants and to the consequent decrease of replicative ability. These results demonstrate **direct (reversible) temperature effects** on the secondary (and tertiary) structures of DNA and RNA.

4.4. The thermodynamic stability hypothesis: protein results

As far as **protein stability** is concerned, one should consider the following points.

(i) In our original paper on the thermodynamic stability hypothesis (Bernardi and Bernardi, 1986a), we observed that "*the aminoacid replacements that accompany GC increases in codon positions comprise those (Argos et al., 1979: Zuber, 1981) that lead to thermodynamically more stable proteins*" (see **Table 12.2**). We also noted that "*the aminoacids that are most frequently acquired in thermophiles and that most contribute to increased stability (alanine and arginine) increase in frequency with increasing exon GC* (**Fig. 12.12**), *whereas those that are correspondingly lost and that diminish stability (serine and lysine) decrease. In the case of compartmentalized genomes, these changes may take place within the same genome; for instance, in the human genome the (Ala + Arg)/(Ser + Lys) molar ratio varies by a factor of four between proteins encoded by the GC-poorest and the GC-richest genes, respectively* (**Fig. 12.12**). *In conclusion, the compositional changes that make DNA*

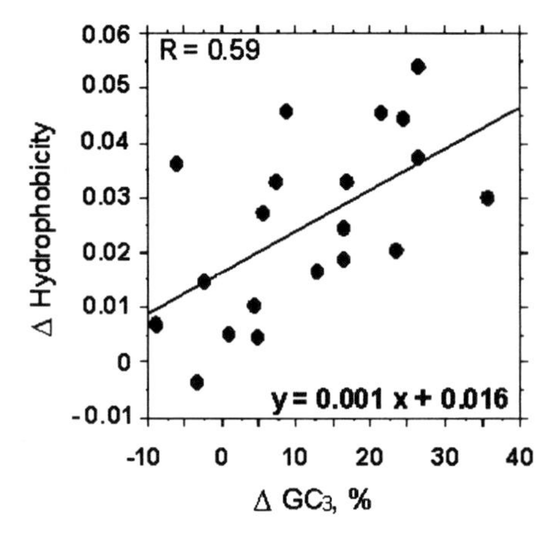

Figure 12.13. Correlation between Δ GC_3 and Δ hydrophobicity of orthologous proteins from human and *Xenopus*. Proteins were first sorted according to the GC_3 values of the corresponding human coding sequences, and then the entire data-set was divided into 20 groups, each comprising 23 proteins. The equation of the linear regression line and the regression coefficient are indicated. The significance of the correlation is p=0.0065. (From Cruveiller et al., 1999).

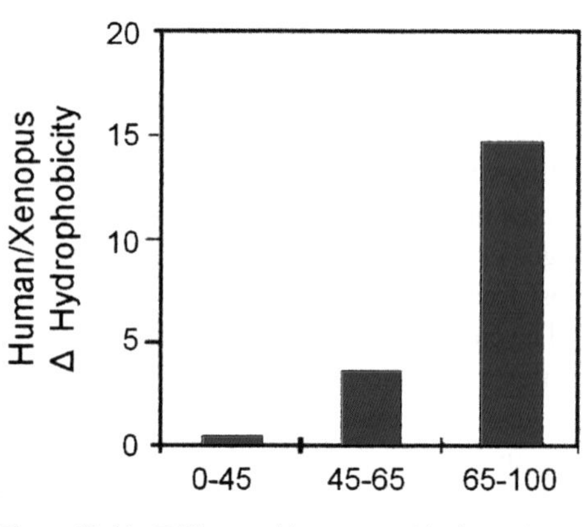

Figure 12.14. Difference histogram of hydropathy values of homologous proteins from human and *Xenopus*. Proteins are partitioned in three groups according to the GC_3 values of the corresponding coding sequences. (From D'Onofrio et al., 1999b).

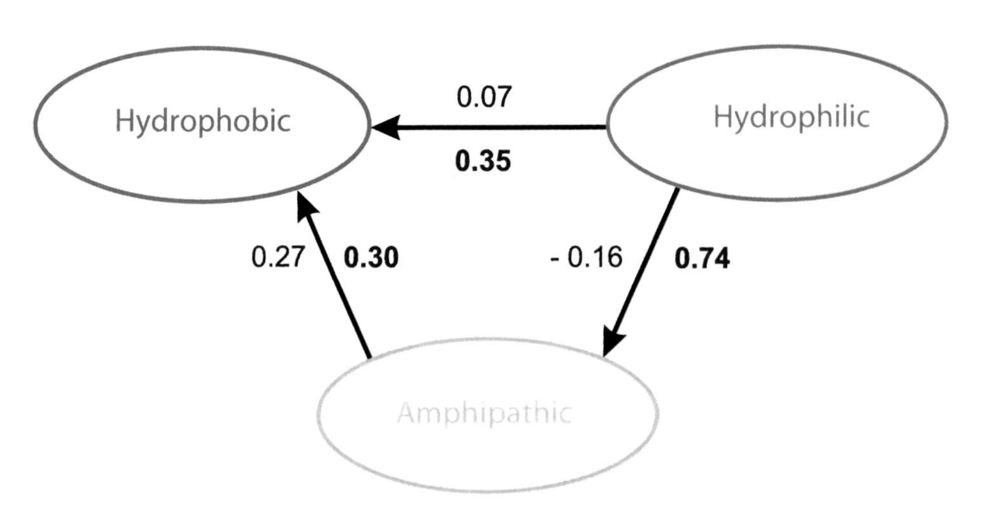

Figure 12.15. Substitution frequency differences among hydrophobic, amphipathic and hydrophilic amino acids from *Xenopus* and man. Values in italics type concern the proteins encoded by genes in the range 0 - 45 % GC_3, whereas values in bold type concern the proteins encoded by genes > 65 % GC_3. (From Cruveiller et al., 1999).

4.5. The primum movens problem

While, on a first approximation, the advantages associated with an increased thermodynamic stability are achieved simultaneously at the DNA, RNA and protein levels (Bernardi and Bernardi, 1986a), one can ask which one is the **driving force**, the *primum movens*, in these thermodynamic stabilizations. It is clear that RNA can be ruled out because the observed compositional changes affect both transcribed and non-transcribed sequences. Proteins can also be ruled out because the compositional changes concerned both translated and non-translated RNA sequences. Indeed, since both coding and noncoding sequences are concerned, the driving force should be DNA. This does not rule out, however, feedbacks from RNA and proteins on DNA. The extremely high GC levels attained by ribosomal RNAs are obviously linked to particular requirements for stem-and-loop systems that have an effect on the composition of the corresponding genes. In the case of proteins, aminoacid changes affecting hydrophobicity may not correspond to increases in codon GC (*e.g.*, Ser→Thr), but this is a relatively rare situation.

CHAPTER 5

Objections to selection

As already mentioned in **Chapter 1**, the compositional changes, the major shifts, that took place between cold- and warm-blooded vertebrates, and the maintenance of the compositional patterns so formed cannot be accounted for by any explanation only or essentially relying on stochastic processes. This led us (Bernardi and Bernardi, 1986a) to propose a **natural selection mechanism acting on the genome phenotype**.

The reason why the natural selection hypothesis met with a strong opposition was, however, its **apparent questioning of the neutral theory**. We have now explained again (see **Chapters 1–3**) that, while our findings cannot be accounted for by that theory, they are compatible with it. A positive result of the objections and alternative explanations published during the past sixteen years is that our proposal underwent an unusual degree of scrutiny, and that the rebuttal of the objections (and the failure of alternative explanations) reinforced our original explanation. In some cases, an analysis of the objections provided further support for the hypothesis (see [vii] below for an example).

In this chapter, we will consider the objections to selection, that were not accompanied by alternative explanations, whereas the following chapter will address the alternative explanations that were proposed. We will begin with the most popular, yet logically weakest, objection to the thermodynamic stability hypothesis.

(i) *While GC levels of ribosomal and transfer RNA stems of prokaryotes are well correlated with optimal growth temperature, no such correlation was found with GC levels of prokaryotic DNAs* (Galtier and Lobry, 1997). These results have been widely accepted as disproving the thermodynamic stability hypothesis (see, for example, Eyre-Walker and Hurst, 2001). Since few issues in evolutionary genomics have been criticized more superficially than the thermodynamic stability hypothesis, we will deal with this point in some detail, starting with some premises: 1) Many factors are collectively responsible for the base composition of a genome. Some of these factors can be recognized because of their specific effects. For example, one can mention the avoidance of the dinucleotide TpT in bacteria exposed to ultraviolet light (which causes the dimerization of TpT with the consequent problems in cell replication), and the avoidance of C in hyperthermophiles (because C is deaminated to dU at high temperature). The majority of factors cannot, however, be identified because of their interplay. 2) Different strategies were developed by different prokaryotes to cope with long-term high temperatures. It is known, for example, that the DNAs of some thermophilic bacteria are strongly stabilized by particular DNA-binding proteins (Robinson et al., 1998) and that, in turn, these proteins can be stabilized by thermostable chaperonins (Taguchi et al., 1991). The emergence of alternative strategies can be understood if one considers that the highest temperatures withstood by some hyperthermophiles are higher than the melting temperature of DNA formed only by GC base pairs. 3) In prokaryotes, most of the DNA is made up of coding sequences. Any compositional change is immediately reflected in changes of protein structure. Different structural constraints on different proteins in prokaryotes living in an extremely wide range of ecological niches will give different feedbacks on genome composition.

What Galtier and Lobry (1997) did was to use a sample of prokaryotes including *Archaea* and *Eubacteria*, mesophiles, thermophiles and hyperthermophiles, aerobes and anaerobes (the latter two groups of bacteria being characterized by different GC levels; Naya et al., 2002), without taking into consideration the points mentioned above. This led to looking at very different and often contrasting inputs as far as genome composition is concerned. **Under these circumstances, plots of prokaryotic GC (or GC_3) *versus* optimal growth temperature are meaningless**, because they mix different effects on genome composition. A similar problem is probably responsible for the lack of a correlation between optimal growth temperature and GC levels in a comparison between *Thermoplasma volcanicum* and seven phylogenetically distant *Archaea* (Kawashima et al., 2000).

In contrast, when a small taxonomic group of organisms is taken into consideration, it is likely that temperature effects can be detected because most other factors influencing base composition are shared. Indeed, 1) when comparisons are made within small taxa, such as *Bacillus* (Claus and Berkeley, 1986) and *Methanobacterium* (Whitman et al., 1992), thermophiles exhibit higher GC levels than mesophiles (see McDonald, 2001, for a detailed discussion); 2) when a pair of closely related eubacteria, such as *Corynebacterium efficiens* and *C. glutamicum,* are taken into consideration, the relationship between optimum growth temperature and GC level is striking (see Nishio et al., 2003); 3) when a transition in body temperature takes place in a small taxonomic group, such as vertebrates, the compositional change between cold- and warm-blooded vertebrates is evident.

To sum up, the results of Galtier and Lobry (1997) have no bearing on our conclusions concerning vertebrates. In fact, they do not even shed light on the problem of the correlation between growth temperature and GC levels of protein-coding sequences in the case of *Eubacteria* and *Archaea* (see note added in proof at the end of this chapter).

(ii) *The compositional changes undergone by GC-rich isochores concerned not only coding sequences, but also non-coding intragenic (introns) and intergenic sequences, and selection on the latter is difficult to visualize.* Indeed, there are two problems that need to be discussed. The first one is the impossibility of selection acting on individual nucleotides of noncoding sequences to form isochores. This problem was solved by the models presented in the preceding **Chapters 2** and **3**. The second problem is the visualization of the function(s) under selection in the noncoding sequences. Some introns are known to be endowed with regulatory roles (in the mammalian α globin gene, for example, a promotor is located in an intron). Intergenic sequences comprise regulatory sequences, CpG islands and conserved sequences (see below). Moreover, they have a role in the expression of the flanking genes. Indeed, integrated viral sequences are optimally transcribed when these sequences are located in compositionally matching chromosomal environments of the host, or, in other words, when their location mimics that of compositionally similar host genes; in contrast, they are not transcribed when they are located in a compositionally non-matching environment (see **Part 6**). Recent results on the conservation of intergenic sequences (Hardison, 2000; Shabalina et al., 2001; Dermitzakis et al., 2002) and of isochores in murids and human (Pavliček et al., 2002b; see **Part 11**) go in the same direction.

(iii) *Some duplicated genes, like the human GC-poor β globin and GC-rich α globin genes, are expressed in the same cells and fulfill the same role* (Wolfe et al., 1989; Li and Graur,

1991; Li, 1997; Graur and Li, 2000). This argument neglects the fundamentally different regulation of these two genes. The latter, in contrast to the former, has an internal promoter and is associated with a different regulatory element, a CpG island instead of a TATA box. This means that two genes can be transcribed in a coordinated way even if their regulatory sequences are completely different. It is, then, difficult to see why a different base composition of two related genes would necessarily be an obstacle to a coordinated expression. It would obviously be of great interest to understand why the compositional fates of the two genes were different (see **Part 6**, **Chapter 3**), but this problem will only be solved by further investigations.

(iv) *The population size of mammalian orders is too small to allow the negative selection process to take place in the time available since the appearance of mammals.* This could be true if, but only if, the formation and maintenance of GC-rich isochores were caused by single-nucleotide changes. We have discussed in **Chapters 1 and 2** why this possibility cannot be taken into consideration.

(v) "*The variation in the pattern of nucleotide substitution seems likely to be due to differences in the underlying mutational process rather than to selection because the results are observed in repetitive elements throughout the genome*" (Lander et al., 2001). Again, this argument holds if, but only if, repetitive elements are "selfish DNA", a hypothesis which is no longer tenable (see **Part 6**).

(vi) "*Reptiles exhibit a warm-blooded isochore structure*" (Hughes et al., 1999). This claim hinged on the demonstration of a unity slope in the GC_3 plot of reptiles *vs.* birds. Unfortunately, the very small sample of largely partial coding sequences from only two reptiles (ten sequences from a crocodile and six from a turtle) cannot define a statistically reliable slope. In addition, the points show no correlation with the orthologous *Xenopus* genes, as they should (see **Fig. 11.7**). This suggests that the points either concern outliers, or that the plots are simply unreliable because of the sample being very small and several of the coding sequences only partial. In spite of the problems just mentioned, these results have been uncritically accepted by a number of authors as showing that reptiles "*are likely to have isochores, just as birds do*" (Eyre-Walker and Hurst, 2001).

Even if the observations of Hughes et al. (1999) were correct, they would only confirm the existence of a limited formation of GC-rich isochores in crocodile and turtle, a point previously shown by Aïssani and Bernardi (1991a,b), but not mentioned by Hughes et al. (1999). However, both the DNA properties (asymmetry, heterogeneity of DNA profiles; see **Part 4**) and the absence of CpG islands in the two species under consideration (see **Part 5**) make them quite different from *bona fide* warm-blooded vertebrates, as well as from reptiles showing a very low asymmetry and heterogeneity of DNA profiles.

Indeed, it was shown in **Fig. 4.9** and pointed out in **Part 4, Chapter 3**, that within the suborder of snakes (which was taken into consideration because it comprises the largest number of analyzed subfamilies or genera; see **Tables 4A and 4B**), the range of asymmetry of CsCl profiles is much wider than those of entire classes of vertebrates, fishes/amphibians or mammals/birds. The lack of similarity in asymmetry (and compositional patterns) in the snakes is a strong indication that compositional patterns developed independently in small taxons (subfamilies, genera). Our suggestion is that this independent development took place under the effect of different body temperatures. Several observations lend support to this explanation: 1) different small taxa of reptiles are characterized by different body

Table 12.3. Differences between the genomes of reptiles and warm-blooded vertebrates

	Reptiles	Warm-blooded vertebrates
Compositional heterogeneity	Low-Medium	High
CpG islands	Absent	Present
T bands	Absent	Present
Karyotype change	Slow	Fast
Speciation	Slow	Fast

temperatures (see, for example, Heatwole and Taylor, 1987); this is in contrast with the essentially uniform body temperatures of warm-blooded vertebrates and with the narrow range of body temperatures of the vast majority of fishes and amphibians; 2) some data (see, for example, Olmo, 2003) indicate a correlation between body temperature and GC levels of reptiles; 3) DNA methylation, a parameter which is very sensitive to body temperature, not only is relatively low in reptiles, but also more variable than in other vertebrate classes (Jabbari et al., 1997; see **Part 5, Section 3.3**).

Next to the lack of the high compositional heterogeneity and lack of CpG islands, other general features distinguish reptilian genomes from those of warm-blooded vertebrates (see **Table 12.3**). Indeed, T bands are absent in chromosomes from turtles, lizards and snakes (M. Schmid, personal comm.), and karyotypic change and species formation are much slower in this vertebrate class (as they also are in fishes and amphibians) compared to warm-blooded vertebrates (Bush et al., 1977; see also Bernardi, 1993b).

Undoubtedly, the simplest explanation for the striking similarity of mammalian/avian data would be that this similarity is the consequence of a compositional transition in the common ancestor of mammals and birds. The ancient *Haematotherma* hypothesis (Haeckel, 1866; Owen, 1866; a hypothesis revived by Hedges et al., 1990), which puts mammals and birds in sister groups, whose common ancestor might have undergone the transition would have satisfied this requirement, if the hypothesis had not been disposed of (see Caspers et al., 1996, and Janke et al., 2001, for discussions on this point).

The proposal of Hughes et al. (1999) that this single transition took place in the common ancestor of tetrapods (**Fig. 12.16A**) is untenable because there is no evidence for a compositional transition between fishes and the common ancestor of tetrapods (compare the data on the genomes of fishes and amphibians presented in **Part 4**). An alternative hypothesis is that this single transition occurred in the common ancestor of reptiles and mammals (Duret et al., 2002; **Fig. 12.16B**). This different hypothesis, surprisingly presented (see legend of **Fig. 12.16B**) as identical to that of Hughes et al. (1999), is based on the assumption that the "*genomes of mammals, birds and reptiles are generally highly heterogeneous and contain GC-rich isochores*". Such an assumption neglects well-established facts that differentiate warm-blooded vertebrates from reptiles (see **Table 12.3**) and is simply not acceptable. As far as the "erosion" of GC-rich isochores of Cetartiodactyls and Primates is concerned (see **Fig. 12.16B**), the hypothesis of **vanishing isochores** (Duret et al., 2002; Smith et al., 2002; Webster et al., 2003) is backed only by an unreliable reconstruction of ancestral GC levels (Alvarez-Valin et al., 2004).

Moreover, if we consider the "classical" phylogenetic tree of amniotes (approximately

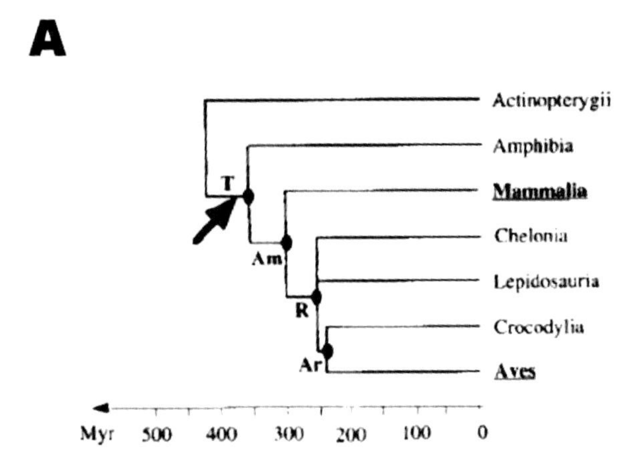

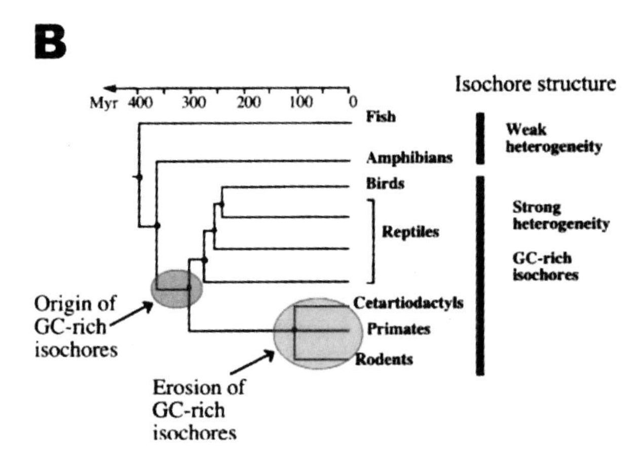

Figure 12.16. **A.** Simplified phylogeny of vertebrates. Homeotherms are boldface and underlined. T = Tetrapoda, Am = Amniota, R = Reptilia, Ar = Archosauria. The arrow indicates the point at which the GC increase of a part of genome could have taken place. (From Hughes et al., 1999). **B.** Phylogenetic distribution of GC-rich isochores in vertebrates. Genomes of fishes and amphibians are weakly heterogeneous in base composition. Genomes of mammals, birds, and reptiles are generally highly heterogeneous and contain GC-rich isochores. (From Hughes et al., 1999). Fig. 12.16B and its legend are from Duret et al. (2002).

represented in **Fig. 12.16A**), the problem with the single transition is that some squamates or *lipidosauria* (snakes and lizards) have genomes that are typically "cold-blooded", namely fairly homogeneous in terms of base composition (Thiery et al., 1976; Bernardi and Bernardi 1990 a,b; Hughes et al., 2002). Since, according to the proposals of Hughes et al. (1999) and Duret et al. (2002), the ancestors of squamates had "warm-blooded", heterogeneous genomes, the genomes of those squamates could then only be the result of a process of extreme compositional homogeneization due to increased substitution rates (as claimed by Hughes and Mouchiroud, 2001, again on a very limited experimental basis, but apparently forgotten by Duret et al., 2002; see **Fig. 12.16B**), for which no comparable example is known among vertebrate genomes. Indeed, only the case of murids is known in which an increased mutation rate led to a decreased genome heterogeneity compared to other mammals, yet this homogeneization did not reach, by far, the remarkable homogeneity of the genomes of some squamates (see Bernardi, 2000a,b). This problem persists if squamates are at the root of the reptilian tree, as suggested by recent data (Hedges and Poling, 1999; see **Fig. 12.10**).

(vii) The objection to the thermal stability hypothesis raised by Belle et al. (2002) has already been rebutted in **Chapter 4**. In fact, the data of Belle et al. (2002) contradict their conclusion and provide, if anything, support to the thermal stability hypothesis.

(viii) According to Hamada et al. (2002), "*The codon usages (and GC_3) in α-globin genes from two heterotherms* (cuckoo and bat) *and three snakes are similar to those in α-globin genes from warm-blooded vertebrates*". On this basis, they concluded that "*these results refute the influence of body temperature pattern upon codon usages (and GC_3) in α-globin genes, and support the hypothesis that the increase in GC content in the genome occurred in the common ancestor of amniotes*".

That α globin from cuckoo and bat show the same GC_3 level and codon usage as other birds and mammals, respectively, may only surprise those who imagine that daily or seasonal torpor (with decreased body temperature) should cause a change in the compositional patterns of the genomes. We made clear (Bernardi 1993a) that, logically, what matters from a thermodynamic point of view is Tmax, the maximal body temperature, tolerated for extended lapses of time. The results on the three snakes are also not surprising in that they apparently support the existence of at least one moderate GC-rich, but not very GC-rich, isochore (embedding the α globin gene) in those reptiles. Under these circumstances, the authors' conclusion is obviously a far-fetched extrapolation.

(ix) Since "*Neither* LDH-A *nor the α-actin gene shows a correlation between GC content and adaptation temperature*", Ream et al. (2003) concluded that "*whereas GC contents of isochores may show variation among different classes of vertebrates, there is no consistent relationship between adaptation temperature and the percentage of thermal stability-enhancing G+C base pairs in protein coding genes*". This is an undue generalization for three different reasons: 1) the major compositional shifts only concerned about 50% of the genes, those present in the gene-dense regions of the ancestral genome core. Unfortunately, one of the two genes (α actin) chosen by Ream et al. (2003) is GC-poor in mammals and did not undergo the compositional transition. This reduces the gene sample to one gene only, LDH-A. 2) The lack of correlation should obviously be shown using orthologous sequences in the 51 vertebrate species investigated. In the case of α actin, finding orthologous sequences is a big challenge. The actins form a large multigene family that is known to have undergone considerable compositional diversification even within the human genome (Dodemont et al., 1982; in fact, there are several paralogs even among α actins). Using paralogous instead of orthologous genes introduces additional noise in the data because of the preferential integration of duplicated genes into the ancestral genome leading to increases in their GC levels in warm-blooded vertebrates (see **Part 6**). Unfortunately, Ream et al. (2003) used paralogous genes not only in the case of the multigene family of α actin, but also in that of the A and B forms of LDH in *Xenopus* (see Clay et al., 2003b). 3) Even a true lack of correlation for two orthologous genes, would not prove the

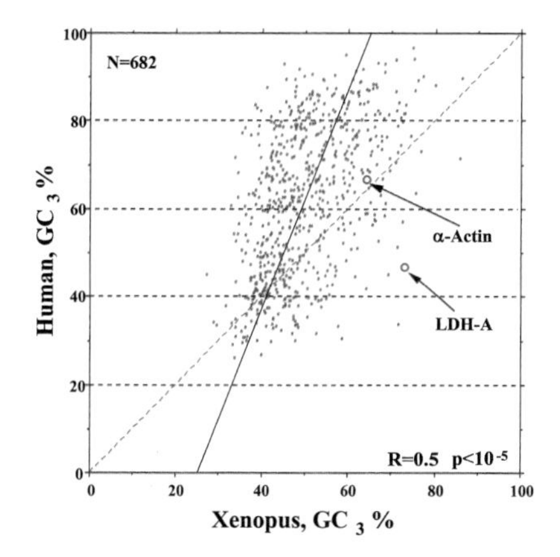

Figure 12.17. Scatterplot of the GC_3 levels of putative orthologous genes from *Xenopus* and human. Each point represents a gene "family" as defined by HOVERGEN (Duret et al., 1994), and shows the GC_3 levels of the sequenced human and mouse genes in that gene family that is most similar (as assessed at the amino acid level). In many, but not all, cases the genes will be orthologs: although the criterion extracts only potential orthologs, it may be the strictest automatic criterion that can be applied, in the absence of complete sequence information for all species. The sequence pairs for LDH-A and α-actin are indicated by red circles. (From Clay et al., 2003b).

absence of a "*consistent relationship between adaptation temperature and the percentage of thermal stability-enhancing G+C bias pairs in protein coding genes*", because the genes tested might be outliers (as they are in the *Xenopus*/human GC_3 plot; see **Fig. 12.17**).

Note added in proof.

The points made under (i) above have been confirmed by Musto et al. (2004), who have investigated the genomic GC level of *Eubacteria* and *Archaea* from families covering an optimal growth temperature range of 15°C to >100°C. When families comprising at least five species (N = 57) were investigated, the majority of them showed a positive correlation between GC and T_{opt}, a minority showing either no correlation or a negative correlation. The significance of the correlation greatly increased when only families comprising at least 10 species (N = 28) were considered (see **Fig. 12.18A** for an example). Furthermore, a strong and significant positive correlation was found between the range of GC levels (ΔGC) and the range of T_{opt} (ΔT; see **Fig. 12.18B**). Needless to say, mixing all GC values under consideration leads to uncorrelated plots.

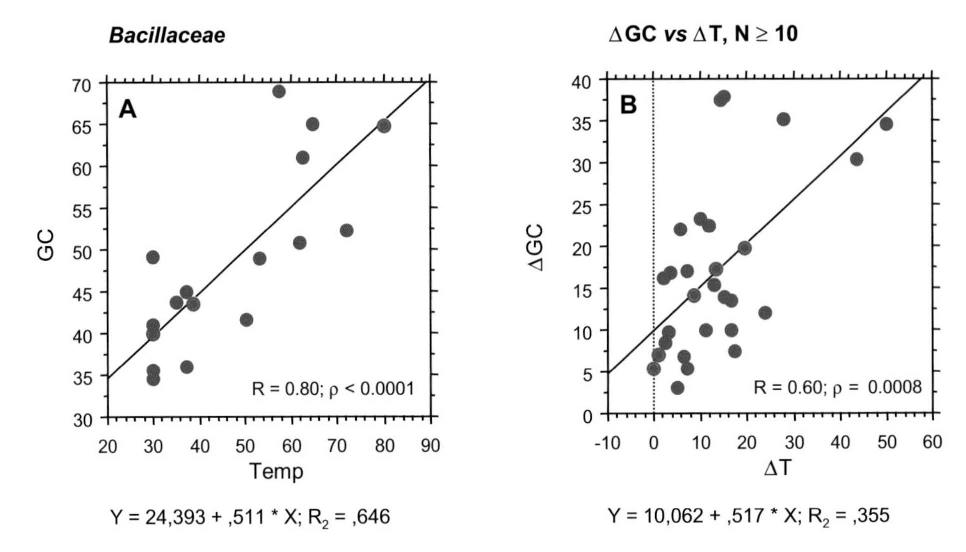

Figure 12.18. **A.** A plot of GC levels of 18 eubacteria from the family of *Bacillaceae* against their optimal growth temperature. **B.** A plot of ΔGC against ΔT (the differences found in GC levels and in optimal growth temperatures) from 28 families of *Eubacteria* and *Archaea*. (From Musto et al., 2004).

Alternative explanations for the major shifts

Several alternative explanations have been proposed to account for the formation and maintenance of GC-rich isochores (see **Table 12.4**).

Table 12.4. Formation and maintenance of GC-rich isochores in the vertebrate genomes

The original explanation:
Natural selection: thermodynamic stability hypothesis (Bernardi & Bernardi, 1986a)

Alternative explanations:
(i) Repair (Filipski, 1987)
(ii) Mutational bias (Sueoka, 1988; Francino and Ochman, 1999)
(iii) Changes in nucleotide pools during DNA replication (Wolfe et al., 1989)
(iv) Recombination (Eyre-Walker, 1993)
(v) Cytosine deamination (Fryxell & Zuckerkandl, 2000)
(vi) "Bendability" of DNA (Vinogradov, 2001)
(vii) Biased gene conversion (Eyre-Walker & Hurst, 2001; Duret et al., 2002; Galtier, 2003; Marais, 2003)

From a general point of view, all the alternative hypotheses of **Table 12.4** do not answer one or more of the following crucial questions which are answered by natural selection:

(i) why regional compositional changes never appeared in the genomes of cold-blooded vertebrates (except, to a limited extent, in some reptiles; see **Chapter 4**), that also are characterized by gene-poor and gene-rich regions, by biphasic (early-late) replication timings and by differential recombination rates;

(ii) why they paralleled each other in the independent lines of mammals and birds;

(iii) why they reached an equilibrium a long time ago, both in mammals and birds, and were maintained since;

(iv) why they were different in exons and introns of the same genes and in their flanking regions (CpG islands, 5' and 3' untranslated sequences; Pesole et al., 2000), as well as in different regions of the coding sequences that correspond to specific protein structures (Chiusano et al., 2000; Alvarez-Valin et al., 2000; D'Onofrio et al., 2002);

(v) why changes were consistently in the direction of an increased thermodynamic stability of DNA and why similar compositional changes that might also be related to thermal stability were found in organisms phylogenetically unrelated to vertebrates;

(vi) why the mutational GC→AT bias of both GC-poor and GC-rich human genes did not lead to a GC-poor compositional pattern in the human genome.

(vii) why an increase in substitution rates (such as that exhibited by murids; see the following section) did not lead to increased GC_3 levels of the genes that underwent the major transition;

Some specific comments on the alternative hypotheses of **Table 12.4** follow:

(i) The **repair hypothesis** (Filipski, 1987) is based on the fact that some types of repair are biased (Brown and Jiricny, 1988) and vary along chromosomes, but suffers from the fact that repair is tightly linked with transcription (Bohr et al., 1985; see also Balajee and Bohr, 2000, for a review). This raises problems with the compositional changes and their maintenance in non-transcribed sequences, as well as with the correlations between coding and noncoding sequences. Moreover, repair has never been shown to vary over the scales needed to generate isochores (Eyre-Walker and Hurst, 2001).

(ii) A **mutational bias** was originally proposed to be responsible for the broad compositional spectrum of bacterial genomes (Freese, 1962; Sueoka, 1962). Mutational biases cannot be denied, since they were convincingly demonstrated by **mutator mutations** in bacteria (see **Part 9** and the following chapter) and since they are also exhibited by the human genome (see **Part 11**). What can be questioned, however, is that they are the cause of the compositional changes and of their maintenance in the genomes of vertebrates. Indeed, one should not forget that mutator mutations are due to changes in some subunits of the replication machinery, and that a single replication machinery is at work in the nucleus of vertebrates. Under the mutational bias explanation, changes should, therefore, not only affect the totality of the genome, but also have the same directionality. Indeed, this is the case for the GC→AT mutational bias found in the human genome (see **Part 11**).

The fact that some genome regions of the ancestors of present-day mammals and birds underwent the transition while others did not, might, however, be explained by "*regional*" mutational biases (Sueoka, 1988). Then additional hypotheses are needed to explain why the changes were regional, instead of concerning the totality of the genome. Such hypotheses have been proposed. Indeed, different chromatin structures were postulated to be responsible for the "*regional*" mutational biases (Sueoka, 1988). These "*plausible explanations*" amount, however, to abandoning the biases of the replication and repair machineries as the cause of the compositional shifts undergone by vertebrate genomes in favor of completely different causes, namely local differences in chromatin structure. Now, either the latter are due to the compositional changes in DNA, and then the argument becomes a circular one (the compositional changes being precisely what is in the need of an explanation); or they are not, and then the reason(s) for the appearance of different chromatin structures should be given. In addition, the analysis of polymorphism datasets (Eyre-Walker, 1999; Smith and Eyre-Walker, 2001) led to the rejection of mutation bias hypotheses.

(iii) **Changes in nucleotide pools**. Wolfe et al. (1989) proposed that isochores arose from mutational biases due to compositional changes in the precursor nucleotide pool during the replication of germ-line DNA. The GC-rich isochores would replicate early in the germ-line cell cycle, when the precursor pool has a high GC content and, thus, a propensity to mutate to GC. The GC-poor isochores, on the other hand, would replicate late in the cell cycle, when the precursor pool has a high AT content and a propensity to mutate to AT. This **mutationist hypothesis** (Graur and Li, 2000) is based on the observations that the composition of the nucleotide precursor pool changes during the cell cycle and that such changes can in fact lead to altered base ratios in the newly synthesized DNA (Leeds et al., 1985).

It has been repeatedly stressed that the depletion hypothesis (still considered as a possibility by Eyre-Walker and Hurst, 2001) was untenable already at the time it was proposed, since the inactive X chromosome and the GC-rich satellite DNAs of mammals were known to replicate at the very end of the cell cycle (see Bernardi et al., 1988; 1993a; see also Graur and Li, 2000). This shows that the GC depletion at the end of the cell cycle, if present at all, is not strong enough to alter the base composition of newly replicated DNA.

(iv) **Recombination**. According to Fullerton et al. (2001), "*the idea that recombination may be involved in determining GC content (Holmquist, 1992; Eyre-Walker, 1993; Charlesworth, 1994) arose from the observation of an association between recombination and GC-rich chromosomal regions. The clustering of chromosomal rearrangements in isochores with high GC contents was described first by Bernardi (1989) and subsequently by Holmquist (1992)*". There is, however, a cause-effect problem in this correlation, in that GC might favour recombination, or recombination might lead to increasing GC. The latter explanation is, however, ruled out by the fact that this phenomenon only occurred in the two independent ancestral lines leading to mammals and birds and not in any other line of cold-blooded vertebrates.

(v) **Cytosine deamination.** It has been claimed that cytosine deamination plays a primary role in the evolution of mammalian isochores (Fryxell and Zuckerkandl, 2000). The predicted two-fold decline of cytosine deamination for each 10% increase in GC content, and its 5.7-fold increase for each 10°C increase in temperature would lead to "*a positive feedback loop between cytosine deamination and GC content*" which "*implies an evolutionary pattern of divergent genetic drift to high or low GC contents*". In spite of the explanations provided, this model can hardly account for the formation and maintenance of GC-rich isochore patterns of vertebrates, with closely similar genome phenotypes in mammals, and birds, respectively.

(vi) **Bendability of DNA.** This is a selectionist hypothesis which is complementary to the thermal stability hypothesis (Vinogradov, 2001). It relies on the statistical significance of a faster increase in "*bendability*" of human exons and introns compared to their melting energy (which reflects thermal stability) with elevation of GC level. Whether the data supporting it are solid enough is not yet clear, because they depend upon two indirect sets of data: the trinucleotide table for bendability based on consensus values obtained from the DNAse I digestion and nucleosome positioning studies of Gabrielian et al. (1996), and the dinucleotide table for free energy of melting (ΔG) obtained from the ultraviolet absorbance and temperature profiles of SantaLucia et al. (1996).

(vii) **Biased gene conversion (BGC).** This alternative hypothesis is best presented and criticized by quoting Eyre-Walker and Hurst (2001). "*BGC is thought to arise during homologous recombination through the formation of heteroduplex DNA. This leads to a base mismatch if the heteroduplex extends across a heterozygous site. These base mismatches are sometimes repaired by the DNA repair machinery, but this process tends to be biased, leading to an excess of one allele in gametes. For example, base mismatches tend to be repaired to GC in mammalian cell lines. So, the variation in recombination rate across a genome will cause variation in GC level if the rate of BGC is sufficiently high.*

The suggestion that BGC might cause isochore formation comes from two observations indicating that a positive correlation exists between the rate of recombination and GC content. First, there is a correlation between the frequency of recombination and G+C content both between and within human chromosomes (Lander et al., 2001; Eyre-Walker, 1993;

Ikemura and Wada, 1991; Fullerton, 2001). Second, sequences that have stopped recombining are either declining in G+C content, or have a lower G+C content than their recombining paralogues (Eyre-Walker, 1993). But these observations do not establish causation..." (our underlining). Indeed, "*the high GC_3 values of some Y-linked genes and the positive correlation between Ks and GC_3 seems to be evidence against the BGC hypothesis*" (Eyre-Walker and Hurst, 2001).

Unfortunately, there are more problems with BGC, in spite of this "**strange phenomenon**" (Marais, 2003) having a very attractive feature, a "regional effect", namely the fact that compositional changes concern whole genome regions: the sheer lack of any quantitative estimation, the short tract size of gene conversion (hundreds of base pairs), the need for high homology, and the unlikelihood that it can more than compensate the mutational AT bias (a phenomenon of a higher order of magnitude, which has no preference for coding or noncoding regions) to form and maintain the GC-rich isochores of warm-blooded vertebrates. Moreover, as a completely neutral process, BGC would have had no way of forming and maintaining very similar genome phenotypes, the isochore patterns, along the independent lines of mammalian and avian orders.

Note added in proof

A very recent paper by Vinogradov (2003) reported data of oligonucleotide microarray experiments. This report is of interest here for three different reasons: (i) It reached the conclusion that "*The housekeeping (ubiquitously expressed) genes in the mammal genome were shown to be on average slightly GC-richer than tissue-specific genes. Both housekeeping and tissue-specific genes occupy similar ranges of GC content, but the former tend to concentrate in the upper part of the range. In the human genome, tissue-specific genes show two maxima, GC-poor and GC-rich. The strictly tissue-specific human genes tend to concentrate in the GC-poor region; their distribution is left-skewed and thus reciprocal to the distribution of housekeeping genes. The intermediately tissue-specific genes show an intermediate GC content and the right-skewed distribution.*" This conclusion is at a slight variance with previous results (D'Onofrio, 2002; see **Fig. 5.4**). However, both D'Onofrio's and Vinogradov's conclusions are in agreement with our previous suggestions (Mouchiroud et al., 1987; Bernardi, 1993, 1995; Pesole et al., 1999) and contradict the conclusions of Gonçalves et al. (2000) and Ponger et al. (2001; see **Part 5, Section 1.2**). Moreover, "*Both for the total data set and for the most part of tissues taken separately, a weak positive correlation was found between gene GC content and expression level.*" This is in agreement with a similar conclusion independently reached (using EST instead of microarray data) by Arhondakis et al. (2003; see **Fig. 12.19**) and by Lercher et al. (2003). Incidentally, the claim by the latter authors that their "*results provide the first direct systematic evidence of a general relationship between expression patterns and chromatin structures and base composition*" (our underlining) is an exaggeration. While this link was first proposed at least ten years ago (Bernardi, 1993) on the basis of the very widely different gene densities in H3 and L isochores, evidence has been recently provided about the correlation between gene-dense (GC-rich) regions and expression on the one hand (Caron et al., 2001; see **Part 7**) and open chromatin on the other (Saccone et al., 2002; see **Part 7**).

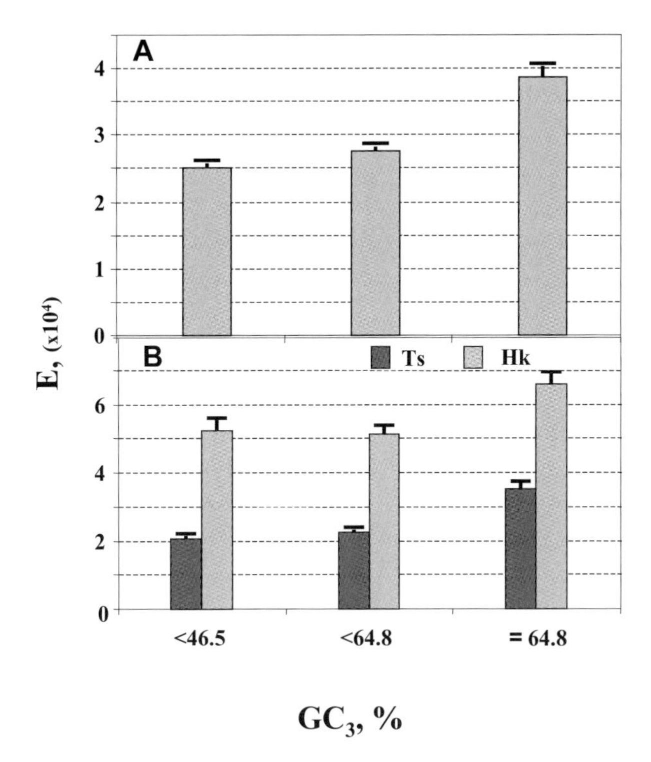

Figure 12.19. **A.** Histogram of the average gene expression level (E) of the genes in the three compositional classes. The boundaries between classes are from D'Onofrio (2002). **B.** Histogram of the average gene expression level (E) of tissue-specific (Ts) and housekeeping (Hk) genes in the three compositional classes. Standard error is reported for each histogram bar. (From Arhondakis et al., 2003).

(ii) The paper also concluded that "*Both in the human and mouse, gene specific for some tissues (*e.g.*, parts of the central nervous system) have a higher average GC content than housekeeping genes. Since they are not transcribed in the germ line (in contrast to housekeeping genes), and therefore have a lower probability of inheritable gene conversion, this finding contradicts the biased gene conversion (BGC) explanation for elevated GC content in the heavy isochores of mammal genome. Genes specific for germ-line tissues (ovary, testes) show a low average GC content, which is also in contradiction to the BGC explanation.*" This point is of interest because it provides an additional, independent argument against BGC being responsible for the formation of GC-rich isochores.

(iii) The report insists on the author's previous conclusions (Vinogradov, 2001; 2003) that the emergence of GC-rich isochores was due to requirements of transcription. This explanation forgets, however, to consider that such requirements are satisfied by the corresponding isochores of cold-blooded vertebrates which are remarkably less GC-rich. Incidentally, this point was also missed by Lercher et al. (2003).

CHAPTER 7

Natural selection and the "whole genome" shifts of prokaryotes and eukaryotes

The explanation originally proposed to account for the **different base composition of prokaryotic genomes** was that genome composition shifts because of directional mutations due to biases in replication enzymes (Freese, 1962; Sueoka, 1962). Indeed, "**mutator mutations**" may lead not only to highly increased mutation rates, but also to strong biases in base substitutions. It has been argued that while these mutational biases are acceptable as the mechanism of the compositional changes, they are not necessarily its cause (Bernardi et al., 1988). In fact, in the laboratory, such biases have only been detected in mutational hot spots (Nghiem et al., 1988; Wu et al., 1990). Overall changes in genome composition after 1,200–1,600 generations, such as those published by Cox and Yanofsky (1967), although still accepted by many authors, fall, in fact, within experimental error (Bernardi, 1993a) and were never confirmed. In spite of this problem, it is conceivable that in nature overall changes in the base composition of bacterial genomes can be achieved through mutational biases. The question remains, however, whether the resulting "whole-genome" compositional changes are only determined by the vagaries of **random mutations** in the genes encoding the protein sub-units of the replication machinery, or are under the control of **natural selection**. Two facts point toward the second conclusion.

The first one is that changes in nucleotide composition correspond to changes in amino acids; in turn, these correspond to changes in the hydrophobicity and in the secondary structure of proteins. If changes only depended upon mutational bias, then one should accept the untenable viewpoint that the corresponding functional changes are selectively irrelevant. Now, that some degree of neutrality may be associated with non-synonymous and synonymous changes is perfectly acceptable, but **the correlations discussed in Part 10, between not only nonsynonymous but also synonymous changes and protein structure, put strong limits on the neutrality of those changes**. Neutrality of changes was probably overestimated because early comparisons concerned orthologous mammalian genes, which are compositionally very similar to each other and mainly undergo conservative changes. This situation is obviously very different in prokaryotes.

The second fact is that GC spectra cover a much wider range in the genomes of prokaryotes and unicellular eukaryotes than in those of invertebrates, cold-blooded vertebrates and warm-blooded vertebrates, these three groups showing a progressively narrower range. Since the potentially relevant mutations in the replication machinery responsible for different mutational biases presumably are comparable in all classes of organisms, and since the spread of GC levels is so different, the only reasonable explanation for the different ranges of **Fig. 12.20** is that there is a different selection on the mutations in the replication machinery of these different classes of organisms, some very deleterious mutations being quickly eliminated by negative selection.

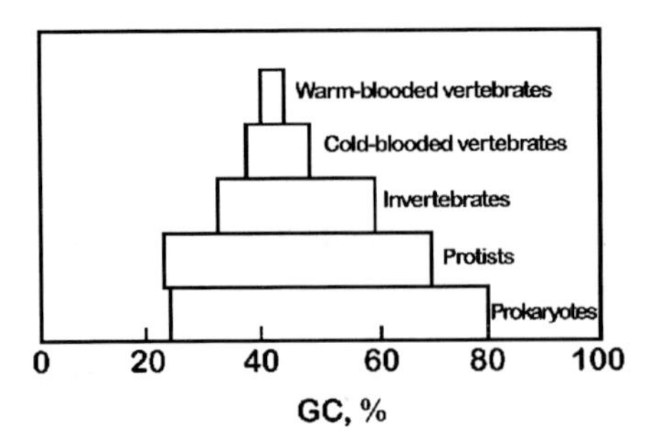

Figure 12.20. GC level ranges of DNAs from warm- and cold-blooded vertebrates, invertebrates, protists, and prokaryotes. (From Bernardi and Bernardi, 1990b).

Needless to say, natural selection does not only act on the mutations occurring in the genes encoding the proteins of the replication machinery. As already mentioned, C is counterselected in hyperthermophilic prokaryotes because of its thermal deamination; TT are counterselected in bacteria exposed to ultraviolet light, because of their dimerization, and aerobic prokaryotes are GC-richer than anaerobic ones (Naya et al., 2002) to quote just three examples. Moreover, compositional heterogeneity of a number of bacterial genomes may also be due to selection, as already mentioned in **Part 9, Chapter 5**. It is then difficult to escape the conclusion that, even if originally developed for the vertebrate genomes to designate the compositional features of the genome, **the concept of genome phenotype should be extended to prokaryotes and generalized to all living organisms**.

What is proposed is that **genome composition** in prokaryotes is under the control of natural selection which optimizes the adaptation of genome composition to the environmental conditions (temperature being only one among many factors). Because of the compactness of the prokaryotic genomes, which are essentially made up of coding and regulatory sequences, selection (mainly negative selection) operates on single nucleotide changes, whereas in vertebrates selection operates also on the very abundant noncoding sequences that characterize those genomes.

This proposal reverses a trend that has dominated the literature for the past 40 years. The reasoning was that since the different base compositions of prokaryotic genomes were due to biases in replication enzymes (Freese, 1962; Sueoka, 1962), the same process should also explain the formation and maintenance of isochores in vertebrate genomes. The results presented in this book in favour of natural selection being responsible for the isochores of vertebrates and the data of **Fig. 12.20** lead, in contrast, to explaining the different compositions of prokaryotic genomes according to the natural selection schemes proposed for the isochores of vertebrates.

Recapitulation*

* In this section, some key figures from the main text (numbers in bold type with an R prefix) are represented on a small scale, with an abbreviated legend (carrying the original figure number) and no reference. See the main text figures for further information. Some other figures and tables from the main text are only referred to (they are in plain type).

1. Structural genomics of warm-blooded vertebrates

Our main conclusions on the structural genomics of vertebrates can be summarized as follows (see Table 12.1).

(i) The **genome compartmentalization**, *i.e.*, a genome's discontinuous compositional heterogeneity, is best illustrated by describing the **organization of the genomes of warm-blooded vertebrates**. We can take as an example the **human genome,** since we have shown that it is typical for most mammalian genomes and similar to the avian genomes. The first property that we discovered was its **compositional compartmentalization (Fig. R1)**. When human DNA is randomly broken down during preparation to an average size of 100 kb, the fragments were shown to cover a broad compositional spectrum (30-60% GC) and to be distributed in a small number of **families** (the "major DNA components"), each one of which is comparable in heterogeneity to the least heterogeneous genomes, those of prokaryotes (Figs. 3.10, 3.11). These families, L1, L2, H1, H2, H3, represent about 33%, 30%, 24%, 7.5%, 5% of the genome. We could also show that the 100-kb DNA fragments derived from much longer, fairly homogeneous chromosomal segments (at least 300 kb in average size), which we called **isochores** for (compositionally) equal landscapes. Incidentally, satellite and ribosomal DNAs could also be visualized as isochores, because of their compositional homogeneity. To sum up, the genomes of warm-blooded vertebrates can be described as **mosaics of isochores (Fig. R1A)**.

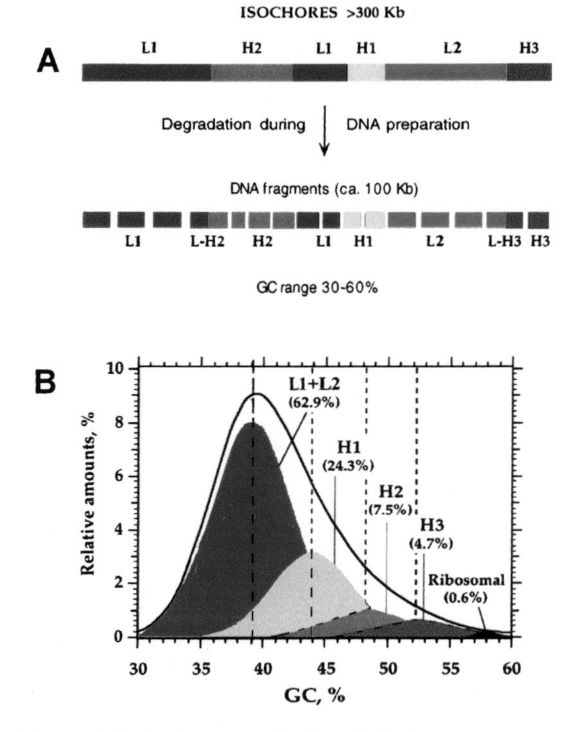

Figure **R1** (3.9). **A.** Scheme of the isochore organization of the human genome. **B.** The CsCl profile of human DNA (ca. 100 kb) is resolved into its major DNA components.

(ii) Isochore (or compositional) patterns of vertebrate genomes were called **genome phenotypes**, because they show phylogenetic differences. The major difference was observed between the genomes of warm- and cold-blooded vertebrates, the latter consisting of isochores that cover a much narrower compositional range and attain lower GC levels. Lesser yet significant differences were found between the isochore patterns of birds (which comprise an additional very small, very GC-rich isochore family H4), or murids (which do not exhibit a distinct H3 isochore family), and the general mammalian pattern presented by the human genome (see **Fig. R2A**). The compositional distributions of coding sequences (or of their third codon positions, **Fig. R2B**) from vertebrates belonging to different classes mimic their isochore patterns in that, compared to human, *Xenopus* shows a narrower spectrum centered on low values, whereas chicken and mouse extend more and less, respectively, towards high values.

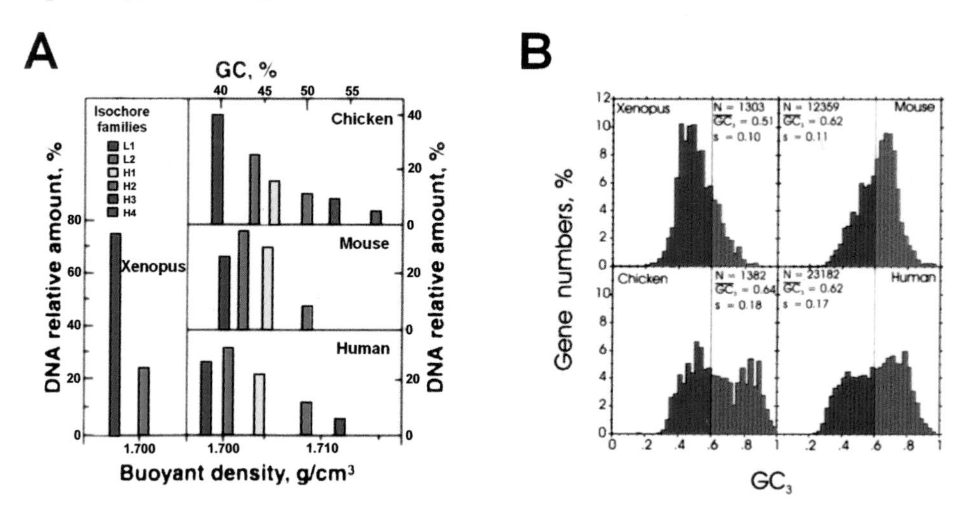

Figure **R2** (3.17). **A.** Isochore families as deduced from density gradient centrifugation. **B.** Compositional patterns of third codon positions (GC_3).

(iii) **The genomic code** comprises three kinds of relations: 1) **the compositional correlations between coding and noncoding sequences** which were found for all vertebrate genomes explored (see **Figs. R3A** and 3.18, 3.19). This was a striking result since coding sequences only represent a minute amount of those genomes (2-3% in the case of the human genome), but also provided the strongest support for the genome as **a coordinated ensemble**. 2) The functional compatibility or incompatibility between coding and noncoding sequences indicated the existence of a **genomic fitness**. This can be exemplified by the effect of compositionally matching *vs.* non-matching contiguous coding sequences on the transcription of integrated viral sequences (**Fig. R4**). 3) **A universal correlation** (**Fig. R5**) was found among the compositions of the three codon positions, both within and among genomes, ranging from bacteria to human.

(iv) **The distribution of genes** in the human genome is remarkably non-uniform, about

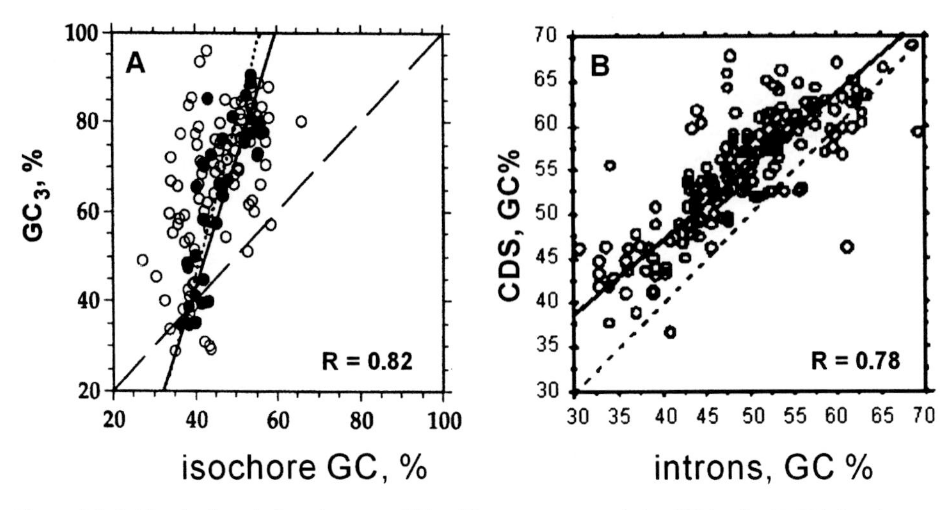

Figure **R3** (3.20). **A.** Correlations between GC_3 of human genes and the GC level of DNA fractions or YACs in which the genes were localized (filled circles), or of 3' flanking sequences (open circles); **B.** Correlations between GC levels of human coding sequences (CDS) and of the corresponding introns.

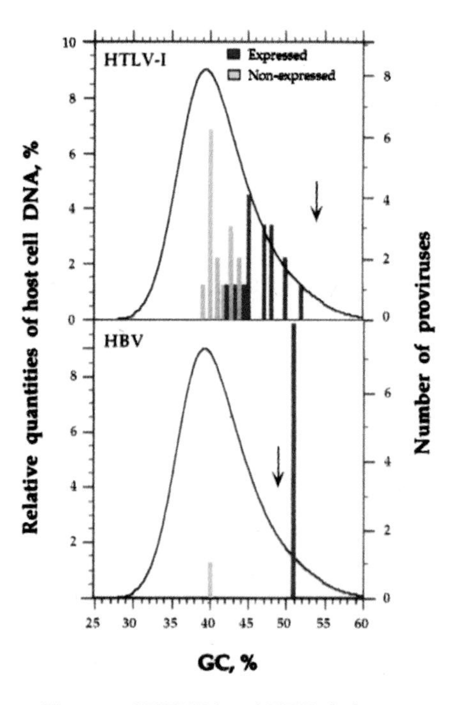

Figure **R4** (6.4). Distribution of integrated HTLV-1 and HBV viral sequences in the human genome. The CsCl profile of human DNA is shown. Arrows indicate the GC level of viral sequences. Expressed sequences are in red, non-expressed sequences in yellow.

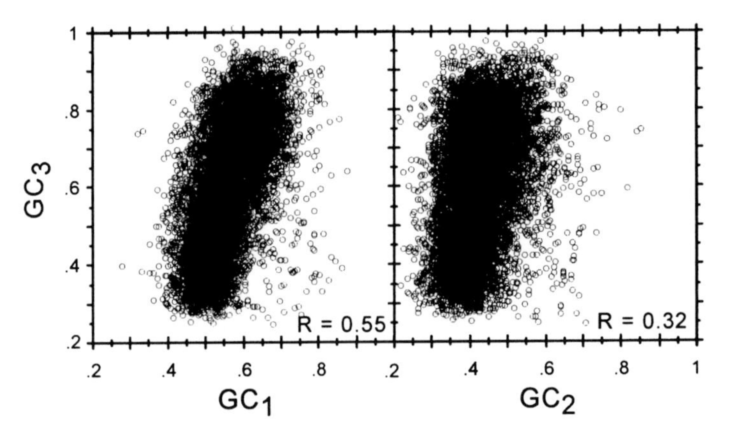

Figure **R5** (from Fig. 10.9). Correlations between GC_3 and GC_1 or GC_2 for 20,148 human genes.

50% of the genes being located in H2 and H3 isochores, which only represent about 12% of the genome, the other 50% or so being present in the remaining 88% of the genome formed by the GC-poorer L1, L2 and H1 isochores (**Fig. R6**). The gene distribution defines, in fact, **two gene spaces**, the small **genome core** and the large **genome desert** (or empty quarter, from the classical name of the Arabian desert).

(v) The genome core and the genome desert exhibit **distinct structural and functional properties** (Table 5.1; see also **Section 2** below). The genome core is characterized not only by a **high gene density,** but also by **abundant CpG islands, short introns and short untranslated sequences, early replication, high recombination, high transcription, open chromatin structures and distal chromosomal locations**, whereas the genome desert is characterized by opposite features. Moreover, the **open chromatin** of the genome core is preferentially targeted by **transposed duplicated genes** and by **viral sequences**, which are, however, stable and transcriptionally active only in **isopycnic isochores** (*i.e.*, isochores having a GC level

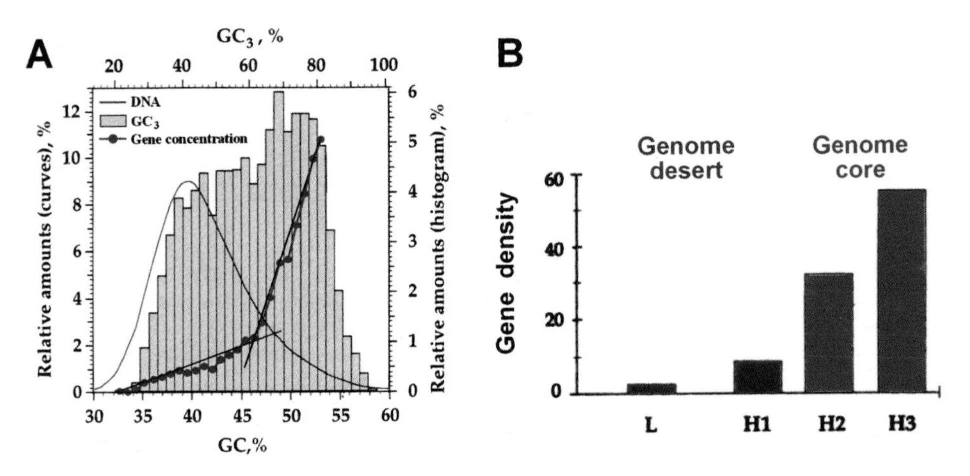

Figure **R6** (5.1). **A.** Profile of relative gene concentration (red dots) in the human genome. **B.** Density of genes in isochore families.

approximating that of the proviral sequence under consideration; see **Fig. R5** and Part 6). Likewise, GC-rich SINE sequences and GC-poor LINEs have their highest densities and stabilities in isopycnic isochores.

2. Chromosomes and interphase nuclei

Compositional maps of human chromosomes (**Figs. R7** and 3.13-3.16) allow visualizing of the isochores and of the gene densities in them. As far as the distribution of isochores in **prometaphase chromosomes** is concerned, four subsets of bands could be distinguished (Table 7.1 and **Fig. R8**): two subsets, $H3^+$ and $H3^-$, of Reverse bands and two subsets, $L1^+$ and $L1^-$, of Giemsa bands. $H3^+$ bands are the R bands that hybridize H3 and H2 isochores; they largely correspond to the T bands of Dutrillaux (1973) and are preferentially located at telomeric and distal regions of chromosomes. $L1^+$ bands are the G bands that hybridize L1 isochores; they correspond to the two sets of the most intense staining G bands of Francke (1994) and tend to be located in the central parts of chromosomes. GC levels decrease from $H3^+$ to $H3^-$ R bands and then to $L1^-$ and $L1^+$ G bands, in this order. $H3^-$ and $L1^-$ bands are, however, not very different compositionally, (in fact, $H3^-$ bands exist that are GC-poorer than non contiguous $L1^-$ bands), but G bands are always flanked by GC-richer R bands, and R bands by GC-poorer G bands in chromosomes 21 and 22. Moreover, DNA is more tightly packed in G bands than in R bands.

In **interphase nuclei**, H3 isochores are in a much more **expanded chromatin** configuration compared to the **compact chromatin** of L1 isochores (**Figs. R9**, 7.27). Moreover, H3 isochores are located in the center of the nucleus, whereas L1 isochores are packed against the nuclear membrane.

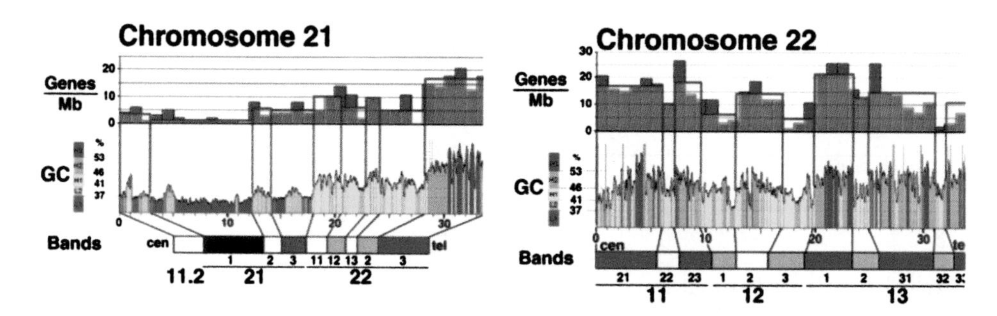

Figure **R7** (7.13A). Bands, isochores and gene concentration of human chromosomes 21 and 22. **Bands** comprise four classes of G bands characterized by different staining intensity (from black to pale grey) and two classes of R bands ($H3^+$, red; $H3^-$, white); the two chromosomes are represented according to their cytogenetic size. **GC** profiles are shown through 100 kb windows. **Genes/Mb**, the blue bar plot concerns chromosomal bands, the red plot 1-Mb segments .

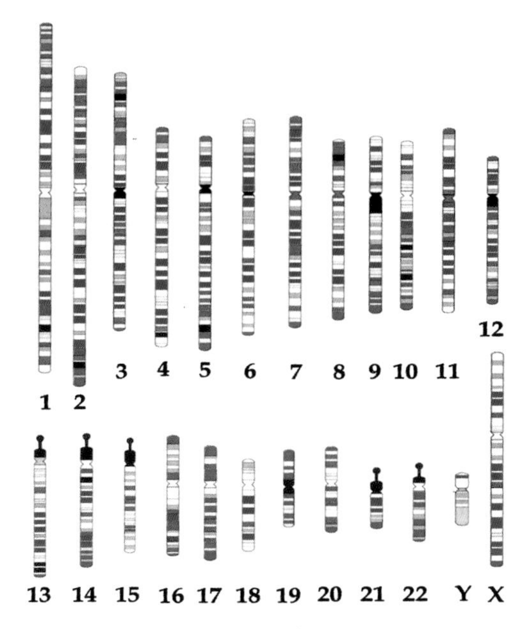

Figure **R8** (7.12). Identification of the GC-poorest ($L1^+$, blue bars) and the GC richest ($H3^+$, red bars) human chromosomal bands. The $H3^-$ R bands are in white. The grey scale of the $L1^-$G bands is according to Francke (1994).

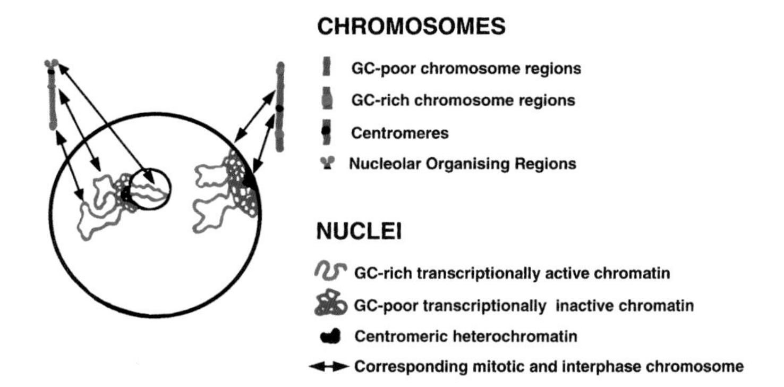

Figure **R9** (7.27). A cartoon showing that the centromeric-proximal regions of chromosomes tend to be present in compact chromatin structures at the periphery of the interphase nucleus (blue blocks), whereas the distal regions are in an open chromatin structure (red filaments) in the center of the nucleus. Only two chromosomes are depicted (including one acrocentric).

3. Comparative and evolutionary genomics of vertebrates

(i) Several basic structural and functional properties are shared by the genomes of all vertebrates. These properties include a **bimodal gene density** (see **Sections 1** and **2**), a **biphasic** (early-late) **replication**, a **higher recombination rate in telomeric regions** of chromosomes, and **a higher level of transcription in GC-rich regions**.

threshold and cause a "**phase transition**" leading to structural changes in DNA and in the chromatin, which affect the expression of genes located within or near the isochore under consideration.

It should be noted that, until regional selection has acted, only neutral and nearly neutral changes will be detected (see **Fig. R13**). In other words, the **neutral and nearly neutral theories account for features of molecular evolution that take place within certain compositional thresholds**. They represent, therefore only one mode, the **neutral mode**, of genome evolution, which is checked by the predominant and pervasive control of selection on compositional patterns, the **selective mode** of genome evolution.

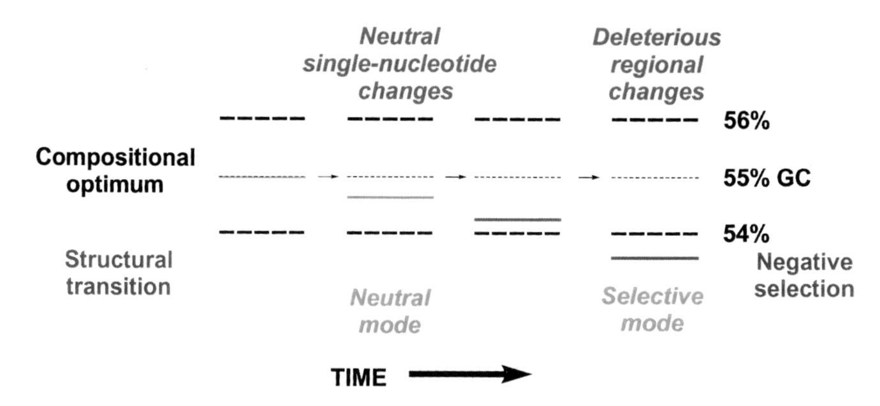

Figure **R13** (12.4). Time course of compositional evolution of a GC-rich noncoding intergenic region from a warm-blooded vertebrate in the conservative mode. It is assumed that the overall compositional trend is dominated by an AT bias. In an early phase, the average GC of the region, initially visualized at its compositional optimum (set at 55% GC), is decreasing but remains within a tolerated range (whose arbitrary thresholds are indicated by the horizontal heavy broken lines). In a late phase, the average GC trespasses the lower threshold, the region undergoes a structural change which is deleterious for gene expression and is subject to negative selection. Until then, the changes are neutral or nearly neutral.

It should be stressed that the drift pushing a region beyond the acceptable thresholds is not only influenced by mutational biases but also by recombination events and transposon insertions. While these factors may slow down, accelerate or alter the direction of the compositional drift, the final control is still exerted by natural selection.

While the stress has been put so far on negative selection, **positive selection** may also play a role in coding sequences.

In summary, natural selection can explain the maintenance of isochores and is pervasive at the genome level. Natural selection (essentially negative, or purifying, selection) acts both on single nucleotide changes in coding and regulatory sequences, and on regional changes in noncoding sequences (through critical changes).

An obvious implication of selection acting on noncoding sequences is that the latter fulfil a function. Their interpretation as **junk DNA** can be rejected on several grounds. First of all, on a gross scale, genome size is highly conserved in both mammals and birds. Since noncoding sequences represent 97-98% of mammalian genomes, this means that size changes in intergenic sequences are generally modest, on the average. Second, regulatory sequences are present in noncoding sequences, sometimes dozens of kb away from the genes that they control. Third, there is increasing evidence for long stretches of sequence conservation in noncoding sequences. Fourth, the expression of integrated viral sequences depends upon the matching or non-matching compositional properties of the flanking host sequences (**Fig. R4**). Fifth, compositional (isochore) patterns are conserved even when sequence homology is lost, as illustrated by the human/mouse comparison of the MHC locus (Fig. 11.6). **In conclusion, genome (and isochore) sizes appear to be under selection.** This is also true for the mitochondrial genome, which covers a five-fold size range in eukaryotes.

(iv) If we now consider the **transitional, or shifting, mode** of evolution (**Fig. R10**) and compare third codon positions of human (or chicken) genes with orthologous *Xenopus* genes, GC-poor genes are compositionally similar, whereas GC-rich genes show increasing differences (**Fig. R14**). In contrast, orthologous human/chicken genes are compositionally similar (Fig. 11.10). Likewise, if we compare the average codon frequencies of human (or chicken) genes with orthologous *Xenopus* genes, we find no difference for GC-poor genes and large differences for GC-rich genes (Fig. 11.9), whereas no differences are found in either case for orthologous human and chicken genes (Fig. 11.11). If the emergence of homeothermy was accompanied by an increasing optimal GC level shifting upward the lower threshold of acceptable average composition of the ancestral genome core (corresponding to an upper optimal level), then the compositional transition would strictly require only negative selection, although a contribution by positive selection is likely (**Fig. R15**). **In conclusion**, negative selection, far from simply maintaining the *status quo* (and being "*boring*", Kondrashov, 2000), can help genome adaptation (and be "*creative*", Schmalhausen, 1949).

The **independent compositional transitions** that occurred in the ancestral lines of mammals and birds were accompanied by other functionally important changes: 1) CpG and

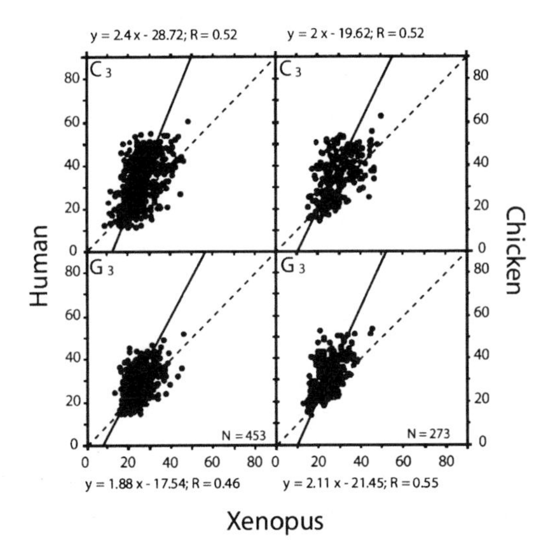

Figure **R14** (11.8). Correlation between C_3 and G_3 values of orthologous genes from human, or chicken, and Xenopus. The orthogonal regression lines are shown together with the diagonals (dashed lines).

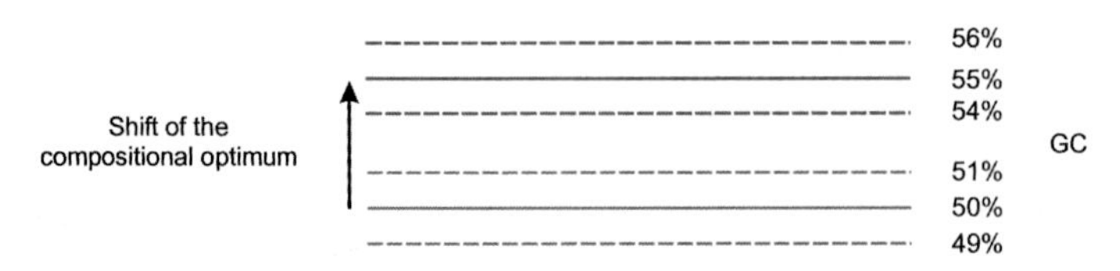

Figure **R15** (12.5). A scheme of the shift of compositional optimum caused by a gradual increase in body temperature. At the same time thresholds within which changes are tolerated (broken lines) are shifted upwards. GC values are only indicative.

methylation levels decreased by a factor of two in mammals/birds compared to fishes/amphibians, but, remarkably, remained stable in each of the two groups (Figs. 11.12 and 11.13), indicating two equilibria separated by a transition; interestingly, reptiles show a wider DNA methylation range (and a wider compositional heterogeneity) compared to the other classes of vertebrates. 2) CpG islands were independently formed at the emergence of mammals and birds; indeed, they are only found in warm-blooded vertebrates (Figs. 5.8-5.10). 3) The nucleotide context of the ATG initiator codons showed an increase in its GC level, at least in mammals (Table 5.1). 4) The formation of a strong banding pattern in warm-blooded vertebrates was paralleled by a higher level of karyotype changes; the disadvantages associated with this increased chromosome instability were apparently compensated by the increased speciation, which, in turn, facilitated the invasion of new ecological niches. 5) Finally, hydrophobicity increased in the encoded proteins (Fig. 12.3), entailing a faster folding of highly expressed proteins.

(v) Natural selection implies the elimination of disadvantages and/or the acquisition of advantages. As far as the compositional transition is concerned, we originally proposed that the selective advantages were the higher thermal stabilities simultaneously achieved at the DNA, RNA and protein levels to cope with the higher body temperature of warm-blooded vertebrates. This can be understood because in a small taxon, such as vertebrates, most factors acting on the compositional pattern of the genome are likely to be similar and because body temperature is the major physiological property that underwent a change at the emergence of warm-blooded vertebrates. The situation of vertebrates is in sharp contrast with that of prokaryotes in which different factors have an input on genome composition of different taxa, making it very difficult, if not impossible, to identify them and assess their influence. However, when closely related bacterial genomes were compared, a higher optimal growth temperature was observed to be accompanied by a higher GC level (Part 12).

The evidence supporting the **thermodynamic stability hypothesis** comes from the following observations. As far as DNA is concerned, the appearance of GC-rich isochores is correlated with body temperature in vertebrates (see Part 12, Chapter 4). The reasons why only one small part (10-15%) of the genome, the gene-dense part of it, underwent the GC increase in both its coding and noncoding sequences can now be understood. Indeed, in interphase nuclei the genome core is present in very open chromatin structures (**Fig. R16**), and DNA stabilization may have become necessary when body temperatures increased with the appearance of homeothermy. This stabilization was not required in the genome desert, where the compact chromatin structure could provide the necessary stabilization.

The reasons why only one small part (10-15%) of the genome, the gene-dense part of it, underwent the GC increase in both its coding and noncoding sequences can now be understood. Indeed, in interphase nuclei the genome core is present in very open chromatin structures (**Fig. R16**), and DNA stabilization may have become necessary when body temperatures increased with the appearance of homeothermy. This stabilization was not

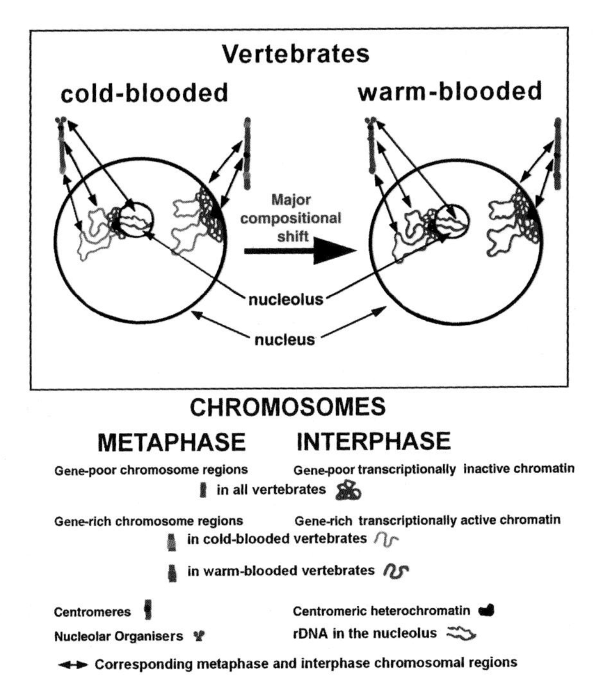

Figure **R16** (12.9). A model for the major compositional transition (see text).

required in the genome desert, where the compact chromatin structure could provide the necessary stabilization.

In the case of RNA, the GC increase leads to the stabilization of stem-and-loop structures against higher temperature.

As far as proteins are concerned, we demonstrated an increase in hydrophobicity with increasing GC_3. This can be understood because quartet and duet codons are positively and negatively correlated with GC_3, respectively, and quartets comprise three abundant hydrophobic amino acids (leucine, alanine and valine), whereas duets comprise six abundant hydrophilic amino acids (lysine, asparagine, glutamine, histidine, glutamic acid and aspartic acid). This increased hydrophobicity was possibly advantageous not only for protein stability but also for increasing the folding rate of the highly expressed proteins that are encoded in the genome core. Finally, as far as the *primum movens* problem is concerned, clearly the driver was DNA. This is the simplest way to account for the fact that both non-transcribed and non-translated sequences of the ancestral genome core also underwent the change.

At this point, it should be mentioned that several objections have been raised against the thermodynamic stability hypothesis, but these have been rebutted (see Part 12, Chapter 5). Moreover, a number of **alternative hypotheses** have been put forward to account for the formation of GC-rich isochores and their maintenance (Table 12.4). While some of them may explain one or two of the reported observations, none can account for most of the other known facts, whereas natural selection can. Indeed, these alternative hypotheses cannot explain why regional compositional changes 1) never occurred in cold-blooded vertebrates living at moderate or low temperatures, 2) were similar in the independent lines of mammals and birds, 3) reached an equilibrium at least 100 million years ago in mammals and were maintained since then, 4) were different in exons, introns, intergenic sequences, and in coding sequences corresponding to different secondary structures of proteins, 5) affected only a small part of the genome (the ancestral genome core) in both coding and noncoding sequences. Other questions could also not be answered, *e.g.* why the mutational AT bias of the human genome did not lead to the formation of a GC-poor genome, why the increased substitution rates of murids did not increase rather than decrease compositional contrasts, why a decrease in CpG and methylation occurred at the transition from cold- to warm-blooded vertebrates, or why CpG islands were formed and the pre-ATG nucleotides were changed.

4. The eukaryotic genome

Although our investigations were mainly focused on the vertebrate genome, we have explored other eukaryotic genomes, ranging from organelle genomes to protists, invertebrates and plants, and found that some of the basic compositional features present in vertebrate genomes are remarkably widespread.

Indeed, **genome compartmentalization** was not only found in warm-blooded, and to a smaller extent, in cold-blooded vertebrates, but also in other eukaryotes, ranging from trypanosomes to *Drosophila*. At least in some cases, this compartmentalization seemed also to be linked to the thermodynamic stability of DNA.

The existence of two gene spaces, the **gene-dense** (genome core) **and the gene-poor** (genome desert) **regions** also appears to be quite widespread. Indeed, such regions were not only found in vertebrates, but also in insects (*Drosophila, Anopheles*) and plants. **Two classes of genes**, characterized by different intron sizes and different expression levels, are associated with the two gene spaces, short introns and high expression leves being found in the gene-dense genome core, long introns and low expression levels in the gene-poor genome desert. In compositionally compartmentalized genomes, gene density was correlated with GC levels. At least in vertebrates and plants, gene-dense regions tend to be located near telomeres in chromosomes, and to be characterized by early replication, high recombination and strong transcription.

At first sight, the **gene distribution of** *Gramineae* is completely different from that of vertebrates, because genes are concentrated in regions of the genome (collectively called the "**gene space**") characterized by a very narrow compositional range. This situation is, however, due to a recent massive integration of nested transposons in the gene-dense regions of the genome. These transposons then dictated the overall compositional properties of the invaded regions.

Linear correlations were found between GC (and GC_3) levels of genes and GC levels of the corresponding introns and flanking sequences.

Finally, **the mitochondrial genome of yeast** taught us several lessons concerning the **reversible transconformations** undergone by stem-and-loop systems of *ori* sequences under the effect of temperature, the **formation of coding sequences** (the *var*1 gene) from noncoding sequences, the **contraction-expansion phenomena** of noncoding sequences, the existence of **genomes without genes**, as well as that of **dispensable genomes.** As discussed in Part 2, the abundant noncoding sequences of the large mitochondrial DNA from yeast cause genome instability, but this is compensated by a finer modulation of crucial functions, such as replication, transcription and recombination, compared to its compact counterpart in animals.

5. The prokaryotic genome

The concept of **genome phenotype** can be extended to prokaryotes. In this case, genome composition, although not showing the large internal heterogeneity of many eukaryotic genomes, covers a very wide spectrum. The original explanation for this finding was that genome composition shifts because of directional mutations due to biases in replication enzymes (Freese, 1962; Sueoka, 1962). It was argued, however, that, while these **mutational biases** are acceptable as the mechanism of the compositional changes, they are not necessarily its cause (Bernardi et al., 1988). The question is whether the resulting overall compositional changes are only determined by the vagaries of **random mutations**, the **mutator mutations**, in the genes encoding the protein sub-units of the replication machinery, or are under the control of **natural selection**. Two facts favour the second conclusion.

The first one is that changes in nucleotide composition correspond to changes in amino acids and, in turn, to changes in the hydrophobicity and in the secondary structure of proteins. If changes only depended upon mutational bias, then one should accept the untenable viewpoint that the corresponding functional changes are selectively irrelevant.

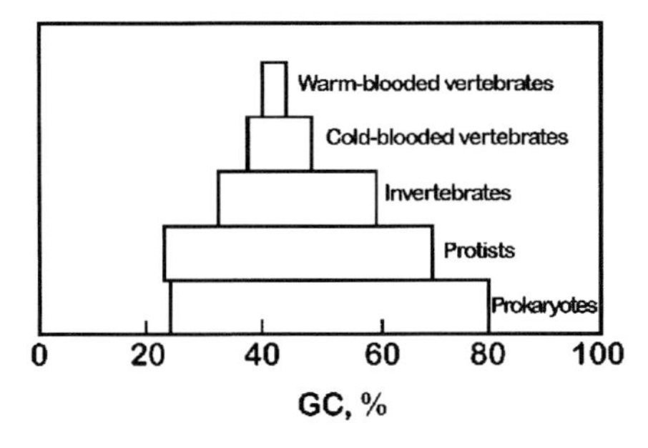

Figure **R17** (12.19). GC level ranges of DNA from warm- and cold blooded vertebrates, invertebrates, protists and prokaryotes.

The second fact is that GC changes cover a much wider range in the genomes of prokaryotes and unicellular eukaryotes than in those of invertebrates, cold-blooded vertebrates and warm-blooded vertebrates, these three groups showing a progressively narrower range. Since the potentially relevant mutator mutations in the replication machinery responsible for different mutational biases presumably are comparable in all classes of organisms, and since the spread of GC levels is so different, the only explanation for the different ranges of **Fig. R17** is that there is a different selection on the mutations in the replication machinery of these different classes of organisms, some changes being quickly eliminated by negative selection.

What we propose is that **genome composition** in prokaryotes is under the control of natural selection which optimizes the adaptation of genome composition to the environmental conditions (temperature being only one among many factors). Because of the compactness of the prokaryotic genomes, which are essentially made up of coding and regulatory sequences, natural selection (mainly negative selection) operates on single nucleotide changes, whereas in vertebrate genomes regional changes also play an important role.

Since the different base compositions of prokaryotic genomes were thought to be due to biases in replication enzymes, it was supposed that the same process should also explain the formation and maintenance of isochores in vertebrate genomes. In contrast, the results presented in this book in favour of natural selection as the factor responsible for the isochores of vertebrates and the data of **Fig. R17** lead to explaining the different compositions of prokaryotic genomes according to the natural selection scheme proposed here for the isochores, thus reversing a trend that has dominated the literature for the past 40 years.

Conclusions

According to **the classical neo-Darwinian selection theory**, natural selection, "*the preservation of favourable variations and the rejection of injurious variations*" (Darwin, 1859), or the differential multiplication of mutant types, essentially occurs through the elimination of the progeny of individual organisms carrying deleterious mutations (**negative, or purifying, selection**), and only very rarely *via* the preferential propagation of organisms with advantageous mutations (**positive selection**).

Twenty years ago, Motoo Kimura formulated the "**mutation-random drift theory**" according to which "*the main cause of evolutionary change at the molecular level – change in the genetic material itself – is random fixation of selectively neutral or nearly neutral mutants*". However, "*The neutral theory is not antagonistic to the cherished view that evolution of form and function is guided by Darwinian selection, but brings out another facet of the evolutionary process by emphasizing the much greater role of mutation pressure and random drift at the molecular level*", in which "*increases and decreases in the mutant frequencies are due mainly to chance*" (Kimura, 1983).

So far, the classical selection and the neutral theories have been generally visualized as two opposite viewpoints. However, these theories may be seen as complementary. Indeed, one could consider that the selection theory applies, to a large extent, to coding and regulatory sequences, but not to the vast majority of the noncoding sequences that represent 97-98% of a typical mammalian genome. *Vice versa*, the neutral theory may apply to the vast majority of noncoding sequences, but only partially to coding and regulatory sequences. This picture is, however, only valid **on a first approximation**. Indeed, both theories only took into consideration single-nucleotide changes. Unfortunately, whether the result of random fixation or selection, single nucleotide changes, cannot account for genome properties such as the genome compartmentalization into isochores, the genome phenotype and the genomic code.

Obviously, a "**paradigm shift**" was needed to reconcile the indisputable abundance of neutral changes with our results, which inescapably implied natural selection (a conclusion strengthened by sixteen years of critical scrutiny; see Part 12). The first step toward this paradigm shift was made by postulating that changes in base composition could lead to **regional phase transitions** in DNA/chromatin structure that were deleterious for the expression of the genes located in or near the regions under consideration. This **isochore selection** proposal (Bernardi et al., 1988) eliminated the need for selection on individual nucleotides in the long non-coding regions of the genome and provided a first model for our previous suggestion (Bernardi and Bernardi, 1986a) that natural selection was the basic explanation for the formation and maintenance of GC-rich isochores. A further step was made here by proposing an improved model in which the accumulation of nucleotide changes (known to be characterized by a strong AT bias) leads to **critical changes** that, although being single-nucleotide changes, cause the average GC of the region to trespass the tolerated compositional range. The consequence is a **transition in DNA/chromatin structure**, that is deleterious for the expression of the embedded or flanking genes and affects the fitness of the individual. These compositionally altered regions are then removed in the progeny by **negative (purifying) selection**, which is also operative, at the individual nucleotide level, in coding and regulatory sequences. This model accounts for the **conservative mode of evolution** of vertebrate genomes. In fact, only negative selection is strictly required also in the **shifting mode of evolution**. In this case, the gene-dense regions of the ancestors of warm-blooded vertebrates increased their GC level in response to the upward shift of the compositional optimum associated with the emergence of homeothermy and the consequent need for stabilizing thermodynamically the DNA present in the open chromatin of the gene-dense isochores.

Our DNA/chromatin transition model, which we propose to call the **neo-selectionist model**, has the merit of reconciling the prevalence in numbers of neutral, nearly neutral and slightly deleterious mutations with the pervasive control exerted at the regional level by natural selection (essentially negative selection). If one looks at a broader perspective, this "**reconciliation**" should not hide, however, the fact that the neutral theory and the neo-selectionist model proposed here lead to **two very different general views of genome evolution**. This contrast can be best visualized by focusing on three points.

(i) Kimura realized that his neutral theory raised a dilemma, "*why natural selection is so prevalent at the phenotype level and yet random fixation of selectively neutral and nearly neutral alleles prevails at the molecular level*". Although Kimura showed a remarkable insight by stating "*The answer to this question, I think, comes from the fact that the most common type of natural selection at the phenotype level is stabilizing selection*", he concluded that "*laws governing molecular evolution are clearly different from those governing phenotypic evolution*". A beginning to the solution of this serious dichotomy was resolved by our proposal of 1986 that natural selection controls not only the "**classical phenotype**" of form and function (or, in molecular terms, of the proteins and of their expression), as generally accepted, but also the "**genome phenotype**", the compositional properties of the genome and all their functional implications. In other words, our proposal is that **natural selection extends to the molecular level, to the genome itself, and to its compositional patterns** (and not only to the single nucleotides of coding or regulatory sequences). This natural selection operating on the genome phenotype could be called **compositional selection** (which is regional in vertebrates, but typically genome-wide in prokaryotes), and would act in addition to the **classical selection**, namely natural selection operating on the classical phenotype, and to the neutral changes.

(ii) When the compositional selection has not yet acted, obviously only neutral and nearly neutral changes will be detected. This is to say that the neutral and nearly neutral theories account for features of molecular evolution which take place within certain compositional thresholds, and represent one mode, the **neutral mode**, of genome evolution. This neutral mode is, however, dominated by the pervasive control of natural selection in the **selection mode**. As already mentioned, the overall model could be called **the neo-selectionist model** of evolution.

(iii) Kimura's view that "*increases and decreases in the mutant frequencies are due mainly to chance*" led him, logically, to the idea of the "**survival of the luckiest**". Obviously, the pervasive control of natural selection on the "genome phenotype" not only solves Kimura's dilemma, but leads us back to the "**survival of the fittest**". Although these expressions have been criticized as tautologous, they summarize very different assessments of the role of chance in evolution. It is then difficult not to conclude that, **even at the molecular**

level, "*it may be said that natural selection is daily and hourly scrutinizing, throughout the world, every variation, even the slightest; rejecting that which is bad, preserving and adding up all that is good; silently and insensibly working, whenever and wherever opportunity offers, at the improvement of each organic being in relation to its organic and inorganic conditions of life*" (Darwin, 1859).

Abbreviations and acronyms

- **A**, CsCl band asymmetry, $<\rho>-\rho_0$
- **BAC(s),** Bacterial Artificial Chromosome(s)
- **BAMD**, bis(acetomercurimethyl)dioxane
- **BLV**, Bovine Leukaemia Virus
- **bp**, base pairs(s)
- **cM,** centimorgan(s)
- **C_0t**, initial DNA concentration x reannealing time
- ***cox 1,*** citochrome oxidase sub-unit 1
- ***c* value,** genome size
- ***cyt b,*** apocytochrome b
- **DAPI,** 4,6 diamino-2-phenylindole
- **DNase,** deoxyribonuclease
- **ESAG(s)**, Expression Site Associated Gene(s)
- **EST(s)**, Expressed Sequence Tag(s)
- **FITC,** fluorescein isothiocyanate
- **Gb**, gigabase(s), or billions of base pairs
- **GC**, molar ratio of guanosine and cytidine in DNA
- **GC_1, GC_2, GC_3,** GC levels in first, second and third codon position
- **GTG bands,** G bands by trypsin using Giemsa
- **H**, intermolecular compositional heterogeneity
- **HBV,** Hepatitis B Virus
- **HIV-1,** Human Immunodeficiency Virus 1
- **HTLV-1,** Human T-cell Leukaemia Virus 1
- **IRES,** Internal Ribosome Entry Site
- **kb**, kilobase(s), or thousand(s) of base pairs
- **LINE(s),** Long Interspersed Repeat(s)
- **LTR(s)**, Long Terminal Repeat(s)
- **Mb**, Megabase(s), or millions of base pairs
- **MHC,** Major Histocompatibility locus
- **MMTV**, Mouse Mammary Tumour Virus
- **MoMLV**, Moloney Murine Leukemia Virus
- **MuLV**, Murine Leukemia Virus
- **Mya**, Million(s) years ago
- **Myr(s)**, Million year(s)
- **N(s)**, undetermined nucleotide(s)
- **ORF**, Open Reading Frames
- ***ori***, origin of DNA replication
- **PCR**, Polynucleotide Chain Reaction
- **pg**, picogram(s)
- **PKR**, Particulate Protein Kinase
- **rf**, ligand/nucleotide molar ratio

- **RIDGE(s)**, Region(s) of Increased Gene Expression
- **RSV**, Rous Sarcoma Virus
- **RFLP(s)**, Restriction Fragment Length Polymorphism(s)
- **SCC**, Sequence Compositional Complexity
- **sd**, standard deviation
- **SINE(s)**, Short Interspersed Repeat(s)
- **SNP(s)**, Single Nucleotide Polymorphism(s)
- **SV40**, Simian virus 40
- ***T*b**, body temperature
- ***T*max**, maximal *T*b sustained for extended periods of time
- ***T*opt**, optimal growth temperature
- **T-DNA**, Transferred DNA (from *Agrobacterium* plasmids)
- **TE**, Transposable Elements
- **TRITC**, tetra methyl rhodamine *(3* isothiocyanate)
- **UTR(s)**, Untraslated Sequence(s)
- **YAC(s)**, Yeast Artificial Chromosome(s)
- **5mC**, 5-methylcytosine
- $<\rho>$, mean buoyant density
- ρ_0, modal buoyant density

References

References are given in alphabetical order for papers with one or two authors, in chronological order for the papers with more than two authors. Only first authors are quoted for sequence papers.

Abe K., Wei J.F., Wei F.S., Hsu Y.C. Uehara H., Artzt K., Bennett D. (1988). Searching for coding sequences in the mammalian genome: the H-2K region of the mouse MHC is replete with genes expressed in embryo. EMBO J. 7: 3441-3449.

Abe T., Kanaya S., Kinouchi M., Ichiba Y., Kozuki T., Ikemura T. (2003). Informatics for unveiling hidden genome signatures. Genome Res. 4: 693-702.

Adams M. et al. (2000). The genome sequence of *Drosophila melanogaster*. Science 287: 2185-2195.

Ahn S., Anderson J.A., Sorrells M.E., Tanksley S.D. (1993). Homoeologous relationships of rice, wheat and maize chromosomes. Mol. Gen. Genet. 241: 483-590.

Aïssani B. and Bernardi G. (1991a). CpG islands: features and distribution in the genome of vertebrates. Gene 106: 173-183.

Aïssani B. and Bernardi G. (1991b). CpG islands, genes and isochores in the genome of vertebrates. Gene 106: 185-195.

Aïssani B., D'Onofrio G., Mouchiroud D., Gardiner K., Gautier C., Bernardi G. (1991). The compositional properties of human genes. J. Mol. Evol. 32: 497-503.

Akashi H. (1994). Synonymous codon usage in *Drosophila melanogaster*: natural selection and translational accuracy. Genetics 136: 927-935.

Akashi H. (1997). Codon bias evolution in *Drosophila*. Population genetics of mutation-selection drift. Gene 205: 269-278.

Akashi H. and Eyre-Walker A. (1998). Translational selection and molecular evolution. Curr. Opin. Genet. Dev. 8: 688-693.

Alvarez-Valin F., Jabbari K., Bernardi G. (1998). Synonymous and nonsynonymous substitutions in mammalian genes: intragenic correlations. J. Mol. Evol. 46: 37-44.

Alvarez-Valin F., Tort J.F., Bernardi G. (2000). Non-random spatial distribution of synonymous substitutions in the *Leishmania* GP63 gene. Genetics 155: 1683-1692.

Alvarez-Valin F., Lamolle G., Bernardi G. (2002). Isochores, GC_3 and mutation biases in the human genome. Gene 300: 161-168.

Alvarez-Valin F., Clay O., Cruveiller S., Bernardi G. (2004). Inaccurate reconstruction of ancestral GC levels gives the illusion of "vanishing isochores". Mol. Phylogenet. Evol. (submitted)

Amann J., Valentine M., Kidd V. J., Lahti J.M. (1996). Localization of chi 1-related helicase genes to human chromosome regions 12p11 and 12p13: similarity between parts of these genes and conserved human telomeric-associated DNA. Genomics 32: 260-265.

Ambros, P.F. and Sumner, A.Y. (1987). Metaphase bands of human chromosomes, and distinctive properties of telomeric regions. Cytogenet. Cell Genet. 44: 223-228.

Ananiev E.V., Phillips R.L., Rines H.W. (1998). Chromosome-specific molecular organization of maize (*Zea mays* L.) centromeric regions. Proc. Natl. Acad. Sci. USA. 95: 13073-13078.

Anderson W.F. (1998). Human gene therapy. Nature 392: 25-30.

Andersson S.G., Zomorodipour A., Andersson J.O., Sicheritz-Ponten T., Alsmark U.C., Podowski R.M., Naslund A.K., Eriksson A.S., Winkler H.H., Kurland C.G. (1998). The genome sequence of *Rickettsia prowazekii* and the origin of mitochondria. Nature 396:133-140.

Andreozzi L., Federico C., Motta S., Saccone S., Sazanova A.L., Sazanov A.A., Smirnov A.F., Galkina S.A., Lukina N.A., Rodionov A.V., Carels N., Bernardi G. (2001). Compositional mapping of chicken chromosomes and identification of the gene-richest regions. Chromosome Res. 9: 521-532.

Antequera F. and Bird A.P. (1988). Unmethylated CpG islands associated with genes in higher plant DNA. EMBO J. 7: 2295-2299.

Aota S. and Ikemura, T. (1986). Diversity in G+C content at the third position of codons in vertebrate genes and its cause. Nucleic Acids Res. 14: 6345-6355.

Aparicio S. et al. (2002). Whole-genome shotgun assembly and analysis of the genome of *Fugu rubripes*. Science 297: 1301-1310.

Arabidopsis Genome Initiative (2000). Analysis of the genome sequence of the flowering plant *Arabidopsis thaliana*. Nature 408: 796-815.

Archibald J.D., Hedges S.B., Kumar S., Rich T.H., Vickers-Rich P., Flannery T. F., Foote M., Hunter J.P., Janis C.M., Sepkoski J.J. (1999). Divergence times of eutherian mammals. Science 285: 2031.

Arcot S.S., Adamson A.W., Risch G.W., LaFleur J., Robichaux M.B., Lamerdin J.E., Carrano A.V., Batzer M.A. (1998). High-resolution cartography of recently integrated human chromosome 19-specific Alu fossils. J. Mol. Biol. 281: 843-856.

Argos P., Rossmann M.G., Grau U.M., Zuber A., Franck G., Tratschin J.D. (1979). Thermal stability and protein structure. Biochemistry 18: 5698-5703.

Arhondakis S., Auletta F., Torelli G., D'Onofrio G. (2004). Base composition and expression level of human genes. Gene (in press).

Athas G.B., Starkey C.R., Levy L.S., (1994). Retroviral determinants of leukemogenesis. Crit. Rev. Oncol. 5: 169-199.

Bachmann K. (1972). Genome size in mammals. Chromosoma. 37: 85-93.

Bailey W.J., Kim J., Wagner G.P., Ruddle F.H. (1997). Phylogenetic reconstruction of vertebrate *Hox* cluster duplications. Mol. Biol. Evol. 14: 843-853.

Bailly S., Guillemin C., Labrousse M. (1973). Comparison du nombre et de la position des zones spécifiques révélées sur les chromosomes mitotiques de l'Amphibien Urodèle *Pleurodeles waltlii Michah* par les techniques de coloration au colorant de Giemsa et à la moutarde de quinacrine. C.R. Acad. Sci. Paris 276: 1867-1869

Baker D. and Sali A. (2001). Protein structure prediction and structural genomics. Science 294: 93-96.

Balajee A.S. and Bohr V.A. (2000). Genomic heterogeneity of nucleotide excision repair. Gene 250: 15-30.

Baldacci G. and Bernardi G. (1982). Replication origins are associated with transcription initiation sequences in the mitochondrial genome of yeast. EMBO J. 1: 987-994.

Baldacci G., Cherif-Zahar B., Bernardi G. (1984). The initiation of DNA replication in the mitochondrial genome of yeast. EMBO J. 3: 2115-2120.

Barakat A., Carels N., Bernardi G. (1997). The distribution of genes in the genomes of *Gramineae.* Proc. Natl. Acad. Sci. USA 94: 6857-6861.

Barakat A., Matassi G., Bernardi G. (1998). Distribution of genes in the genome of *Arabidopsis thaliana* and its implications for the genome organization of plants. Proc. Natl. Acad. Sci. USA 95: 10044-10049.

Barakat A., Tran Han D., Benslimane A., Rode A., Bernardi G. (1999). The gene distribution in the genomes of pea, tomato and date-palm. FEBS Letters 464: 60-62.

Barakat A., Gallois P., Raynal M., Mestre-Ortega D., Sallaud C., Guiderdoni E., Delseny M., Bernardi G. (2000). The distribution of T-DNA in the genomes of transgenic *Arabidopsis* and rice. FEBS Letters 471: 161-164.

Bashirullah A., Cooperstock R.L., Lipshitz H.D. (1998). RNA localization in development. Annu. Rev. Biochem. 67: 335-394.

Batzer M. A. and Deininger P.L. (2002). Alu repeats and human genomic diversity. Nature Reviews Genet. 3: 370-379.

Batzer M.A., Deininger P.L., Hellmann-Blumberg U., Jurka J., Labuda D., Rubin C.M., Schmid C.W., Zietkiewicz E., Zuckerkandl E. (1996). Standardized nomenclature for Alu repeats. J. Mol. Evol. 42: 3-6.

Bailey J.A., Yavor A.M., Massa H.F., Trask B.J., Eichler E.E. (2001). Segmental duplications: organization and impact within the current human genome project assembly. Genome Res. 11: 1005-1017.

Beilharz M.W., Cobon G.S., Nagley P. (1982). Physiological alteration of the pattern of transcription of the *oli2* region of yeast mitochondrial DNA. FEBS Letters 147: 235-238.

Bell M.V., Cowper A.E., Lefranc M.P., Bell J.I., Screaton G.R. (1998). Influence of intron length on alternative splicing of CD44. Mol. Cell. Biol. 18: 5930-5941.

Belle E.M.S., Smith N., Eyre-Walker A. (2002). Analysis of the phylogenetic distribution of isochores in vertebrates and a test of the thermal stability hypothesis. J. Mol. Evol. 55: 356-363.

Belozerski A.N. and Spirin A.S. (1958). A correlation between the compositions of deoxyribonucleic and ribonucleic acids. Nature 182: 111-112.

Bennetzen J.L. and Hall B.D. (1982). Codon selection in yeast. J. Biol. Chem. 257: 3026-3031.

Bentvelzen P., Daams J.H., Hageman P., Calafat J. (1970). Genetic transmission of viruses that incite mammary tumor in mice. Proc. Natl. Acad. Sci. USA 67: 377-384.

Bernaola-Galvan P., Roman-Roldan R., Oliver J.L. (1996). Compositional segmentation and long-range fractal correlations in DNA sequences. Phys. Rev. E. Stat. Phys. Plasmas. Fluids Relat. Interdiscip. Topics 53: 5181-5189.

Bernaola-Galvan P., Carena P., Roman-Roldan R., Oliver J. (2002). Study of statistical correlations in DNA sequences. Gene 300: 105-115.

Bernaola-Galvan P., Oliver J., Carpena P., Clay O., Bernardi G. (2004). Quantifying intragenomic heterogeneity in prokaryotic genomes. Gene (submitted).

Bernardi G. (1962). Chromatography of denatured deoxyribonucleic acid on calcium phosphate. Biochem. J. 83: 32-33.

Bernardi G. (1965a). Dimeric structure and allosteric properties of spleen acid deoxyribonuclease J. Mol. Biol. 13: 603-605.

Bernardi G. (1965b). Chromatography of nucleic acids on hydroxyapatite. Nature (London) 206: 779-783.

Bernardi G. (1971). Spleen acid deoxyribonuclease. In *The Enzymes,* 3rd edition. (P.D. Boyer, ed.) vol. 4, pp. 271-287, Academic Press, New York, NY, USA.

Bernardi G. (1979a). The *petite* mutation in yeast. Trends in Biochem. Sci. 4: 197-201

Bernardi G. (1979b). Organization and evolution of the eukaryotic genome. In *Recombinant DNA and genetic experimentation* (J. Morgan and W.J. Whelan, eds.) pp. 15-20, Pergamon, New York, NY, USA.

Bernardi G. (1982a). The evolutionary origin and the biological role of non-coding sequences in the mitochondrial genome of yeast. In *Mitochondrial Genes* (G. Attardi, P. Borst, P.P. Slonimski, eds.) pp. 269-278, Cold Spring Harbor Laboratory, Cold Spring Harbor, NY, USA.

Bernardi G. (1982b). The origins of replication of the mitochondrial genome of yeast. Trends in Biochem. Sci. 7: 404-408.

Bernardi G. (1983). Genome instability and the selfish DNA issue. Folia Biol. 29: 82-92.

Bernardi G. (1984). Sequence organization of the vertebrate genome. In *Genetic Manipulation : Impact on Man and Society* (W. Arber, K. Illmensee, J. Peacock, P. Starlinger, eds.) pp. 171-178, Cambridge University Press, Cambridge, U.K.

Bernardi G. (1985). The organization of the vertebrate genome and the problem of the CpG shortage. In *Biochemistry and biology of DNA methylation* (G.L. Cantoni and A. Razin, eds.) pp. 3-10, Alan Liss, NY, USA.

Bernardi G. (1989). The isochore organization of the human genome. Ann. Rev. Genet. 23: 637-661.

Bernardi G. (1993a). The vertebrate genome: isochores and evolution. Mol. Biol. Evol. 10: 186-204.

Bernardi G. (1993b). Genome organization and species formation in vertebrates. J. Mol. Evol. 37: 331-337.

Bernardi G. (1995). The human genome: organization and evolutionary history. Annu. Rev. Genet. 29: 445-476.

Bernardi G., (2000a). Isochores and the evolutionary genomics of vertebrates. Gene 241: 3-17.

Bernardi G. (2000b). The compositional evolution of vertebrate genomes. Gene 259: 31-43.

Bernardi G. (2001). Misunderstandings about isochores. I. Gene 276: 3-13.

Bernardi G. and Bach M.L. (1968). Studies on acid deoxyribonuclease. VII. Inactivation of *Haemophilus influenzae* transforming DNA by spleen acid deoxyribonuclease. J. Mol. Biol. 37: 87-98.

Bernardi A. and Bernardi G. (1968). Studies on acid hydrolases. IV. Isolation and properties of spleen exonuclease. Biochim. Biophys. Acta 155: 360-370.

Bernardi A. and Bernardi G. (1976). Cloning of all EcoRI fragments of phage lambda in *E. coli.* Nature 264: 89-90.

Bernardi G. and Bernardi G. (1980). Repeated sequences in the mitochondrial genome of yeast. FEBS Letters 115: 159-162.

Bernardi G. and Bernardi G. (1985). Codon usage and genome composition. J. Mol. Evol. 22: 363-365.
Bernardi G. and Bernardi G. (1986a). Compositional constraints and genome evolution. J. Mol. Evol. 24: 1-11.
Bernardi G. and Bernardi G. (1986b). The human genome and its evolutionary context. In Cold Spring Harbor Symp. Quant. Biol. "Molecular Biology of *Homo sapiens*" vol. 51, pp. 479-487, Cold Spring Harbor, New York, NY, USA.
Bernardi G. and Bernardi, G. (1990a). Compositional patterns in the nuclear genomes of cold-blooded vertebrates. J. Mol. Evol. 31: 265-281.
Bernardi G. and Bernardi G. (1990b). Compositional transitions in the nuclear genomes of cold-blooded vertebrates. J. Mol. Evol. 31: 282-293.
Bernardi G. and Bernardi G. (1991). Compositional properties of nuclear genes from cold-blooded vertebrates. J. Mol. Evol. 33: 57-67.
Bernardi G. and Cook W.H. (1960a). An electrophoretic and ultracentrifugal study on the proteins of the high-density fraction of egg yolk. Biochim. Biophys. Acta 44: 86-96.
Bernardi G. and Cook W.H. (1960b). Separation and characterization of the two high-density lipoproteins of egg yolk, a- and β-lipovitellin. Biochim. Biophys. Acta 44: 96-105.
Bernardi G. and Cook W.H. (1960c). Molecular weight and behavior of lipovitellin in urea solution. Biochim. Biophys. Acta 44: 105-109.
Bernardi G. and Timasheff S.N. (1970). Optical rotatory dispersion and circular dichroism properties of yeast mitochondrial DNA's. J. Mol. Biol. 48: 43-52.
Bernardi G. and Saccone S. (2003). Gene distribution on human chromosomes. In *Encyclopedia of the Human Genome*, 2: 617-620. Nature Publishing Group, London, UK.
Bernardi G., Champagne M., Sadron C. (1960). Enzymatic degradation of deoxyribonucleic acid into sub-units. Nature (London) 188: 228-229.
Bernardi G., Bernardi A., Chersi A., (1966). Studies on acid hydrolases. I. A procedure for the preparation of acid deoxyribonuclease and other acid hydrolase. Biochim. Biophys. Acta 129: 1-11.
Bernardi G., Carnevali F., Nicolaieff A., Piperno G., Tecce G. (1968). Separation and characterization of a satellite DNA from a yeast cytoplasmic *petite* mutant. J. Mol. Biol. 37: 493-505.
Bernardi G., Faurès M., Piperno G., Slonimski P.P. (1970). Mitochondrial DNA's from respiratory-sufficient and cytoplasmic respiratory-deficient mutant yeast. J. Mol. Biol. 48: 23-42.
Bernardi G., Federico C., Saccone S. (2000). The human prometaphase chromosomal bands: compositional features and gene distribution. In *Chromosomes today* Vol. 13 (Olmo E. and Redi C.A., eds.) Birkhäuser Verlag AG, Germany.
Bernardi G., Giro G., Gaillard C. (1972a) Chromatography of polypeptides and proteins on hydroxyapatite columns; further investigations. Biochim. Biophys. Acta 278: 409-420.
Bernardi G., Piperno G., Fonty G. (1972b). The mitochondrial genome of wild-type yeast cells. I. Preparation and heterogeneity of mitochondrial DNA. J. Mol. Biol. 65: 173-189.
Bernardi G., Ehrlich S.D., Thiery J.P. (1973). The specificity of deoxyribonucleases and their use in nucleotide sequence studies. Nature New Biology 246: 36-40.

Bernardi G., Prunell A., Kopecka H. (1975). An analysis of the mitochondrial genome of yeast with restriction enzymes. In *Molecular Biology of Nucleocytoplasmic Relationships* (S. Puiseux-Dao, ed.) pp. 85-90, Elsevier, Amsterdam, The Netherlands.

Bernardi G., Prunell A., Fonty G., Kopecka H., Strauss F. (1976). The mitochondrial genome of yeast : organization, evolution, and the petite mutation. In *Proceedings of the 10th International Bari Conference on the Genetic Function of Mitochondrial DNA* (C. Saccone and A.M. Kroon, eds.) pp. 185-198, Elsevier North-Holland, Amsterdam, The Netherlands.

Bernardi G., Baldacci G., Bernardi G., Faugeron-Fonty G., Gaillard C., Goursot R., Huyard A., Mangin M., Marotta R., de Zamaroczy M. (1980a). The petite mutation: excision sequences, replication origins and suppressivity. In *The organization and expression of the mitochondrial genome* (A.M. Kroon and C. Saccone, eds.) pp. 21-31, Elsevier North-Holland, Amsterdam, The Netherlands.

Bernardi G., Baldacci G., Culard F., Faugeron-Fonty G., Gaillard C., Goursot R., Strauss F., de Zamaroczy M. (1980b). Excision and replication of mitochondrial genomes from spontaneous *petite* mutants of yeast. In *FEBS Symposium on DNA* (J. Sponar and S. Zadrazil, eds.) pp. 77-84, Pergamon Press, New York, NY, USA.

Bernardi G., Olofsson B., Filipski J., Zerial M., Salinas J., Cuny G., Meunier-Rotival M., Rodier F. (1985a). The mosaic genome of warm-blooded vertebrates. In *Proceedings of the FEBS Congress*, Moscow, July 1984 (Yu. A. Ovchinnikov, ed.) VNU Science Press BV, part B, pp. 69-77.

Bernardi G., Olofsson B., Filipski J., Zerial M., Salinas J., Cuny G., Meunier-Rotival M., Rodier F. (1985b). The mosaic genome of warm-blooded vertebrates. Science 228: 953-958.

Bernardi G., Mouchiroud D., Gautier C., Bernardi G. (1988). Compositional patterns in vertebrate genomes: conservation and change in evolution. J. Mol. Evol. 28: 7-18.

Bernardi G., Mouchiroud D., Gautier. C. (1993). Silent substitutions in mammalian genomes and their evolutionary implications. J. Mol. Evol. 37: 583-589.

Bernardi G., Hughes S., Mouchiroud D. (1997). The major compositional transitions in the vertebrate genome: a review. J. Mol. Evol. 44: S41-S51.

Bettecken T., Aïssani B., Mueller C.R., Bernardi G. (1992). Compositional mapping of the human dystrophin gene. Gene 122: 329-335.

Biémont M.C., Laurent C., Couturier J., Dutrillaux B. (1978). Chronology of the replication of sex chromosome bands in lymphocytes of normal subjects and patients. Ann. Genet. 21: 133-141.

Bird A.P. (1980). DNA methylation and the frequency of CpG in animal DNA. Nucleic Acids Res. 8: 1499-1504.

Bird A.P. (1987). CpG islands as gene markers in the vertebrate nucleus. Trends Genet. 3: 342-347.

Bird A.P., Taggart M., Frommer M., Miller O.J., Macleod D. (1985). A fraction of the mouse genome that is derived from islands of nonmethylated, CpG-rich DNA. Cell 40: 91-99.

Birdsell J.A. (2002). Integrating genomics, bioinformatics, and classical genetics to study the effects of recombination on genome evolution. Mol. Biol. Evol. 19: 1181-1197.

Bishop J.M., (1980). Enemies within: the genesis of retrovirus oncogenes. Cell 23: 5-6.

Blanc H. and Dujon B. (1980). Replicator regions of the yeast mitochondrial DNA responsible for suppressiveness. Proc. Natl. Acad. Sci. USA 77: 3942-3946.

Boeke J.D. (1997). LINEs and Alus–the polyA connection. Nature Genet. 6: 6-7.

Bohr V.A., Smith C.A., Okumoto D.S., Hanawalt P.C. (1985) DNA repair in an active gene: removal of pyrimidine dimers from the DHFR gene of CHO cells is much more efficient than in the genome overall. Cell 40: 359-369.

Boivin A. Vendrely R., Vendrely C. (1948). L'acide desoxyribonucléique du noyau cellulaire, dépositaire des caractères héreditaires; arguments d'ordre analytique. C.R. Acad. Sci. 226: 1061-1063.

Bond H.E., Flamm W.G., Burr H.E., Bond S.B. (1967). Mouse satellite DNA. Further studies on its biological and physical characteristics and its intracellular localization. J. Mol. Biol. 27: 289-302.

Botstein D., White R.L., Skolnick M., Davis R.W. (1980). Construction of a genetic linkage in man using restriction fragment length polymorphism. Am. J. Hum. Genet. 32, 314-331.

Bourbaki N. (1969). Intégration. Numéro 35 in Fascicules/Eléments de Mathématique. Hermann, Paris.

Bourgaux P. and Bourgaux-Ramoisy D. (1967). Chromatographic separation of the various forms of polyoma virus DNA J. Gen. Virol. 1: 323-332.

Boyle S., Gilchrist S., Bridger J.M., Mahy N.L., Ellis J.A., Bickmore W.A. (2001). The spatial organization of human chromosomes within the nuclei of normal and emerin-mutant cells. Hum. Mol. Genet. 10: 211-219.

Brack A. and Orgel L.E. (1975). Beta structures of alternating polypeptides and their possible prebiotic significance. Nature 256: 383-387.

Breindl M., Harbers K., Jaenisch R., (1984). Retrovirus induced lethal mutation in collagen 1 gene of mice is associated with an altered chromatin structure. Cell 38: 9-16.

Brenner S., Elgar G., Sandford R., Macrae A., Venkatesh B., et al. (1993). Characterization of the pufferfish (Fugu) genome as a compact model vertebrate genome. Nature 366: 265-268.

Britten R.J. and Kohne D.E. (1968). Repeated sequences in DNA. Hundreds of thousands of copies of DNA sequences have been incorporated into the genomes of higher organisms. Science 161: 529-540.

Britten R.J. and Davidson E.H. (1969). Gene regulation for higher cells: a theory. Science 165: 349-357.

Britten R.J. and Smith J. (1969). A bovine genome. In *Carnegie Inst. Wash. Year Book* 68: 378-386.

Bromhan L. and Penny D. (2003). The modern molecular clock. Nature Rev. Genet. 4: 216-224.

Brookfield J.F.Y. (2001) Selection on Alu sequences? Current Biology 11: R900-R901.

Brosius J. (1999) Genomes were forged by massive bombardments with retroelements and retrosequences. Genetica 107: 209-238.

Brown P.O., (1997). Integration. In *Retroviruses* (J.M. Coffin, S.H. Hughes, H.E. Varmus, eds.) pp. 161-205, Cold Spring Harbor Laboratory Press, New York, NY, USA.

Bucciarelli G., Bernardi G. and Bernardi G. (2002). An ultracentrifugation analysis of 200 fish genomes. Gene 295: 153-162.

Bulmer M. (1991). The selection-mutation-drift theory of synonymous codon usage. Genetics 129: 897-907.

Bulmer M., Wolfe K.H., Sharp P.M. (1991). Synonymous nucleotide substitution rates in mammalian genes: implications for the molecular clock and the relationship of mammalian orders. Proc. Natl. Acad. Sci. USA 88: 5974-5978.

Bünemann M. and Dattagupta N. (1973). On the binding and specificity of 3,6-bis-(acetatomercurimethyl)-dioxane to DNAs of different base composition. Biochim.Biophys. Acta 331: 341-348.

Bush G.L., Case S.M., Wilson A.C., Patton J.L. (1977). Rapid speciation and chromosomal evolution in mammals. Proc. Natl. Acad. Sci. USA 74: 3942-3946.

Bushman F. (1995). Targeting retroviral integration. Science 276: 1443-1444.

Cacciò S., Perani P., Saccone S., Kadi F., Bernardi G. (1994). Single-copy sequence homology among the GC-richest isochores of the genomes from warm-blooded vertebrates. J. Mol. Evol. 39: 331-339.

Cacciò S., Zoubak S., D'Onofrio G., Bernardi G. (1995). Nonrandom frequency patterns of synonymous substitutions in homologous mammalian genes. J. Mol. Evol. 40: 280-292.

Cacciò S., Jabbari K., Matassi G., Guermonprez F., Desgrès J., Bernardi G. (1997). Methylation patterns in the isochores of vertebrate genomes. Gene 205: 119-124

Caenorhabditis elegans Sequencing Consortium (1998). Genome sequence of the nematode *C. elegans*: a platform for investigating biology. Science 282: 2012-2018.

Campbell A. (1969). *Episomes*. Harper & Row, New York, NY, USA.

Capel J., Montero L.M., Martinez-Zapater J.M., Salinas J. (1993). Non-random distribution of transposable elements in the nuclear genome of plants. Nucleic Acids Res. 21: 2369-2373.

Carels N. and Bernardi G. (2000). Two classes of genes in plants. Genetics 154: 1819-1825.

Carels N., Barakat A., Bernardi, G. (1995). The gene distribution of the maize genome. Proc. Natl. Acad. Sci. USA 92: 11057-11060.

Carels N., Hatey P., Jabbari K., Bernardi G. (1998). Compositional properties of homologous coding sequences from plants. J. Mol. Evol. 46: 45-53.

Caron H., van Schaik B., van der Mee M., Baas F., Riggins G., van Sluis P., Hermus M.C., van Asperen R., Boon K., Voute P.A., Heisterkamp S., van Kampen A., Versteeg R. (2001). The human transcriptome map: clustering of highly expressed genes in chromosomal domains. Science 291: 1289-1292.

Carrara M. and Bernardi G. (1968). Separation of nucleosides on polyacrylamide gel columns. Biochim. Biophys. Acta, 155: 1-7.

Carroll R.L. (1987). *Vertebrate paleontology and evolution.* (W.H. Freeman, ed.) New York, NY, USA

Carteau S., Hoffmann C., Bushman F. (1998). Chromosome structure and human immunodeficiency virus type 1 cDNA integration: centromeric alphoid repeats are a disfavored target. J. Virol. 72: 4005-4014.

Carulli J.P., Krane D.E., Hartl D.L, Ochman H. (1993). Compositional heterogeneity and patterns of molecular evolution in the *Drosophila* genome. Genetics 134: 837-845.

Caspers G.J., Reinders G.J., Leunissen J.A.M., Wattel J., de Jong W.W. (1996). Protein sequences indicate that turtles branched off from the amniote tree after mammals. J. Mol. Evol. 42: 580-586.

Caspersson T., Castleman K.R., Lomakka G., Modest E.J., Moller A., Nathan R., Wall R.J., Zech L. (1971). Automatic karyotyping of quinacrine mustard stained human chromosomes. Exp. Cell Res. 67: 233-235.

Cavalier-Smith T. (1985). *The evolution of genome size.* Wiley, Chichester, U.K.

Charlesworth B. (1994). Genetic recombination. Patterns in the genome. Curr. Biol. 4: 182-184.

Chersi A., Bernardi A., Bernardi G. (1966). Studies on acid hydrolases. II. Isolation and properties of spleen acid phosphomonoesterase. Biochim. Biophys. Acta 129: 11-22.

Chevallier M.R. and Bernardi G. (1965). Transformation by heat-denatured deoxyribonucleic acid. J. Mol. Biol. 11: 658-660.

Chevallier M.R. and Bernardi G. (1968). The transforming activity of denatured *Haemophilus influenzae* DNA. Mol. Biol. 32: 437-452.

Chiapello H., Lisacek F. Caboche M., Henaut A. (1998). Codon usage and gene function are related in sequences of *Arabidopsis thaliana.* Gene 209: GC1-GC38.

Chiusano M.L., D'Onofrio G., Alvarez-Valin F., Jabbari K., Colonna G., Bernardi G. (1999). Correlations of nucleotide substitution rates and base composition of mammalian coding sequences with protein structure. Gene 238: 23-31.

Chiusano M.L., Alvarez-Valin F., Di Giulio M., D'Onofrio G., Ammirato G., Colonna, G., Bernardi, G. (2000). Second codon positions of genes and the secondary structures of proteins. Relationships and implications for the origin of the genetic code. Gene 261: 63-69.

Choi Y.C., Henrard D.H., Lee I., Ross S.R. (1987). The mouse mammary tumor virus long terminal repeat directs expression in epithelial and lymphoid cells of different tissues in transgenic mice. J. Virol. 61: 3013-3019.

Chowdhary B.P., Raudsepp T., Fronicke L., Scherthan H. (1998) Emerging patterns of comparative genome organization in some mammalian species as revealed by Zoo-FISH. Genome Res. 8: 577-589.

Chumakov I. et al. (1992). Continuum of overlapping clones spanning the entire human chromosome 21q. Nature 359: 380-387.

Chumakov I. et al. (1995). A YAC contig map of the human genome. Nature 377: 175-297.

Chung H.M., Shea C., Fields S., Taub R.N., Van der Ploeg L.H., Tse D.B. (1991). Architectural organization in the interphase nucleus of the protozoan *Trypanosoma brucei*: location of telomeres and mini-chromosomes. EMBO J. 9: 2611-2619.

Clark S.J., Harrison J., Frommer M. (1995). CpNpG methylation in mammalian cells. Nature Genet. 10: 20-27.

Clark-Walker G.D. and McArthur C.R. (1978). Structural and functional relationships of mitochondrial DNAs from various yeasts. In *Biochemistry and genetics of yeasts* (M. Bacila et al., eds.) p. 225. Academic Press, New York, NY, USA.

Claus D. and Berkeley R.C.W. (1986). Genus *Bacillus* Cohn 1872. In *Bergey's manual of systematic bacteriology* (P.H.A. Sneath, ed.) vol. 2. pp. 1105-1139. Williams and Wilkins, Baltimore, MD, USA.

Clay O. (2001). Standard deviations and correlations of GC levels in DNA sequences. Gene 276: 33-38.

Clay O. and Bernardi G. (2001a). The isochores in human chromosomes 21 and 22. Biochem. Biophys. Res. Commun. 285: 855-856.

Clay O. and Bernardi G. (2001b). Compositional heterogeneity within and among isochores in mammalian genomes. II. Some general comments. Gene 276: 25-31.

Clay O., Cacciò S., Zoubak S., Mouchiroud D., Bernardi G. (1996). Human coding and non-coding DNA: compositional correlations. Mol. Phylogenet. Evol. 5: 2-12.

Clay O., Carels N., Douady C., Macaya G. and Bernardi G. (2001) Compositional heterogeneity within and among isochores in mammalian genomes - I. CsCl and sequence analyses. Gene 276: 15-24.

Clay O., Douady C.J., Carels N., Hughes S., Bucciarelli G., Bernardi G. (2003a). Using analytical ultracentrifugation to study compositional variation in vertebrate genomes. Eur. Biophis. J. 32: 418-426.

Clay O., Arhondakis S., D'Onofrio G., Bernardi G. (2003b). LDH-A and a-actin as tools to assess the effects of temperature on the vertebrate genome: some problems. Gene 317: 157-160.

Colbert E.H. and Morales M. (1991). *Evolution of the vertebrates*, 4th edition. Wiley-Liss, New York, NY, USA.

Cooper D.N. and Krawczak M. (1989). Cytosine methylation and the fate of CpG dinucleotides in vertebrate genomes. Hum. Genet. 83: 181-188.

Cooper D.N., Taggart M.H., Bird A.P. (1983). Unmethylated domains in vertebrate DNA. Nucleic Acids Res. 11: 647-658.

Comings D.E. (1978). Mechanisms of chromosome banding and implications for chromosome structure. Annu. Rev. Genet. 12: 25-46.

Corneo G., Ginelli E., Soave C., Bernardi G. (1968). Isolation and characterization of mouse and guinea pig satellite DNA's. Biochemistry 7: 4373-4379.

Corneo G., Ginelli E., Polli E. (1970). Repeated sequences in human DNA. J. Mol. Biol. 48, 319-327.

Cortadas J., Macaya G., Bernardi G. (1977) An analysis of the bovine genome by density gradient centrifugation: fractionation in Cs_2SO_4/3,6 bis (acetato-mercurimethyl) dioxane density gradient. Eur. J. Biochem. 76: 13-19.

Cortadas J., Olofsson B., Meunier-Rotival M., Macaya G., Bernardi G. (1979). The DNA components of the chicken genome. Eur. J. Biochem. 99: 179-186.

Coruzzi G., Bonitz S.G., Thalenfeld B.E., Tzagoloff A. (1981). Assembly of the mitochondrial membrane system. Analysis of the nucleotide sequence and transcripts in the *oxil* region of yeast mitochondrial DNA. J. Biol. Chem. 256: 12780-12787.

Cosson J. and Tzagoloff A. (1979). Sequence homologies of (guanosine+cytidine)-rich regions of mitochondrial DNA of *Saccharomyces cerevisiae*. J. Biol. Chem. 254: 42-43.

Costantini M. (2003). Genome organization in sponges. Ph.D. Thesis, Open University.

Coulondre C., Miller J.H., Farabaugh P.J., Gilbert W. (1978). Molecular basis of base substitution hotspots in *Escherichia coli*. Nature 274: 775-780.

Cox E.C. and Yanofsky C. (1967). Altered base ratios in the DNA of an *Escherichia coli* mutator strain. Proc. Natl. Acad. Sci. USA 58: 1895-1902.

Craig M. and Bickmore W.A. (1994) The distribution of CpG islands in mammalian chromosomes. Nature Genet. 7: 376-382.

Cremer T. and Cremer C. (2001). Chromosome territories, nuclear architecture and gene regulation in mammalian cells. Nature Rev. Genet. 2: 292-301.

Crews S., Ojala D., Posakony J., Nishiguchi J., Attardi G. (1979). Nucleotide sequence of a region of human mitochondrial DNA containing the precisely identified origin of replication. Nature 277: 192-198.

Croft J.A., Bridger J.M., Boyle S., Perry P., Teague P., Bickmore W.A. (1999). Differences in the localization and morphology of chromosomes in the human nucleus. J. Cell Biol. 145: 1119-1131.

Cruveiller S., D'Onofrio G., Jabbari K., Bernardi G. (1999) Different hydrophobicities of orthologous proteins from *Xenopus* and man. Gene 238: 15-21.

Cruveiller S., D'Onofrio G., Bernardi G. (2000). The compositional transition between the genomes of cold- and warm-blooded vertebrates: codon frequencies in orthologous genes. Gene 261: 71-83.

Cruveiller S., Jabbari K., Clay O., Bernardi G. (2003a). Incorrectly predicted genes in rice? Science (in press).

Cruveiller S., Jabbari K., Clay O., Bernardi G. (2003b). Compositional features of eukaryotic genomes for checking predicted genes. Brief. Bioinfor. 4: 43-52.

Cruveiller S., Jabbari K., Clay O., Bernardi G. (2003c). Compositional gene landscapes in vertebrates. Genome Res. (submitted).

Cuny G., Macaya G., Meunier-Rotival M., Soriano P., Bernardi G. (1978). Some properties of the major components of the mouse genome. In *Genetic Engineering* (W.H. Boyer and S. Nicosia, eds.) pp. 109-115, Elsevier, Amsterdam, The Netherlands.

Cuny G., Soriano P., Macaya G., Bernardi G. (1981). The major components of the mouse and human genomes: preparation, basic properties and compositional heterogeneity. Eur. J. Biochem. 111: 227-233.

Darwin C. (1859). *On the origin of species by means of natural selection, or the preservation of favoured races in the struggle of life*. John Murray, London, U.K.

Davidson E.R. and Britten R.J. (1973). Organization, transcription, and regulation in the animal genome. Quart. Rev. Biol. Dev. 48: 565-613.

Davidson E.R. and Britten R.J. (1979). Regulation of gene expressions: possible role of repetitive sequences. Science 204: 1052-1059.

Dayhoff M.O. (1972). *Atlas of protein sequence and structure.* Vol. 5, Natl. Biomed. Res. Found., Washington, DC.

Dehal V. et al. (2002). The draft genome of *Ciona intestinalis*: insights into chordate and vertebrate origins. Science 298: 2157-2167.

Deininger P.L. and Slagel W.K. (1988). Recently amplified Alu family members share a common parental Alu sequence. Mol. Cell. Biol. 8:4566-4569.

De Luca di Roseto G., Bucciarelli G., Bernardi G. (2002). An analysis of the genome of *Ciona intestinalis*. Gene 295: 311-316.

de Massy B., Rocco V., Nicolas A. (1995). The nucleotide mapping of DNA double-strand breaks at the CYS3 initiation site of meiotic recombination in *Saccharomyces cerevisiae*. EMBO J. 14: 4589-4598.

de Miranda A.B., Alvarez-Valin F., Jabbari K., Degrave W.M., Bernardi G. (2000). Gene expression, amino acid conservation, and hydrophobicity are the main factors shaping codon preferences in *Mycobacterium tuberculosis* and *Mycobacterium leprae*. J. Mol. Evol. 50: 45-55

Dermitzakis E.T., Reymond A., Lyle R., Scamuffa N., Ucla C., Deutsch S., Stevenson B.J., Flegel V., Bucher P., Jongeneel C.V., Antonarakis S. (2002) Numerous potentially functional but non-genic conserved sequences on human chromosome 21. Nature 420: 578-582.

De Sario A., Aïssani B., Bernardi G. (1991) Compositional properties of telomeric regions from human chromosomes. FEBS Letters 295: 22-26.

De Sario A., Geigl E.M. and Bernardi G. (1995). A rapid procedure for the compositional analysis of yeast artificial chromosomes. Nucl. Acids Res. 23: 4013-4014.

De Sario A., Geigl E.M., Palmieri G., D'Urso M., Bernardi G. (1996). A compositional map of human chromosome band Xq28. Proc. Natl. Acad. Sci. USA 93: 1298-1302

De Sario A., Roizès G., Allegre N., Bernardi G. (1997). A compositional map of the cen-q21 region of human chromosome 21. Gene 194: 107-113.

Devillers-Thiery A. (1974). Utilisation des endonucléases dans l'étude des sequences des ADN. Thesis, Université Paris VII.

de Zamaroczy M. and Bernardi G. (1985). Sequence organization of the mitochondrial genome of yeast - a review. Gene 37: 1-17

de Zamaroczy M. and Bernardi G. (1986a). The GC clusters of the mitochondrial genome of yeast and their evolutionary origin. Gene 41: 1-22.

de Zamaroczy M. and Bernardi G. (1986b). The primary structure of the mitochondrial genome of *S. cerevisiae* - a review. Gene 47: 155-177

de Zamaroczy M. and Bernardi G. (1987). The AT spacers and the *var1* genes from the mitochondrial genome of *Saccharomyces cerevisiae* and *Torulopsis glabrata*: evolutionary origin and mechanism of formation. Gene 54: 1-22

de Zamaroczy M., Baldacci G., Bernardi G. (1979). Putative origins of replication in the mitochondrial genome of yeast. FEBS Letters 108: 429-432

de Zamaroczy M., Marotta R., Fonty G., Goursot R. , Mangin M., Baldacci G., Bernardi G. (1981). The origins of replication of the mitochondrial genome of yeast and the phenomenon of suppressivity. Nature 292: 75-78.

de Zamaroczy M., Faugeron-Fonty G., Bernardi G. (1983). Excision sequences in the mitochondrial genome of yeast. Gene 21: 193-202.

de Zamaroczy M., Faugeron-Fonty G., Baldacci G., Goursot R., Bernardi G. (1984). The *ori* sequences of the mitochondrial genome of a wild-type yeast strain : number, location, orientation and structure. Gene 32: 439-457.

Dickerson R.E. (1971). The structures of cytochrome c and the rates of molecular evolution. J. Mol. Evol. 1: 26-45.

Dickerson R.E. and Geis I. (1983). *Hemoglobins: structure, function, evolution and pathology*. Cummings, Menlo Park, CA.

Di Giulio M. (1996). The beta-sheets of proteins, the biosynthetic relationships between amino acids, and the origin of the genetic code. Orig. Life Evol. Biosph. 26: 589-609.

Di Giulio M. (1997). On the origin of the genetic code. J. Theor. Biol. 187: 573-581.

Dodemont H.J., Soriano P., Quax W.J., Ramaekers F., Lenstra J.A., Groenen M.A., Bernardi G., Bloemendal H. (1982). The genes coding for the cytoskeletal proteins actin and vimentin in warm-blooded vertebrates. EMBO J. 1: 167-171.

D'Onofrio G. (2002). Expression patterns and gene distribution in the human genome. Gene 300: 155-160.

D'Onofrio G. and Bernardi G. (1992). A universal compositional correlation among codon positions. Gene 110: 81-88.

D'Onofrio G., Mouchiroud D., Aïssani B., Gautier C., Bernardi G. (1991). Correlations between the compositional properties of human genes, codon usage and aminoacid composition of proteins. J. Mol. Evol. 32: 504-510.

D'Onofrio G., Jabbari K., Musto H., Bernardi G. (1999a). The correlation of protein hydropathy with the composition of coding sequences. Gene 238: 3-14.

D'Onofrio G., Jabbari K., Musto H., Alvarez-Valin F., Cruveiller S. and Bernardi G. (1999b). Evolutionary genomics of vertebrates and its implications. Ann. N.Y. Acad. Sci. 18: 81-94.

D'Onofrio G., Ghosh T.C., Bernardi G. (2002). The base composition of the genes is correlated with the secondary structures of the encoded proteins. Gene 300: 179-187.

Doolittle R.F. (2002). The grand assault. Nature 419: 493-494.

Doolittle W.F. and Sapienza C. (1980). Selfish genes, the phenotype paradigm and genome evolution. Nature 284: 601-603.

Doskocil J. and Sorm F. (1962). Distribution of 5-methylcytosine in pyrimidine sequences of deoxyribonucleic acids. Biochim. Biophys. Acta 55: 953.

Douady C., Carels N., Clay O., Catzeflis F. and Bernardi G. (2000). Diversity and phylogenetic implications of CsCl profiles from rodent DNAs. Mol. Phylogen. Evol. 17: 219-230.

Drouin R., Holmquist G., Richer C.L. (1994) High resolution replication bands compared with morphologic G- and R-bands. Adv. Hum. Genet. 22: 47-115.

Dubcovsky J., Ramakrishna W., SanMiguel P.J., Busso C.S., Yan L., Shiloff B.A., Bennetzen J.L. (2001). Comparative sequence analysis of collinear barley and rice bacterial artificial chromosomes. Plant Physiol. 125: 1342-1353.

Dujon B. (1996). The yeast genome project: what did we learn? Trends Genet. 12: 263-270.

Dunham I. et al. (1999). The DNA sequence of human chromosome 22. Nature 402: 489-495.

Duret L. and Mouchiroud D. (2000). Determinants of substitution rates in mammalian genes: expression pattern affects selection intensity but not mutation rate. Mol. Biol. Evol. 17: 68-74.

Duret L., Mouchiroud D., Gouy M. (1994). HOVERGEN: A database of homologous vertebrate genes. Nucleic Acids Res. 22: 2360-2365.

Duret L., Mouchiroud D., Gautier C. (1995). Statistical analysis of vertebrate sequences reveals that long genes are scarce in GC-rich isochores. J. Mol. Evol. 40: 308-317.

Duret L., Semon M., Piganeau G., Mouchiroud D, Galtier N. (2002). Vanishing GC-rich isochores in mammalian genomes. Genetics 162: 1837-1847.

Dutrillaux B. (1973) Nouveau système de marquage chromosomique: les bandes T. Chromosoma 41: 395-402.

Dutrillaux B., Rethoré M.O., Lejeune J. (1975). Comparison of the karyotype of the orangutan (*Pongo pygmaeus*) to those of man, chimpazee, and gorilla. Ann.Genet. 18: 153-161.

Dutrillaux B., Couturier J., Richer C.-L., Viegas-Pequinot E. (1976). Sequence of DNA replication in 277 R- and Q-bands of human chromosomes using a BrdU treatment. Chromosoma 58: 51-61.

Easteal S. (1990). The pattern of mammalian evolution and the relative rate of molecular evolution. Genetics 124: 165-173.

Edmondson D.G. and Roth S.Y. (1996). Chromatin and transcription. FASEB J. 10: 1173-1182.

Ehrlich S.D., Thiery J.P., Bernardi G. (1972). The mitochondrial genome of wild-type yeast cells. III. The pyrimidine tracts of mitochondrial DNA. J. Mol. Biol. 65: 207-212.

Ehrlich M., Gama-Sosa M.A., Huanh L.H., Midgett R.M., Kuo K.C. McCune RA, Gehrke C. (1982). Amount and distribution of 5-methylcytosine in human DNA from different types of tissues of cells. Nucleic Acids Res. 10: 2709-2721.

Eichler E.E., Archidiacono N., Rocchi M. (1999). CAGGG repeats and the pericentromeric duplication of the hominoid genome. Genome Res. 9: 1048–1058.

Eigner J. and Doty P. (1965). The native, denatured and renatured states of deoxyribonucleic acid. J. Mol. Biol. 12: 549-580.

Eizirik E., Murphy W.J., O'Brien S.J. (2001). Molecular dating and biogeography of early placental mammal radiation. J. Hered. 92: 212-219.

Elleder D., Pavlicek A., Paces J., Hejnar J. (2002). Preferential integration of human immunodeficiency virus type 1 into genes, cytogenetic R bands and GC-rich DNA regions: insight from the human genome sequence. FEBS Lett. 517: 285-286.

Elton R.A. (1974). Theoretical models for heterogeneity of base composition in DNA. J. Theor. Biol. 45: 533-553.

Endo T., Imanishi T., Gojobori T., Inoko H. (1997). Evolutionary significance of intragenome duplications on human chromosomes. Gene 205: 19-27.

Engelman D.M., Steitz T.A., Goldman A. (1986). Identifying nonpolar transbilayer helices in amino acid sequences of membrane proteins. Annu. Rev. Biophys. Biophys. Chem. 15: 321-353.

Ephrussi B. (1949). Action de l'acriflavine sur les levures. In *Unités biologiques douées de continuité génétique*. Publications du CNRS, Paris, France.

Ephrussi B. (1953). *Nucleo-cytoplasmic relations in micro-organisms – Their bearing on cell heredity and differentiation.* Oxford at the Clarendon Press, Oxford, UK.

Ephrussi B., de Margerie-Hottinguer H., Roman H. (1955). Suppressiveness: a new factor in the genetic determinism of the synthesis of respiratory enzymes in yeast. Proc. Nat. Acad. Sci. USA 41: 1065-1071.

Eyre-Walker A. (1993). Recombination and mammalian genome evolution. Proc R. Soc. Lond. B. Biol. Sci. 252:237-243.

Eyre-Walker A. (1999) Evidence of selection on silent site base composition in mammals: potential implications for the evolution of isochores and junk DNA. Genetics 152:675-683.

Eyre-Walker A. and Hurst L.D. (2001). The evolution of isochores. Nature Rev. Genet. 2: 549-555.

Faugeron-Fonty G., Culard F., Baldacci G., Goursot R., Prunell A. and Bernardi G. (1979). The mitochondrial genome of wild-type yeast cells. VIII. The spontaneous cytoplasmic *petite* mutation. J. Mol. Biol. 134: 493-537.

Faugeron-Fonty G., Le Van Kim C., de Zamaroczy M., Goursot R., Bernardi G. (1984).

A comparative study of the *ori* sequences from the mitochondrial genomes of 20 wild-type yeast strains. Gene 32: 459-473.

Feder M.E. (1996). Ecological stress and evolutionary physiology of the stress proteins and the stress response: the *Drosophila melanogaster* model in animals and temperature. In *Phenotypic and Evolutionary Adaptation* (I.A. Johnston and A.F. Bennett, eds.) Cambridge University Press, Cambridge, UK.

Federico C., Saccone S., Bernardi G. (1998) The gene-richest bands of human chromosomes replicate at the onset of the S-phase. Cytogenet. Cell Genet. 80: 83-88

Federico C., Andreozzi L., Saccone S., Bernardi G. (2000). Gene density in the Giemsa bands of human chromosomes. Chromosome Res. 8: 737-746.

Federico C., Saccone S., Scavo C., Motta S., Bernardi G. (2003). Identification of the gene-richest regions in the chromosomes and nuclei of *Rana esculenta* by *in situ* hybridization with the chicken gene-rich isochores. Gene (submitted).

Fennoy S.L. and Bailey-Serres J. (1993). Synonymous codon usage in *Zea Mays* L. nuclear genes is varied by levels of C and G-ending codons. Nucleic Acids Res. 21: 5294-5300.

Ferreira J., Paolella, G., Ramos C., Lamond A.I. (1997). Spatial organization of large-scale chromatin domains in the nucleus: a magnified view of single chromosome territories. J. Cell Biol. 139: 1597-1610.

Feuillet C. and Keller B. (1999). High gene density is conserved at syntenic loci of small and large grass genomes. Proc. Natl. Acad. Sci. USA 96: 8265-8270.

Fickett J.W., Torney D.C., Wolf D.R. (1992). Base compositional structure of genomes. Genomics 13: 1056-1064.

Fields C., Adams M.D., White O., Venter J.C. (1994). How many genes in the human genome? Nature Genet. 7: 345-346.

Filipski J. (1987). Correlation between molecular clock ticking, codon usage fidelity of DNA repair, chromosome banding and chromatin compactness in germline cells. FEBS Lett. 217:184-186.

Filipski J. and Mucha M. (2002). Structure, function and DNA composition of *Saccharomyces cerevisiae* chromatin loops. Gene 300, 63-68.

Filipski J., Thiery J.P., Bernardi G. (1973). An analysis of the bovine genome by Cs_2SO_4-Ag+ density gradient centrifugation. J. Mol. Biol. 80: 177-197.

Fincham V.J. and Wyke J.A., (1991). Differences between cellular integration sites of transcribed and non transcribed Rous sarcoma proviruses. J. Virol. 65: 461-463.

Fiser A., Simon I., Barton G.J. (1996) Conservation of amino acids in multiple alignments: aspartic acid has unexpected conservation. FEBS Lett. 397: 225-229.

Flamm W.G., McCallum M., Walker P.M.B. (1967). The isolation of complementary strands from a mouse DNA fraction. Proc. Natl. Acad. Sci. USA 57: 1729.

Flavell R.B., Bennett M.D., Smith J.B., Smith D.B. (1974). Genome size and the proportion of repeated nucleotide sequence DNA in plants. Biochem. Genet. 12: 257-269.

Flavell R.B., Gale M.D., O'Dell M., Murphy G., Moore G. (1993). Molecular organization of genes and repeats in the large cereal genomes and implications for the isolation of genes by chromosome walking. In *Chromosomes Today* (A.T. Sumner and A.C. Chandley, eds.) vol. 11: 199-213.

Fonty G., Crouse E.J., Stutz E., Bernardi G. (1975). The mitochondrial genome of *Euglena gracilis*. Eur. J. Biochem. 54: 367-372.

Fonty G., Goursot R., Wilkie D., Bernardi G. (1978). The mitochondrial genome of wild-type yeast cells. Recombination in crosses. J. Mol. Biol. 119: 213-235.

Force A., Lynch M., Pickett F.B., Amores A., Yan Y.L., Postlethwait J. (1999). Preservation of duplicate genes by complementary, degenerative mutations. Genetics 151: 1531-1545.

Forget B.G., Cavallesco C., DeRiel J.K., Spritz R.A., Choudary P.V., Wilson J.T., Reddy V.B., Weissman S.M. (1979). Structure of the human globin genes. In *Eukaryotic Gene Regulation*. ICN-UCLA Symposium on Molecular and Cellular Biology XIV, p. 367-381 (R. Axel, T. Maniatis, C.F. Fox, eds.) Academic Press, New York, NY, USA.

Foury F., Roganti T., Lecrenier N., Purnelle B. (1998) The complete sequence of the mitochondrial genome of *Saccharomyces cerevisiae*. FEBS Lett. 440: 325-331.

Francino M.P. and Ochman H. (1999) Isochores result from mutation not selection. Nature 400: 30-31.

Francke W. (1994) Digitized and differentially shaded human chromosome ideograms for genomic applications. Cytogenet. Cell Genet. 6: 206-219.

Freese E. (1962). On the evolution of base composition of DNA. J. Theor. Biol. 3: 82-101.

Freund A.M. and Bernardi G. (1963). Viscosity of deoxyribonucleic acid solutions in the "sub-melting" temperature range. Nature (London), 200: 1318-1319.

Friedman R. and Hughes A.L. (2001). Pattern and timing of gene duplication in animal genomes. Genome Res. 11: 1842-1847.

Frischmeyer P.A., van Hoof A., O'Donnell K., Guerreiro A.L., Parker R., Dietz H.C. (2002). An mRNA surveillance mechanism that eliminates transcripts lacking termination codons. Science 295: 2258-2261.

Froelich D., Strazielle C., Bernardi G., Benoit H. (1963) Low-angle light-scattering of deoxyribonucleic acid solutions. Biophys. J. 3: 115-125.

Froese R. and Pauly D. (2002). FISHBASE. World Wide Web electronic publication. www.fishbase.org, version 26/12/2002 (pag 342).

Fryxell K.J. and Zuckerkandl E. (2000). Cytosine deamination plays a primary role in the evolution of mammalian isochores. Mol. Biol. Evol. 17: 1371-1383.

Fukagawa T., Sugaya K., Matsumoto K., Okumura K., Ando A., Inoko H., Ikemura T. (1995) A boundary of long-range G + C% mosaic domains in the human MHC locus: pseudoautosomal boundary-like sequence exists near the boundary. Genomics 25: 184-191.

Fullerton S.M., Carvalho A.B., Clark A.G. (2001). Local rates of recombination are positively correlated with GC content in the human genome. Mol. Biol. Evol. 18: 1139-1142.

Furst A, Brown E.H., Braunstein J.D., Schildkraut C.L. (1981). Alpha-globulin sequences are located in a region of early-replicating DNA in murine erythroleukemia cells. Proc. Natl. Acad. Sci. USA 78: 1023-1027.

Gabrielian A., Simoncsits A., Pongor S. (1996). Distribution of bending propensity in DNA sequences. FEBS Letters 393: 124-130.

Gaillard C. and Bernardi G. (1979). The nucleotide sequence of the mitochondrial genome of a spontaneous *petite* mutant in yeast. Mol. Gen. Genet. 174: 335-337.

Gaillard C., Strauss F., Bernardi G. (1980). Excision sequences in the mitochondrial genome of yeast. Nature 283: 218-220.

Gaillard C., Doly J., Cortadas J., Bernardi G. (1981). The primary structure of bovine satellite 1.715. Nucleic Acids Res. 9: 6069-6082.

Galtier N. (2003). Gene conversion drives GC content evolution in mammalian histones. Trends Genet. 19: 65-68.

Galtier N. and Lobry J.R. (1997) Relationships between genomic G+C content, RNA secondary structures, and optimal growth temperature in prokaryotes. J. Mol. Evol. 44: 632-636.

Galtier N. and Mouchiroud D. (1998) Isochore evolution in mammals: a human-like ancestral structure. Genetics 150: 1577-1584.

Galtier N., Piganeau G., Mouchiroud D., Duret L. (2002). GC-content evolution in mammalian genomes: the biased gene conversion hypothesis. Genetics 159: 907-911.

Gama-Sosa M.A., Midgett R.M., Slagel V.A., Githens S., Kuo K.C., Gehrke C.W., Ehrlich M. (1983a). Tissue-specific differences in DNA methylation in various mammals. Biochim. Biophys. Acta 740: 212-219.

Gama-Sosa M.A., Wang R.Y., Kuo K.C., Gehrke C.W., Ehrlich M. (1983b). The 5-methylcytosine content of highly repeated sequences in human DNA. Nucleic Acids Res. 11: 3087-3095.

Gardiner K., Watkins P., Munke M., Drabkin H., Jones C., Patterson D. (1988). Partial physical map of human chromosome 21. Somat. Cell. Mol. Genet. 14: 623-637.

Gardiner K., Aïssani B., Bernardi G. (1990). A compositional map of human chromosome 21. EMBO J. 9: 1853-1858.

Gardiner-Garden M. and Frommer M. (1987). CpG islands in vertebrate genomes. J. Mol. Biol. 196: 261-282.

Gardiner-Garden M., Sved J., Frommer M. (1992). Methylation sites in angiosperm genes. J.Mol. Evol. 34: 219-230.

Garel A. and Axel R. (1976) Selective digestion of transcriptionally active ovalbumin genes from oviduct nuclei. Proc. Natl. Acad. Sci. USA 11: 3966-3970.

Gill K.S., Gill B.S., Endo T.R., Mukai Y. (1993). Fine physical mapping of Ph1, a chromosome pairing regulator gene in polyploid wheat. Genetics 134: 1231-1236.

Gill K.S., Gill B.S., Endo T.R., Boyko E.V. (1996a). Identification and high-density mapping of gene-rich regions in chromosome group 5 of wheat. Genetics 143: 1001-1012.

Gill K.S., Gill B.S., Endo T.R., Taylor T. (1996b). Identification and high-density mapping of gene-rich regions in chromosome group 1 of wheat. Genetics 143: 1883-1891.

Gillespie J.H. (1991). *The Causes of Molecular Evolution.* Oxford Univ. Press, Oxford.

Gillum A.M. and Clayton D.A.(1979). Mechanism of mitochondrial DNA replication in mouse L-cells: RNA priming during the initiation of heavy-strand synthesis. J. Mol. Biol. 135: 353-368.

Ginatulin A.A. (1984). *Structure, organization, evolution of vertebrate genome.* Nauka, Moscow.

Gleba Y.Y., Parokonny A., Kotov V., Negrutiu I., Momot V. (1987). Spatial separation of parental genomes in hybrids of somatic plant cells. Proc. Natl. Acad. Sci. USA 84: 3709-3713.

Glukhova L.A., Zoubak S.V., Rynditch A.V., Miller G.G., Titova I.V., Vorobyeva N., Lazurkevitch Z.V., Graphodatskii A.S., Kushch A.A. and Bernardi G. (1999). Localization of HTLV-1 and HIV-1 proviral sequences in chromosomes of persistently infected cells. Chromosome Res. 7: 177-183.

Goff S.P. (1992). Genetics of retroviral integration. Annu. Rev. Genet. 26: 527-544.

Goff, S. A. et al. (2002) A draft sequence of the rice genome (*Oryza sativa L. ssp. japonica*). *Science* 296: 92-100.

Goffeau A., Barrell B.G., Bussey H., Davis R.W., Dujon B., Feldmann H., Galibert F., Hoheisel J.D., Jacq C., Johnston M., Louis E.J., Mewes H.W., Murakami Y., Philippsen P., Tettelin H., Oliver S.G. (1996). Life with 6000 genes. Science 274: 546-567.

Gojobori T., Li W.H., Graur D. (1982) Patterns of nucleotide substitution in pseudogenes and functional genes. J. Mol. Evol. 18: 360-369.

Goldman M.A., Holmquist G.P., Gray M.C., Caston L.A., Nag A. (1984). Replication timing of genes and middle repetitive sequences. Science 224: 686-692.

Goldring E.S., Grossman L.I., Krupnick D., Cryer D.R., Marmur J. (1970). The petite mutation in yeast. Loss of mitochondrial deoxyribonucleic acid during induction of petites with ethidium bromide. J. Mol. Biol. 52: 323-335.

Gonçalves I., Duret L., Mouchiroud D. (2000). Nature and structure of human genes that generate retropseudogenes. Genome Res. 10: 672-678.

Goodman M. and Moore G.W. (1977). Use of Chou-Fasman amino acid conformational parameters to analyze the organization of the genetic code and to construct protein genealogies. J. Mol. Evol. 10: 7-47.

Gould S.J. and Lewontin R.C. (1979) The spandrels of San Marco and the Panglossian paradigm: a critique of the adaptationist programme. Proc. R. Soc. Lond. B. Biol. Sci. 205: 581-598.

Goursot R., de Zamaroczy M., Baldacci G., Bernardi G. (1980). Supersuppressive *petite* mutants in yeast. Current Genet. 1: 173-176.

Goursot R., Mangin M., Bernardi G. (1982). Surrogate origins of replication in the mitochondrial genome of *ori* petite mutants of yeast. EMBO J. 1: 705-711.

Goursot R., Goursot R., Bernardi G. (1988). Temperature can reversibly modify the structure and the functional efficiency of *ori* sequences of the mitochondrial genome from yeast. Gene 69: 141-145.

Gouy M. and Gautier C. (1982). Codon usage in bacteria: correlation with gene expressivity. Nucleic Acids Res. 10: 7055-7074.

Gouy M., Gautier C., Attimonelli M., Lanave C., di Paola G. (1985). ACNUC – a portable retrieval system for nucleic acid sequence databases: logical and physical designs and usage. Comput. Appl. Biosci. 1: 167-172.

Grandgenett D.P. and Mumm S.R. (1990). Unraveling retrovirus integration. Cell 60: 3-4.

Grantham R. (1980). Workings on the genetic code. Trends Biochem Sci. 5: 327-333.

Grantham R., Gautier C., Gouy M., Mercier R., Pavé A. (1980). Codon catalog usage and the genome hypothesis. Nucleic Acids Res. 8: r49-r62.

Grantham R., Gautier C., Gouy M., Jacobzone M., Mercier R. (1981). Codon catalog usage is a genome strategy modulated for gene expressivity. Nucleic Acids Res. 9: r43-r74.

Graur D. and Li W.H. (2000) *Fundamentals of Molecular Evolution*, 2nd edition. Sinauer, Sunderland, MA, USA.

Grippo P., Iaccarino M., Parisi E., Scarano E. (1968). Methylation of DNA in developing sea urchin embryos. J. Mol. Biol. 36: 195-208.

Groisman E., Sturmoski M., Solomon F., Lin R., Ochman H. (1993). Molecular, functional, and evolutionary analysis of sequences specific to salmonella. Proc. Natl. Acad. Sci. USA 90: 1033-1037.

Grosjean H., Sankoff D., Min Jou W., Fiers W., Cedergren R.J. (1978). Bacteriophage MS2 RNA: a correlation between the stability of the codon/anticodon interaction and the choice of code words. J. Mol. Evol. 12: 113-119.

Gruenbaum Y., Stein R., Cedar H., Razin A. (1981). Methylation of CpG sequences in eukaryotic DNA. FEBS Lett. 124: 67-71.

Gu X. and Huang W. (2002). Testing the parsimony test of genome duplications: a counterexample. Genome Res. 12: 1-2.

Gu X. and Li W.H. (1992). Higher rates of amino acid substitution in rodents than in humans. Mol. Phylogenet. Evol. 1: 211-214.

Gu X., Hewett-Emmett D., Li W.H. (1998). Directional mutational pressure affects the amino acid composition and hydrophobicity of proteins in bacteria. Genetica 102/103: 383-391.

Gu X., Wang Y., Gu J. (2002). Age distribution of human gene families shows significant roles of both large- and small-scale duplications in vertebrate evolution. Nat. Genet. 31: 205-209.

Guild W. (1963). Evidence for intramolecular heterogeneity in pneumococcal DNA. J. Mol. Biol. 6: 214-229.

Gurdon J.B. (1962). The developmental capacity of nuclei taken from intestinal epithelium cells of feeding tadpoles. J. Embryol. Exp. Morph. 10: 622-640.

Haag J., O'Huigin C., Overath P. (1998). The molecular phylogeny of trypanosomes: evidence for an early divergence of the *Salivaria*. Mol. Biochem. Parasitol. 91: 37-49.

Haeckel E. (1866). *Generelle Morphologie der Organismen*. Reimer, Berlin, Germany.

Hake S. and Walbot V (1980). The genome of *Zea mays*, its organization and homology to related grasses. Chromosoma 79: 251-270.

Hamada K., Horiike T., Kanaya S., Nakamura H., Ota H., Yatogo T., Okada K., Nalamura H., Shinozawa T. (2002). Changes in body temperature pattern in vertebrates do not influence the codon usages of α-globin genes. Genes Genet. Syst. 77: 197-207

Hardison R.C. (2000). Conserved noncoding sequences are reliable guides to regulatory elements. Trends Genet. 16: 369-372.

Häring D. and Kypr J. (2001). No isochores in human chromosomes 21 and 22? Biochem. Biophys. Res. Commun. 280: 567-573.

Hartl D. and and Clark A.G. (1997). *Principles of population genetics*. Sinauer Associates, Sunderland MA, USA.

Harvey P.H. and Pagel M.D. (1991). *The comparative method in evolutionary biology*. Oxford University Press, Oxford, UK.

Hasegawa M., Yasunaga T., Miyata T. (1979). Secondary structure of MS2 phage RNA and bias in code word usage. Nucleic Acids Res. 7: 2073-2079.

Hattori M., Toyoda A., Ishikawa H., Ito T., Ohgusu H., Oishi N., Kano T., Kuhara S., Ohki M., Sakaki Y. (1993). Sequence-tagged *Not*I sites of human chromosome 21. Sequence analysis and mapping. Genomics 17: 39-44.

Hattori M. et al. (2000). The DNA sequence of human chromosome 21. Nature 405: 311-319.

Heatwole H. and Taylor J. (1987). *Ecology of reptiles*. Surrey Beatty & Sons, Chipping Norton, N.S.W., Australia.

Hedges S.B. and Poling L.L.(1999). A molecular phylogeny of reptiles. Science 283: 998-1001.

Hedges S.B., Moberg K.D., Maxson L.R. (1990). Tetrapod phylogeny inferred from 18S and 28S ribosomal RNA sequences and a review of the evidence for amniote relationships. Mol. Biol. Evol. 7: 607-633.

Heilig R. et al. (2003). The DNA sequence and analysis of human chromosome 14. Nature 421: 601-607.

Heizmann P., Doly J., Hussein Y., Nicolas P., Nigon V., Bernardi G. (1981). The chloroplast genome of bleached mutants of *Euglena gracilis*. Biochim. Biophys. Acta 653: 412-415.

Hellmann-Blumberg U., Hintz M.F., Gatewood J.M., Schmid C.W. (1993). Developmental differences in methylation of human Alu repeats. Mol. Cell. Biol. 13: 4523-4530.

Hennig W. and Walker P.M. (1970). Variations in the DNA from two rodent families (*Cricetidae* and *Muridae*). Nature 225: 915-919.

Henrard D. and Ross S.R. (1988). Endogenous mouse mammary tumor virus is expressed in several organs in addition to the lactating mammary gland. J. Virol. 62: 3046-3049.

Heslop-Harrison J.S. and Bennett M.D. (1990). Nuclear architecture in plants. Trends Genet. 6: 401-405.

Hethcote H. (2000). The mathematics of infectious diseases. SIAM Review 42: 599-653.

Hey J. (1999). The neutralist, the fly and the selectionist. Trends Ecol. Evol. 14: 35-38.

Hilleren P., McCarthy T., Rosbash M., Parker R., Jensen T.H. (2001). Quality control of mRNA 3'-end processing is linked to the nuclear exosome. Nature 413: 538-542.

Hinegardner R. (1968). Evolution of cellular DNA content in teleost fishes. Am. Nat. 102: 517-523.

Hinegardner R. (1976). The cellular DNA content of sharks, rays and some other fishes. Comp. Biochem. Physiol. 55: 367-370.

Hodgson C.P. (1996). *Retro-vectors for human gene therapy*. Landes, Austin, TX, USA.

Hoenika J., Arrasate M., Garcia de Yebenes J., Avila J. (2002). A two-hybrid screening of human Tau protein: interactions with Alu derived domain. Neuroreport 13: 343-349.

Holland P.W., Garcia-Fernandez J., Williams N.A., Sidow A. (1994). Gene duplications and the origins of vertebrate development. Development Suppl. 125-133

Holliday R. (1995). *Understanding Ageing*. Cambridge University Press, Cambridge, U.K.

Holmes-Son M.L., Appa R.S., Chow S.A. (2001). Molecular genetics and target site specificity of retroviral integration. Adv. Genet. 43: 33-69.

Holmquist G.P. (1992). Chromosome bands, their chromatin flavors, and their functional features. Am. J. Hum. Genet. 51: 17-37.

Holt R.A. et al. (2002). The genome sequence of the Malaria Mosquito *Anopheles gambiae*. Science 298: 129-149.

Hori T., Suzuki Y., Solovei I., Saitoh Y., Hutchison N., Ikeda J.E., Macgregor H., Mizuno S. (1996). Characterization of DNA sequences constituting the terminal heterochromatin of the chicken Z chromosome. Chromosome Res. 4: 411-426.

Horvath J., Viggiano L., Loftus B., Adams M., Rocchi M., Eichler E. (2000). Molecular structure and evolution of an alpha/non-alpha satellite junction at 16p11. Hum.Mol.-Genet. 9: 113-123.

Houck C.M., Rinehart F.P., Schmid C.W. (1979). A ubiquitous family of repeated DNA sequences in the human genome. J. Mol. Biol. 132: 289-306.

Huchon D., Catzeflis F.M., Douzery E.J. (2000). Variance of molecular datings, evolution of rodents and the phylogenetic affinities between *Ctenodactylidae* and *Hystricognathi*. Proc. R. Soc. Lond. B. Biol. Sci. 267: 393-402.

Hudson A.P., Cuny G., Cortadas J., Haschemeyer A.E.V., Bernardi G. (1980). An analysis of fish genomes by density gradient centrifugation. Eur. J. Biochem. 112: 203-210.

Hudspeth M.E., Ainley W.M., Shumard D.S., Butow R.A., Grossman L.I. (1982). Location and structure of the *var1* gene on yeast mitochondrial DNA: nucleotide sequence of the 40.0 allele. Cell 30: 617-626.

Hugenholtz P. (2002). Exploring prokaryotic diversity in the genomic era. Genome Biol., 3: REVIEWS0003.

Hughes S.H. (1983). Synthesis, integration and transcription of the retroviral provirus. Curr. Top. Microbiol. Immunol. 103: 23-51.

Hughes S. and Mouchiroud D. (2001). High evolutionary rates in nuclear genes of squamates. J. Mol. Evol. 53: 70-76.

Hughes S., Zelus D., Mouchiroud D. (1999). Warm-blooded isochore structure in Nile crocodile and turtle. Mol. Biol. Evol. 16:1521-1527.

Hughes S., Clay O. and Bernardi G. (2002). Compositional patterns in reptilian genomes. Gene 295: 323-329.

Hurst L.D. and Merchant A.R. (2001). High guanine-cytosine content is not an adaptation to high temperature: a comparative analysis amongst prokaryotes. Proc. R. Soc. Lond. B. Biol. Sci. 268: 493-497.

Hyde J.E. and Sims P.F. (1987). Anomalous dinucleotide frequencies in both coding and non-coding regions from the genome of the human malaria parasite *Plasmodium falciparum*. Gene 61: 177-187.

Iborra F.J., Jackson D.A., Cook P.R. (2001). Coupled transcription and translation within nuclei of mammalian cells. Science 293: 1139-1142.

Ichikawa H., Shimizu K., Saito A., Wang D., Oliva R., Kobayashi H., Kaneko Y., Miyoshi H., Smith C.L., Cantor C.R., Ohki M. (1992). Long-distance restriction mapping of the proximal long arm of human chromosome 21 with *Not*I linking clones. Proc. Natl. Acad. Sci. USA 89: 23-27.

Ifft J.B., Voet D.M., Vinograd J. (1961). The determination of density distributions and density gradients in binary solutions at equilibrium in the ultracentrifuge. J. Phys. Chem. 65: 1138-1145.

Ikemura T. (1981a). Correlation between the abundance of Escherichia coli transfer RNAs and the occurrence of the respective codons in its protein genes. J. Mol. Biol. 146: 1-21.

Ikemura T. (1981b). Correlation between the abundance of Escherichia coli transfer RNAs and the occurrence of the respective codons in its protein genes: a proposal for a synonymous codon choice that is optimal for the *E. coli* translational system. J. Mol. Biol. 151: 389-409.

Ikemura T. (1982). Correlation between the abundance of yeast transfer RNAs and the occurrence of the respective codons in protein genes. Differences in synonymous codon choice patterns of yeast and *Escherichia coli* with reference to the abundance of isoaccepting transfer RNAs. J. Mol. Biol. 158: 573-597.

Ikemura T. (1985). Codon usage and tRNA content in unicellular and multicellular organisms. Mol. Biol. Evol. 2: 13-34.

Ikemura T. and Aota S. (1988). Alternative chromatic structure at CpG islands and quinacrine-brightness of human chromosomes. Global variation in G+C content along vertebrate genome DNA. Possible correlation with chromosome band structures. J. Mol. Biol. 60: 909-920.

Ikemura T. and Wada K. (1991). Evident diversity of codon usage patterns of human genes with respect to chromosome banding patterns and chromosome numbers: relation between nucleotide sequence data and cytogenetic data. Nucleic Acid Res. 16: 4333-4339.

Ikemura T., Wada K., Aota S. (1990). Giant G+C% mosaic structures of the human genome found by arrangement of GenBank human DNA sequences according to genetic positions. Genomics 8: 207-216.

Ina Y. (1995). New methods for estimating the numbers of synonymous and nonsynonymous substitutions. J. Mol. Evol. 40: 190-226.

Ingle J., Pearson G.G., Sinclair J. (1973). Species distribution and properties of nuclear satellite DNA in higher plants. Nature New Biol. 242: 193-197.

International Human Genome Sequencing Consortium (2001). Initial sequencing and analysis of the human genome. Nature 409: 860-921.

Isacchi A., Bernardi G., Bernardi G. (1993). Compositional compartmentalization of the nuclear genomes of *Trypanosoma brucei* and *Trypanosoma equiperdum*. FEBS Letters 335: 181-183.

Jabbari K. and Bernardi G. (1998). CpG doublets, CpG islands and Alu repeats in long human DNA sequences from different isochore families. Gene 224: 123-128.

Jabbari K. and Bernardi G. (2000). The distribution of genes in the *Drosophila* genome. Gene 247: 287-292.

Jabbari K. and Bernardi G. (2004a). A compositional analysis of the sequences of three fish genomes. Gene (in press)

Jabbari K. and Bernardi G. (2004b). On the relation between cytosine methylation and CpG shortage, TpG (CpA) excess and TpA Loss. Gene (in press).

Jabbari K., Cacciò S., Païs de Barros J.-P., Desgrès J. and Bernardi G. (1997). Evolutionary changes in CpG and methylation levels in vertebrate genomes. Gene 205: 109-118

Jabbari K., Cruveiller S., Clay O., Bernardi G. (2003a). The correlation between GC_3 and hydropathy in human genes. Gene 317: 137-140.

Jabbari K., Rayko E., Bernardi G. (2003b). The major shifts of human duplicated genes. Gene 317: 203-208.

Jabbari K., Clay O., Bernardi G. (2003c). GC_3 heterogeneity and body temperature in vertebrates. Gene 317: 161-163.

Jabbari K., Cruveiller S., Clay O., Le Saux J., Bernardi G. (2004). The new genes of rice: a closer look. Trends Plant Sci. (submitted).

Jackson D.A., Pombo A., Iborra F. (2000). The balance sheet for transcription: an analysis of nuclear RNA metabolism in mammalian cells. FASEB J. 14: 242-254.

Janke A., Erpenbeck D., Nilsson M., Arnason U. (2001). The mitochondrial genomes of the iguana (*Iguana iguana*) and the caiman (*Caiman crocodylus*): implications for amniote phylogeny. Proc. R. Soc. Lond. B. Biol. Sci. 268: 623-631.

Jankers J. and Berns A. (1996) Retroviral insertional mutagenesis as a strategy to identify cancer genes. Biochem. Biophys. Acta 1287: 29-57.

Jeffreys A.J. and Neumann R. (2002). Reciprocal crossover asymmetry and meiotic drive in a human recombination hot spot. Nature Genet. 31: 267-271.

Jensen R.H. and Davidson N. (1966). Spectrophotometric, potentiometric, and density gradient ultracentrifugation studies of the binding of silver ion by DNA. Biopolymers 4: 17-32.

Johannsen W. (1909). *Elemente der exakten Erblichkeitslehre*. Fischer, Jena, Germany.

Jolicoeur P. (1990). Bivariate allometry: Interval estimation of the slopes of the ordinary and standardized normal major axes and structural relationship. J. Theor. Biol. 144: 273-285.

Jordan I.K., Rogozin I.B., Glazko G.G., Koonin E.V. (2003). Origin of a substantial fraction of human regulatory sequences from transposable elements. Trends Genet. 19: 68-72

Josse J., Kaiser A.D., Kornberg A. (1961). Enzymatic synthesis of deoxyribonucleic acid. VIII. Frequencies of nearest neighbor base sequence in deoxyribonucleic acid. J. Biol. Chem. 236: 864-875.

Jukes T.H. and Bhushan V. (1986). Silent nucleotide substitutions and G + C content of some mitochondrial and bacterial genes. J. Mol. Evol. 24: 39-44.

Jurka J. (2000). Repbase update: a database and an electronic journal of repetitive elements. Trends Genet. 16: 418-420.

Jurka J. and Smith T.F. (1987). Beta turns in early evolution: chirality, genetic code, and biosynthetic pathways. Cold Spring Harbor Symp. Quant. Biol. 52: 407-410. Cold Spring Harbour, NY, USA.

Kadi F., Mouchiroud D., Sabeur G., Bernardi G. (1993). The compositional patterns of the avian genomes and their evolutionary implications. J. Mol. Evol. 37: 544-551.

Kagawa Y., Nojima H., Nukima N., Ishizuka M., Nakajima T., Yasuhara T., Tanaka T., Oshima T. (1984). High guanine plus cytosine content in the third letter codons of an extreme thermophile. J. Biol. Chem. 259: 2956-2960.

Kanaya S. Yamada Y., Kudo Y., Ikemura T. (1999). Studies of codon usage and tRNA genes of 18 unicellular organisms and quantification of *Bacillus subtilis* tRNAs: gene expression level and species-specific diversity of codon usage based on multivariate analysis. Gene 238: 143-155.

Kaneko T. et al. (2000). Complete genome structure of the nitrogen-fixing symbiotic bacterium *Mesorhizobium loti*. DNA Res. 6: 331-338.

Karlin S., Doerfler W., Cardon L.R. (1994). Why is CpG suppressed in the genomes of virtually all small eukaryotic viruses but not in those of large eukaryotic viruses? J. Virol. 68: 2889-2897.

Karlin S., Blaisdell B.E., Sapolsky R.J., Cardon L., Burge C. (1993). Assessments of DNA inhomogeneities in yeast chromosome III. Nucleic Acids Res. 21:703-11.

Katsanis N., Fitzgibbon J., Fisher E.M. (1996). Paralogy mapping: identification of a region in the human MHC triplicated onto human chromosomes 1 and 9 allows the prediction and isolation of novel PBX and NOTCH loci. Genomics 35: 101-108.

Kawarabayasi Y. et al. (1999). Complete genome sequence of an aerobic hyper-thermophilic crenarchaeon, *Aeropyrum pernix K1*. DNA Res. 6: 83-101, 145-152.

Kawasaki T. (2003). *Theory of chromatography*. Liguori, Naples, Italy (in press)

Keith P., Allardi J., Moutou B. (1992). Livre rouge des espèces menacées de poissons d'eau douce de France et bilan des introductions. Museum National d'Histoire Naturelle. Secretariat de la Faune et de la Flore, Conseil Supérieur de la Peche, CEMAGREF and Ministère de l'Environment, p. 111.

Kemp D., Thompson J., Walliker D., Corcoran L. (1987). Molecular karyotype of *Plasmodium falciparum* conserved linkage groups and expendable histidine-rich protein genes. Proc. Natl. Acad. Sci. USA 84: 7672-7676.

Kerem B.S., Goiten R., Diamond G., Cedar H., Marcus M. (1984). Mapping of DNAase I sensitive regions of mitotic chromosomes. Cell 38: 493-499.

Kerr A., Peden J., Sharp P. (1997). Systematic base composition variation around the genome of *Mycoplasma genitalium*, but not *Mycoplasma pneumoniae*. Mol. Microbiol. 25:1177-1184.

Kettmann R., Meunier-Rotival M., Cortadas J., Cuny G., Ghysdael J., Mammerickx M., Burny A., Bernardi G., (1979). Integration of bovine leukemia virus DNA in the bovine genome. Proc. Natl. Acad. Sci. USA 76: 4822-4826.

Kettmann R., Cleuter Y., Mammerickx M., Meunier-Rotival M., Bernardi G., Burny A., Chantrenne H., (1980). Genomic integration of bovine leukemia provirus: comparison of persistent lymphocytosis with lymph node tumor form of enzootic bovine leukemia. Proc. Natl. Acad. Sci. USA 77: 2577-2581.

Kimura M. (1968). Evolutionary rate at the molecular level. Nature 217: 624-626.

Kimura M. (1983). *The Neutral Theory of Molecular Evolution*. Cambridge University Press, Cambridge, U.K.

Kimura M. (1986). DNA and the neutral theory. Philos. Trans. R. Soc. Lond. B. Biol. Sci. 312: 343-354.

King J.L. and Jukes T.H. (1969). Non-Darwinian evolution. Science 164: 788-798.

Kirsch I.R., Green E.D., Yonescu R., Strausberg R., Carter N., Bentley D., Leversha M.A., Dunham I., Braden V.V., Hilgenfeld E., Schuler G., Lash A.E., Shen G.L., Martelli M., Kuehl W.M., Klausner R.D., Ried T. (2000). A systematic, high-resolution linkage of the cytogenetic and physical maps of the human genome. Nature Genet. 24: 339-340.

Kit S. (1960). Compositional heterogeneity of normal and malignant tissue deoxyribonucleic acids (DNA). Biochem. Biophys. Res. Comm. 3: 361-367.

Kit S. (1961). Equilibrium sedimentation in density gradients of DNA preparation from animal tissues. J. Mol. Biol. 3: 711-716.

Kit S. (1962). Species differences in animal deoxyribonucleic acids as revealed by equilibrium sedimentation in density gradients. Nature 193: 274-275.

Kondrashov A.S. (2000). Molecular darwinism. Trends Genet. 16, 580.

Kong A., Gudbjartsson D.F., Sainz J., Jonsdottir G.M., Gudjonsson S.A., Richardsson B., Sigurdardottir S., Barnard J., Hallbeck B., Masson G., Shlien A., Palsson S.T.,

Frigge M.L., Thorgeirsson T.E., Gulcher J.R., Stefansson K. (2002). A high-resolution recombination map of the human genome. Nature Genet. 31: 241-247.

Kopecka H., Chevallier M.R., Prunell A., Bernardi G. (1973). Degradation of transforming *Haemophilus influenzae* DNA by deoxyribonucleases. Biochim. Biophys. Acta 319: 37-47.

Kopecka H., Macaya G., Cortadas J., Thiery J.P., Bernardi G. (1978). Restriction enzyme analysis of satellite DNA components from the bovine genome. Eur. J. Biochem. 84: 189-195.

Korenberg J.R. and Engels W.R. (1978). Base ratio, DNA content, and quinacrine-brightness of human chromosomes. Proc. Natl. Acad. Sci. USA 75: 3382-3386.

Korenberg J. and Rikowski M. (1988). Human molecular organization: alu, Lines, and the molecular structure of metaphase chromosome bands. Cell 53: 391-400.

Kornberg A., Bertsch L.L., Jackson J.F., Khorana H.H. (1964). Enzymatic synthesis of DNA. XVI. Oligonucleotides as template and the mechanism of their replication. Proc. Natl. Acad. Sci. USA 51: 315.

Kramer B., Kramer W., Fritz H.J. (1984). Different base/base mismatches are corrected with different efficiencies by the methyldirected DANN mismatch repair system of *E. coli*. Cell 38: 879-887.

Krawczak M. and Cooper D.N. (1996). Single base-pair substitutions in pathology and evolution: two sides to the same coin. Hum. Mutat. 8: 23-31.

Kumar S. and Hedges S.B. (1998). A molecular time scale for vertebrate evolution. Nature 392: 917-920.

Kung H.J., Baerkoel C., Carters T.H. (1991). Retroviral mutagenesis of cellular oncogenes: a review with insights into the mechanisms of insertional activation. Curr. Top. Microbiol. Immunol. 171: 1-25.

Kunnath L. and Locker J. (1982). Characterization of DNA methylationn in the rat. Biochim. Biophys. Acta 699: 264-271.

Kyte J. and Doolittle R.F. (1982). A simple method for displaying the hydropathic character of a protein. J. Mol. Biol. 157: 105-132.

Lambert P.F., Kawashima E., Reznikoff W.S. (1987). Secondary structure at the bacteriophage G4 origin of complementary strand DNA synthesis: in vivo requirements. Gene 53: 257-264.

Lander E.S. et al., (2001). Initial sequencing and analysis of the human genome. Nature 409: 860-921.

Laskowski M. Sr. (1971). Deoxyribonuclease I. In *The enzymes* (P.D. Boyer, ed.) vol. IV, pp. 289-311. Academic Press, New York, NY, USA.

Laskowski M., Sr. (1982). Nucleases: historical perspectives. In *Nucleases.* (S.M. Linn and R.J. Roberts, eds.) pp. 1-22. Cold Spring Harbor Laboratory, NY, USA.

Lawrence J.G. and Ochman H. (1997). Amelioration of bacterial genomes: rates of change and exchange. J. Mol. Evol. 44: 383-397.

Lawrence S., Collins A., Keats B.J., Hulten M., Morton N.E. (1993). Integration of gene maps: chromosome 21. Proc. Natl. Acad. Sci. USA 90: 7210-7214.

Leclercq I., Mortreux F., Cavrois M., Leroy A., Gessain A., Wain-Hobson S., Wattel E. (2000). Host sequences flanking the human T-cell leukemia virus type 1 provirus in vivo. J. Virol. 74: 2305-2312.

Lee K.Y., Wahl R., Barbu E. (1956). Contenu en bases puriques et pyrimidiques des acides désoxyribonucléiques des bactéries. Ann. Inst. Pasteur 91: 212-224.

Leeds J.M., Slabaugh M.B., Mathews C.K. (1985). DNA precursor pools and ribonucleotide reductase activity: distribution between the nucleus and cytoplasm of mammalian cells. Mol. Cell. Biol. 5: 3443-3450.

Lennon G.G. and Fraser N.W. (1983). CpG frequency in large DNA segments. J. Mol. Evol. 19: 286-288.

Lercher M., Smith N.G.C, Eyre-Walker A., Hurst L. (2002). The evolution of isochores: evidence from SNP frequency distributions. Genetics 162: 1805-1810.

Lercher M., Urrutia A. O., Pavlicek A., Hurst L.D. (2003). A unification of mosaic structures in the human genome. Hum. Mol. Genet. 12: 2411-2415.

Levens D., Ticho B., Ackerman E., Rabinowitz M. (1981). Transcriptional initiation and 5' termini of yeast mitochondrial RNA. J. Biol. Chem. 256: 5226-5232.

Lewin A., Morimoto R., Rabinowitz M. (1978). Restriction enzyme analysis of mitochondrial DNAs of petite mutants of yeast: classification of petites, and deletion mapping of mitochondrial genes. Mol. Gen. Genet. 163: 257-275.

Leutwiler L.S., Hough-Evans B.R., Meyerowitz E.M. (1984). The DNA of *Arabidopsis thaliana*. Mol.Gen. Genet. 194: 15-23.

Li W. (2001). Delineating relative homogeneous C+G domains in DNA sequences. Gene 276: 57-72.

Li W.-H. (1997). *Molecular evolution*. Sinauer Associates, Sunderland, MA, USA.

Li W.-H and Graur D. (1991). *Fundamentals of molecular evolution.* Sinauer, Sunderland, MA, USA.

Li W.-H., Gouy M., Sharp P.M., D'Huigin C., Yang Y-.W. (1990). Molecular phylogeny of *Rodentia*, *Lagomorpha*, *Primates*, *Artiodactyla*, and *Carnivora* and molecular clocks. Proc. Natl. Acad. Sci. USA 87: 6703-6707.

Li W.-H., Gu Z., Wang H., Nekrutenko A. (2001). Evolutionary analyses of the human genome. Nature 409: 847-849.

Li W., Bernaola-Galvan P., Carpena P., Oliver J.L. (2003). Isochores merit the prefix 'iso'". Computational Biology and Chemistry 27: 5-10.

Lima-de-Faria A., Isaksson M., Olsson E. (1980). Action of restriction endonucleases on the DNA and chromosomes of *Muntiacus muntjak*. Hereditas 92: 267-73.

Lobry J.R. and Gautier C. (1994). Hydrophobicity, expressivity and aromaticity are the major trends of amino-acid usage in 999 *Escherichia coli* chromosome-encoded genes. Nucleic Acids Res. 22: 3174-3180.

Locker J., Rabinowitz M., Getz G.S. (1974). Tandem inverted repeats in mitochondrial DNA of petite mutants of *Saccharomyces cerevisiae*. Proc. Natl. Acad. Sci. USA 71: 1366-1370.

Lundin L.G. (1993). Evolution of the vertebrate genome as reflected in paralogous chromosomal regions in man and the house mouse. Genomics 16: 1-19.

Macaya G., Thiery J.P., Bernardi G. (1976). An approach to the organization of eukaryotic genomes at a macromolecular level. J. Mol. Biol. 108: 237-254.

Macaya G., Cortadas J., Bernardi G. (1978) An analysis of the bovine genome by density gradient centrifugation. Eur. J. Biochem. 84: 179-188.

Machon O., Hejnar J., Hajkowa P., Geryk J., Svoboda J. (1996). The LTR, *v-src*, LTR provirus in H-19 hasmter tumor cell line is integrated adjacent to the negative regulatory region. Gene 174: 9-17.

Macino G. and Tzagoloff A. (1979). Assembly of the mitochondrial membrane system: partial sequence of a mitochondrial ATPase gene in *Saccharomyces cerevisiae*. Proc. Natl. Acad. Sci. USA 76: 131-135.

Maquat L.E. and Carmichael G.G. (2001). Quality control of mRNA function. Cell 104: 173-176.

Marais G. (2003). Biased gene conversion: implications for genome and sex evolution. Trends Genet. 19: 330-338.

Marcus S.L., Smith S.W., Sarkar N.H., (1981). Quantitative of murine mammary tumor virus-related RNA in mammary tissues of low- and high-mammary-tumor-incidence mouse strains. J. Virol. 40: 87-95.

Margulis L. and Sagan D. (2002). *Acquiring genomes. A theory of the origins of species.* Basic Books, New York, NY, USA.

Marmur J. and Doty P. (1962). Determination of the base composition of deoxyribonucleic acid from its thermal denaturation temperature. J. Mol. Biol. 5: 109-118.

Marotta R., Colin Y., Goursot R., Bernardi G. (1982). A region of extreme instability in the mitochondrial genome of yeast. EMBO J. 1: 529-534.

Matassi G., Melis R., Kuo K.C., Macaya G., Gehrke C.W., Bernardi G. (1992). Large-scale methylation patterns in the nuclear genomes of plants. Gene 122: 239-245.

Matassi G., Melis R., Macaya G., Bernardi G. (1991). Compositional bimodality of the nuclear genome of tobacco. Nucleic Acids Res. 19: 5561-5567.

Matassi G., Montero L.M., Salinas J., Bernardi G. (1989). The isochore organization and the compositional distribution of homologous coding sequences in the nuclear genome of plants. Nucleic Acids Res. 17: 5273-5290.

Matsuo K., Clay O., Takahashi T., Silke J., Schaffner W. (1993). Evidence for erosion of mouse CpG islands during mammalian evolution. Somat. Cell Mol. Genet. 19: 543-555.

Mayr E. (1976). *Evolution and the diversity of life.* Harvard Univ. Press., Cambridge, MA, USA.

Mayr E. (1988). *Toward a new philosophy of biology. Observation of an evolutionist.* Harvard Univ. Press, Cambridge, MA, USA.

Mazrimas J.A. and Hatch F. T. (1972). A possible relationship between satellite DNA and the evolution of kangaroo rat species (genus *Dipodomys*). Nature New Biol. 240: 102-105.

McCarthy J.E. and Kollmus H. (1995). Cytoplasmic mRNA-protein interactions in eukaryotic gene expression. Trends Biochem. Sci. 20: 191-197.

McCutchan T.F., Dame J.B., Miller L.H., Barnwell J. (1984). Evolutionary relatedness of *Plasmodium* species as determined by the structure of DNA. Science 225: 808-811.

McCutchan T.F., Dame J.B., Gwadz R.W., Vernick K.D. (1988). The genome of *Plasmodium cynomolgi* is partitioned into separable domains which appear to differ in sequence stability. Nucleic Acids Res. 16: 4499-4510.

McDonald J.H. (2001). Patterns of temperature adaptation in proteins from the bacteria *Deinococcus radiodurans* and *Thermus thermophilus*. Mo. Biol. Evol. 18: 741-749.

McGeoch D.J. (1970). Some base sequence charachteristics of deoxyribonucleic acid. PhD thesis. University of Glasgow. U.K.

McKenna M.C. and Bell S.K. (1997). *Classification of mammals above the species level.* Columbia Univ. Press, New York, NY, USA.

McLysaght A., Hokamp K., Wolfe K.H. (2002). Extensive genomic duplication during early chordate evolution. Nature Genet. 31: 200-204.

Medrano L., Bernardi G., Couturier J., Dutrillaux B., Bernardi G. (1988). Chromosome banding and genome compartmentalization in fishes. Chromosoma 96: 178-183.

Mehrotra B.D. and Mahler H.R. (1968). Characterization of some unusual DNAs from the mitochondria from certain "petite" strains of *Saccharomyces cerevisiae*. Arch. Biochem. Biophys. 128: 685-703.

Melville S.E., Leech V., Navarro M., Cross G.A.M. (2000). The molecular karyotype of the megabase chromosomes of *Trypanosoma brucei* stock 427. Mol. Biochem. Parasit. 111: 261-273.

Mery-Drugeon E., Crouse E.J., Schmitt J., Bohnert H.J., Bernardi G. (1981). The mitochondrial genomes of *Ustilago cynodontis* and *Acanthamoeba castellanii*. Eur. J. Biochem. 114: 577-583.

Meselson M., Stahl F.W., Vinograd J. (1957). Equilibrium sedimentation of macromolecules in density gradients. Proc. Nat. Acad. Sci. USA 43: 581-588.

Meunier-Rotival M., Cortadas J., Macaya G., Bernardi G. (1979). Isolation and organization of calf ribosomal DNA. Nucleic Acids Res. 6: 2109-2123.

Meunier-Rotival M., Soriano P., Cuny G., Strauss F., Bernardi G., (1982). Sequence organization and genomic distribution of the major family of interspersed repeats of mouse DNA. Proc. Natl. Acad. Sci. USA 79: 355-359.

Meyer A. and Schartl M. (1999). Gene and genome duplications in vertebrates: the one-to-four (-to-eight in fish) rule and the evolution of novel gene functions. Curr. Opin. Cell Biol. 11: 699-704.

Meyerowitz E.M. (1992). Introduction to the *Arabidopsis* genome. In *Methods in Arabidopsis Research*, (C. Koncz, N. Chua and J. Shell, eds.) pp. 100-118, World Scientific Publishing, Singapore.

Meyers B.C., Tingey S.V., Morgante M. (2001). Abundance, distribution and transcriptional activity of repetitive elements in the maize genome. Genome Res. 11: 1660-1676.

MHC Sequencing Consortium (1999). Complete sequence and gene map of a human major histocompatibility complex. Nature 401: 921-923.

Michels P.A. (1987). Evolutionary aspects of trypanosomes: analysis of genes. J. Mol. Evol. 24: 45-52.

Mills D.R., Peterson R.L., Spiegelman S. (1967). An extracellular Darwinian experiment with a self-duplicating nucleic acid molecule. Proc. Natl. Acad. Sci. USA 58: 217

Mirsky A. and Ris H. (1949). Varable and constant components of chromosomes. Nature 163: 666-667.

Mirsky A. and Ris H. (1951). The deoxyribonucleic acid content of animal cells and its evolutionary significance J. Gen. Physiol. 34: 451-462.

Montero L.M., Salinas J., Matassi G., Bernardi G. (1990). Gene distribution and isochore organization in the nuclear genome of plants. Nucleic Acids Res. 18: 1859-1867.

Moore M.J. (2002). Nuclear RNA turnover. Cell 108, 431-434.

Moore G., Abbo S., Cheung W., Foote T., Gale M., Koebner R., Leitch A., Leitch I., Money T., Stancombe P., Yano M., Flavell R. (1993). Key features of cereal genome organization as revealed by the use of cytosine methylation-sensitive restriction endonucleases. Genomics 15: 472-482.

Mooslehner K., Karl U., Harbers K. (1990). Retroviral sites in transgenic *Mov* mice frequently map in the vicinity of transcribed DNA region. J. Virol. 64: 3056-3058.

Morescalchi M.A., Schempp W., Consigliere S., Bigoni F., Wienberg J., Stanyon R. (1997). Mapping chromosomal homology between humans and the black-handed spider monkey by fluorescence in situ hybridization. Chromosome Res. 5: 527-536.

Moriyama E.N. and Gojobori T. (1992). Rates of synonymous substitution and base compostion of nuclear genes in *Drosophila*. Genetics 130: 855-864.

Mouchiroud D. and Bernardi G. (1993). Compositional properties of coding sequences and mammalian phylogeny. J. Mol. Evol. 37: 109-116

Mouchiroud D., Fichant G., Bernardi G. (1987). Compositional compartmentalization and gene composition in the genome of vertebrates. J. Mol. Evol. 26: 198-204

Mouchiroud D., Gautier C., Bernardi G. (1988) The compositional distribution of coding sequences and DNA molecules in humans and murids. J. Mol. Evol. 27: 311-320

Mouchiroud D., D'Onofrio G., Aïssani B., Macaya G., Gautier C., Bernardi G. (1991). The distribution of genes in the human genome. Gene 100: 181-187.

Mouchiroud D., Gautier C., Bernardi G. (1995) Frequencies of synonymous substitutions in mammals are gene-specific and correlated with frequencies of non-synonymous substitutions. J. Mol. Evol. 40: 107-113.

Mouse Genome Sequencing Consortium (2002). Initial sequencing and comparative analysis of the mouse genome. Nature 420: 520-562.

Mungall A.J. (2003). The DNA sequence and analysis of human chromosome 6. Nature 425: 805-811.

Murphy W.J., Eizirik E., O'Brien S.J., Madsen O., Scally M., Douady C.J., Teeling E., Ryder O.A., Stanhope M.J., de Jong W.W., Springer M.S. (2001). Resolution of the early placental mammal radiation using Bayesian phylogenetics. Science 294: 2348-2351.

Musto H., Cacciò S., Rodriguez-Maseda H., Bernardi G. (1997). Compositional constraints in the extremely GC-poor genome of *Plasmodium falciparum*. Mem. Inst. Oswaldo Cruz 92: 835-841.

Musto H., Rodriguez-Maseda H., Bernardi G. (1994). The nuclear genomes of African and American trypanosomes are strikingly different. Gene 141: 63-69.

Musto H., Rodriguez-Maseda H., Bernardi G. (1995). Compositional properties of nuclear genes from *Plasmodium falciparum.* Gene 152: 127-132.

Musto H., Romero H., Zavala A., Jabbari K., Bernardi G. (1999). Synonymous codon choices in the extremely GC-poor genome of *Plasmodium falciparum*: compositional constraints and translational selection. J. Mol. Evol. 49: 27-35.

Musto H., Cruveiller S., D'Onofrio G., Romero H. and Bernardi G (2001). Translational selection on codon usage in *Xenopus laevis*. Mol. Biol. Evol. 18: 1703-1707.

Musto H., Zavala A., Bernardi G. (2004). Genomic GC levels are correlated with optimal growth temperature in families of prokaryotes. Proc. Natl. Acad. Sci. USA (submitted).

Muto A. and Osawa S. (1987). The guanine and cytosine content of genomic DNA and bacterial evolution. Proc. Natl. Acad. Sci. USA 84: 166-169.

Myers E.W. et al., (2000). A whole-genome assembly of *Drosophila*. Science 287: 2196-2204.

Nachman M. (2002). Variation in recombination rate across the genome: evidence and implications. Curr. Opin. Genet. Develop. 12: 657-663.

Nagley P. and Linnane A.W. (1970). Mitochondrial DNA deficient petite mutants of yeast. Biochem. Biophys. Res. Commun. 39: 989-996.

Naya H., Romero H., Zavala A., Alvarez B., Musto H. (2002). Aerobiosis increases the genomic guanine plus cytosine content (GC%) in prokaryotes. J. Mol. Evol. 55: 260-264

Naylor G.J., Collins T.M., Brown W.M. (1995). Hydrophobicity and phylogeny. Nature 373: 565-566.

Nei M. (1987). *Molecular Evolutionary Genetics.* Columbia University Press, New York, NY, USA.

Neil J.C. (1983). Defective Avian Sarcoma Viruses. Curr. Top. Microbiol. Immunol. 103: 51-75.

Nekrutenko A. and Li W.H. (2000). Assessment of compositional heterogeneity within and between eukaryotic genomes. Genome Res. 10: 1986-1985.

Nekrutenko A. and Li W.H. (2001). Transposable elements are found in a large number of human protein-coding genes. Trends Genet. 17: 619-621.

Nelson J.S. (1994). *Fishes of the world,* 2nd edition. Wiley, New York, NY, USA.

Nghiem Y., Cabrera M., Cupples C.G., Miller J.H. (1988). The *mutY* gene: a mutator locus in Escherichia coli that generates GC→TA transversions. Proc. Natl. Acad. Sci. USA 85: 2709-2713.

Nishio Y., Nakamura Y., Kawarabayasi Y., Usuda Y., Kimura E., Sugimoto S., Matsui K., Yamagishi A., Kikuchi H., Ikeo K., Gojobori T. (2003). Comparative complete genome sequence analysis of the amino acid replacements responsible for the thermostability of *Corynebacterium efficiens.* Genome Res. 13: 1572-1579.

Norell M., Ji Q., Gao K, Yuan C., Zhao Y., Wang L. (2002). 'Modern' feathers on a non-avian dinosaur. Nature 416: 36-37.

Nowak R.M. and Paradiso J. (1983). *Walker's mammals of the world,* 4th edition. Johns Hopkins University Press, Baltimore and London.

Ochman H. (2001). Lateral and oblique gene transfer. Curr. Opin. Genet. Dev. 11: 616-619.

Ochman H., Lawrence J.G., Groisman E.A.(2000). Lateral gene transfer and the nature of bacterial innovation. Nature 405: 299-304.

Ohno S. (1970). *Evolution by gene duplication.* Springer, Berlin, Germany.

Ohno S. (1972). So much "junk" DNA in our genome. Brookhaven Symp. Biol. 23: 366-370.

Ohno S. (1999). The one-to-four rule and paralogues of sex-determining genes. Cell. Mol. Life Sci. 55: 824-830.

Ohta T. (1972). Population size and rate evolution. J. Mol. Evol. 1: 150-157

Ohta T. (1992). The nearly neutral theory of molecular evolution. Annu. Rev. Ecol. Syst. 23: 263-286.

Ohta T. (2002). Near-neutrality in evolution of genes and gene regulation. Proc. Natl. Acad. Sci. USA 99: 16134-16137.

Ohta T. and Ina Y. (1995). Variation in synonymous substitution rates among mammalian genes and the correlation between synonymous and nonsynonymous divergences. J. Mol. Evol. 41: 717-720.

Okamuro J.K. and Goldberg R.B. (1985). Tobacco single-copy DNA is highly homologous to sequences present in the genomes of its diplois progenitors. Mol. Gen. Genet. 198: 290-298.

Oliver J.L., Bernaola-Galvan P., Carpena P., Roman-Roldan R. (2001). Isochore chromosome maps of eukaryotic genomes. Gene 276: 47-56.

Oliver J.L., Carpena P., Roman-Roldan R., Mata-Balaguer T., Mejias-Romero A., Hackenberg M., Bernaola-Galvan P. (2002). Isochore chromosome maps of the human genome. Gene 300: 117-127.

Olmo E. (1981). Evolution of genome size and DNA base composition in reptiles. Genetica 57: 39-50.

Olmo E. (2003). Reptiles: a group of transition in the evolution of genome size and of the nucleotypic effect. Cytogenet. Genome Res. (in press).

Olmo E., Capriglione T., Odierna G. (1989). Genome size evolution in vertebrates: trends and constraints. Comp. Biochem. Physiol. B. 92:447-53.

Olmo E., Capriglione T., Odierna G. (2002). Different genomic evolutionary rates in the various reptile lineages. Gene 295: 317-321.

Olofsson B. and Bernardi G. (1983a). Organization of nucleotide sequences in the chicken genome. Eur. J. Biochem. 130: 241-245

Olofsson B. and Bernardi G. (1983b). The distribution of CR1, an Alu-like family of interspersed repeats in the chicken genome. Biochim. Biophys. Acta 740: 339-341

Orgel L.E. and Crick F.H. (1980). Selfish DNA: the ultimate parasite. Nature 284: 604-607.

Osinga K.A. and Tabak H.F. (1982). Initiation of transcription of genes for mitochondrial ribosomal RNA in yeast: comparison of the nucleotide sequence around the 5'-ends of both genes reveals a homologous stretch of 17 nucleotides. Nucleic Acids Res. 10: 3617-3626.

Ostashevsky J. (1998). A polymer model for the structural organization of chromatin loops and minibands in interphase chromosomes. Mol. Biol. Cell 9: 3031-3040.

Ostertag E.M. and Kazazian H.H. (2001). Biology of mammalian L1 retrotransposons. Annu. Rev. Genet. 35: 501-538.

Owen R. (1866). *On the anatomy of vertebrates*, Vol. I. Longmans, Green, London, U.K.

Pačes J., Zíka R., Pačes V., Pavlíček A., Clay O., Bernardi G. (2004). Representing GC variation along eukaryotic chromosomes. Gene (in press).

Panstruga R., Buschges R., Piffanelli P., Schulze-Lefert P. (1998). A contiguous 60 kb genomic stretch from barley reveals molecular evidence for gene islands in a monocot genome. Nucleic Acids Res. 26: 1056-1062.

Pavliček A, Jabbari K, Pačes J, Pačes V, Hejnar J, Bernardi G (2001) Similar integration but different stability of Alus and LINEs in the human genome. Gene 276: 39-45.

Pavliček A., Paces J., Clay O., Bernardi G. (2002a). A compact view of isochores in the draft human genome sequence. FEBS Letters 511: 165-169.

Pavliček A., Clay O., Jabbari K., Pačes J., Bernardi G. (2002b). Isochore conservation between MHC regions on human chromosome 6 and mouse chromosome 17. FEBS Letters 511: 175-177.

Pearson W.R., Wood T., Zhang Z., MillerW.(1997). Comparison of DNA sequences with protein sequences. Genomics 46: 24-36

Pebusque M.J., Coulier F., Birnbaum D., Pontarotti P. (1998). Ancient large-scale genome duplications: phylogenetic and linkage analyses shed light on chordate genome evolution. Mol. Biol. Evol. 15: 1145-1159.

Pech M., Streeck R.E., Zachau H.G. (1979). Patchwork structure of a bovine satellite DNA. Cell 18: 883-93.

Perrin P. and Bernardi G. (1987). Directional fixation of mutations in vertebrate evolution. J. Mol. Evol. 26: 301-310.

Perrins C.M. and Middleton A.L.A. (eds) (1985). *The encyclopedia of birds*. Allen & Unwin, London, U.K.

Pesole G., Liuni S., Grillo G., Saccone C. (1997). Structural and compositional features of untranslated regions of eukaryotic mRNAs. Gene 205: 95-102.

Pesole G., Bernardi G., Saccone C. (1999). Isochore specificity of AUG initiator context of human genes. FEBS Letters 464: 60-62.

Pesole G., Gissi C., Grillo G., Licciulli F., Liuni S., Saccone C, (2000). Analysis of oligonucleotide AUG start codon context in eukariotic mRNAs. Gene 261: 85-91.

Pesole G., Mignone F., Gissi C., Grillo G., Licciulli F., Liuni S. (2001). Structural and functional features of eukaryotic mRNA untranslated regions. Gene 276:73-81.

Pesole G., Liuni S., Grillo G., Licciulli F., Mignone F., Gissi C, Saccone C. (2002). UTRdb and UTRsite: specialized databases of sequences and functional elements of 5' and 3' untranslated regions of eukaryotic mRNAs. Update 2002. Nucleic Acids Res. 30: 335-340.

Petranovic M., Vlahovic K., Zahradka D., Dzidic S., Radman M. (2000). Mismatch repair in xenopus egg extracts is not strand-directed by DNA methylation. Neoplasma 47: 375-381.

Petrov D. A. (2001). Evolution of genome size: new approaches to an old problem. Trends Genet. 17: 23-28.

Pilia G., Little R.D., Aïssani B., Bernardi G., Schlessinger D. (1993) Isochores and CpG islands in YAC contigs in human Xq26.1-qter. Genomics 17: 456-462

Piperno G. and Bernardi G. (1971) Separation of nucleosides on polyacrylamide gel columns. Further developments. Biochim. Biophys. Acta 238: 388-396.

Pizon V., Cuny G., Bernardi G. (1984). Nucleotide sequence organization in the very small genome of a tetraodontid fish, *Arothron diadematus*. Eur. J. Biochem. 140: 25-30.

Pollak Y., Katzen A., Spira D, Golenser J. (1982). The genome of *Plasmodium falciparum*. I: DNA composition. Nucleic Acids Res. 10: 539-546.

Pollock Jr.J.M., Swihart M., Taylor J.H. (1978). Methylation of DNA in early development: 5-methyl cytosine content of DNA in sea urchin sperm and embryos. Nucleic Acids Res. 5: 4855-4861.

Ponger L., Duret L., Mouchiroud D. (2001). Determinants of CpG islands: expression in early embryo and isochore structure. Genome Res. 11: 1854-1860.

Poso D., Sessions R.B., Lorch M., Clarke A.R. (2000). Progressive stabilization of intermediate and transition states in protein folding reactions by introducing surface hydrophobic residues. J. Biol. Chem. 275: 33723-33726.

Pradet-Balade B., Boulmé F., Beug H., Müllner E.W., Garcia-Sanz J.A. (2001). Translation control: bridging the gap between genomics and proteomics? Trends Biochem. 26: 225-229.

Prunell A. and Bernardi G. (1977). The mitochondrial genome of wild-type yeast cells. VI. Organization and genetic units. J. Mol. Biol. 110: 53-74.

Prunell A., Kopecka H., Strauss F., Bernardi G. (1977). The mitochondrial genome of wild-type yeast cells. V. Size, homogeneity and evolution. J. Mol. Biol. 110: 17-52.

Raudsepp T., Fronicke L., Scherthan H., Gustavsson I., Chowdhary B.P. (1996) Zoo-FISH delineates conserved chromosomal segments in horse and man. Chromosome Res. 4: 218-225.

Rayko E., Goursot R., Cherif-Zahar B., Melis R., Bernardi G. (1988). Regions flanking *ori* sequences affect the replication efficiency of the mitochondrial genome of *ori* + petite mutants from yeast. Gene 63: 213-226.

Rayko E., Jabbari K., Bernardi G. (2003). Introns and CpG-islands features of human duplicated genes. Gene (submitted).

Ream R., Johns G., Somero G. (2003). Base compositions of genes encoding a-actin and lactate dehydrogenase-A from differently adapted vertebrates show no temperature-adaptive variation in G+C content. Mol. Biol. Evol. 20: 105-110.

Reyes A., Pesole G., Saccone C. (1998). Complete mitochondrial DNA sequence of the fat dormouse, Glis glis: further evidence of rodent paraphyly. Mol. Biol. Evol. 15: 499-505

Rice Chromosome 10 Sequencing Consortium (2003). In-depth view of structure, activity and evolution of Rice Chromosome 10. Science 300: 1566-1569.

Ridley M. (2004). *Evolution.* Third Edition. Blackwell Publishing, Oxford, U.K.

Robinson H.L. and Gagnon G., (1986). Patterns of proviral insertion and deletion in avian leukosis virus-induced lymphomas. J. Virol. 57: 28-36.

Robinson H., Gao Y., Mccray B.S., Edmondson S.P., Shriver J.W., Wang A.H.J. (1998). The hyperthermophyle chromosomal protein *Sac7d* sharply kinks DNA. Nature 392: 202-205.

Roelofs J. and Van Haaster M. (2001) Genes lost during evolution. Nature 411: 1013-1014.

Rohdewohld H., Weinher H., Reik W., Jaenisch R., Breindl M. (1987). Retrovirus integration and chromatine structure: Moloney murine leukemia proviral integation sites map near DNase I-hypersensitive sites. J. Virol. 61: 336-343.

Rolfe R. and Meselson M. (1959). The relative homogeneity of microbial DNA. Proc. Natl. Acad. Sci. USA 45: 1039-1043.

Romani M., Casciano J., Querzola F., De Ambrosis A., Siniscalco M. (1993). Analysis of a viral integration event in a GC-rich region at the 1p36 human chromosomal site. Gene 135: 153-160.

Romanov G.A. and Vanyushin B.F. (1981). Methylation of reiterated sequences in mammalian DNAs. Effects of the tissue type, age, malignancy and hormonal induction. Biochim. Biophys. Acta 653: 204-218.

Roman-Roldan R., Bernaola-Galvan P., Oliver J. (1998). Sequence compositional complexity of DNA through an entropic segmentation algorithm. Phys. Rev. Letts. 80: 1344-1347.

Romero H., Zavala A., Musto H., Bernardi G. (2003). The influence of translational selection on codon usage in fishes from the family *Cyprinidae.* Gene 317: 141-147.

Rose G.D., Geselowitz A.R., Lesser G.J., Lee R.H., Zehfus M.H. (1986). Hydrophobicity of amino acid residues in globular proteins. Science 229: 834-838.

Ross M.T. (2003). L isochore map: gene-poor isochores. In *Nature Encyclopedia of the Human Genome* (D.N. Cooper, ed.) vol. 3, pp. 729-733, Nature Publishing Group. London, UK

Ruelle D. (1989). *Chaotic evolution and strange attractors: the statistical analysis of time series for deterministic nonlinear systems*. Cambridge University Press, Cambridge, U.K.

Russell G.J. (1974). Characterization of deoxyribonucleic acids by doublet frequency analysis. Ph.D. Thesis. University of Glasgow, U.K.

Russell G.J. and Subak-Sharpe J.H. (1977). Similarity of the general designs of protochordates and invertebrates. Nature 266: 533-536.

Russell G.J., Walker P.M., Elton R.A., Subak-Sharpe J.H. (1976). Doublet frequency analysis of fractionated vertebrate nuclear DNA. J. Mol. Biol. 108:1-23.

Rynditch A., Kadi F., Geryk J., Zoubak S., Svoboda J., Bernardi G., (1991). The isopycnic, compartmentalized integration of Rous. Gene 106: 165-172.

Rynditch A., Zoubak S., Tsyba L., Tryapitsina-Guley N., Bernardi G. (1998). The regional integration of retroviral sequences into the mosaic genomes of mammals. Gene 222: 1-16.

Sabeur G., Macaya G., Kadi F., Bernardi G. (1993). The isochore patterns of mammalian genomes and their phylogenetic implications. J. Mol. Evol. 37: 93-108.

Saccone S. and Bernardi G. (2001). Human chromosomal banding by *in situ* hybridization of isochores. Methods in Cell Science 23: 7-15.

Saccone C., Cacciò S., Perani P., Andreozzi L., Rapisarda A., Motta S., Bernardi G. (1997). Compositional mapping of mouse chromosomes and identification of the gene-rich regions. Chromosome Res. 5: 293-300.

Saccone S., De Sario A., Della Valle G., Bernardi G. (1992) The highest gene concentrations in the human genome are in T bands of metaphase chromosomes. Proc. Natl. Acad. Sci. USA 89: 4913-4917

Saccone S., De Sario A., Wiegant J., Rap A.K., Della Valle G., Bernardi G. (1993). Correlations between isochores and chromosomal bands in the human genome. Proc. Natl. Acad. Sci. USA 90: 11929-11933.

Saccone S., Cacciò S., Kusuda J., Andreozzi L., Bernardi G. (1996) Identification of the gene-richest bands in human chromosomes. Gene 174: 85-94.

Saccone S.., Cacciò S., Perani P., Andreozzi L., Rapisarda A., Motta S., Bernardi G. (1997). Compositional mapping of mouse chromosomes and identification of the gene-rich regions. Chromosome Res. 5: 293-300.

Saccone S., Federico C., Solovei I., Croquette M.F., Della Valle G., Bernardi G. (1999). Identification of the gene-richest bands in human prometaphase chromosomes. Chromosome Res. 7: 379-386.

Saccone S., Pavlicek A., Federico C., Paces J., Bernardi G. (2001). Genes, isochores and bands in human chromosomes 21 and 22. Chromosome Res. 9: 533-539.

Saccone S., Federico C., Andreozzi L., D'Antoni S., Bernardi G. (2002). Localization of the gene-richest and the gene-poorest isochores in the interphase nuclei of mammals and birds. Gene 300: 169-178.

Sadoni N., Langer S., Fauth C., Bernardi G., Cremer T., Turner B.M., Zink D. (1999). Functional nuclear organization of mammalian genome. J. Cell Biology 146: 1211-1226.

Saito A., Abad J.P., Wang D., Ohki M., Cantor C., Smith C.L. (1991). Construction and characterization of a *Not*I linking library of human chromosome 21. Genomics 10: 618-630.

Salemme F.R., Miller M.D., Jordan S.R.. (1977). Structural convergence during protein evolution. Proc. Natl. Acad. Sci. USA 74: 2820-2824.

Salinas J., Zerial M., Filipski J. and Bernardi G. (1986). Gene distribution and nucleotide sequence organization in the mouse genome. Eur. J. Biochem. 160: 469-478.

Salinas J., Zerial M., Filipski J., Crépin M., Bernardi G., (1987). Non-random distribution of MMTV proviral sequences in the mouse genome. Nucleic Acids Res. 15: 3009-3022.

Salinas J., Matassi G., Montero L.M., Bernardi G. (1988) Compositional compartmentalization and compositional patterns in the nuclear genomes of plants. Nucleic Acids Res. 16: 4269-4285.

Salser W. (1977). Globin mRNA sequences: analysis of base pairing and evolutionary implications. Cold Spring Harbour Symp. Quant. Biol. 40: 985-1002. Cold Spring Harbour, NY, USA.

Samejima T. and Yang J.T. (1965). Optical rotatory dispersion and conformation of deoxyribonucleic and ribonucleic acids from various sources. J. Biol. Chem. 240: 2094-2100.

Samonte R.V. and Eichler E.E. (2002). Segmental duplications and the evolution of the primate genome. Nature Rev. Genet. 3: 65-72.

Sandhu D. and Gill K.S. (2002). Gene-containing regions of wheat and the other grass genomes. Plant Physiol. 128: 803-811

SanMiguel P., Tikhonov A., Jin Y.K., Motchoulskaia N., Zakharov D., Melake-Berhan A., Springer P.S., Edwards K.J., Lee M., Avramova Z., Bennetzen J.L. (1996). Nested retrotransposona in the intergenic regions of the maize genome. Science 274: 765-768.

SanMiguel P., Gaut B.S., Tikhonov A., Nakajima Y., Bennetzen J.L. (1998). The paleontology of intergene retrotranspons of maize. Nature Gen. 20: 43-45.

SanMiguel P.J., Ramakrishna W., Bennetzen J.L., Busso C.S., Dubcovsky J. (2002). Transposable elements, genes and recombination in a 215-kb contig from wheat chromosome 5A(m). Funct. Integr. Genomics 2: 70-80.

SantaLucia J., Allawi H., Seneviratne P.A. (1996). Improved nearest-neighbor parameters for predicting DNA duplex stability. Biochemistry 35: 3555-3562.

Sasaki T. et al. (2002). The genome sequence and structure of rice chromosome 1. Nature 420: 312-316.

Satoh G, Takeuchi JK, Yasui K, Tagawa K, Saiga H, Zhang P, Satoh N. (2002). Amphi-Eomes/Tbr1: an amphioxus cognate of vertebrate Eomesodermin and T-Brain1 genes whose expression reveals evolutionarily distinct domain in amphioxus development. J. Exp. Zool. 294: 136-45.

Sawyer J.R. and Hozier J.C. (1986). High resolution of mouse chromosomes: banding conservation between man and mouse. Science 232: 1632-1635.

Scherdin V., Rhodes K., Brendl M. (1990). Transcriptionally active genome regions and preferred targets for retrovirus integration. J. Virol. 64: 907-912.

Schattschneider D. (1990). *Vision of symmetry. Notebooks, periodic drawings and related work of M.C. Escher*. W.H. Freeman & Company, New York, USA.

Scherf A., Hilbich C., Sieg K., Mattei D., Mercereau-Puijalon O., Müller-Hill B. (1988). The 11-1 gene of *Plasmodium falciparum* codes for distinct fast evolving repeats. EMBO J. 7: 1129-1137.

Scherthan H., Cremer T., Arnason U., Weier H.U., Lima-de-Faria A., Fronicke L. (1994). Comparative chromosome painting discloses homologous segments in distantly related mammals. Nature Genet. 6: 342-347.

Schildkraut C.L., Marmur J., Doty P. (1962). Determination of the base composition of deoxyribonucleic acid from its buoyant density in CsCl. J. Mol. Biol. 4: 430-443.

Schimke R.T. (ed.) (1982). *Gene amplification*. Cold Spring Harbor, New York, NY, USA

Schmalhausen I.I. (1949). *Factors of evolution. The theory of stabilizing selection*. The Blakiston Company, Philadelphia, PA, USA.

Schmid C.W. (1998). Does SINE evolution preclude Alu function? Nucleic Acids Res. 26: 4541-4550.

Schmid C.W. and Hearst J.E. (1972). Sedimentation equilibrium of DNA samples heterogeneous in density. Biopolymers 11: 1913-1918.

Schmid M. (1978). Chromosome banding in amphibians. Chromosoma 68, 131-148.

Schmid M. and Guttenbach M. (1988) Evolutionary diversity of reverse (R) fluorescent chromosome bands in vertebrates. Chromosoma 97: 101-114.

Schmitt J.M., Bohnert H.-J., Gordon K.H.J., Herrmann R., Bernardi G., Crouse E.J. (1981). Compositional heterogeneity of the chloroplast DNAs from *Euglena gracilis* and *Spinacea oleracea*. Eur. J. Biochem. 114: 375-382.

Schröder A.R.W., Shinn P., Chen H., Berry C., Ecker J.R., Bushman F. (2002). HIV-1 integration in the human genome favors active genes and local hotspots. Cell 110: 165-172.

Schubach W. and Groudine M., (1984). Alteration of c-myc chromatin structure by avian leukosis virus integration. Nature 307: 702-708.

Schweizer D. (1977). R-banding produced by DNase I digestion of chromomycin-stained chromosomes. Chromosoma 64: 117-124.

Seebacher F., Grigg G.C., Beard L.A. (1999). Crocodiles as dinosaurs: behavioural thermoregulation in very large ectotherms leads to high and stable body temperatures. J. Experim. Biolog. 202: 77-86.

Serrano J., Kuehl D.W., Naumann S. (1993). Analytical procedure and quality assurance criteria for the determination of major and minor deoxynucleosides in fish tissue DNA by liquid chromatography-ultraviolet spectroscopy and liquid chromatography-thermo. J. Chromatogr. 615: 203-213.

Shabalina S.A., Ogurtsov A.Y., Kondrashov V.A., Kondrashov A.S. (2001). Selective constraint in intergenic regions of human and mouse genomes. Trends Genet. 17: 373-376.

Shapiro H.S. (1976). Distribution of purines and pyrimidines in deoxyribonucleic acids. In *Handbook of biochemistry and molecular biology*. (G.D. Fasman, ed.) vol. II, 3rd edition, pp. 241-281, CRC Press, Cleveland, OH, USA.

Sharp P.M. and Li W.H. (1986). An evolutionary perspective on synonymous codon usage in unicellular organisms. J. Mol. Evol. 24: 28-38.

Sharp P.M. and Li W.H. (1989). On the rate of DNA sequence evolution in Drosophila. J. Mol. Evol. 28: 398-402

Sharp P.M. and Lloyd A.T. (1993). Regional base composition variation along yeast chromosome III: evolution of chromosome primary structure. Nucleic Acids Res. 21: 179-183.

Sharp P.M. and Matassi G. (1994). Codon usage and genome evolution. Curr. Opin. Genet. Dev. 4: 851-860.

Sharp P.M., Cowe E., Higgins D.G., Shields D.C., Wolfe K.H., Wright F. (1988). Codon usage patterns in *Escherichia coli, Bacillus subtilis, Saccharomyces cerevisiae, Schizosaccharomyces pombe, Drosophila melanogaster* and *Homo sapiens*; a review of the considerable within species diversity. Nucleic Acids Res. 16: 8207-8211.

Sharp P.M., Averof M., Lloyd A.T., Matassi G., Peden J.F. (1995). DNA sequence evolution: the sounds of silence. Philos. Trans. R. Soc. Lond. B Biol. Sci. 349: 241-247.

Shen J.C., Rideout W.M., Jones P.A. (1994). The rate of hydrolytic deamination of 5-methylcytosine in double stranded DNA. Nucleic Acids Res. 22: 972-976.

Shields R. (1993). Pastoral synteny. Nature 365: 297-298.

Shields D.C., Sharp P.M., Higgins D.G., Wright F. (1988). "Silent" sites in *Drosophila* genes are not neutral: evidence of selection among synonymous codons. Mol. Biol. Evol. 5: 704-716.

Shimizu M. (1982). Molecular basis for the genetic code. J. Mol. Evol., 18: 297-303.

Shiryayev, A. (1984). *Probability*. Springer-Verlag, New York, NY, USA.

Sibley C.G and Monroe Jr. B.L. (1990). *Distribution and taxonomy of birds of the world.* Yale University Press, New Haven, CT, USA.

Sidow A. (1996). Gen(om)e duplications in the evolution of early vertebrates. Curr. Opin. Genet. Dev. 6: 715-722.

Siegfried Z. and Cedar H. (1997). DNA methylation: a molecular lock. Curr. Biol. 7: R305-R307.

Singer M and Berg P. (1991). *Genes and Genomes. A changing perspective*. p.622. University Science Books, Mill Valley, CA, USA.

Sinsheimer R.L. (1955). The action of pancreatic deoxyribonuclease II. Isomeric dinucleotides. J. Biol. Chem. 215: 579.

Skalka A., Fowler A.V., Hurwitz J. (1966). The effect of histones on the enzymatic synthesis of ribonucleic acid. J. Biol. Chem. 241: 588-596.

Smit A.F. (1996). The origin of interspersed repeats in the human genome. Curr. Opin. Genet. Dev. 6: 743-748.

Smit A.F. (1999). Interspersed repeats and other mementos of transposable elements in mammalian genomes. Curr. Opin. Genet. Dev. 9: 657-663.

Smith H.O. and Wilcox K.W. (1970). A restriction enzyme from *Haemophilus influenzae.* I. Purification and general properties. J. Mol. Biol. 51: 379-391.

Smith K.N. and Nicolas A. (1998). Recombiantion at work for meiosis. Curr. Opin. Genet. Dev. 8: 200-211.

Smith N.G.C. and Eyre-Walker A. (2001) Synonymous codon bias is not caused by mutation bias in G+C rich genes in humans. Mol.Biol.Evol. 18: 982-986.

Smith N.G.C and Eyre-Walker A. (2003). Partitioning the variation in mammalian substitution rates. Mol. Biol. Evol. 20: 10-17.

Smith N.G.C., Knight R., Hurst L.D. (1999). Vertebrate genome evolution: a slow shuffle or a big bang? Bioessays 21: 697-703.

Smith N., Webster M., Ellegren H. (2002). Deterministic mutation rate variation in the human genome. Genome Res. 12: 1350-1356.

Sogin M., Gunderson J., Elwood H.J., Alonso R.A., Peattie D.A. (1989). Phylogenetic meaning of the kingdom concept: an unusual ribosomal RNA from *Giardia lamblia*. Science 243: 75-77.

Sonenberg N. (1994). mRNA translation: influence of the 50 and 30 untranslated regions. Curr. Opin. Gen. Dev. 4: 310-315.

Sor F. and Fukuhara H. (1982). Nature of an inserted sequence in the mitochondrial gene coding for the 15S ribosomal RNA of yeast. Nucleic Acids Res. 10: 1625-1633.

Sorenson J.C. (1984). The structure and expression of nuclear genes in higher plants. Adv. Genetics 22: 109-144.

Soriano P., Macaya G., Bernardi G. (1981). The major components of the mouse and human genomes : reassociation kinetics. Eur. J. Biochem. 115: 235-239.

Soriano P., Meunier-Rotival M., Bernardi G. (1983). The distribution of interspersed repeats is non-uniform and conserved in the mouse and human genomes. Proc. Natl. Acad. Sci. USA 80: 1816-1820.

Spring J. (1997). Vertebrate evolution by interspecific hybridisation - are we polyploid? FEBS Lett. 400: 2-8.

Stenico M., Lloyd A.T., Sharp P.M. (1994). Codon usage in *Caenorhabditis elegans*: delineation of translational selection and mutational biases. Nucleic Acids Res. 22: 2437-46.

Stephens R., Horton R., Humphray S., Rowen L., Trowsdale J., Beck S. (1999). Gene organization, sequence variation and isochore structure at the centromeric boundary of the human MHC. J. Mol. Biol. 291: 789-799.

Stevanovic S. and Bohley P. (2001). Proteome analysis by three-dimensional protein separation: turnover of cytosolic proteins in hepatocytes. J. Biol. Chem. 382: 677-682.

Stevens J.R., Noyes H.A., Dover G.A., Gibson W.C. (1999). The ancient and divergent origins of the human pathogenic trypanosomes, *Trypanosoma brucei* and *T. cruzi*. Parasitology 118: 107-116.

Stevens R.C., Yokoyama S., Wilson I.A. (2001). Global efforts in structural genomics. Science, 294: 89-92

Stewart W.N. and Rothwell G.W. (1993) *Paleobotany and the evolution of plants*, Cambridge University Press, U.K.

Stock A.D. and Mengden G.A. (1975). Chromosome banding pattern conservatism in birds and nonhomology of chromosome banding patterns between birds, turtles, snakes and amphibians. Chromosoma 50: 69-77.

Strouboulis J. and Wolffe A. P. (1996). Functional compartmentalization of the nucleus. J. Cell Sci. 109:1991-2000.

Stutz E. and Bernardi G. (1972). Hydroxyapatite chromatography of deoxyribonucleic acids from *Euglena gracilis*. Biochimie 54: 1013-1021.

Subak-Sharpe H., Burk R.R., Crawford L.V., Morrison J.M., Hay J., Keir H.M. (1966). An approach to evolutionary relationships of mammalian DNA viruses through ana-

lysis of the pattern of nearest neighbor base sequences. Cold Spring Harbor Symp. Quant. Biol. 31: 737-486. Cold Spring Harbor, NY, USA.
Subak-Sharpe H., Elton R.A., Russell G.J. (1974). Evolutionary implications of doublet analysis. Symp. Soc. Gen. Microbiol. 24: 131-150.
Sueoka N. (1959). A statistical analysis of deoxyribonucleic acid distribution in density gradient centrifugation. Proc. Nat. Acad. Sci. USA 45: 1480-1490.
Sueoka N. (1961). Variation and heterogeneity of base composition of deoxyribonucleic acids: a compilation of old and new data. J. Mol. Biol. 3: 31-40.
Sueoka N. (1962). On the genetic basis of variation and heterogeneity of DNA base composition. Proc. Natl. Acad. Sci. USA 48: 582-592.
Sueoka N. (1988). Directional mutation pressure and neutral molecular evolution. Proc. Natl. Acad. Sci. USA 85: 2653-2657.
Sueoka N. (1992). Directional mutation pressure, selective constraints, and genetic equilibria. J Mol Evol. 34: 95-114.
Sueoka N., Marmur J., Doty P. (1959). Heterogeneity in deoxyribonucleic acids: II. Dependence of the density of deoxyribonucleic acids on guanine-cytosine content. Nature 183: 1429-1433.
Sved J. and Bird A. (1990). The expected equilibrium of the CpG dinucleotide in vertebrate genomes under a mutation model. Proc. Natl. Acad. Sci. USA 87: 4692-4696.
Swartz M.N., Trautner T.A., Kornberg A. (1962). Enzymatic synthesis of deoxyribonucleic acid. XI Further studies on nearest neighbour base sequences in deoxyribonucleic acid. J. Biol. Chem. 237: 1961-1967.
Swift H. (1950). The costancy of desoxyribose nucleic acid in plant nuclei. Proc. Natl. Acad. Sci. USA 36: 643-654.
Syvanen M. (1994). Horizontal gene transfer: evidence and possible consequences. Annu. Rev. Genet. 28: 237-261.
Szybalski W. (1968). Use of cesium sulfate for equilibrium density gradient centrifugation. In *Methods in Enzymology* (L. Grossman and K. Moldave, eds.) vol. 12, part B, pp. 330-360, Academic Press, New York, NY, USA.
Taguchi H., Konishi J., Ishii N., Yoshida M. (1991). A chaperonin from a thermophilic bacterium, *Thermus thermophilus*, that controls refoldings of several thermophilic enzymes. J. Biol. Chem. 266: 22411-22418.
Tajbakhsh J., Luz H., Bornfleth H., Lampel S., Cremer C., Lichter P. (2000). Spatial distribution of GC- and AT-rich DNA sequences within human chromosome territories. Exp. Cell Res. 255: 229-237.
Tazi J. and Bird A.P. (1990). Alternative chromatin structure at CpG islands. Cell 60: 909-920.
Temin H.M. (1976). The DNA provirus hypothesis. Science 192: 1075-1080.
Tenzen T., Yamagata T., Fukagawa T., Sugaya K., Ando A., Inoko H., Gojobori T., Fujiyama A., Okumura K., Ikemura T. (1997). Precise switching of DNA replication timing in the GC content transition area in the human major histocompatibility complex. Mol. Cell Biol. 17: 4043-4050.
Thiery J.P., Macaya G., Bernardi G. (1976). An analysis of eukaryotic genomes by density gradient centrifugation. J. Mol. Biol. 108: 219-235.
Thompson H.L., Schmidt R., Dean C. (1996). Identification and distribution of seven

classes of middle-repetitive DNA in the *Arabidopsis thaliana* genome. Nucleic Acids Res. 24: 3017-3022.

Tiselius A., Hjertén S., Levin Ö. (1956). Protein chromatography on calcium phosphate columns. Arch. Biochem. Biophys. 65: 132-155

Trask B.J., Massa H., Brand-Arpon V., Chan K., Friedman C., Nguyen O.T., Eichler E.E., van den Engh G., Rouquier S., Shizuya H., et al. (1998). Large multi-chromosomal duplications encompass many members of the olfactory receptor gene family in the human genome. Hum. Mol. Genet. 7: 2007-2020.

Tsichlis P.N. and Lazo P.A. (1991). Virus-host interactions and the pathogenesis of murine and oncogenic retroviruses. Curr. Top. Microbiol. Immuna 171: 95-171.

Tsyba L., Rynditch A., Boeri E., Jabbari K., Bernardi G. (2004). Distribution of HIV-1 in the genomes of infected individuals: localization in GC-poor isochores and high viremia are correlated. FEBS Letters (submitted).

Tugendreich S., Feng Q., Kroll J., Sears D.D., Boeke J.D., Hieter P. (1994). Alu sequences in RMSA-1 protein? Nature 370: 106.

Tyler J.C. (1980). Osteology, phylogeny and higher classification of the fishes of the order *Plectognathi* (*Tetraodontiformes*). NOAA Technical Report NMFS Circular 434: 1-422.

Ullu E. and Tschudi C. (1984). Alu sequences are processed 7SL RNA genes. Nature 312: 171-172.

Van der Ploeg L.H. and Flavell R.A. (1980). DNA methylation in the human gamma delta beta-globin locus in erythroid and nonerythroid tissues. Cell 19:947-958.

Van der Velden A.V. and Thomas A.A. (1999). The role of the 5' untranslated region of an mRNA in translation regulation during development. Int. J. Biochem. Cell Biol. 31:87-106.

Van Nie R. and Verstraeten A.A., (1975). Studies of genetic transmission of mammary tumor virus of C3Hf mice. Int. J. Cancer 16: 922-931.

Vanyushin B.F., Tkacheva S.G., Belozersky A.N. (1970). Rare bases in animal DNA. Nature 225: 948-949.

Vanyushin B.F., Mazin A.L., Vasilyev V.K., Belozersky A.N. (1973). The content of 5-methylcytosine in animal DNA: the species and tissue specificity. Biochim Biophys Acta. 299: 397-403.

Varmus H.E., (1984). The molecular genetics of cellular oncogenes. Ann. Rev. Genet. 18: 553-612.

Varmus H.E. and Brown P., (1989). Retroviruses. In: *Mobile DNA* (D.E. Berg and M.M. Howe, eds.) pp. 53-108. American Society for Microbiology, Washington, DC, USA.

Velculescu V.E., Zhang L., Vogelstein B., Kinzler K.W. (1995). Serial analysis of gene expression. Science 270: 484-487.

Vendrely R. and Vendrely C. (1948). La teneur de noyau cellulaire en acide désoxyribo-nucléique à travers les organes, les individus et les espèces animales. Experientia 4: 434-436.

Venter C. et al. (2001). The sequence of the human genome. Science 291: 1304-1351.

Venturini G., D'Ambrogi R., Capanna E. (1986). Size and structure of the bird genome-I. DNA content of 48 species of *Neognathae*. Comp. Biochem. Physiol. B. 85: 61-65.

Viegas-Pequignot E. and Dutrillaux B. (1978). Une méthode simple pour obtenir des prophases et des prometaphases. Ann. Genet. 21: 122-125.

Vijaya S., Steffen D.L., Kozak C., Robinson H.L. (1986). Acceptor sites for retroviral integrations map near DNA I-hypersensitive sites in chromatin. J. Virol. 60: 683-692.

Vinogradov A.E. (2001). Bendable genes of warm-blooded vertebrates. Mol Biol Evol. 18: 2195-2200.

Vinogradov A.E. (2003). DNA helix: the importance of being GC-rich. Nucleic Acids Res. 31: 1838-1844.

Vinogradov A.E. (2003). Isochores and tissue-specificity. Nuckeic Acids Res. 31: 5212-5220.

Volpi E.V., Chevret E., Jones T., Vatcheva R., Williamson J., Beck S., Campbell R.D., Goldsworthy M., Powis S.H., Ragoussis J., Trowsdale J., Sheer D. (2000). Large-scale chromatin organization of the major histocompatibility complex and other regions of human chromosome 6 and its response to interferon in interphase nuclei. J. Cell Sci. 113:1565-1576.

Wada A. and Suyama A. (1985). Third letters in codond counterbalance the (G+C) content of their first and second letters. FEBS Letters 188: 291-294.

Wada A. and Suyama A. (1986). Local stability of DNA and RNA secondary structure and its relation to biological functions. Progr. Biophys. Mol. Biol. 47:113-157.

Wallon G., Kryger G., Lovett S.T., Oshima T., Ringe D., Petsko G.A. (1997). Crystal structures of *Escherichia coli* and *Salmonella typhimurium* 3-isopropylmalate dehydrogenase and comparison with their thermophilic counterpart from *Thermus thermophilus*. J. Mol. Biol. 266: 1016-1031.

Waring M. and Britten R.J. (1966). Nucleotide sequence repetition: a rapidly reassociating fraction of mouse DNA. Science 154: 791-794.

Watanabe Y., Fujiyama A., Ichiba Y., Hattori M., Yada T., Sakaki Y, Ikemura T. (2002). Chromosome-wide assessment of replication timing for human chromosomes 11q and 21q: disease-related genes in timing-switch regions. Hum. Mol. Gen. 11: 13-21.

Watson J. and Crick F. (1953). Molecular structure of nucleic acids: a structure for deoxyribonucleic acid. Nature 117: 737-738.

Weber L. (1988). A review: molecular biology of malaria parasites. Exp. Parasitol. 66: 143-170.

Webster M., Smith N., Ellegren H. (2003). Compostional evolution of noncoding DNA in the human and chinpanzee genomes. Mol. Biol. Evol. 20: 278-286.

Weintraub H. and Groudine M. (1976). Chromosomal subunits in active genes have an altered conformation. Science 193:848-856.

Weisblum B. and de Haseth P.L. (1972). Quinacrine, a chromosome stain specific for deoxyadenylate-deoxythymidylaterich regions in DNA. Proc. Natl. Acad. Sci. USA 69:629-632.

Wellems T.E., Walliker D., Smith C.L., do Rosario VE., Maloy W.L., Howard R.J., Carter R., McCutchan T.F. (1987). A histidine-rich protein gene marks a linkage group favored strongly in a genetic cross of *Plasmodium falciparum*. Cell 49: 633-42.

Wells R.D. and Blair J.E. (1967). Studies on polynucleotides. LXXI. Sedimentation and buoyant density studies of some DNA-like polymers with repeating nucleotide sequences. J. Mol. Biol. 27: 273-88.

Whitelegge, J. P. (2003) Plant proteomics: BLASTing out of a MudPIT. Proc. Natl. Acad. Sci. USA 99:11564–11566.

Whitman W.B., Bowen T.L., Boone D.R. (1992). The methanogenic bacteria. In *The prokaryotes* (A. Balows, H.G. Trueper, M. Dworkin, W. Harder , K.H. Schleifer, eds.) pp. 719-767. Springer-Verlag, New York, NY, USA.

Wilson D.E. and Reeder D.M. (1993). *Mammal species of the world. A taxonomic and geographic reference.* Random House, Smithsonian Inst. Press, Washington, DC, USA.

Wilusz C.J., Wang W., Peltz S.W. (2001). Curbing the nonsense: the activation and regulation of mRNA surveillance. Genes Dev. 15: 2781-2785.

Winkler H. (1920). *Verbreitung und Ursache der Parthenogenesis im Pflanzen- und Tierreich.* Fischer, Jena, Germany.

Wobus U. (1975). Molecular characterization of an insect genome: *Chironomus thummi.* Eur. J. Biochem. 59: 287-93.

Woese C.R. (1967). The fundamental nature of the genetic code: prebiotic interactions between polynucleotides and polyamino acids or their derivatives. Proc. Natl. Acad. Sci. USA 59: 110-711.

Wolfe K.H. (2001). Yesterday's polyploids and the mystery of diploidization. Nature Rev. Genet. 2: 333-241.

Wolfe K.H. and Sharp P.M. (1993). Mammalian gene evolution: nucleotide sequence divergence between mouse and rat. J. Mol. Evol. 37: 441-456.

Wolfe K.H., Sharp P.M., Li W.H. (1989) Mutation rates differ among regions of the mammalian genome. Nature 337: 283-285.

Woodcock D.M., Crowther P.J., Diver W.P. (1987). The majority of methylated deoxycytidines in human DNA are not in the CpG dinucleotide. Biochem. Biophys. Res. Comm. 145: 888-894.

Wright F. (1990). The "effective number of codons" used in a gene. Gene 87 :23-29.

Wright J., Steer E., Hailey A. (1988). Habitat separation in tortoises and the consequences for activity and thermoregulation. Can. J. Zool. 66: 1537-1544.

Wu C.I. and Li W. (1985). Evidence for higher rates of nucleotide substitution in rodents than in man. Proc. Natl. Acad. Sci. USA 82: 1741-1745.

Wu J., Maehara T., Shimokawa T., Yamamoto S., Harada C., Takazaki Y., Ono N., Mukai Y., Koike K., Yazaki J., Fujii F., Shomura A., Ando T., Kono I., Waki K., Yamamoto K., Yano M., Matsumoto T., Sasaki T. (2002). A comprehensive rice transcript map containing 6591 expressed sequence tag sites. Plant Cell 14: 525-535.

Wu T.H., Clarke C.H., Marinus M.G. (1990). Specificity of *Escherichia coli mutD* and *mutL* mutator strains. Gene 87: 1-5.

Wurster-Hill D.H. and Gray C.W. (1979). The interrelationships of chromosome banding patterns in procyonids, viverrids, and felids. Cytogenet. Cell. Genet. 15: 306-331.

Yamagishi H. (1970). Nucleotide distribution in the DNA of *Escherichia coli.* J. Mol. Biol. 49: 603-608.

Yamagishi H. (1974). Nucleotide distribution in bacterial DNA's differing in G plus C content. J. Mol. Evol. 3: 239-242.

Yang A.S., Gonzalgo M.L., Zingg J.M., Millar R.P., Buckley J.D., Jones P.A. (1996). The rate of CpG mutation in Alu repetitive elements within the p53 tumor suppressor gene in the primate germline. J. Mol. Biol. 258: 240-250.

Yarus M. (1991). An RNA-amino acid complex and the origin of the genetic code. New Biol. 3: 183-189.

Yokota H., Singer M.J., van den Engh G.J., Trask B.J. (1997). Regional differences in the compaction of chromatin in human G0/G1 interphase nuclei. Chromosome Res. 5: 157-166.

Yonenaga-Yassuda Y., Kasahara S., Chu T.H., Rodriguez M.T. (1988). High-resolution RBG-banding pattern in the genus *Tropidurus* (*Sauria, Iguanidae*). Cytogenet. Cell Genet. 48: 68-71.

Yu J. et al. (2002). A draft sequence of the rice genome (*Oryza sativa* L. ssp. *indica*). Science 296: 79-92.

Yuhki N., Beck T., Stephens R.M., Nishigaki Y., Newmann K., O'Brien S.J. (2003). Comparative genome organization of Human, Murine and Feline MHC class II region. Genome Res. 13: 1169-1179.

Yunis J.J. (1976). High resolution of human chromosomes. Science 191: 1268-70.

Yunis J.J. (1981). Mid-prophase human chromosome. The attainment of 2,000 bands. Hum. Genet. 56: 291-298.

Zenvirth D., Arbel T., Sherman A., Goldway M., Klein S., Simchen G. (1992). Multiple sites for double-strand breaks in whole meiotic chromosomes of *Saccharomyces cerevisiae*. EMBO J. 11: 3441-3447.

Zerial M., Salinas J., Filipski J., Bernardi G. (1986a). Gene distribution and nucleotide sequence organization in the human genome. Eur. J. Biochem. 160: 479-485.

Zerial M., Salinas J. Filipski J., Bernardi G. (1986b). Genomic localization of hepatitis B virus in a human hepatoma cell line. Nucleic Acids Res. 14: 8373-8386.

Zhang C.T. and Zhang R. (2003a). An isochore map of the human genome based on the Z curve method. Gene 317: 127-135.

Zhang C.T. and Zhang R. (2003b). Isochore structures in the mouse genome. Genome Res. (in press).

Zietkiewicz E., Makalowski W., Mitchell G.A., Labuda D. (1994). Phylogenetic analysis of a reported complementary DNA sequence. Science 265: 1110-1111.

Zoubak S., Rynditch A., Bernardi G. (1992). Compositional bimodality and evolution of retroviral genomes. Gene 119: 207-213.

Zoubak S., Richardson J., Rynditch A., Hillsberg P., Hafler D., Boeri E., Lever A.M.L., Bernardi G., (1994). Regional specificity of HTLV-I proviral integration in the human genome. Gene 143: 155-163.

Zoubak S., D'Onofrio G., Cacciò S., Bernardi G., Bernardi G. (1995). Specific compositional patterns of synonymous positions in homologous mammalian genes. J. Mol. Evol. 40: 293-307.

Zoubak S., Clay O., Bernardi G. (1996). The gene distribution of the human genome. Gene 174: 95-102.

Zuber H. (1981). Structure and function of thermophilic enzymes. In *Structural and functional aspects of enzyme catalysis* (H. Eggerer and R. Huber, eds.) pp. 114-127. Springer-Verlag, Berlin, Germany.

Zuckerkandl E. and Pauling L. (1962). Molecular disease, evolution, and genetic heterogeneity. In *Horizons in Biochemistry* (M. Kasha and B. Pullman, eds.) pp. 189-225. New York Academic Press, NY, USA

Zuckerkandl E. (1975). The appearance of new structures and functions in proteins during evolution. J. Mol. Evol. 7: 1-57.

Zuckerkandl E. (1976). Evolutionary processes and evolutionary noise at the molecular level. II. A selectionist model for random fixations in proteins. J. Mol. Evol. 7: 269-311.
Zuckerkandl E. (1986). Polite DNA: functional density and functional compatibility in genomes. J. Mol. Evol. 24: 12-27.